Springer Series in Advanced Manufacturing

The **Springer Series in Advanced Manufacturing** includes advanced textbooks, research monographs, edited works and conference proceedings covering all major subjects in the field of advanced manufacturing.
The following is a non-exclusive list of subjects relevant to the series:

1. Manufacturing processes and operations (material processing; assembly; test and inspection; packaging and shipping).
2. Manufacturing product and process design (product design; product data management; product development; manufacturing system planning).
3. Enterprise management (product life cycle management; production planning and control; quality management).

Emphasis will be placed on novel material of topical interest (for example, books on nanomanufacturing) as well as new treatments of more traditional areas.

As advanced manufacturing usually involves extensive use of information and communication technology (ICT), books dealing with advanced ICT tools for advanced manufacturing are also of interest to the Series.

Springer and Professor Pham welcome book ideas from authors. Potential authors who wish to submit a book proposal should contact Anthony Doyle, Executive Editor, Springer, e-mail: anthony.doyle@springer.com.

More information about this series at http://www.springer.com/series/7113

Surjya Kanta Pal · Debasish Mishra · Arpan Pal ·
Samik Dutta · Debashish Chakravarty · Srikanta Pal

Digital Twin – Fundamental Concepts to Applications in Advanced Manufacturing

Springer

Surjya Kanta Pal
Department of Mechanical Engineering
Indian Institute of Technology
Kharagpur, India

Arpan Pal
Embedded Systems and Robotics
Tata Consultancy Services
Kolkata, India

Debashish Chakravarty
Department of Mining Engineering
Indian Institute of Technology
Kharagpur, India

Debasish Mishra
Advanced Technology Development Centre
Indian Institute of Technology
Kharagpur, India

Samik Dutta
Surface Engineering and Tribology
CSIR–Central Mechanical Engineering
Research Institute
Durgapur, India

Srikanta Pal
Department of Electronics
and Communication Engineering
Birla Institute of Technology Mesra
Ranchi, India

ISSN 1860-5168 ISSN 2196-1735 (electronic)
Springer Series in Advanced Manufacturing
ISBN 978-3-030-81814-2 ISBN 978-3-030-81815-9 (eBook)
https://doi.org/10.1007/978-3-030-81815-9

This Springer imprint is published by the registered company Springer Nature Switzerland AG
The registered company address is: Gewerbestrasse 11, 6330 Cham, Switzerland

This book is dedicated to all Industry–Academia Collaborative Ecosystem

Preface

Errors using inadequate data is much less than those using no data at all.
—Charles Babbage

The advent of the *Internet of Things (IoT)* and *Cyber-Physical System (CPS)*, coupled with signal processing and machine learning techniques, including artificial intelligence has been some of the key enablers to advanced automation in industries. This is aimed at reducing machine downtime, enhancing product quality, and improving the efficacy of the manufacturing systems. This ecosystem consisting of the physical devices and cyber components requires infrastructures such as cloud and fog computing, encrypted with a communication protocol. This will ensure seamless network connectivity for automated and precise manufacturing, leading to economic utilization of resources, and sustainable supply chain management through the interaction of physical and virtual resources. These techniques embody the concept of *Industry 4.0*, i.e., the fourth industrial revolution.

In general, it becomes difficult to perform new tests on a physical device without prior information, optimize the test parameters, and identify the wellness of the physical device. To address this issue, a digital representation of the physical device is the need of the day, which is termed as a *digital twin*. It is defined as a virtual representation of a physical device, under the umbrella of Industry 4.0. The digital twin portrays the life of the physical device, with the help of data gathered from the physical device in real time. The digital twin can be considered as a package of different modules, wherein each of the modules performs an intended task that provides specific information about the physical device. The tasks performed by various modules identify the disturbances in the process, assess the health of critical components of the physical device, and sends alarms; all of which occur in real time. The modules include techniques for processing incoming real-time data from the physical device, techniques for prediction of the faults by machine learning, and intelligent techniques for control of the physical device.

This book volume aims at touring the readers with these advanced technologies of Industry 4.0 to ultimately develop a digital twin from the fundamental level. This shall be achieved using the technologies in a consolidated form for a manufacturing process. Several case studies and schematic representations are included all through the discussions to provide the readers with a comprehensive understanding of these techniques. This volume shall be of immense help to academicians for understanding the nitty-gritty of the digital tools of Industry 4.0 through the case studies. The industry personnel will also be able to appreciate the advantages which can be achieved by transforming their production system to be compliant with Industry 4.0, through upscaling the concept of the digital twin. The book includes seven chapters, which are mentioned below with a short description and a number of illustrations.

Chapter 1 sets the prologue of industrial revolutions, starting from the first to the fourth. It aims at making the readers aware of the transformations, happened so far. The tools of Industry 4.0 are introduced, and amongst them, the attention of this volume is focused on the digital twin. Its importance along with the present and future perspectives are highlighted.

Chapter 2 gives an overview of sensor electronics applicable in manufacturing, and their integration with the manufacturing processes, systems, and sub-systems. For each sensor, a case study in manufacturing is included for a practical understanding.

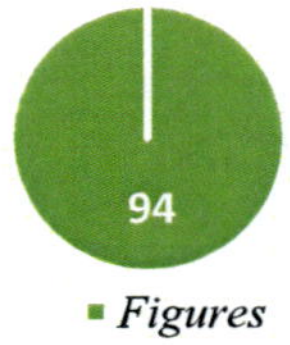

Chapter 3 deals with the acquisition of signals, their processing, and feature extraction, especially useful in various manufacturing processes, systems, and sub-systems. Several examples in the field of manufacturing have been elaborated for each signal processing technique.

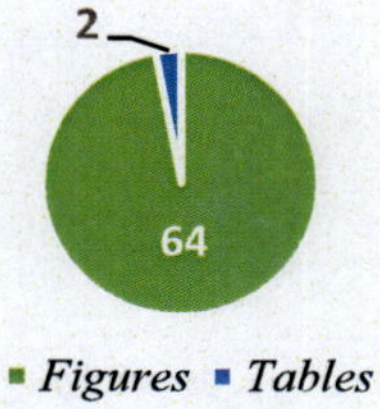

Chapter 4 discusses the fundamental digital image processing techniques along with suitable case studies used in creating digital twin models for real-time product quality assurance in the manufacturing pipeline.

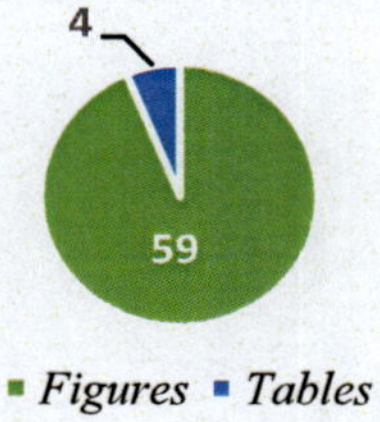

Chapter 5 discusses the seamless data transmission and computing required for developing a digital twin application for the real-time diagnosis and prognosis of manufacturing systems. The chapter includes a thorough discussion regarding the edge, fog, and cloud computing; data transmission; and temporal capacity of data transmission with case studies.

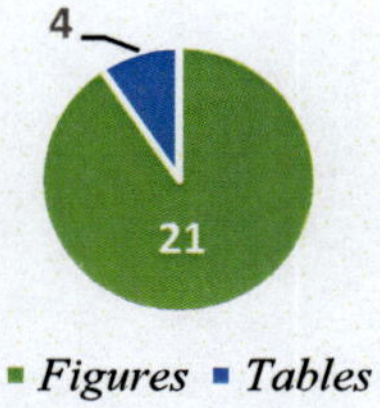

Chapter 6 discusses the artificial intelligence and machine learning techniques required for the diagnosis of manufacturing processes, systems, and sub-systems and prediction of machine health for corrective measures. Several examples in the field of manufacturing are discussed to provide a practical understanding.

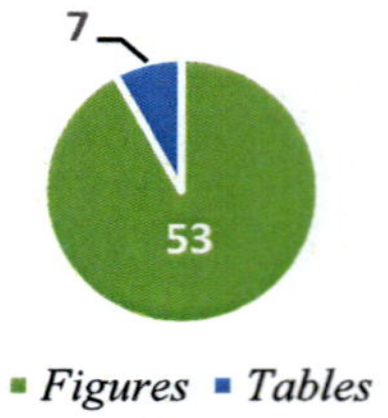

Chapter 7 provides a comprehensive detailing of the digital twin application. It discusses the techniques involved in the building of a digital twin, benefits, and associated challenges. A few case studies are also included in the chapter wherein digital twins have been built for process and health monitoring.

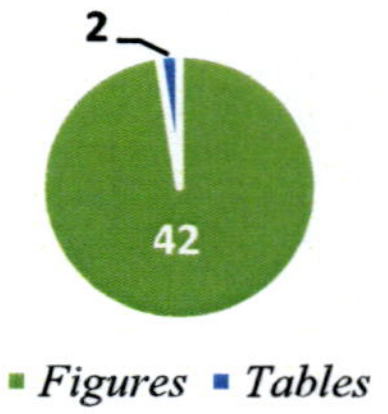

Kharagpur, India — Surjya Kanta Pal
Kharagpur, India — Debasish Mishra
Kolkata, India — Arpan Pal
Durgapur, India — Samik Dutta
Kharagpur, India — Debashish Chakravarty
Ranchi, India — Srikanta Pal

Acknowledgements

The authors have inherited the experiences of writing a book by reading a large number of research articles published across the globe. Hence, a special thanks to all the research contributions.

We would like to thank the *Indian Institute of Technology Kharagpur, India (IIT Kharagpur)*, for equipping us with the state-of-the-art infrastructure using which the case studies were prepared. A special thanks to the *Mechanical Engineering Department* of *IIT Kharagpur* for providing the infrastructure to the *Friction Stir Welding laboratory*. Thanks to all the doctoral scholars, *Karali, Sukhomay, Probir, Kamal, Amrita, Soumya, Harshadeep, Ravi Shankar, Ramkumar, Kanchan, Rahul, Matruprasad, Tiara, Ishita, Samarshi, Raju, Vyas, Abhijit, Iman, Omkar, Suryakanta, Perwej, Debolina, Soumya Sangita, Pooja, Ananta, Rishab, Noorul, Ranjan, Suparna, Avishek, Souvik, Pallab, and Riya*. They have been a constant source of inspiration in this journey by creating a research atmosphere. Also, thanks to the MTech students *Raj, Aditya*, and *Nandikesh* for their efforts on digital twin and robotics work.

A special thanks to *Chandan* from *Forming laboratory*, *Pravanjan* and *Jyotirmoy* from the *Centre of Excellence in Advanced Manufacturing Technology*, and *Prajeesh* from the *Metrology laboratory* of *IIT Kharagpur*. The authors duly acknowledge their help in collecting images of the equipment and instruments.

A special note of thanks to all the interns belonging to different institutes of India, *Nagraj, Ujjwal, Inderjeet, Shilpi, Ravi Ranjan, Chiraag, Aaquib, Nitin, Rohan, Alekhya, Soham, Ankit, Arkadeep, Aman, Riya, Sananya, Abhinav, Pranav, Saad, Sarthak, Ashok, Sristi, Abhishek, Shubranil, Sayani, and Ayan,* and undergraduate students of *IIT Kharagpur*, *Anurup* and *Sandip*. This book would not have been possible without the case studies that they have provided.

A special thanks to the *Department of Heavy Industry (DHI)* of *Ministry of Heavy Industries and Public Enterprises (HI&PE),* Government of India, for funding a research project, which is being executed with *Tata Consultancy Services (TCS)* under the *Centre of Excellence in Advanced Manufacturing Technology* at *IIT Kharagpur*. The authors express a special thanks to the *TCUP* and *PREMAP* teams of *TCS*. The successful collaboration with *TCS* is one of the inspirations behind this book.

Special gratitude to *Dr. Tapas Chakravarty, Mr. Prateep Misra, Dr. Tanushyam Chattopadhyay, Dr. B. P. Gautam, Dr. B. Sreeharikumar, Mr. Satanik Panda, Mr. Arijit Sinharay, Mr. Asif Sumair, Dr. M. Girish Chandra, Dr. A. Anil Kumar, Mr. Raj Rakshit, Mr. Chirabrata Bhaumik, Mr. Suvra Dutta, Mr. Amit Swain, Ms. Kriti Kumar, Dr. Riddhi Panse, and Mr. Ankur Krishna.* A special thanks to *Professor Sudip Misra* and doctoral scholar *Anandaroop* of *IIT Kharagpur* for their contributions to the research project with *TCS*.

Finally, the authors would like to thank *Springer* for accepting the proposal and publishing the textbook on Digital Twin.

From the Authors

I have not failed. I've just found 10,000 ways that won't work.
—Thomas A. Edison

People say *Engineering comes from common sense.* One might ponder what that *sense* is and how *common* is it?

Imagine you are driving down a lane and suddenly your vehicle breaks down. You open the engine cover and look at the messy components. What do you find? Some heat fumes, some oil splashes, and you have an inherent urge to try and fix it. Back in the mind, numerous thoughts flow simultaneously, what the problem could have been. You start to perform a few basic things and ultimately figure out the problem. The approach you followed for arriving at the solution is a feeling of your sense, and it is common because you may not be an expert such as an automobile engineer, yet you could gaze at the problem and find a solution to it. The whole idea is coming up with solutions, does not matter whether they are brilliant. For every problem, there would be numerous ways to solve but ultimately the best amongst them is selected.

If we look back at history, the innovations did not pop over a night. There might have been several failures that must be summed up to report the final achievement. *Industry 4.0*, the fourth industrial revolution, is not about a machine or a system or a style. The components which have taken aback the internet with thousands of posts are also like that. Go to a market and ask for the *Internet of Things*, *Cyber-Physical System*, *Digital Twin*, etc., the store owner will gaze at you with bewilderment. These are items that one needs to develop as they are not readily available. Needless to mention, the real challenge here is to develop an *Internet of Things* or *Cyber-Physical System* framework, a *digital twin* model, etc. that makes our lives easier, and we need the best of the engineering minds to accomplish the same.

As the dependence on technology kept growing, engineers became increasingly popular. Engineers and their efforts are responsible for transforming the theory to practice for the overall progress. One might be curious to know the reasons as to what makes *technology* an essential ingredient in business success. The response is quite simple. Technology saves on the time and effort required to do a job that otherwise would have been a Herculean task. It also provides a sense of security to an organization. If one starts to list the reasons which highlight the importance of technology in business success, it would be an unending process. The role it plays is irreplaceable for the smooth running of business operations. The cut-throat competition in the corporate world is quite aware of the benefits that technology will provide for business excellence. Business operations are equally challenging to every corporate, the one which creates a difference is the one that adapts to the latest technological revolution. It is an uncontested conclusion that to keep the workflow streamlined, businesses need to implement advanced technology like the *Internet of Things*, *Cyber-Physical Systems*, *Digital Twins*, etc. that have been designed specifically for various sectors.

Now the question to ponder upon is *where lies the problem?* The answer goes Industry 4.0 or its concepts, the most famous one called *Industrial Internet of Things*, are often presented at a high level of abstraction by personnel and consultants. The dilemma faced by the industry personnel is to select between the similar but different, i.e., Internet of Things and Industrial Internet of Things. They are not two different entities, but they differ in terms of their application areas. The latter is named so because of its application to core industry-specific areas. So, most of the concepts of Industry 4.0 are presented from a business perspective. The focus of the consultants is mostly on the business models, operational efficiency, and profitable decisions, which are lucrative and are readily understandable.

Well, we, the authors, who have a theoretical as well as application-based knowledge of the concepts of Industry 4.0, have penned down our thoughts, experiences, and insights in this book in the best possible way. We have been thinking of this textbook for a time, and finally, conceptualized, worked on it, and prepared the manuscript. This textbook attempts to address the gap between conceptualizing the application of Industry 4.0 and implementing it from the scratch. While doing so, we have paid special attention to the shop floor related technical difficulties involved in the implementation of Industry 4.0 concepts. With this book, we present you with several case studies that are included in every chapter, which provides a deeper understanding of the concepts from the implementation perspective.

We hope you have a great time reading it.

Please write to us in case of queries. We would also love to know your valuable feedback.

Contents

About the Authors

Prof. Surjya Kanta Pal graduated from Government Engineering College, Jalpaiguri, India, in 1991, M.Tech. from Indian Institute of Technology Kanpur, India, in 1993, and Ph.D. from Indian Institute of Technology Kharagpur, India, in 1999, all in Mechanical Engineering. He did Postdoctoral research from The University of Sheffield, the UK, in Automatic Control and Systems Engineering (1999–2002). He worked with Grasim Industries Limited in Engineering and Development Division from 1994 to 1995. He started teaching at Indian Institute of Technology Guwahati, India, from July 2002 to July 2004 and then joined Indian Institute of Technology Kharagpur where he is presently working as a Professor in the Department of Mechanical Engineering. His specialization is Manufacturing Science. He is the Founder Chairperson of the Centre of Excellence in Advanced Manufacturing Technology at Indian Institute of Technology Kharagpur. This Centre fosters collaborative research between industry and academia in advanced manufacturing topics. He has vast experience in the application of signal and image processing, optimization, and machine learning techniques to manufacturing since 2004. He has been awarded the prestigious "Lord Kumar Bhattacharyya Chair Professor" for his excellent contributions to research in manufacturing.

Debasish Mishra is pursuing doctoral research at Indian Institute of Technology Kharagpur, India, on the application of Industry 4.0 concepts for real-time monitoring and control of friction stir welding process using multiple sensors. He did M.Tech. in Manufacturing Science and Engineering from Veer Surendra Sai University of Technology (VSSUT), Burla (2016), and B.Tech. in Mechanical Engineering from Biju Patnaik University of Technology, India (2013). He has filed four patents on the application of signal processing, machine learning, and artificial intelligence techniques in manufacturing.

Dr. Arpan Pal received B.Tech. and M.Tech. from Indian Institute of Technology Kharagpur, India, and Ph.D. from Aalborg University, Denmark. He has rich experience in signal processing, communication, embedded systems, and robotics. Currently, as Distinguished Chief Scientist and Research Area Head, Embedded Devices and Intelligent Systems, he is leading a team of 75+ scientists of TCS Research in the areas of connected health, smart cities, smart enterprises, smart manufacturing, and remote sensing. His research interests include affective computing, edge computing, intelligent sensing, signal and image processing and artificial intelligence and Internet of Things applications and systems. He is the author of a book on IoT. Dr. Pal also worked for DRDO, India, as a Scientist for Missile Seeker Systems and in Rebeca Technologies (erstwhile Macmet Interactive Technologies) as their Head of Real-time Systems.

Samik Dutta did his BE and M.Tech. in Mechanical Engineering from Regional Engineering College, Durgapur, presently known as National Institute of Technology Durgapur, India, in 2002 and 2004, respectively. He has pursued his Ph.D. in Tool Condition Monitoring using Digital Image Processing, from Indian Institute of Technology Kharagpur, India, in 2015. He is currently working at CSIR-Central Mechanical Engineering Research Institute, Durgapur, India as a Principal Scientist. He has rich experience in the application of digital image processing techniques and sensorial monitoring in manufacturing processes.

Prof. Debashish Chakravarty has completed his studies at Indian Institute of Technology Kharagpur, India (B.Tech. in 1993, M.Tech. in 1995, and Ph.D. in 2001), all in Mining Engineering. He did postdoctoral research in Jülich Research Institute, Germany, in Collaborative and Distributive Virtual Reality (2003). Currently, he is serving as a Professor in the Department of Mining Engineering at Indian Institute of Technology Kharagpur. He has completed numerous research funded projects in the area of autonomous navigation, image perception, SLAM, controls, ML/DL for different tasks in autonomous vehicular motion and trajectory planning in addition to conducting several high-end research-oriented projects in the areas of Mining Engineering. He has worked on VR/AR/XR applications in various industrial problems. He has extensively worked on different image processing projects using underground, terrestrial, airborne, and satellite imageries. Working experience in optical, multispectral, hyperspectral, and radar imaging systems and products makes his expertise unique.

Prof. Srikanta Pal did his graduation from Regional Engineering College Warangal presently known as National Institute of Technology Warangal, India (1990), Masters in Microwave Engineering from Jadavpur University (1992) (Gold Medalist), and D.Phil. in Superconducting Filters for Wireless Communications from Oxford University, UK (2001–2003), and Postdoctoral research in Superconducting Filters in Radio Astronomy from University of Birmingham, UK (2006–2008). He worked as an R&D Engineer in Bharat Electronics Limited (Ministry of Defence, Govt. of India, 1992–2000), in Indian Institute of Technology Delhi, India, CARE as visiting faculty (2003–2005), as Assistant Professor in Department of Electronics and Computer Engineering, Indian Institute of Technology Roorkee, India (2005–2006), and currently as a Professor in Department of Electronics and Communication Engineering of Birla Institute of Technology Mesra since 2008. His research interests include antennas, microwave/terahertz imaging/detectors, statistical signal processing, high-temperature superconductors and application of signal and image processing in manufacturing processes.

Abbreviations

Chapter—1

IT	Information technology
NC	Numeric controlled
CNC	Computer numeric control
CIM	Computer integrated manufacturing
CAD	Computer-aided design/ computer-aided designing
CAM	Computer-aided manufacturing
IoT	Internet of Things
CPS	Cyber-Physical systems
PLC	Programmable logic controller
IIoT	Industrial Internet of Things
SCADA	Supervisory control and data acquisition

Chapter—2

LVDT	Linear variable differential transformer
EMF	Electromotive force
NPN, PNP	Proximity sensor configurations
AC	Alternating current
DC	Direct current
FSW	Friction stir welding
NTC	Negative temperature coefficient
PTC	Positive temperature coefficient
RTD	Resistance temperature detector
RMS	Root mean square

Chapter—3

DAQ	Data acquisition card
ADC	Analogue-to-digital converter
VI	Virtual instrument
IDE	Integrated development environment
I/O	Input/output module
SNR	Signal to noise ratio
RMSE	Root mean square error
CC	Correlation coefficient
SKEW	Skewness
KURT	Kurtosis
FT	Fourier transform
DFT	Discrete Fourier transform
FFT	Fast Fourier transform
GMAW	Gas metal arc welding
P-GMAW	Pulsed- Gas metal arc welding
STFT	Short-time Fourier transform
CT-scan	Computer tomography scanning

Chapter—4

MV	Machine vision
RoI	Region of interest
μ-EDM	Micro-electric discharge machining process
ECM	Electrochemical machining
SMEFC	Scanning micro electrochemical flow cell
IR	Infrared camera
DED	Direct energy deposition
TIG	Tungsten inert gas
GTAW	Gas tungsten arc welding
MAG	Metal arc gas
FPGE	Field programmable gate array
RUL	Remaining useful life
$VB_{average}$	Average tool flank wear width
PCB	Printed circuit board
CPPS	Cyber physical production system
GPU	Graphical processing unit
FSW	Friction stir weld
FOV	Field-of-view
CCD	Charged couple device
CMOS	Complementary metal oxide semiconductor

CTE	Charge transfer efficiency
EHD	Electrohydrodynamic
D_A	Iris opening diameter
DOV	Depth of view
MTF	Modulation transfer function
LED	Light emitting diodes
FWHM	Full width half maximum
ND	Neutral density
LPF	Low-pass filter
CLAHE	Contrast limited histogram equalization
LoG	Laplacian of Gaussian
GLCM	Gray level co-occurrence matrix
RLS	Run length statistical
PPS	Pixel pair spacing
RLM	Run length matrix
DWT	Discrete wavelet transform
STFT	Short time Fourier transform
HT	Hough transform
LBP	Local binary pattern
ANN	Artificial neural network

Chapter—5

WPAN	Wireless personal area networks
WLAN	Wireless local area networks
WWAN	Wireless wide area networks
LLC	Logical link control
PHY	Physical layer
MAC	Media access control
UDP	User datagram protocol
BLE	Bluetooth low energy
L2CAP	Control adaptation protocol layer
RFComm	Radio frequency communication protocol
SDP	Service discovery protocol
BNEP	Bluetooth network encapsulation protocol
TCP/IP	Transmission control protocol/internet protocol
ISM	Industrial, scientific and medical band
CSMA/CA	Carrier sense multiple access
6LoWPAN	Ipv6 over low power wireless personal area network
IETF	Internet engineering task force
LPWAN	Low-power wide-area networks
MTC	Machine-type communications
NB-IoT	Narrow-band IoT

HTTP	Hypertext transfer protocol
MQTT	Message queue telemetry transport
CoAP	Constrained application protocol
DTN	Delay-tolerant-Networking
OFDM	Orthogonal frequency division multiplexing
mmWave	Millimetric wave
MIMO	Multiple input multiple output
mMIMO	Massive multiple input multiple output
eMBB	Enhanced mobile broadband
URLLC	Ultra-reliable low latency communications
mMTC	Massive machine type communications
CoMP	Coordinated multiple access points

Chapter—6

RUL	Remaining useful life
SNR	Signal to noise ratio
FIR	Finite impulse response
IIR	Infinite impulse response
PAA	Piecewise aggregation approximation
SAX	Symbolic aggregate approximation
FDR	Fisher discriminant ratio
LSE	Least Square Estimation
MLE	Maximum Likelihood Estimation
BMSE	Bayesian Minimum Mean Squared Error
KF	Kalman filtering
RoI	Return-on-investment
OR	Operations research
SVM	Support vector machine
TMAZ	Thermo-mechanically affected zone
HAZ	Heat affected zone
FPR	Feed per revolution
UTS	Tensile strength
MSE	Mean square error
LMS	Least-mean-square
RLS	Recursive-least-square
KAF	Kernel adaptive filters
UCR	University of California, Riverside
DTW	Dynamic time warping
MIC	Maximal information coefficient
mRMR	Minimum redundancy maximum relevance
DNN	Deep neural networks
YOLO	You Only Look Once

OLS	Ordinary least square
GAN	Generative adversarial neural network

Chapter—7

NASA	National Aeronautics and Space Administration
PLM	Product lifecycle management
CNC	Computer numeric controlled
RUL	Remaining useful life
DROTA	Doctor's Rule-out Test Approach
WPS	Welding procedure specification
PQR	Procedure qualification record
D1	Wavelet coefficients at the first level
TCP/IP	Transmission control protocol/internet protocol
SVM	Support vector machine
RoI	Region of interest
GUI	Graphical user interface

Symbols

Chapter—2

V_{S1}	Induced EMF in secondary coil (1)
V_{S2}	Induced EMF in secondary coil (2)
k_1	Constant
k_2	Constant
$\varnothing$	Coupling between primary and secondary coils
R	Resistance
ρ	Resistivity
l	Length of a conductor
A	Area of a conductor
V_{out}	Output voltage
V_{in}	Input voltage
ϵ	Strain
Δl	Change in length
$\frac{\Delta l}{l}$	Mechanical strain
ΔR	Change in resistance
$\frac{\Delta R}{R}$	Electrical strain
G	Gauge factor
C	Capacitance
ε_r	Relative permittivity of the dielectric
ε_0	Permittivity of free space
d	Separation between the plates
x	Change in the separation distance between the two plates
ω	Angular speed of a shaft
v	Velocity
α	Position
t	Time
σ	Stress
E	Young's modulus

F	Force
Q	Charge
k	Piezoelectric coefficient
F	Force
n	Number of piezoelectric crystals
p	Charge sensitivity
g	Voltage sensitivity
T_0, T	Temperature
R_T	Resistance at temperature T
R_0	Resistance at temperature T_0
β	Constant
α_T	Temperature coefficient
m	Mass
a	Acceleration
X	Deflection of the spring-mass system
K	Stiffness of the spring
λ	Wavelength of a wave
ϑ	Velocity of a sound wave
f	Frequency of a wave
ϑ_0	Velocity of an observer
ϑ_s	Velocity of a source
f'	Observed frequency
I	Current
V	Voltage
EMF	Electromotive force
N	Number of turns in a coil
φ_B	Magnetic flux
$\frac{d\varphi_B}{dt}$	Change of magnetic flux with time
M	Mutual inductance
$\frac{dI}{dt}$	Rate of change of current
P	Pressure
F_v	Volumetric flow rate
F_m	Mass flow rate
ρ	Density
B	Magnetic field strength
d_i	Internal diameter of a pipe
Δt	Difference in time interval
UC	Uncertainty

Chapter—3

LS	Least significant bit
n	Resolution of an ADC

F_s	Sampling frequency
T_s	Sampling period
W	Highest frequency component
SNR	Signal to noise ratio
$RMSE$	Root mean square error
r	Correlation coefficient
x	Original signal
x_{den}	Denoised signal
N	Number of samples of a signal
σ	Standard deviation
$\bar{x}$	Sample mean
Err	Error
μ	Mean
RMS	Root mean square
$SKEW$	Skewness
$KURT$	Kurtosis
μ_4	Fourth moment
$f(t)$	Periodic function
a_0, A_k, B_k	Constant term
$x(t)/x_n$	Time domain signal
$X(F)$	Fourier transform
k	Wave number/ frequency bins
n	Sample number
Δt	Time of sampling
T	Time period
x_k	Discrete signal
X_k	Discrete Fourier transform of discrete signal
Q	Arc energy
I	Welding current
E	Arc voltage
v	Welding speed
$g(t)$	Window function
$STFT(\tau, F)$	Short time Fourier transform
τ	Time variable of the spectrogram
σ_t	Standard deviation in time
σ_f	Standard deviation in frequency
$\Psi(t)$	Base wavelet
$\Psi^*(t)$	Complex conjugate of the base wavelet
τ	Shift parameter
s	Scale parameter
s_0, τ_0	Coefficients in discrete wavelet transform
$g[n]$	High-pass filter
$h[n]$	Low-pass filter

Chapter—4

t_e	Exposure time
D_B	Blur circle diameter
v	Relative speed between camera and object
α	Angle between the image plane and direction of motion
M	Magnification
F	Focal length
d_o	Distance of object from principal plane of the lens
d_I	Distance of image from principal plane of the lens
D_B	Allowable blur circle diameter
i	Image
f	Filter
o	Output image
σ	Standard deviation
f_g	Gaussian filter
$I(u, v)$	Discrete Fourier transform of an image
u, v	Discrete variables
$D(u, v)$	Distance of point (u, v) from central point i.e. (0,0) in frequency axis
$F(u, v)$	Butterworth filter
n	Order of the filter
$F_{HP}(u, v)$	High-pass filter function
$F_{LP}(u, v)$	Lowpass filter function
∇i_x	Gradient along x-direction
∇i_y	Gradient along y-direction
∇i	Gradient of image
$m_\nabla(x, y)$	Magnitude of the gradient
(x, y)	Target pixel
$\alpha(x, y)$	Direction of the edge at a pixel
$\nabla^2 i_g$	Laplacian operator
i_g	Smoothened image
(e_h)	High threshold
(e_l)	Low threshold
t	Threshold level
$[0, L-1]$	Gray level range
s_R	Between-class-variance
$P_1(t)$	Cumulative summation of probability of occurrence of pixels up to the gray level
m_i	Average intensity of the image
E	Structuring element
E_p	Translation of E by p amount
RP	Run percentage
SRE	Short run emphasis
$R(g, r)$	Element of RLM

n_R, n_p	Total number of runs and pixels
LRE	Long run emphasis
GLN	Gray-level non-uniformity
RLN	Run-length non-uniformity
$LGRE$	Low gray-level run emphasis
$HGRE$	High gray-level run emphasis
$SRLGE$	Short run low gray-level emphasis
$SRHGE$	Short run high gray-level emphasis
$LRLGE$	Long run low gray-level emphasis
$LRHGE$	Long run high gray-level emphasis
q	Dilation level
2^q	Scaling factor
$2^q k$	Translation factor
RMS	Root mean square
ENE	Energy
AVE	Average value of the coefficients

Chapter—6

m_1 and m_2	Mean of the image
$\sigma_1$and σ_2	Variance of the image
FDR	Fisher discriminant ratio
X	Feature
i and j	Two classes
σ	Standard deviation
G_i	Gini impurity
$p_{i,k}$	Ratio of k instances amongst the training instances in the ith node
N	Number of terminal nodes
D	Depth of the tree
H_i	Informational entropy
x_1, y_1, x_2, y_2	Number of instances
$G_{i(left)}$	Gini impurity index of the left leaf node
$G_{i(right)}$	Gini impurity index of the right leaf node
$G_{i(rootnode)}$	Gini impurity index of the root node
ω	Rotational speed
v	Welding speed
α	Tilt angle
r	Pearson correlation coefficient
x_i	Values of the x variable
y_i	Values of the y variable
$\overline{x}$	Mean of the values of the x sample

$\bar{y}$	Mean of the values of the y sample
μ_k	Centroid of the clusters
J	Objective function
T_i	Target output of the ith dataset
O_i^k	Output of the neural network model on the kth interaction when the ith dataset is considered
x_1 and x_2	Inputs
h_1 and h_2	Hidden neurons
w_1, w_2, w_3, and w_4	Weights
o_1	Output
α	Learning rate
s_1 and s_2	Input images
y	Binary indicator
α and β	Constants
D_w	Euclidean distance
f	Function
T	Transform
X	Original data
m	Number of features
N	Length of the features
K	Number of atoms of transform
Z	Coefficient matrix
λ, γ, α	Hyperparameters

Chapter—7

r	Pearson correlation coefficient
x_i	Values of the x variable
y_i	Values of the y variable
$\bar{x}$	Mean of the values of the x sample
$\bar{y}$	Mean of the values of the y sample
∇	Divergence
H	Magnetic field intensity
J	Current density
B	Magnetic flux density
A	Magnetic vector potential
E	Applied electric field
σ	Conductivity
N	Number of turns
I_{coil}	Current flowing in the coil
e_{coil}	Induced voltage
A_{coil}	Cross-sectional area

Chapter 1
Evolution of Manufacturing and Its Journey Towards Digital Twin

1.1 Overview

This chapter sets the prologue of industrial revolutions, starting from the first to the fourth. It aims at making the readers aware of the transformations happened so far. The tools of Industry 4.0 have been introduced; and "*digital twin*", which is one amongst them is the focus of this chapter. Its importance, along with the present and future potential has been highlighted.

1.2 A Few Fascinating Facts

Often, the mind wanders about how the historical monuments would have been erected without the modern-day construction technologies. Taking a glance at one of the world's most popular pre-historic monuments, the "*Stonehenge*" shown in Fig. 1.1, raises a fundamental question, '*how*'? It is the world's most famous stone circle located in England, the United Kingdom. As per archaeologists, the construction period lies between 3000 and 2000 BC [1]. It is fascinating to imagine the erection of this architecture. The modes of transport used to build it remains a mystery.

Another example of such historic architecture is the "*Ellora Temple*" situated in India (Fig. 1.2). It is believed that the entire temple has been carved out from a gigantic piece of rock around the period of 756–773 CE [2]. The beauty of the temple lies in the precision with which the rocks have been cut. Taking a closer look at the entire temple would leave us with some basic questions such as, how was the entire design first visualized and several such related questions which remain unanswered. These examples show that manufacturing existed with the very existence of humans.

S. K. Pal et al., *Digital Twin – Fundamental Concepts to Applications in Advanced Manufacturing*, Springer Series in Advanced Manufacturing,
https://doi.org/10.1007/978-3-030-81815-9_1

Fig. 1.1 Picture of "Stonehenge" located in England [1]

Fig. 1.2 A Picture of "Ellora Temple" located in India [2]

1.3 Manufacturing and the Industrial Revolution

The historic monuments have witnessed the human potential to start from scratch and create spectacular designs. They can create the best solutions to meet their requirements from time to time. For example, it is assumed that when man could not solely survive on plant-based foods, he longed for meat, and hence, he invented tools with sharp edges that can hunt animals for food. Thus, manufacturing in its most nascent stage began with the very existence of humans.

The word "*manufacture*" comes from two Latin words, "*Manu*" which means by hands, and "*Factus*" which means to make. Manufacturing is the process of working on raw materials for coming up with finished goods. It is a process of value addition by the use of tools, machines, and also human labour. Value addition has always been the prime objective of engineering and will continue through the coming ages. But, the thing that has got changed over the years is the way of manufacturing. The industrial revolution has brought this change. It is referred to as a revolution because it brought a significant change in technology and social life as well. The following section discusses the industrial revolutions to help in appreciating the changes that have occurred in the way of manufacturing over the years till the current scenario. To provide the readers with the real sense of revolution, examples of inventions in manufacturing processes occurring in each revolution are discussed.

1.3.1 The First Industrial Revolution

During the pre-industrialization period, manufacturing mostly occurred at home, and was performed by skilled artisans. It also involved the usage of animals' and human's muscular strength. This continued for several years until the introduction of the first-ever steam engine. The engine had a piston-driven motion, a non-rotary type, where the steam moved in one direction under power and returned to the start by gravity. The power that was generated by driving this steam led to a paradigm shift in the industrial context by replacing the animals' and human's muscle power. This power was utilized to run machines that emulated the work done for centuries by skilled artisans. People realized that it became much less expensive in the long run to operate the steam-driven machine instead of paying for animals and workers. The idea that a machine could cost less to operate while doing the same amount of work introduced an element of '*productivity*'. This is referred to as the first industrial revolution. It led to the same amount of work with a few men operating the machine for few hours than the previous arrangement of doing that work with a large number of men, animals, and for several days and nights. The first industrial revolution sustained from 1750 to 1850. The factories erected during this revolution consisted of general-purpose machines which were driven by mechanical prime movers. Neither the factory had a specific design, nor did it have a scientific layout. The factories had to manufacture a wide range of products as per their customers' demands, which thereby affected the rate of production and profitability. Furthermore, the jobs were highly labour-oriented.

Citing a few examples of manufacturing, '*machining*' in the pre-industrialization period meant the way of working on the metals by the use of handmade tools. It took several hours to complete one job, and no two jobs were alike. The concept of standardization was not prevalent during those times, i.e., there was no uniformity in the dimensions. The lathe is believed to be one of the most ancient tools, and it is considered as one of the significant factors of the industrial revolution. It has been termed as the "*mother of machine tools*" because it was the only machine tool available which led to the invention of other machine tools. An improved turning lathe entirely made of metallic structure, with guides for sliding the carriage holding the tool was introduced in this era. After the arrival of the steam engine, the lathe machine was powered by steam flowing through a shaft. The early machine tools were utilized for boring cylinders for steam engines. The milling machines also appeared in this era, where the tool with multiple cutting edges rotates and the job is fixed. Earlier, in the pre-industrialization period, the hand-operated files, which have multiple teeth or cutting edges were utilized for working on the jobs. The milling machine was later invented to serve this purpose.

Another example is the welding of metals. In this era of the first industrial revolution, joining occurred by heating the two metals together and then forging the same. It is not the way how conventional welding works today. The joining process in this era resembles the forging technology practised at the blacksmith shop, which is termed as "forge welding". Thus, the jobs were labour-intensive and inefficient. The technological and social changes that occurred with this revolution are mentioned below.

- **Technological change**: With the advent of the steam engine, the rate of production increased rapidly. This contributed significantly to the world economy. It enabled the miner to remove water and dig deeper which led to the discovery of various minerals. These minerals were the key contributors to the progress of the industrial age. Thus, the invention and implementation of the steam engine brought a significant change leading to amassing of the material and generation of power.
- **Social change**: The steam engines rendered the factory owners the ability to move from the conventional isolated riverside locations to the outskirts of the-then growing cities. It also led a few families to move to a town from the countryside to work in the factories. Gradually, the industrial revolution touched the lives of the people.

1.3.2 The Second Industrial Revolution

Several innovations followed after the steam engine such as petroleum (oil), internal combustion engine, automobile, electricity as the power source, telephone, electric motors, and the list follows. These inventions further replaced the use of muscular strength. In the context of manufacturing, the second revolution occurred on a much bigger scale with the introduction of the '*moving assembly line*' concept. The system consisted of a transfer line carrying the product to be manufactured. The workstations were divided, where each of the workstations performed one specific job only. Manufacturing in the first industrial revolution can be thought of as job-centric, i.e., the job stood still at one fixed location, while all the parts were brought up to the job by the labours for assembling. This concept took a turn in the second industrial revolution by moving the job and fixing the labours' location. The labours were divided into groups, which is termed, labour division. Each group was engaged to perform one specific work. This led to higher productivity. The second industrial revolution is identified between the period from 1850 to 1940.

Continuing with our discussion on advancement in manufacturing, an important contribution was the concept of '*interchangeable parts*' in the field of machining. The parts manufactured in the previous era did not have any standard. Thus, it was a tedious job when it came to replacement/repairing. The interchangeable parts refer to identical parts, which were manufactured with a particular standard. Thus, they fit a large variety of products, which intend to use that part. This enabled mass production. The electric motors were now utilized in the machines for moving the drives. These introduced the concepts of '*accuracy and precision*', which are now the crucial factors in manufacturing. Another remarkable invention was the '*punched card*' which was utilized for recording and storing information by the machines. It was a stiff card, where the data in binary format was entered by punching holes in it. Thus, a stack of such cards was built and used in the machine for performing specific operations.

With the discovery of the electric arc, the hammer of the blacksmith got replaced by a welding gun and electric grinders. The first-ever fusion welding was carried out by employing carbon electrodes for joining two lead plates. Also, the bare metal electrodes got utilized for joining the metals, which were used up in the molten metal pool. However, the quality of the weldment was found to be poor, because of the oxides and nitrides that were formed during the welding. This was because of the environmental attack which happened with the use of bare metal electrodes. This led people to realize the importance of shielding the arc and the weld pool from the atmosphere during the welding. This led to the advent of coated metal electrodes. This technique is known as "shielded metal arc welding". The technological and social changes that occurred with this revolution are mentioned below.

- **Technological changes**: The concept of the assembly line and the use of interchangeable parts reduced the manufacturing lead-time substantially. They also resulted in higher productivity and further reduced the usage of labour to some extent. Mass production of goods started from this era. With the internal combustion engine, automobiles were manufactured. The ironworks and coal mines were used to produce raw materials for the machinery.
- **Social changes**: The innovations further made a positive impact on the lives of the people. It led aspiring entrepreneurs to now invest their capital in factories to utilize new machines and workers to work on them. It also led people to move beyond cities, across countries, and continents. Similarly, mechanization occurred in other industrial sectors as well.

1.3.3 The Third Industrial Revolution

The previous two revolutions brought innovative power sources, the first is the steam engine, and the second is the electricity. These innovations helped humans in increasing productivity and efficiency by reducing the usage of animal and manual labour, and labour division. The incorporation of electronic components and various information technology (IT) enabled systems constituted the third revolution. This resulted in large-scale automation of manufacturing processes, reduced human involvement, and also the lead-time substantially. It also brought the computer to be a core part of the manufacturing operation for programming, planning, and controlling the activities. Other advancements in this era include the software for writing programs, data storage systems, sensors, etc. These increased the productivity of the factories further.

It is believed that the first numeric controlled (NC) machines were built in this era. NC represents a control system in which a machine is controlled using numbers. These numbers specify the action to be performed and were provided on the punched tapes. The objective was to control the movement of the cutting tool relative to the workpiece during machining to achieve precise jobs. To ensure the correctness of the position, a feedback system is also present in the machine. Over time, the computers have replaced the punched tapes, and they are now used as the controller in the NC machines. The programs are written on the computer, and hence, it is referred to as computer numeric control (CNC). Another concept, computer integrated manufacturing (CIM) came in soon after, which includes technologies such as computer-aided design (CAD) and computer-aided manufacturing (CAM). To provide the readers with a crisp definition at this point in the discussion, CAD refers to the use of computers for drafting, modifying, analyzing, and optimizing the design of a product, and CAM refers to the use of computers for managing the production of the product. CIM refers to the use of computers both for engineering technologies i.e., CAD and CAM, and business information such as, managing the inventory of raw materials, spare parts, machinery on the production shop floor, manufactured jobs, and so on.

The third industrial revolution witnessed several inventions in welding technology as well. In the case of the shielded metal arc welding technique, the electrodes get consumed. As an alternative, the "gas tungsten arc welding" technique was invented in which, the arc is delivered by tungsten metal. Tungsten melts at a very high temperature around ~3420 °C. Hence, it does not get consumed during welding. However, a suitable filler material is required to fill the gap between the metals to be joined. The invention of the "gas metal arc welding" technique was a further improvisation as it utilizes a spool of wire acting as the electrode. This wire gets fused with the metals to be joined. Thus, it eradicated the problem of frequent change of electrodes which is in the case of shielded metal arc welding technique and filler material in the case of the gas tungsten arc welding. Other techniques such as resistance spot welding, plasma arc welding, electron beam welding, etc. were also introduced. These inventions mark the steps taken towards higher productivity and automation in each invention, starting from the idea of forge welding to the use of conventional welding techniques. A few major inventions in welding technology also happened towards the end of the third industrial revolution namely, friction welding, friction stir welding, and laser welding, which are presently referred to as the advanced techniques for welding. These techniques have enabled the joining of various light-weight materials, and have brought new capabilities such as elimination of the use of filler material and joining of dissimilar materials.

Robots were also introduced in this era and were utilized in automobile manufacturing factories, mostly for welding. It is believed that the one reason which led to the use of robots was to perform jobs that are risky, accident-prone, and also repetitive. Nevertheless, the other reason was to obtain a product in time with consistent quality. These robots are fed with programs that are triggered one after the other in a particular sequence. So, they keep on operating following the sequence and do not stop until they are asked to. In a way, they foster industrial growth because of their ability to perform the same task over and over again with the same consistency, which is crucial in manufacturing. However, manufacturing kept evolving, and robots during this stage did not incorporate the intelligence of their own. The technological and social changes that occurred with this revolution are mentioned below.

- **Technological change**: From the business point of view, the investors witnessed massive automation and the requirement of hydraulics for driving factories. Technological innovations kept flowing to aid automation. There was a constant increase in the efficiency, production rate by the monitoring and control of the machines.
- **Social change**: A demand for trained and skilled labour arrived. The emergence of the automated machinery raised the usage of electronic components and electrical elements which also created several job opportunities.

1.3.4 A Summary of the Three Industrial Revolutions

A summary of the major inventions in the field of machining and welding in the three industrial revolutions is depicted in Fig. 1.3. This highlights the evolution of factories with technologies and innovations. It began with the use of handmade products until

	1st *Industrial revolution* Advent of steam engine	2nd *Industrial revolution* Moving assembly line	3rd *Industrial revolution* Electronics and IT systems
Advancements in machining	• *Use of handmade tools* • *Steam powered turning lathe* • *Steam powered milling machine*	• *Interchangeable parts* • *Conceptualization of punched tapes* • *Use of motors for drives in lathe and milling machines*	• *Introduction of NC machines* • *Inclusion of feedback in NC machines* • *CNC control* • *Computers in manufacturing* • *CIM, CAD, and CAM concepts*
Advancements in welding	• *Forge welding process*	• *Use of carbon electrodes for fusion welding* • *Use of bare metal electrodes for fusion welding*	• *Conventional techniques (SMAW, GTAW, and GMAW)* • *Resistance, Electron beam, and Plasma arc welding techniques* • *Advanced techniques (Friction, laser, and Friction stir welding)*

Fig. 1.3 Advancements in manufacturing (machining and welding) in each industrial revolution

the first industrial revolution came into the picture wherein water and steam power were employed to manufacture products. Later, the usage of electricity and moving assembly line concept for mass production, and then the integration of electronics, electrical, and mechanical systems came in through computers. The inventions in the three revolutions dictate the very fundamental questions of manufacturing that are: how can the process be made faster, better, and cheaper. With each stage, the milestones of manufacturing were realized which are productivity, quality, accuracy, precision, and low production cost. Thus, the world is adapting to innovative technologies and consumer demands.

Figure 1.4 shows a qualitative description of the three industrial revolutions. The major trend in each revolution is also shown in the picture. It can be summarized that with each passing revolution, the use of labour has decreased. This has been possible with the invention of machines, electricity, electronics, IT systems, and robots. These inventions have brought several innovative concepts such as, the use of transfer lines, CIM, etc., whose implementation has reduced the usage of labour. Machines have brought the importance of production in manufacturing, and thus, it has been increasing with each revolution. Finally, this increase in productivity with less usage of labour is the reason for the decrease in the cost of production.

The successor of the third industrial revolution has already stepped in, which is popularly known as "*Industry 4.0*". The fourth industrial revolution is the ultimate result of the fast and rapidly developing digital life as witnessed by us today. The following section discusses the details of this revolution.

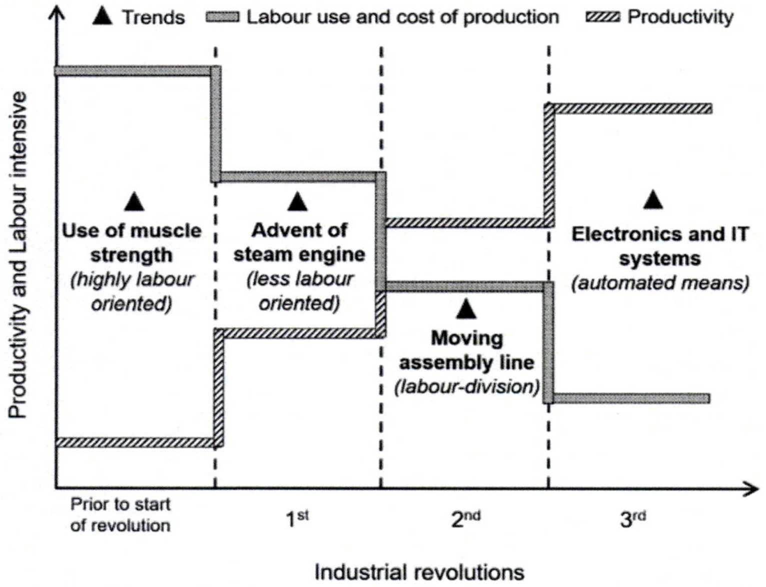

Fig. 1.4 A qualitative description of the industrial revolutions

1.3.5 The Fourth Industrial Revolution

The term Industry 4.0 has been coined by Germany, and it refers to the digitalization of manufacturing. The major trends in this revolution include the "*Internet of Things (IoT)*", "*Cyber-Physical Systems (CPS)*", "*Cloud Computing*" etc. [3–5]. These trends will be discussed eventually in the later sections. The following paragraphs discuss Industry 4.0 in more detail.

From the Industry 4.0 point of view, there exist two worlds, one is a "*physical world*" and the other is a "*digital world*". The first one comprises of the physical things, i.e., the tangible things (which can be seen with bare eyes and can be touched). Citing the examples of the physical world, it could be a machine or a part, a tool, a sensor, any device inside a machine, etc. The second one includes all digital things such as data, algorithms, processing applications, and mathematical models, etc.

It can be an advocate, these things existed long before, so what is it innovative? The difference is in '*connectivity*'. Industry 4.0 aims to merge the two worlds, i.e., establishing real-time communication between the two worlds. This connection would enable them to exchange information amongst themselves in real-time. This information refers to the manufacturing data. The objective is to replace the traditional mechanical chains as used during the nineteenth century with the digital chains where data is the building block. The machines on the manufacturing shop floor are intended to be connected with various sensors and actuators which will facilitate the digitalization of different activities. Sensors will be utilized for collecting the data enabling a means of decision making and actuators for controlling the machine as feedback. In simple terms, the sensors and actuators will become the ear and eyes of the machines/manufacturing system enabling real-time monitoring and control. This would create a flexible and autonomous means of manufacturing.

1.3.6 Advancement of Industry 4.0 Over the Third Industrial Revolution

Automation in manufacturing started in the third industrial revolution with the application of electronics, computers, and IT systems. Automation is also a key aspect of the fourth industrial revolution. To clear this ambiguity, it is essential to note that the objective of automation in both revolutions remains the same. However, the difference lies in the '*way*' or manner used to fulfil this objective.

For instance, let us consider a robot employed for performing some shop floor activity, as shown in Fig. 1.5. The robot is expected to shift a job kept at the 'initial location 1' which is incoming from a transporting unit. For this, the robot follows the 'path' to transport the job and place it at the 'final location'. To perform this operation, the robot is trained with some logic using the programmable logic controller (PLC). The process is automated by sequences pre-set by the operator. This largely remained the idea of automation in the third industrial revolution. The robots were

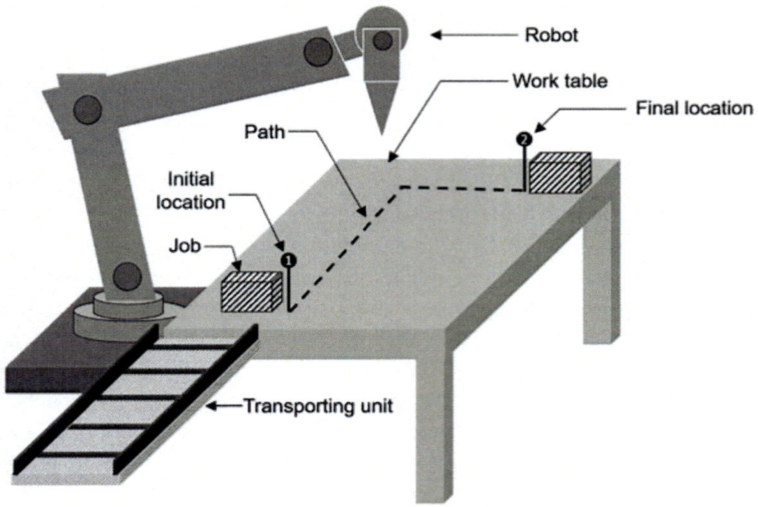

Fig. 1.5 Robot in Industry 3.0 era

also integrated with a few sensors, mostly for determining the position and speed. The robot keeps operating until it is not stopped by the user. Thus, in case of any failure, the activity of the robot, gets halted. The health of the robot is purely dependent on the scheduled maintenance scheme as well. Collecting all these facts, it shows that the robot lacked intelligence with regards to its way of operation. Next, the idea of Industry 4.0 is explained in this same situation.

As highlighted earlier, Industry 4.0 is about connectivity, i.e., connecting the machines (robots) with sensors, digital systems etc. so that real-time insights into the critical processes can be derived. Figure 1.6 attempts to explain this connectivity. The robot shown in this picture will be integrated with various sensors, which will be crucial for the production process being performed and the health of the robot. As the operation begins, the data acquired by the sensors will be transferred to a digital system (computer as shown in the picture) via the internet. This data will be utilized for online descriptive analysis of the process, feedback to the robot for process control, performance alarms, health prediction by historical trend analysis, uptime and downtime overviews, etc. In this way, the level of automation will be higher. This interconnection between the physical machine, digital systems, and human is referred to as "*interoperability*", which is one of the five design and operating principles of Industry 4.0. The second principle is termed "*real-time capability*", which is the transfer of the data from the physical machine to the digital systems and vice versa.

Robots in the fourth industrial revolution will also have the 'autonomy' feature, in addition to the automation achieved through the use of PLCs. This feature is explained as follows. Consider an obstacle in the path of the robot. It is expected that the robot must be able to identify the obstacle, if any, in its path, and accordingly choose the path to reach the dispatch location. But how will a robot achieve this, if

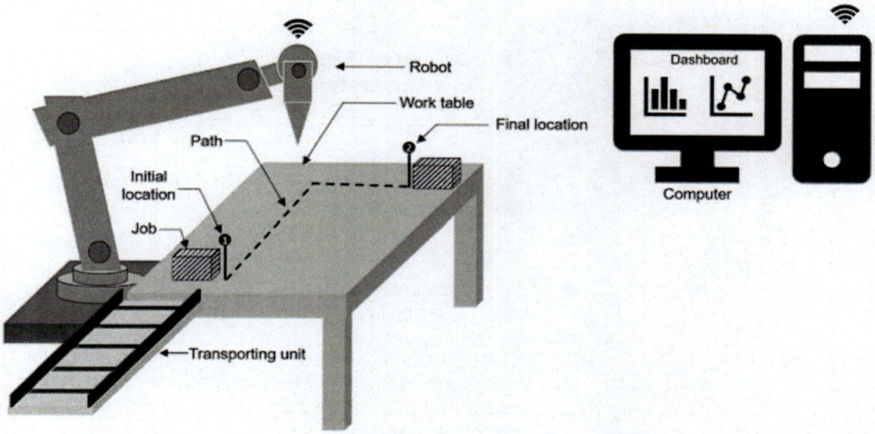

Fig. 1.6 Robots in Industry 4.0 era

it has been only programmed to follow a single path? The concern now is that the robot needs to be intelligent for this purpose. The engagement of sensors into the robot will provide it 'eyes' to look around itself, and identify the obstacles, if any, in its path. Moreover, once the robot identifies the obstacle, it has to choose a new path for which, it has to perform a few computations in real-time. These computations are performed on the sensory data for real-time decisions. The autonomy feature also demands the robots to be 'adaptable' to the working environment. For instance, there would be several robots working in a production line, which may be engaged to perform a similar type of operation. If one of them detects an error and configures itself, other robots also adapt themselves accordingly. This is termed as "machine-to-machine" communication, where the machines are connected via the internet and are communicating amongst themselves by exchanging data. In this way, machines themselves can identify the maintenance requirement by comparing their data with that of the others.

Thus, Industry 4.0 aims at integrating the internet into manufacturing activities for attaining higher benefits. In this revolution, data is the building block that will connect devices over a network. Considering the industries in this era, several machines (robot-based) are needed in a production line. The real-time insights about the manufacturing process and the health of the machine will be highly beneficial as it will help in making more profitable decisions. In a nutshell, it will increase competitiveness and will allow unmatched levels of productivity. From these discussions, it can be observed that the success lies in coming up with the digital counterpart corresponding to a physical machine. This representation is termed as "digital twin", which is a component under the umbrella of Industry 4.0. The details of the digital twin have been introduced in the following section. Figure 1.7 pictorially summarizes the contributions of the four industrial revolutions, i.e., from the first to the

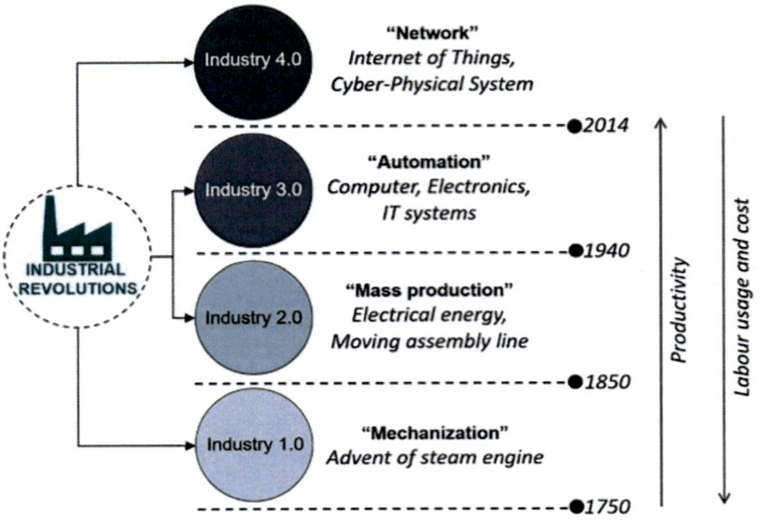

Fig. 1.7 Evolution of the industrial revolutions from Industry 1.0 to 4.0

fourth. The major trend in each revolution such as mechanization, mass production, automation, and network in Industry 1.0–4.0, respectively is also shown. In the previous section, Fig. 1.4 summarizes the achievements starting from Industry 1.0 to 3.0. The inventions in the three revolutions dictate the very fundamental questions of manufacturing that are: how can the process be made faster, better, and cheaper. As Industry 4.0 has arrived, productivity will further increase lowering the cost of production because of the real-time insights of the manufacturing activities enabling decision making and control.

1.4 Introduction to Digital Twin

A digital twin is one of the concepts of Industry 4.0. It refers to the creation of a virtual model which would correspond to a physical system. To simplify the concept, the term "*twin*" signifies a replica. This replica could be of a human, product, process, or machine. The other term "*digital*" signifies its mode of existence. Hence, digital twin refers to a model that can be built and deployed in a computer and can replicate a physical entity. Factors to be considered here are: how to replicate, and to what extent? Further, one could also see this idea similar to a CAD model. This is so because a CAD model is also built and deployed on a computer and it resembles a product. However, this is not the objective of a digital twin. A CAD model is responsible for product design, and it is heavily based on the designer's experience. A digital twin is expected to serve throughout the life cycle of the product. It is intriguing to think about 'throughout the life cycle', but one needs to understand the purpose.

Let us consider the case of purchasing a new automobile. The automobile manufacturer must have performed several tests to ensure the integrity and performance of this product before introducing it in the market. Here, the manufacturer would have utilized the CAD models only. Suddenly, at some point, the product starts malfunctioning or produces some unknown errors, it may be brought to the service centre. The service centre identifies the possible errors and rectifies them. This is a regular practice. Often, this practice introduces several problems because of no prior information on the fault that kept growing in the product. This may lead to repair expenses and may consume a huge time as well. The problem can be even bigger considering the example of a machine, or a manufacturing system consisting of several machines. This sort of problem will introduce heavy downtime, which might prove devastating to the industry. Industry 4.0 provides a new outlook to this practice by introducing the concept of the digital twin, which is an internet-based "product-service facility". The digital twin will also have the CAD models of the physical entity, but at the same time, it is also provided with the latest sensory data. The twin will utilize the sensory data to update its model, and provide real-time information such as present condition, remaining useful life, suggestions to improve the present condition, etc. about the physical entity. Thus, the twin attempts to replicate the life of the physical entity by mirroring itself through the sensors' data. Thus, a digital twin of an automobile is expected to keep a track of the critical components inside the engine, predict the failure of components, dynamic modelling for fault diagnosis, etc.

1.5 Key-Aspects of a Digital Twin Model

Figure 1.8 pictorially shows a digital twin residing in the virtual machine for the descriptive and prescriptive analyses of a physical machine. The data acquired from the physical machine through the sensors ($sensor_1$, $sensor_2$, $sensor_3$, …, $sensor_n$) will be transferred to a virtual machine via the internet in real-time. The virtual machine will contain the trained simulation and data-driven models which will act on the received data to derive information about the physical machine. The collected information may be fused to derive other associated information as well. The insights will be made available in a dashboard, as shown in the figure. The real-time insights will assist the manufacturer in tracking the operation of the components, the health condition of critical components, suggestions to improve the present performance of the components, etc. In accordance to the above, certain initiatives may be adopted for rectification of the product while incorporating the feedback into the process as well. Thus, the usefulness of a digital twin model will depend upon the analytics that will be performed based on the data acquired from the sensors engaged with the physical devices. The supporting analytics will only form the intelligence of a digital twin model. These discussions pertain to the third principle of Industry 4.0 termed "*virtualization*", which aims at the creation of virtual models and simulations for the

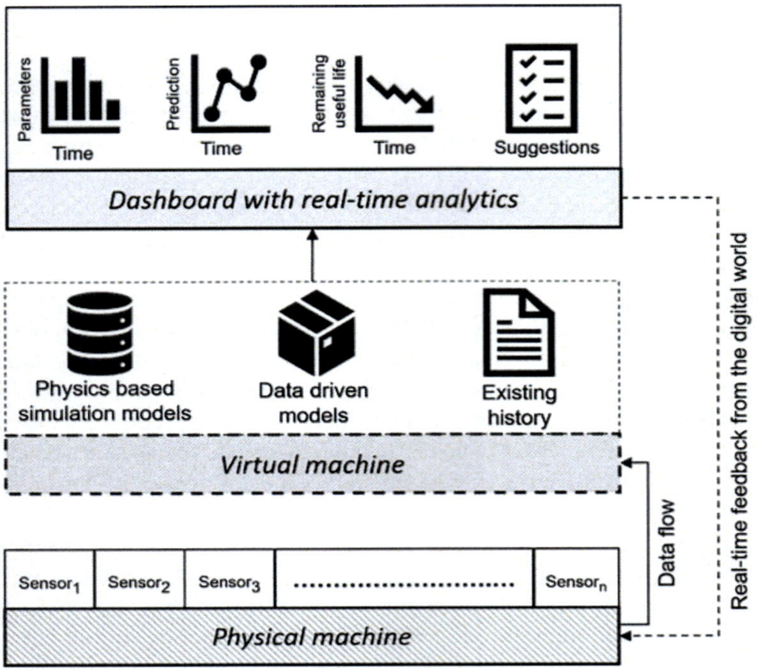

Fig. 1.8 Pictorial representation of a digital twin for descriptive and prescriptive analyses of a physical system

physical factory. The next essential thing in this discussion is to understand how to channelize this entire model.

1.6 Components for Envisaging the Digital Twin

This section will discuss the essential things required to channelize the flow of information to envisage a digital twin model. Figure 1.9 depicts the components required for envisaging the digital twin. The attributes have been shown to the left of the figure. For better clarification, these attributes have been shown in the same representation which is depicted in Fig. 1.8. The five attributes are the components of Industry 4.0. Each attribute is discussed in the following sections to understand its purpose in detail.

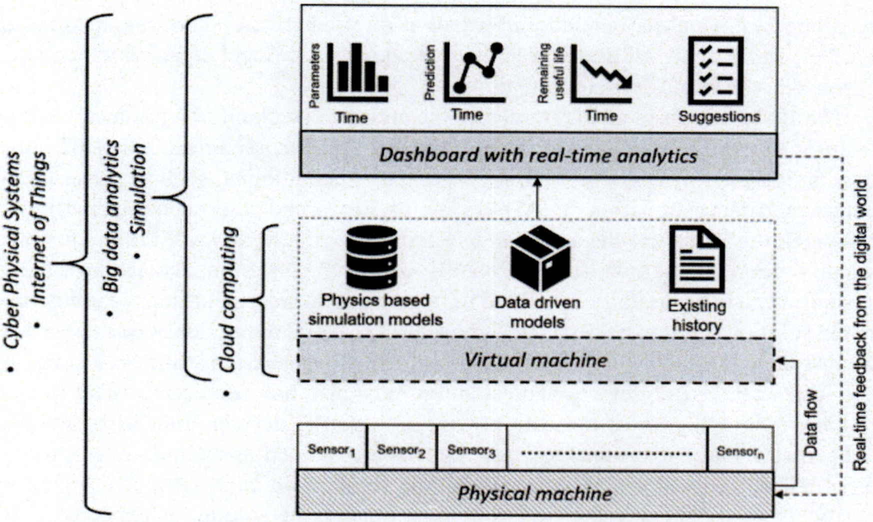

Fig. 1.9 Attributes for envisaging digital twin

1.6.1 Internet of Things

As per a recent definition, "*The Internet of Things can be defined as a world of interconnected things that are capable of sensing, actuating, and communicating among themselves and with the environment (*i.e., *smart things or smart objects)*" [6, 7]. The term IoT specifies all the smart things around us. A common example of daily life could be a smartphone. The reason a phone is being referred to as smart is that it is capable to communicate with other devices over the internet. It can send data, and also receive data. In simple terms, when a user opens "*YouTube*" from his smartphone with internet connectivity, it requests the website to play a song, and the website responds by playing that song. To build a digital twin, smart sensors will be required to be engaged with the physical machine. These sensors will sense, collect, and transfer the information wirelessly (i.e., over the internet) to the virtual machine. These smart sensors, the machine, and the connected virtual machine can be referred to as the IoT.

The IoT in application to the industrial sector is specifically referred to as the "Industrial Internet of Things" (IIoT). The term 'smart' is encapsulated with four features namely, *monitoring, control, optimization,* and *autonomy*. Thinking from the perspective of manufacturing, '*monitoring*' refers to the knowledge of surroundings i.e., the environmental data, collecting the process data, data about the health of the machine, etc. This is aimed at enabling real-time alerts and decisions about the manufacturing process. '*Control*' refers to the use of software residing in the cloud enabling the control of the process, as and when required, to improve the quality of the product. By enabling monitoring and control, the said manufacturing activity can be *optimized*. This may include information on predictive diagnostics, repair, etc. Further, while doing all these three, *autonomous control* can be gained which aims at self-diagnosis and self-coordination.

The IIoT is a subset of IoT, targeting the entire value chain of a product. Certain features that differentiate the IIoT from the IoT are discussed next. The IIoT solution is expected to co-exist with several legacy technologies such as supervisory control and data acquisition (SCADA) systems, PLCs, and other such systems, based on which the industrial machines and systems are existing today. Many industries, which account for the small-scale economy, consist of conventional machines. These systems need to be digitally retrofitted for making them smart. Further, the industrial-based solution needs to be capable of considering several thousands of devices. Thus, it must include suitable communication protocols to ensure the seamless transfer of data across thousand devices. The solution must also have dedicated check-points to ensure the integrity of the data flowing into digital devices. Industries are also concerned about cyber-attacks. These attacks are defined as an attempt to expose, alter/delete, steal etc. the manufacturing data resulting in mishaps. Thus, "cybersecurity" is very critical in the IIoT solutions. Finally, the solution is expected to be flexible enough to cater to a large number of manufacturing processes and problems, and robust in terms of accuracy and performance. This is so because it will be handling an entire industrial infrastructure.

1.6.2 Cyber-Physical Systems

As per a recent definition, "*Cyber-Physical Systems has emerged as a unifying name for systems where the cyber parts,* i.e., *the computing and communication parts, and the physical parts are tightly integrated, both at the design time and during operation*" [2]. As discussed above, Industry 4.0 is all about the physical and digital worlds. The CPS is a term used for the integration of the physical world with the digital world. The digital world processes the data received from the physical world using computer algorithms, processing applications, and models to derive useful information about the component in the physical world in real-time. This digital world forms the cyber component. Decisions are taken based on this derived information which is sent as feedback to the physical machine from the digital world or it could also be in the form of suggestions for improving the condition of the machine.

1.6.3 Cloud Computing

Earlier, there was frequent use of hard drive, pen drive, etc. for storing the data. The practice has now shifted to services such as "*pCloud*", "*OneDrive*", "*Box*",

"*Dropbox*", "*Google Drive*", etc. This shift in practices is possible because of digitalization, which has led people wish to access their data at all times, from any location, and be secured. These aforementioned platforms are termed cloud services, and they have storage, application, and processing power. The major features of a cloud service are providing: (a) access whenever it is needed and through the web, (b) flexibility for expanding or shrinking a drive space, (c) flexibility of sharing a space, and (d) usage statistics, etc. In the context of building the digital twin, a cloud server is the virtual machine, which will be required to manage the manufacturing data. This server will contain the process applications and digital models which will operate on the data. More details on cloud computing are discussed in Chap. 5.

1.6.4 Simulation

Manufacturing industries always aim at improving the quality of their products, lowering the production expenses, increasing the efficiency of the process, etc. To achieve these objectives, research is conducted on new, advanced, and innovative materials, innovations in tooling utilized in manufacturing, extensive optimization of production parameters, and many such fundamental studies. This requires investment in procurement, design, experimentation, validation, prototyping, etc. Instead, if all these activities can be performed by simulating a real-time environment, the investment can be significantly brought down.

Simulation can be defined as the imitation of a real-world process or system over time [8–10]. It helps in understanding the process or the system by imitating the activities of that process or the system in a controlled environment. It also provides flexibility to ensure the results by repeating the simulated conditions. Once built, it acts as a digital model of that process or the system. While simulation is popular for designing and its verification, it can bring larger benefits to manufacturing. As the digital twin is expected to provide the complete picture of a physical system, the sensors' data alone cannot fulfil this objective. This is because a product or a system is made up of numerous sub-components, and a sensor cannot acquire all the information. However, owing to the importance of these entities in the process or the system, information of the same may be obtained by creating a virtual environment for that process or the system. The virtual environment is the simulation. The benefits of the simulation model include: identifying opportunities that could lead to cost savings, identifying the manufacturing bottlenecks, determining the variation of physical parameters, verifying the suitability and efficiency of new tools, materials, etc. in a process.

1.6.5 Big Data Analytics

The term "data analytics" or "data science" refers to the process of playing with the data. Of course, the objective is to harness useful information from the data. This is quite straightforward, but, before one starts playing with data, it is imperative to understand the following: What is the source of the data? What the data corresponds to? In which form the data is available? What is to be derived? Are there any dependencies in the data? What will be the usage of the result that would be derived from the data? It is worth mentioning that data is the fuel for the successful implementation of Industry 4.0 in manufacturing. With the rapid digitalization, there will be a few billions of devices connected to digital platforms, which will be continuously sharing and receiving data. Likewise, for an industry, thousands of sensors may be engaged with the physical machine, which might be having different sampling rates, and different output formats. Thus, there would be an enormous amount of data at each instance of time. The term 'big data' refers to this large volume of the dataset and is characterized by four features namely: volume, velocity, variety, and veracity [11–13]. With billions of digital assets and thousands of sensors in machinery, there will be a large volume of the dataset, with different velocities of acquisition, variety of formats, and veracity referring to the amount of useful information. The term 'big data analytics' is about collecting and analysing this huge dataset to discover valuable information and hidden insights. For industrial applications, the tasks included are real-time processing of the data for identification of faults, predicting the quality of the product in real-time, predicting the optimized parameters for better quality, predicting the health of machine components, scheduling maintenance of the manufacturing machine, etc. The analytics will apply to both data-driven and physics-based approaches. For a manufacturing process, both these approaches are important because the data-driven is built on the historical data, and the physics-based is built on mathematical models. Hence, coming up with a hybrid model by combining the two approaches will be more beneficial as it would increase reliability.

1.7 Utilities of the Digital Twin in Manufacturing

The digital twin will become the most powerful and intelligent consultants in the industry. Figure 1.10 depicts the major utilities of a digital twin. The following discusses these utilities along with their benefits.

The digital twin will be powerful because, with the help of its constituents i.e., physics-based simulation and data-driven models, it will operate on the real-time data collected from the ongoing operations or systems, and will predict the performance or errors, if any. Thus, there will be complete transparency between the ongoing operation and its quality. Further, if any defect/error is identified, the digital twin will suggest corrective measure in real-time as well. With this information, the investment in offline quality inspection shall be eradicated, and machine downtime can also be prevented. The digital twin is expected to become intelligent with continuous access to sensory data and analytics. The digital twin can predict the remaining useful life of an asset by analysing the sensors' real-time data along with the historical information. The aforementioned points make way for a higher level of machine availability by reducing the downtime, leading to an increase in the efficiency of the

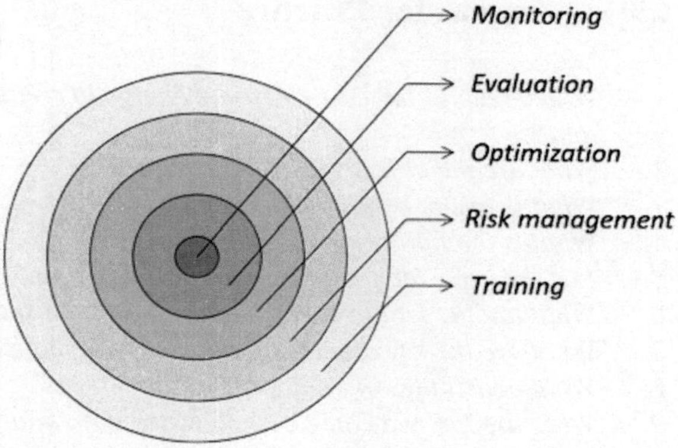

Fig. 1.10 Utilities of a digital twin

production unit and the profit. The digital twin can also be utilized for optimization and risk management. The twin can be used to simulate a real-world situation, and the outcome of the same can be visualized in the virtual platform. This will help in the planning of the innovation that could be successfully applied in the physical system by avoiding the occurrence of defects and errors. It could also be utilized for training the employees in an organization about the operation of the critical components. The conventional practice includes the building of prototypes, which are first tested to determine the flaws, and accordingly, the necessary changes are made. This may prove unsafe because of the associated industrial hazards. A digital twin will provide a virtual platform for learning. This will facilitate the rectification of the problems faced by the operator in the model itself eliminating the processes of backtracking or redesigning of the prototypes for further testing. Thus, a digital twin is expected to help industries in upskilling their workforce at minimum costs.

1.8 Summary

This chapter introduced the industrial revolutions, starting from the first to the fourth. A philosophical approach could be seen throughout the discussions, along with some real-world examples. It discussed the major trends in each revolution, along with the technological and social changes brought by them. The journey of the fourth industrial revolution to its present status has also been discussed with a suitable example. These discussions have been converged towards the importance of digital twin in Industry 4.0, where its role has been discussed in detail. This has been the prime objective of this chapter. For enabling a digital twin, the associated technologies such as the IoT, CPS, cloud computing, big data analytics, and simulation have been explained as well. The chapter is concluded by mentioning the benefits of the digital twin. The next chapter will discuss the sensors, which are a crucial part of the physical world, as observed in the discussions made in this chapter.

1.9 Questions for Practice

1. *Which revolution used electrical energy to create mass production for the first time?*
2. *What are the objectives of Industry 4.0?*
3. *What is meant by autonomy?*
4. *What is the full form of IIoT?*
5. *List the expected features of an IIoT solution.*
6. *What are the features which make a product smart?*
7. *What are the key aspects of a digital twin model?*
8. *What constitutes big data?*
9. *What are the two categories to which data analytics can be applied?*
10. *List the benefits of a digital twin.*

References

1. Wikipedia Stonehenge. https://en.wikipedia.org/wiki/Stonehenge. Accessed 6 Jan 2021
2. Wikipedia Kailasa Temple, Ellora. https://en.wikipedia.org/wiki/Kailasa_Temple,_Ellora. Accessed 6 Jan 2021
3. Gilchrist A (2016) Industry 4.0. Dyna 90:16–17. https://doi.org/10.6036/7392
4. Mishra D, Gupta A, Raj P et al (2020) Real time monitoring and control of friction stir welding process using multiple sensors. CIRP J Manuf Sci Technol. https://doi.org/10.1016/j.cirpj.2020.03.004
5. Roy RB, Mishra D, Pal SK et al (2020) Digital twin: current scenario and a case study on a manufacturing process. Int J Adv Manuf Technol 107:3691–3714. https://doi.org/10.1007/s00170-020-05306-w
6. Press I, Perez R (2018) Internet of Things A to Z. https://doi.org/10.1002/9781119456735
7. Khanna A, Kaur S (2020) Internet of things (IoT), applications and challenges: a comprehensive review. Springer, US
8. Rout M, Pal SK, Singh SB (2016) Finite element simulation of a cross rolling process. J Manuf Process 24:283–292. https://doi.org/10.1016/j.jmapro.2016.09.012
9. Rout M, Pal SK, Singh SB (2017) Finite element modelling of hot rolling. In: Computational methods and production engineering. Elsevier, pp 83–124
10. Jain R, Pal SK, Singh SB (2018) Finite element simulation of pin shape influence on material flow, forces in friction stir welding. Int J Adv Manuf Technol 94:1781–1797. https://doi.org/10.1007/s00170-017-0215-3
11. Da XuL, He W, Li S (2014) Internet of things in industries: A survey. IEEE Trans Ind Inform 10:2233–2243. https://doi.org/10.1109/TII.2014.2300753
12. Lee J, Davari H, Singh J, Pandhare V (2018) Industrial Artificial Intelligence for industry 4.0-based manufacturing systems. Manuf Lett 18:20–23. https://doi.org/10.1016/j.mfglet.2018.09.002
13. Tao F, Sui F, Liu A et al (2019) Digital twin-driven product design framework. Int J Prod Res 57:3935–3953. https://doi.org/10.1080/00207543.2018.1443229
14. Matruprasad, Rout Surjya K, Pal Shiv B, Singh (2016) Finite element simulation of a cross rolling process. J Manuf Process 24283–292. https://doi.org/10.1016/j.jmapro.2016.09.012

Chapter 2
Sensor Electronics for Digital Twin

2.1 Overview

For building a digital twin, sensors and associated electronic components will be at the core as they would sense and gather necessary information from the physical machine. As pointed out in the previous chapter, the sensors and actuators will be the 'ear' and 'eyes' of the manufacturing systems enabling real-time monitoring and control. Some sensors are to monitor the process while the rest are for monitoring the health of the machine. This chapter will give an exhaustive overview of the sensor electronics applicable in the manufacturing processes and systems. The chapter is also filled with several real-world case studies, wherein, the engagement of sensors in the manufacturing processes is described and shown pictorially. Further, the data collected from these sensors are also shown in the form of plots to mark their importance in the respective manufacturing processes/systems. Discussions are also highlighted on how to decide the selection of sensors for an application.

2.2 The Need for Electronics in Manufacturing

The term 'electronics' has been conceived from the word 'electron'. The electrons are charged particles which upon flowing through a conductor generate an electric current. This section will highlight the need for electronic components in manufacturing. Let us consider the example of an automated production line for the manufacturing of automobiles. This production line has a set of operations that are being performed automatically maintaining a sequence. The operations may consist of manufacturing of the chassis, assembling of the sub-components, welding, painting, and aesthetic enhancement. These operations will be performed in dedicated workstations, which are interconnected employing a transfer system. To channelize the transfer system automatically, servo motors are engaged, a few robots are installed

S. K. Pal et al., *Digital Twin – Fundamental Concepts to Applications in Advanced Manufacturing*, Springer Series in Advanced Manufacturing,
https://doi.org/10.1007/978-3-030-81815-9_2

in each work-station for operating on the automobile, and some instruments are engaged to track the position of the automobile in the production line, and so on. These components i.e., motor, actuator, the instrument for sensing, etc. are all electronic components that enable the automation of the manufacturing process. The engagement of the mechanical system with the electronic components makes the process more efficient and reliable. As automation is at the core of the fourth industrial revolution, thinking, framing, and implementing Industry 4.0 without electronics will be next to impossible. The next section introduces the sensor and elaborates its role in manufacturing.

2.3 Introduction to Sensor and Transducer

Figure 2.1 is a schematic diagram for understanding the sensor and the transducer. A module with an ability to sense and respond to changes in a system can be called a sensor. The changes in a system refer to the various physical quantities, for example, during a machining process, they can be force, torque, vibration, position, etc. The sensor module consists of elements that can capture the occurring change. This is referred to as the sensing element. This response of the sensing element is sent to a device that can show the detected information. Many times, it is essential to convert one form of energy into another to extract and read the information. This conversion of energy is termed transduction and the operation is served by a device called a transducer. It converts the non-electrical quantities into an equivalent electrical quantity. A sensor might come with a transducer connected to it to convert the information.

Figure 2.2 depicts the components of industrial automation, which are sensors, controllers, and actuators. The information acquired by a sensor is utilized for controlling a process via controllers and actuators. The actuators are devices that convert an electrical signal into an action. This art of measuring and controlling via instruments is referred to as "instrumentation".

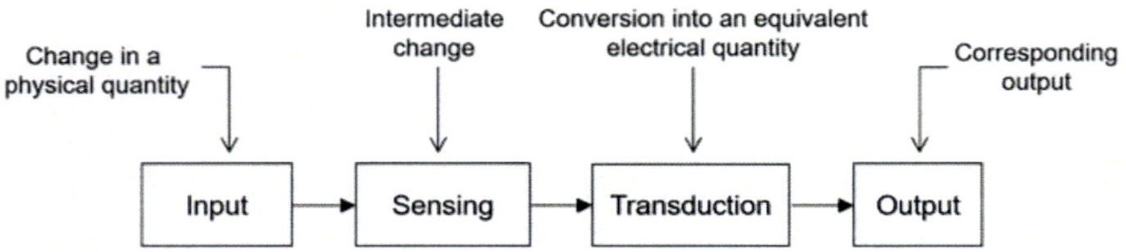

Fig. 2.1 Schematic diagram for a sensor and a transducer

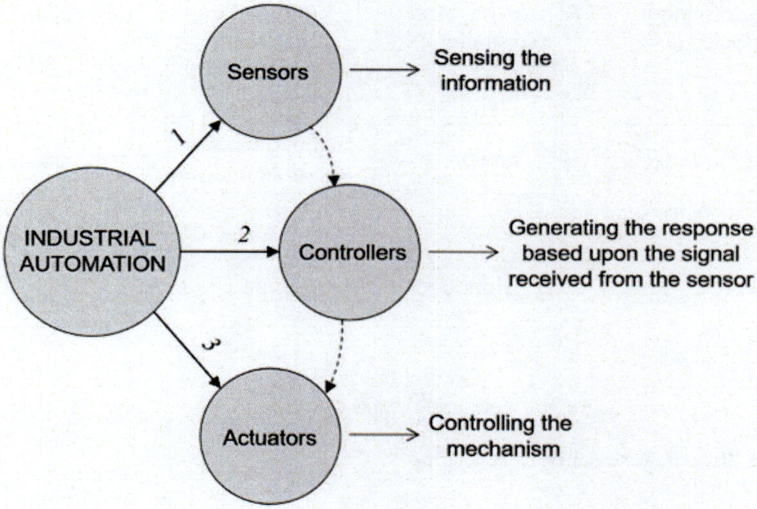

Fig. 2.2 Components of industrial automation

2.4 Role of a Sensor in Manufacturing

The role of a sensor is to sense and provide the information in a readable format. To begin with, let us consider an instrument 'stethoscope', which doctors use during the diagnosis of patients. The doctor puts one end of the stethoscope on the patient's body, and the other end has earpieces. This instrument transmits the low-volume sounds (such as heartbeat) to the ear of the doctor. The listener (doctor) identifies the abnormalities in those sounds, if any, and accordingly diagnoses the patient and prescribes necessary medication. In the previous chapter, sensors are framed to be the 'eyes' and 'ears' of the manufacturing machine/equipment. The stethoscope is the 'eye' and 'ear' for a doctor which is used for sensing the human body. Thus, the existence of sensors roots back to the very old days.

Similarly, sensors in a manufacturing process are also utilized for diagnosing the process. For instance, considering a machining process, one of the objectives will be to track the condition of the cutting tool during the process. Figure 2.3 shows a block diagram for illustrating the role of a sensor in manufacturing. The process begins with the engagement of a sensor, and the selection of the sensor depends upon the objective. For monitoring the condition of a cutting tool, the prevailing force or the current being consumed during the machining process could be crucial, and for that, one can use a force sensor and a current sensor. So, the sensor senses the physical information, which in this example, will be the force generated or the current consumed during machining. The signal, which is the output of the sensor is processed via signal processing and data analysis techniques to derive useful information. This information indicates identifying instances from the signal, which

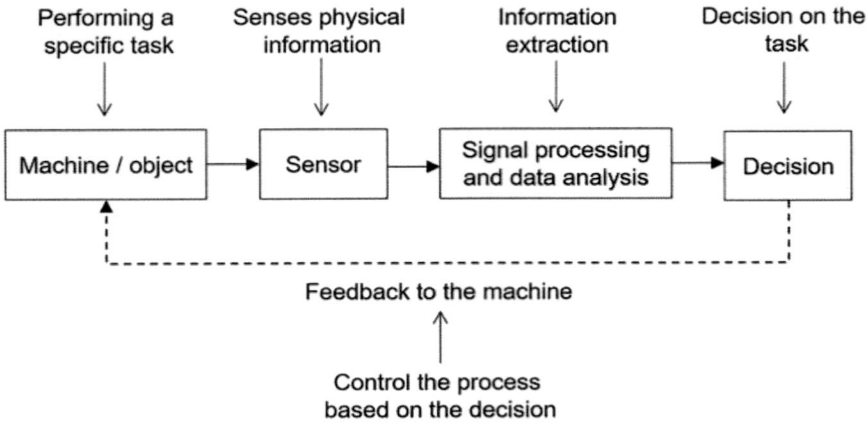

Fig. 2.3 Role of the sensor in manufacturing

resembles the failure of the cutting tool, and similarly, the next step may be to identify the instances of the signal which corresponds to degradation of the cutting tool, and so on. Finally, this derived information will help in deciding on the life of the cutting tool, and accordingly, the feedback can be sent to the machine.

Thus, the sensor forms an integral part in the automation of the manufacturing process by sensing the physical underlying parameters and providing it as a signal which could be processed. To imprint the usage of the sensor in manufacturing (considering machining process) on the example of doctor's stethoscope, the machine/equipment is the patient whose cutting tool is being diagnosed, the instrument is the force or the current sensor, the signal processing and data analysis techniques become the doctor who will derive the information, and the decision becomes the medication for the cutting tool. The following section describes the various types of sensors being employed to carry out different intelligent activities.

2.5 Types of Sensors

This section will elaborate on various types of sensors used for industrial automation. The discussion will include the working principle of the sensor, along with the advantages and limitations of the sensor. In the end, the illustration of real-world data collected by that sensor is also shown. This section intends to make the readers familiar with the various types of sensors. The understanding of the sensor is like keeping the stepping stone for implementation of Industry 4.0 in manufacturing. It will also help in identifying the type of sensor required for a particular application.

2.5.1 *Position and Displacement Sensors*

Position sensors determine the position of a body or component they are attached to. The position is identified concerning a reference point. The displacement sensor measures the amount by which a body has been moved. Both position and displacement sensors are available in two typical configurations namely, contact and non-contact. The former category refers to the one where there is physical contact between the component being sensed and the sensor, while there is no physical contact between the component and the sensor in the latter category. The manufacturing machines comprise heads over which the components are mounted. It is crucial to track the position and displacement of the heads during the manufacturing process, which enables feedback to the controller. The following are some of the examples of sensors used for determining the position and the amount of displacement of a component.

2.5.1.1 Differential Transformer

The differential transformer is also referred to as the linear variable differential transformer (LVDT), which is a transducer and is widely used in determining the linear position and displacement [1]. Figure 2.4 depicts the schematic picture of an LVDT. It consists of a hollow metallic cylinder and a push rod which is of smaller diameter than that of the hollow cylinder. The push rod can freely move back and forth along the long axis of the hollow cylinder. It is named LVDT because it measures the linear displacement, it has two coils, one is the primary and the other secondary which is similar to a transformer, and the output corresponds to the voltage difference of the secondary windings.

The working principle of this sensor is depicted in Fig. 2.5. It measures the displacement along a single axis of the object. The sensor is a mutual inductance device and it carries three coils, positioned symmetrically in an insulated tube. Out of the three coils, the central coil is the primary coil, and the other two are secondary coils connected in series with the primary coil. There is a core magnetic element, free to move through the tube along its linear axis. The linkage between the component being measured and the core element is made through a push rod, such that, any displacement of that component would move the core element from its position. The primary coil is driven by applying an alternating current, which induces the voltage

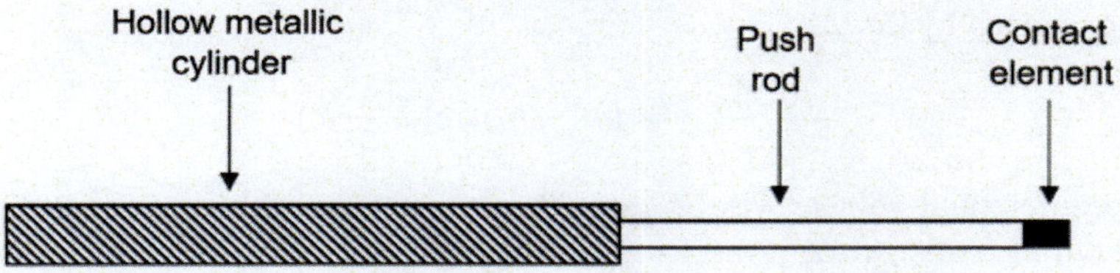

Fig. 2.4 LVDT sensor

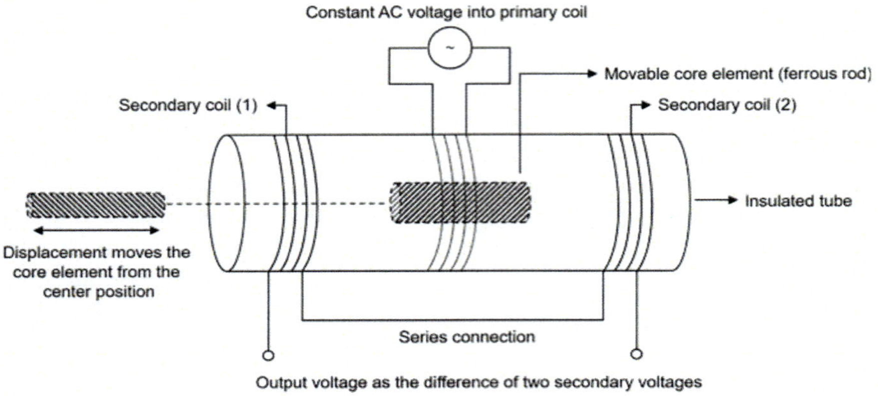

Fig. 2.5 Woking principle of LVDT

in the two secondary coils connected to the primary one. As the core element moves, a differential voltage is induced, the magnitude of which is determined linearly by the magnitude of the elements' extension from the centre position and the phase by the direction of elements' movement.

When the core element is at the centre position, the voltage induced in the secondary coils is equal, and thus, the difference in output voltage will be null. Let us consider the sinusoidal input current to the primary coil is given as (Eq. 2.1):

$$i = I\sin\omega t \tag{2.1}$$

The electromotive force (EMF) induced in the secondary coils due to this current can be given by (Eqs. 2.2 and 2.3):

$$V_{s1} = k_1\sin(\omega t - \emptyset) \tag{2.2}$$

$$V_{s2} = k_2\sin(\omega t - \emptyset) \tag{2.3}$$

where V_{s1} and V_{s2} denote the induced EMF in secondary coils (1) and (2), respectively, k_1, k_2, and Ø depends on the coupling between the primary and secondary coils (1) and (2) as shown in the picture) for a particular position. The differential voltage is given by Eq. 2.4:

$$V_{s1} - V_{s2} = (k_1 - k_2)\sin(\omega t - \emptyset) \tag{2.4}$$

The differential voltage will be zero when the core would be at equal distance from both the coils, while it results in a value as the core displaces either towards any of the secondary coils.

2.5.1.2 Potentiometer Sensor

It is a position transducer, which converts the mechanical displacement into electrical signals, and thus, it is referred to as an electromechanical transducer [1]. A schematic view of this transducer is shown in Fig. 2.6, which consists of a rotating shaft, three electrical terminals, and the casing of the sensor. Out of the three terminals, two are fixed, and the remaining one is variable. It also has a resistor element that is connected with the fixed terminals. Two types of potentiometer are widely used, one is the linear, and the other is a rotary type.

The basic construction of this device has been shown in Fig. 2.7. The resistive element is in the resistive track that is inside the body of the sensor. The resistive element is having a sliding contact that can be moved over the length of the element. The resistive element is a wire-wound track made of conductive plastic, cermet, etc. The track comprises a large number of closely packed turns of a resistive wire. Conductive plastic is made up of plastic resin embedded with carbon powder. Cermet is a type of material having a high resistance to temperature because of its low temperature coefficient. During sensing, a voltage is applied across the resistive element.

Referring to the circuit diagram shown in Fig. 2.8 depicts three terminals (Terminal 1, 2, and 3) as shown in Figs. 2.6 and 2.7. A voltage divider circuit is formed when the slider comes into contact with the wire. The 'Terminal 1' and 'Terminal 3' are connected with the resistive element, and 'Terminal 2' is connected to the wiper. The input voltage is supplied across 'Terminal 1' and 'Terminal 3'. As shown in the picture, the wiper divides the resistive element into two resistors. The one with a longer resistive path will offer higher resistance as per Eq. 2.5. The output voltage is measured as shown in Eq. 2.6. If the wiper position is at 'Terminal 1', 'R_1' becomes null, the output voltage will be equal to the input voltage. Thus, the output voltage is proportional to the displacement of the slider over the wire.

$$R = \rho \frac{l}{A} \tag{2.5}$$

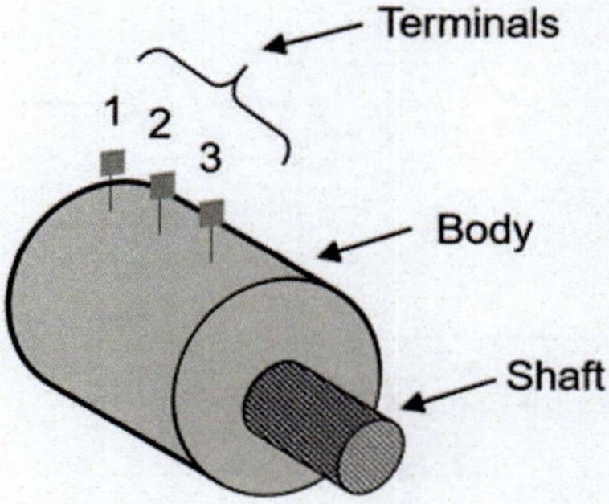

Fig. 2.6 Schematic picture of the potentiometer sensor

$$V_{out} = \frac{R_2}{(R_1 + R_2)} V_{in} \tag{2.6}$$

where l is the length of the conductor, A is the area of the conductor, ρ is a proportionality constant called resistivity and R is the resistance offered by the conductor to the flow of the current. In Eq. 2.6, V_{in} indicates the input voltage, R_1 and R_2 are the resistances, and V_{out} is the output voltage.

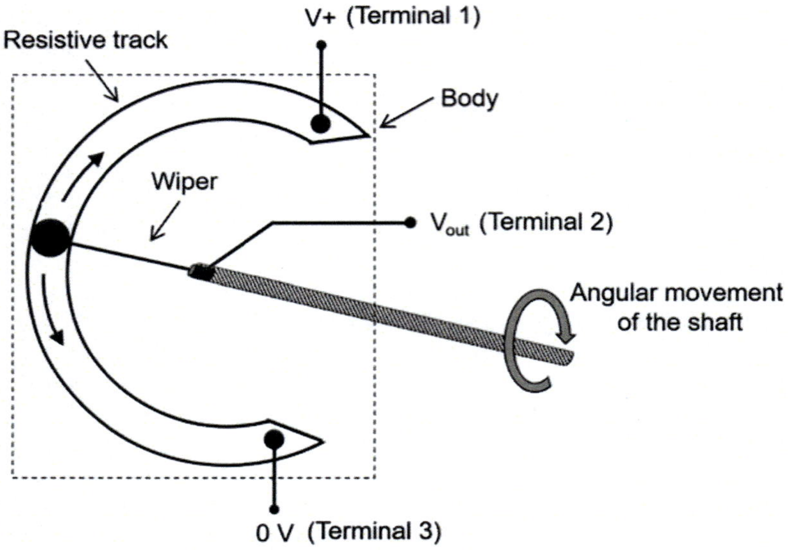

Fig. 2.7 Construction of a potentiometer

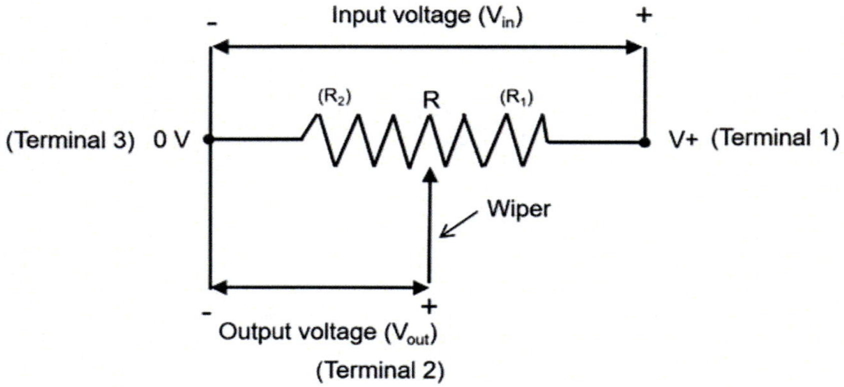

Fig. 2.8 Circuit diagram of the potentiometer

2.5.1.3 Strain Gauges

Stress and strain are two crucial quantities of engineering. The stress refers to the force applied per unit area, and the strain is the ratio of change in length to the original length measured in the direction of an applied load. Thus, whenever, a force is applied to a material, it will give rise to stress, which will induce a strain in the material. The strain gauge sensor employs a mechanism to measure the strain induced in a material. It consists of a metal wire or foil or a semiconductor strip and this element is directly stuck on to the surface of the component whose displacement is to be measured. Any strain that is generated in the component being measured causes a change in the resistance of the element, and the fractional change in the resistance is proportional to the strain.

This sensor depends upon the electrical resistivity of a conductor, which depends upon the change in its length and cross-sectional area [1]. This refers to Eq. 2.5, which states the direct dependency of the resistance on the length, and an inverse dependency on the area. Consider a conductor with length l. This is also referred to as the unstrained condition. Let us assume that an external load is applied to the conductor, which brings a change in its length i.e., Δl. Thus, the strain in the conductor can be represented as given in Eq. 2.7.

$$\epsilon = \frac{\Delta l}{l} \tag{2.7}$$

where Δl is the change in the length, l is the original length, and ϵ is the strain. $\frac{\Delta l}{l}$ is also referred to as the mechanical strain. As the resistance is having a direct dependency on the length, the application of the external load will tend to change the electrical resistance. This is represented in Eq. 2.8. Similar to the mechanical strain, the $\frac{\Delta R}{R}$ is referred to as the electrical strain.

$$\frac{\Delta R}{R} \propto \varepsilon$$

$$\Rightarrow \frac{\Delta R}{R} = G\varepsilon \tag{2.8}$$

where R is the resistance of the metal strip, ΔR is the change in the resistance, and G is the constant of proportionality and is termed as the gauge factor. Thus, the gauge factor is the ratio of the electrical strain to the mechanical strain. The change in the resistance gives a measure of the displacement of the component with which the sensor has been attached. To measure the strain or the change in the resistance, a Wheatstone Bridge is used, which is shown in Fig. 2.9. Due to the arrangement of the four resistors in a bridge fashion, it is named so. It acts like a voltage divider where, R_1 and R_2 are on one side, and R_3 and R_4 on the other. The circuit has an input voltage connected across the bridge.

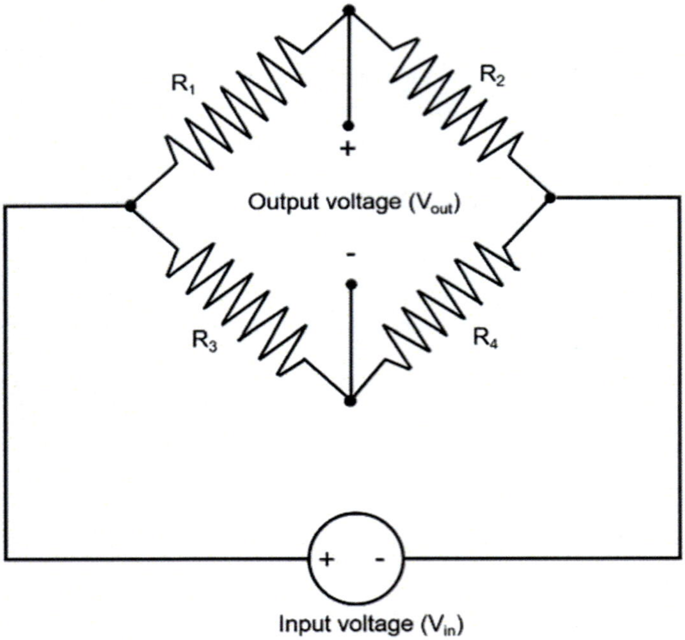

Fig. 2.9 Picture of a Wheatstone Bridge

During the unstrained condition, there is no change in the resistance of the four resistors, thus, the output voltage is null. This is represented in Eq. 2.9. In this situation, the bridge is referred to be in a balanced state, and the output of the bridge is mentioned in Eq. 2.10.

$$R_1 = R_2 = R_3 = R_4 = R \tag{2.9}$$

$$V_{\text{out}} = \left(\frac{V_{in} R_2}{R_1 + R_2} - \frac{V_{in} R_3}{R_3 + R_4} \right) \tag{2.10}$$

During a strained condition i.e., in the presence of an external load, or, if there occurs a change in the physical parameters, there would be a change in the resistance. Let us assume that a strain element is connected in the circuit, as shown in Fig. 2.10. Let the change in the resistance be ΔR (Eq. 2.11).

$$R_2 = R + \Delta R \tag{2.11}$$

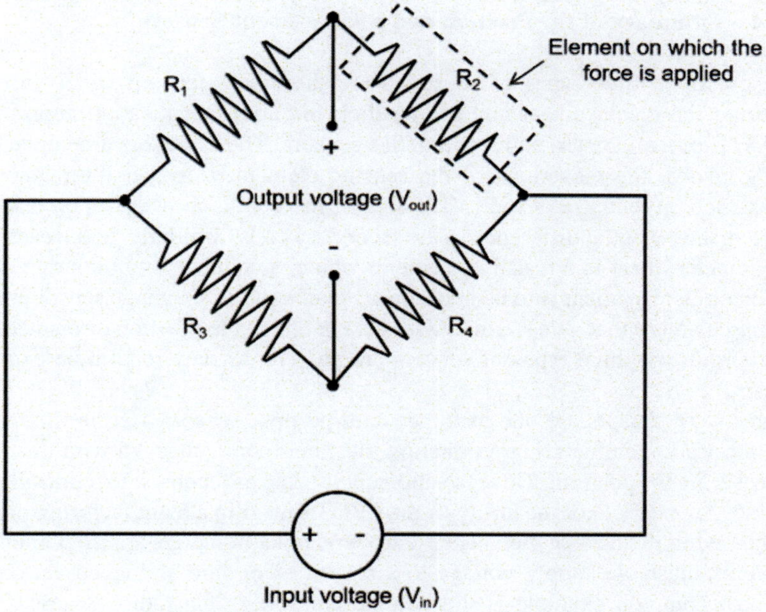

Fig. 2.10 Picture of a strain element in the Wheatstone Bridge

With the change in the resistance, the bridge becomes unbalanced, and the output voltage will be the quantity as shown in Eq. 2.12, which is simplified, and the final output voltage is shown in Eq. 2.13 (assuming ΔR to be a very small quantity when compared to R).

$$V_{\text{out}} = \frac{V_{\text{in}}[R + \Delta R]}{R_1 + R_2} \quad \text{(using Eqs. 2.9–2.11)} \tag{2.12}$$

$$\Rightarrow V_{\text{out}} = \frac{V_{\text{in}}[R + \Delta R]}{R + R + \Delta R} - \frac{V_{\text{in}} R}{R + R} \quad \text{(using Eq. 2.11)}$$

$$\Rightarrow V_{\text{out}} = \frac{V_{\text{in}} \Delta R}{4R + 2\Delta R}$$

$$\Rightarrow V_{\text{out}} = \frac{V_{\text{in}} \Delta R}{4R} \tag{2.13}$$

From Eq. 2.13, it can be said that during the unbalanced condition, the output voltage is one-fourth times the input voltage multiplied with the ratio of change in the resistance to the original resistance. Thus, the output voltage is directly dependent upon the change in the resistance.

2.5.1.4 Summary of the Position and Displacement Sensors

In this section, three types of sensors were discussed that are widely used for measuring the displacement and the position in various application areas. Those are LVDT, potentiometer, and strain gauge sensors. These sensors are contact-type as they involve the movement of the sensing element to trigger the phenomena. For instance, in the case of the LVDT sensor, the core is moved, which leads to a change in the potential difference in the secondary coils. Similarly, in the case of the potentiometer, there is a resistive element, which is connected to the target whose position or displacement is to be measured. Finally, the strain gauge sensor employs a sensing element that gets strained with the change in the position or displacement of the target. All three types of sensors are passive, as they require an excitation voltage.

The LVDT sensors are one of the general-purpose sensors used in the industry and almost all manufacturing/production machines come engaged with this sensor for measuring the position. This measurement is taken as feedback for controlling the machine. An example of the utility of the LVDT sensor in a manufacturing machine is discussed in the next section. The rotary type potentiometers are used mainly for obtaining adjustable supply voltage to a part of electronic and electrical circuits. The most common example is the volume controller of a radio, where a rotary potentiometer is employed to control the volume of the device. It is controlled by moving the rotary knob, which controls the supply to the amplifier. The linear types are employed in measuring the voltage across a branch of a circuit. For example, the equalizer of the sound system employs linear potentiometers for changing the music configurations. The strain gauge sensor is employed in measuring the displacement, load, pressure, etc. A few examples include the application of this sensor in bridges for monitoring the strain on the cables, displacement, and load prevailing because of the running vehicles. It is also used in manufacturing machines for measuring the displacement, load, and torque in rotating parts.

2.5.1.5 Real-Life Example in Manufacturing

As an example, a machine performing an extrusion process is considered. Extrusion is one of the metal forming processes. It is a plastic deformation process in which a block of the metal is forced to flow by compressing through the die opening of a smaller cross-sectional area than that of the original metal block. The extrusion machine is driven by hydraulic power. During the forming process, the metal block is placed inside a die and punch arrangement that is manufactured to fit the original metal block. Compressive load is applied on the punch, which plastically deforms the metal block and fills the die container, and finally, the deformed material goes out through the die opening in a regular shape. An LVDT sensor is used here to measure the displacement of the punch from the start position to the end of the process. A picture of the machine, consisting of the die and punch arrangement, and the LVDT sensor is shown in Fig. 2.11. The schematic for data acquisition from the LVDT sensor is shown in Fig. 2.12.

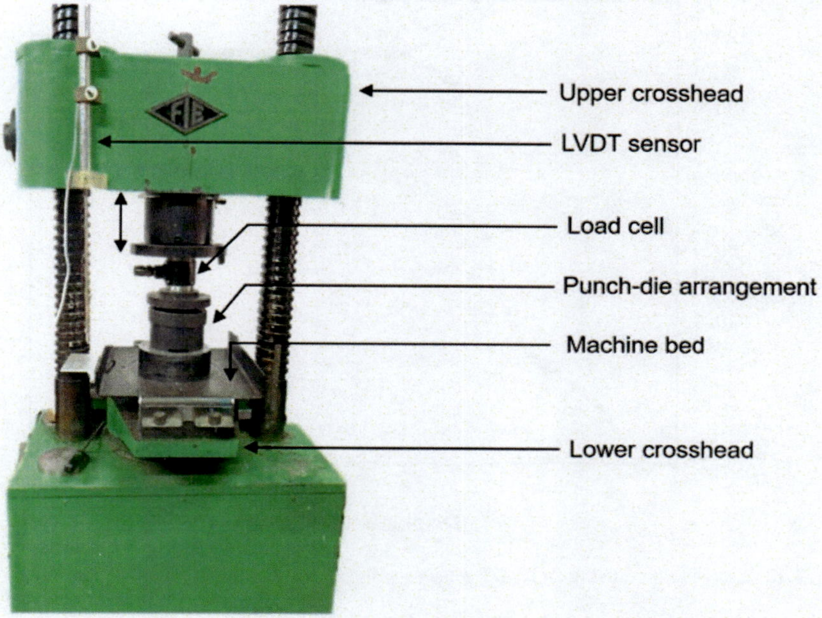

Fig. 2.11 Picture of a compression testing machine fitted with an LVDT sensor

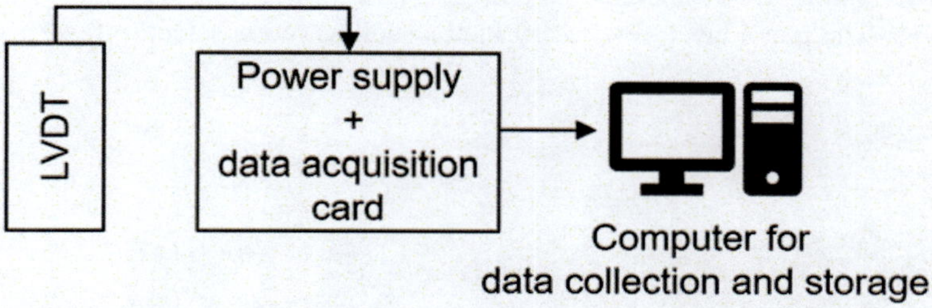

Fig. 2.12 Schematic diagram for data acquisition from an LVDT sensor

Figure 2.13 shows a plot depicting the displacement of the punch concerning the applied load during the extrusion process. From the plot, it can be inferred how the change in the applied load has brought the change in the displacement. As the load increases from 1000 to 10,000 Kgf, the change in the displacement observed is small as compared to the change in the displacement once this upper limit of the load is crossed. Through this data, the operator gains knowledge about the amount of load required to deform the material. This process also requires several trial experiments to finally determine the optimum values. With the optimum values known, a suitable monitoring system can be built which will keep a track of the displacement, which is being measured through the LVDT.

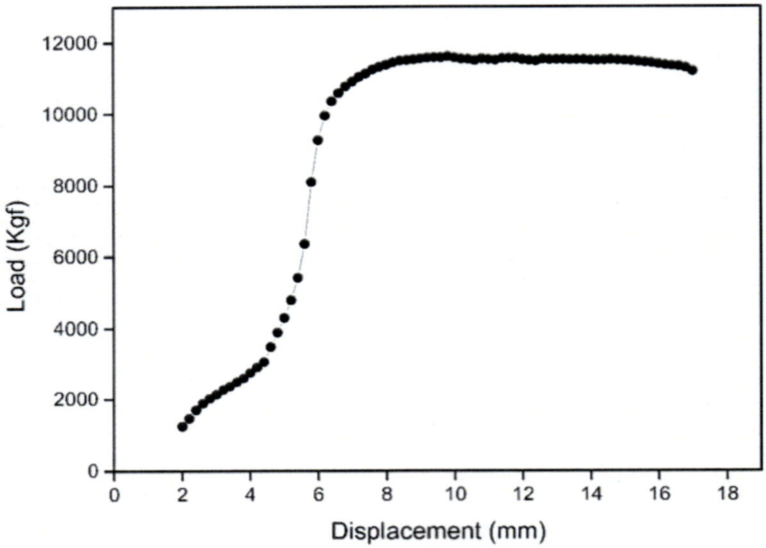

Fig. 2.13 Data captured using an LVDT sensor

The use of a potentiometer can be found in the control panels of the manufacturing machines. Figure 2.14 shows a computer numeric controlled (CNC) turning machine *(Ace Micromatic, LT-20 XL)* and an enlarged view of the control panel is shown in Fig. 2.15. The panel has a few rotary knobs such as mode selector, feed override,

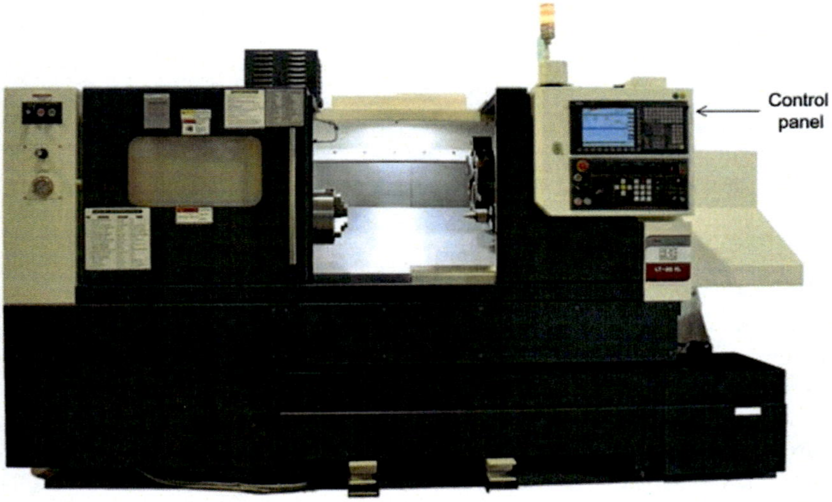

Fig. 2.14 Picture of a CNC turning machine

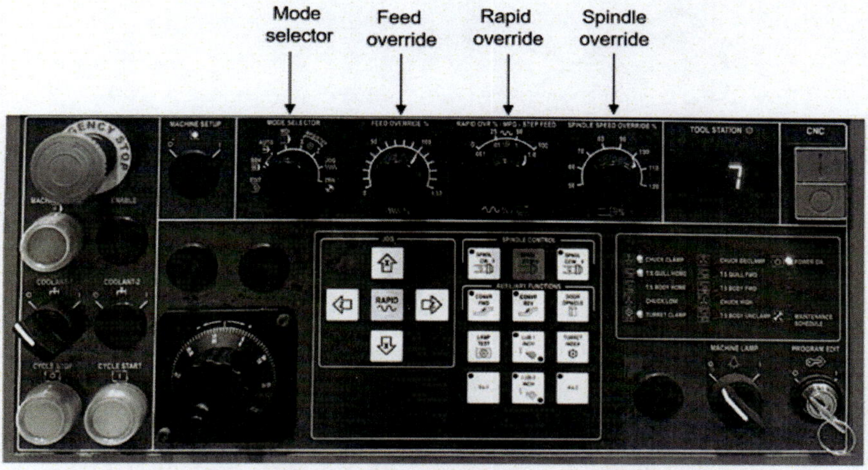

Fig. 2.15 Picture of the control panel of the CNC turning machine

rapid override, and spindle override. A user can rotate the mode selector knob to switch amongst the modes for operating the machine such as manual, auto, and semi-auto. The feed override knob is for increasing or decreasing the speed of the bed movement, while the rapid override is for rapid movement. Finally, the spindle override is for changing the speed of the spindle. All these knobs are potentiometers that are programmed suitably that the machine recognizes once the user rotates the switches. This is because as the switch is rotated, the output voltage will vary, and the machine is already set with those corresponding measurements.

2.5.2 *Proximity Sensors*

The term proximity has a close meaning to the position. This sensor is used to determine when an object has moved to some critical distance from the sensor. These sensors are non-contact type and have found a wide range of applications starting from household electronics, smartphones, shopping malls, parking lots, automobiles, manufacturing machines, etc. The discussions in this section will include two types of sensors namely inductive and capacitive proximity sensors.

2.5.2.1 Inductive Proximity Sensor

This sensor is typically used for the detection of conductive materials, usually ferrous metals [2]. A schematic picture of this sensor is shown in Fig. 2.16, which resembles

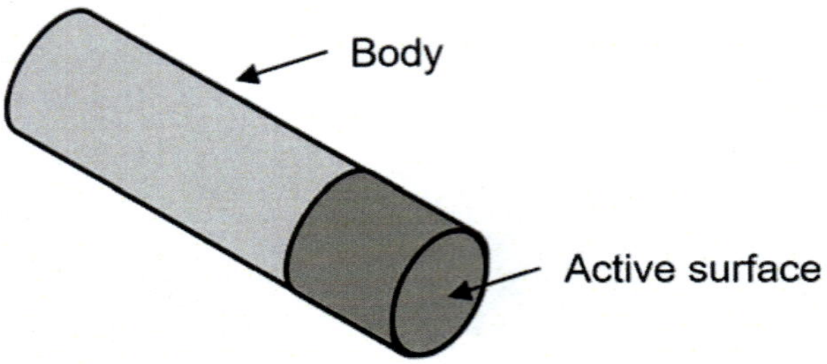

Fig. 2.16 A picture of an inductive type proximity sensor

a cylindrical shape. The active surface is the one, which is exposed to the object to be detected.

The basic principle behind this sensor is the use of an electromagnetic field. The construction of this sensor includes an inductive coil, an oscillator, a detector, and a triggering switch, as shown in Fig. 2.17. The coil consists of turns and also includes a capacitor for storing the electrical charge. An alternating current passes through the coil, which is provided by an external excitation. This, in turn, results in the generation of an alternating magnetic field, which is at the active surface of the sensor. This active surface is exposed to the target. This is the reason why the inductive coil is placed just beside the surface so that the generated electromagnetic field exists there at the active surface. On bringing the conductive element (target) near this electromagnetic field, the inductive characteristics of the metal change the properties of this field. A part of the energy i.e., the electrical current is induced in the target, which is referred to as eddy current. The eddy current can be defined as loops of current that is induced within a conductor because of a time-varying magnetic field, which is popularly referred to as *Faradays' law of induction.* The magnetic field generated by the eddy current opposes the source of the primary magnetic field and hence, results in a change in the impedance of the coil or the natural frequency of the circuit. This, in turn, reduces the amplitude of the current. Because of this, there

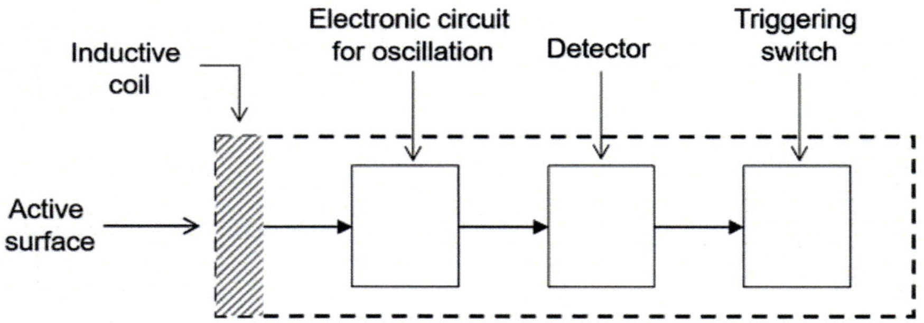

Fig. 2.17 Elements of an inductive proximity sensor

is a loss of power i.e., the intensity of the oscillations decreases. Below a pre-set value, this change is detected by the detector which triggers a switch.

2.5.2.2 Capacitive Proximity Sensor

A capacitor can store electrical charge, and the capacitance describes how the space between two conductors affects an electric field between them [2]. This sensor looks similar to the inductive type proximity sensor shown in Fig. 2.16. However, it differs in the sensing mechanism, which in this case is the electrostatic field. Figure 2.18 shows a schematic picture to understand the basic principle behind a capacitance sensor. It consists of two conductive plates separated by air or a dielectric medium. The plates are conductive i.e., they can get electrically charged when kept in an electrical field, and the separation medium acts as an insulator between them, which may be ceramic or plastic. The basic principle behind the operation of this sensor is to play with the capacitance or the amount of charge.

The capacitance can be altered by changing the spacing between the plates or the surface area, or by changing the dielectric medium. The following discusses each of these techniques. The expression for the capacitance of a parallel plate capacitor is shown in Eq. 2.14.

$$C = \frac{\varepsilon_r \varepsilon_0 A}{d} \tag{2.14}$$

where ε_r is the relative permittivity of the dielectric between the plates, ε_0 is the permittivity of free space, A is the overlap area between the plates, d is the separation between the plates, and C is the capacitance. The schematic diagram of the aforementioned techniques to alter the capacitance is shown in Figs. 2.19 and 2.20, respectively.

As the separation distance between the plates changes due to the displacement of the component being monitored, the capacitance changes. The expression for the capacitance of this new configuration is shown in Eq. 2.15.

$$C + \Delta C = \frac{\varepsilon_r \varepsilon_0 A}{d - x} \tag{2.15}$$

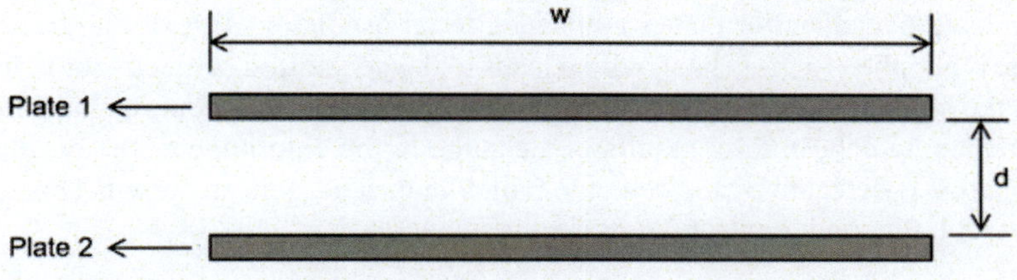

Fig. 2.18 Schematic view of a capacitance sensor

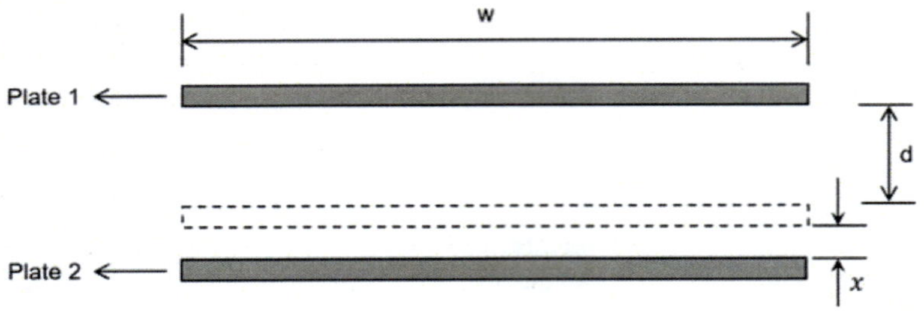

Fig. 2.19 Measurement by the sensor through the change in the separation between the plates

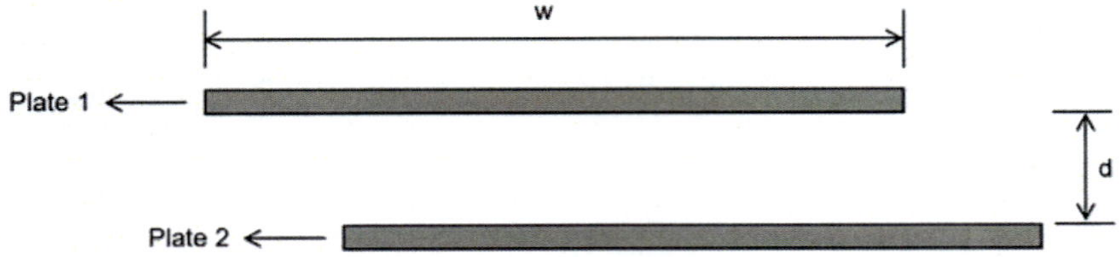

Fig. 2.20 Measurement by the sensor through the change in the overlapping area

where x is the change in the separation distance between the two plates. Similarly, the other configuration is the change in the overlapping area of the two plates, as shown in Fig. 2.20. If one of the plates is displaced in a direction parallel to the other plate, the effective area of the plates will change proportionally to the value of capacitance. The expression for the capacitance of this new configuration is shown in Eq. 2.16.

$$C - \Delta C = \frac{\varepsilon_r \varepsilon_0 (A - \Delta A)}{d} \tag{2.16}$$

In actual application, it refers to detecting the change in the capacitance between a sensing object and the sensor itself. Out of the two plates, as discussed above, one of the plates is the sensing surface, and the other is the object to be sensed. The amount of capacitance varies depending on the size and distance of the object being sensed. The construction of the sensor is shown in Fig. 2.21, which includes a conductive plate, an oscillator, a detector, and a triggering switch. The conductive plate acts like a capacitor that is connected to an oscillator. An external excitation voltage is supplied to the sensor, which leads to the generation of electrostatic forces. When the object to be sensed comes into the target area, there occurs a change in the capacitance, which, in turn, results in a change in the amplitude of the oscillation. This change is detected by the detector, which in turn triggers an output. This sensor can detect both conductive and non-conductive targets.

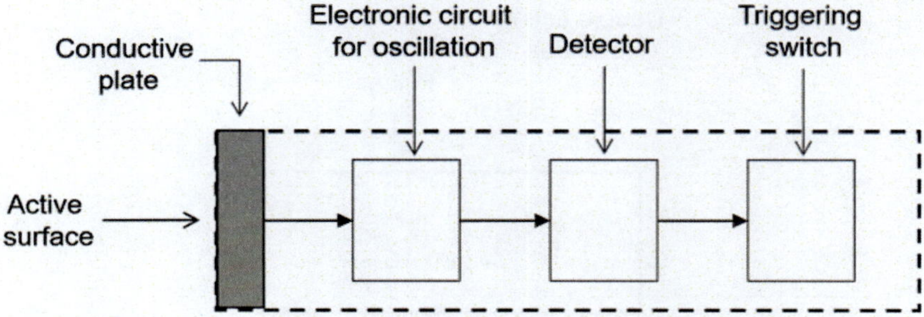

Fig. 2.21 Elements of a capacitive proximity sensor

2.5.2.3 Summary of the Proximity Sensors

In this section, two types of sensors were discussed, which are widely used in various application areas. Those are inductive and capacitive type proximity sensors. Both the sensors are passive i.e., they require an external excitation voltage for sensing. Both are also non-contact type sensors, which makes them attractive to be used in various areas as the chances of breakage or getting damaged over time because of contact with the target is not the case. However, the inductive type sensor can only detect ferrous metals because of the inclusion of an electromagnetic field, whereas the capacitive type sensor is suitable to detect both metallic and non-metallic objects. The inductive type sensor can measure in the range up to 100 mm whereas the capacitive type has a smaller range going up to 50 mm. The applications of the inductive sensor include elevators in malls, overhead cranes for identifying the start and end positions, manufacturing machines, etc. The capacitive sensors are employed for liquid level monitoring, conveyor belt, filling of bottles, etc.

2.5.2.4 Real-Life Example in Manufacturing

For an illustration of the proximity sensor, an example of controlling a cylinder's actuation is shown. The schematic diagram for the actuation of the cylinder is shown in Fig. 2.22. The cylinder has two inlets, one for forward and the other for a backward motion of the piston. Because of the two motions, it is referred to as double-acting. A solenoid coil is employed for switching in between the forward and backward movements of the piston of the cylinder. This coil is operated manually. When 'A' of the solenoid coil is powered, the piston will move forward, else if 'B' is powered, the piston will move backwards. The double-acting cylinder is utilized as an actuator in production machines. To switch between these movements, a user has to continuously operate the coil, which is a tedious task. Thus, the objective is to automate this operation i.e., to move the piston automatically.

For this purpose, in the present case, two inductive proximity sensors are employed. The sensors have been engaged in the experimental set-up, as shown

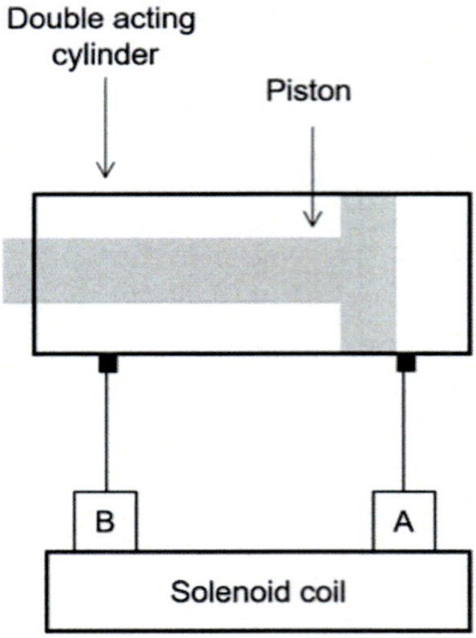

Fig. 2.22 Schematic diagram for mechanical actuation of a double-acting cylinder

in Fig. 2.23. This sensor is available in two configurations namely 'NPN' and 'PNP', where N stands for negative and P stands for positive. These configurations indicate the type of output one needs to have from the sensor. As discussed, the detector in the proximity sensors triggers the output. Depending upon whether this output is positive or negative, one selects NPN or PNP configuration, respectively.

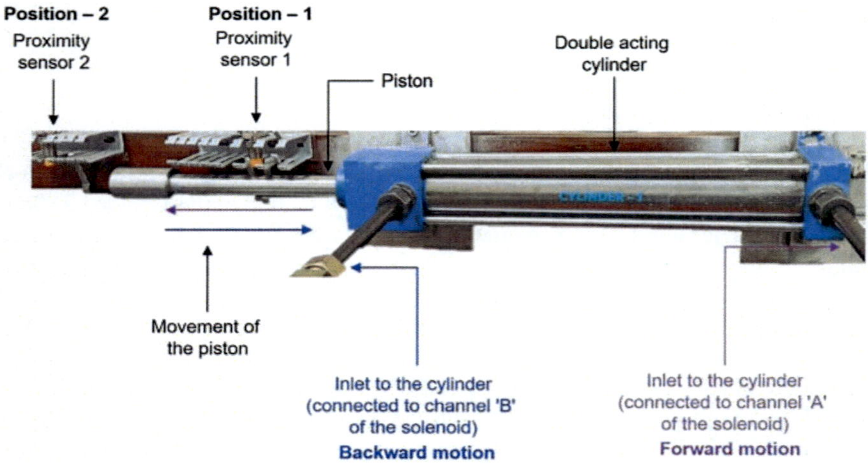

Fig. 2.23 Experimental set-up for sensor-based actuation of the cylinder

One proximity sensor is fixed at the 'Position 1', and the other is fixed at the 'Position 2' as shown in Fig. 2.23. These positions refer to the minimum and maximum displacements of the piston, respectively. Both proximity sensors are provided with an input voltage from a suitable source. The feedback of the sensor at 'Position 1' is connected to 'A' of the solenoid coil, and that of the sensor at 'Position 2' is connected to 'B' of the coil. In this example, the proximity sensor needs to provide a positive feedback signal to the solenoid coil, thus, the sensors are of NPN configuration. As the sensors are switched on, the electromagnetic field is generated, which detects the presence of the target (piston of the cylinder), thus it triggers 'A' of the coil, which is its feedback. This in turn moves the piston in the forward direction. As the piston reaches 'Position 2', it gets detected by the proximity sensor located in this position. However, this sensor triggers output to switch 'B' of the solenoid coil, which retracts the piston back to its original position. In this way, the actuation process is automated. A software named "AUTOMGEN" was utilized for making the circuit.

Figure 2.24 shows another application of a proximity sensor in a CNC turning machine *(Ace Micromatic, LT-20 XL)*. The application area is the ball screw drive of the machine which is guided through a motor. Here the objective of the proximity sensor is to sense the position of the drive. In the case of looseness or some other fault, this distance between the sensor and the mechanical arrangement will change. As this occurs, the sensor triggers a signal as an alarm for the feedback. In this way, the sensor helps in detecting the faults in the machine.

Fig. 2.24 Use of proximity sensor in CNC turning machine for sensing the axial position of a shaft

2.5.3 *Velocity and Motion*

Manufacturing machines involve numerous rotating and translating components. After identifying the position and displacement through the sensors discussed above, the rate of displacement forms another crucial quantity to be measured. For instance, consider a CNC turning machine, where the objective is to perform operations such as turning, threading, finishing, etc. on a block of metal. For these operations, the job, which is mounted over the spindle needs to be rotated at different speeds. Thus, it becomes important to ensure that the spindle is rotating at the instructed speed, which is achieved by using a sensor that can detect linear and angular velocities and motion. The following discusses the typical sensors under this category.

2.5.3.1 Tacho-Generator

The tacho-generator is a contact-type sensor that is used to measure the angular velocity [3]. It converts mechanical energy into an equivalent electrical quantity. Figure 2.25 shows a schematic picture of this sensor, which has a shaft that can be seen on the external body. This shaft is connected to the mechanical component whose rotational speed is to be measured.

This sensor works opposite to an electrical motor. The motor is given electrical energy as the input that is converted to mechanical energy. However, in the case of the tacho-generator, the input is the mechanical energy of a component, which is the target. The output will be an electrical signal, which will be equivalent to the rotation of the mechanical component. The tacho-generator is also referred to as the dual of an electrical motor, as it also works on the principle of *Faraday's law of electromagnetic induction.* It is available in two forms namely; alternating current (AC) generator and direct current (DC) tacho-generator.

Figure 2.26 shows the elements of a DC tacho-generator, which are a stator, permanent magnet, a rotating armature circuit, and a commutator and brush. The housing consists of the rotating armature which is in the field of a permanent magnet. The movement of the shaft when it is connected with the rotating mechanical component

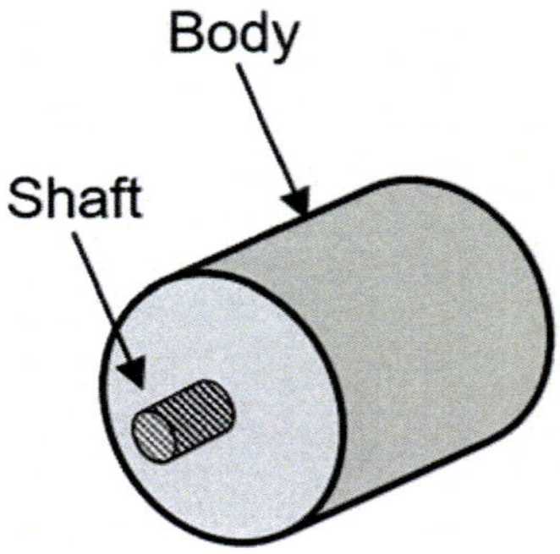

Fig. 2.25 Picture of a tacho-generator

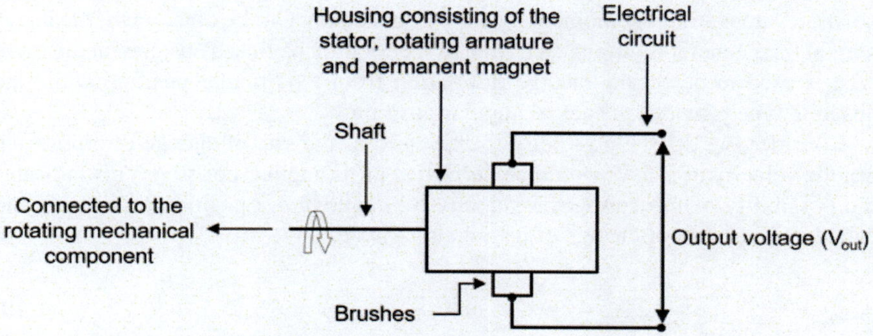

Fig. 2.26 Components of a DC tacho-generator

drives the armature inside the housing. This rotation induces electromagnetic force, whose magnitude is proportional to the speed of the shaft. The commutator and brush assembly convert this AC of the armature coil to a DC, which is obtained as the output voltage. Further, the polarity of the output voltage indicates the direction of rotation.

As discussed, the output voltage is proportional to the speed of the shaft as shown in Eq. 2.17.

$$V_{\text{out}} \propto \omega$$

$$\Rightarrow V_{\text{out}} = K\omega \tag{2.17}$$

where V_{out} is the output voltage obtained from the sensor, ω is the speed of the shaft, and K is the proportionality constant and depends upon factors such as temperature, maximum load, maximum velocity, etc. The value of this constant is usually provided by the manufacturer. The tacho-generator is an active sensor i.e., it does not require any external supply for the operation. However, because of the presence of the commutator, large fluctuations are observed in the output voltage. This is taken care of by the slip rings present in the case of the AC generator.

2.5.3.2 Encoder

An encoder is a sensing device that measures linear and angular displacement [3]. It converts the motion into an electrical signal that can be read by another device. The one measuring the linear displacement is called the linear encoder and the rotational motion is measured by the rotary encoder. Each of these two types is also available in two forms namely absolute and incremental. The absolute encoders provide unique position values from the moment they are switched on. On the other hand, the incremental type generates output each time as the rotation begins. Thus, the incremental type is used to detect the change in rotation from a particular position while the absolute type provides the actual angular position.

Consider Eq. 2.18, the velocity is expressed as the rate of change of a position. So, the velocity of a shaft would be derivative of its angular position. This indicates the possibility of measuring the velocity by employing a position transducer and calculating the rate of change of its output variable.

$$v = \frac{d\alpha}{dt} \tag{2.18}$$

where v is the velocity, α denotes the position, and t is the time. The position transducers have already been introduced in this chapter. If one intends to measure the velocity by using a potentiometer, an additional differentiator circuit would be required to be interfaced for calculating the derivative of the position. However, the stabilization of the circuit poses a problem. Besides, the capacitive and inductive type have a poor angular resolution. In contrast to these, the optical system is one of the popular methods for determining the velocity. Thus, the optical type encoders are the most widely used ones.

In the case of an optical type of encoder, there is a light source that generates the light, and a detector that receives the light. Together, they are termed as the optoelectronic pair. There is a track consisting of opaque and transparent sections arranged in a series as shown in Fig. 2.27. This track moves in between the optoelectronic pair with the change in the position of the target. The light source emits a beam of radiation that is focused on the track by using a suitable arrangement such as a focusing lens. Based on the presence of an opaque or a transparent section, the light is detected by the detector. The detector which is a photosensor produces a sinusoidal

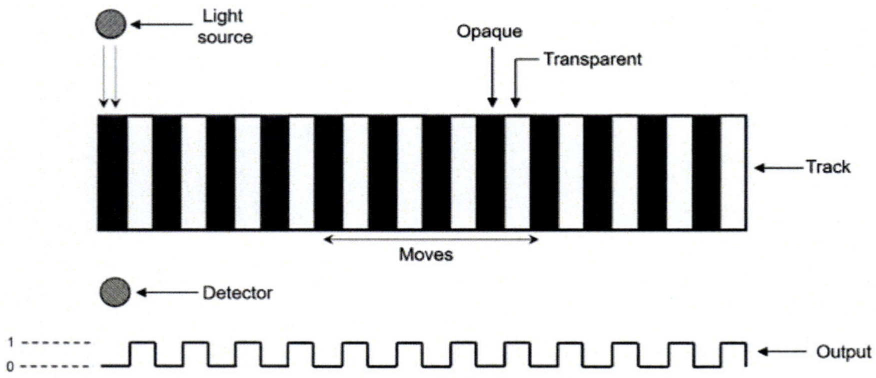

Fig. 2.27 Encoder components

waveform as a response to the light beam, which is then transformed into a square wave or pulse train as shown in Fig. 2.27.

Considering the output shown in the picture, a signal of zeros and ones is generated. Zero corresponds to an opaque section while one corresponds to a transparent section. If the machine element is connected with the track, then the moment it starts moving, these pulses are generated. Technically, these pulses are referred to as counts. The pulsed output produced is so that the number of pulses is proportional to the movement of the track. The counts are used to calculate the displacement and speed. This discussion corresponds to what is referred to as an incremental type of encoder. It is named so because the counts which are generated as the track moves are increments that are encoded to provide the desired output. However, it can be observed that one cannot infer the direction of the movement from the output pulse train. This is because the output will only be a pulse train from either direction. To serve the above purpose, two tracks are utilized and are arranged in a manner that one of the pulse trains leads while the other lags. This is named the quadrature encoder. A schematic of the same is depicted in Fig. 2.28.

As shown in Fig. 2.28, the encoder is having two tracks named 'A' and 'B'. Each of the two tracks is having a few contact zones, which are marked as opaque and transparent zones. This again means that an opaque zone will not allow the light to pass through it, while it could pass through a transparent zone. These zones are arranged in a manner that one lags the other. A rotating contact is shown for demonstration of the working principle. This can move in both clockwise and anticlockwise directions. This sensor when connected to a machine, the element which will move is connected to this rotating contact. Let us look into the working principle of this sensor and why it is named the quadrature encoder. The working principle is shown schematically in Fig. 2.29.

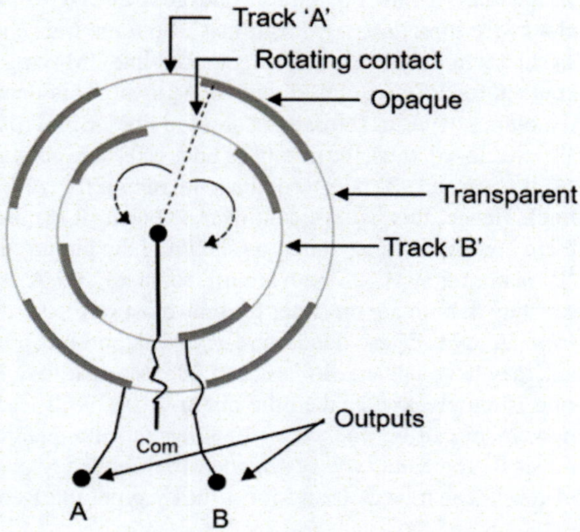

Fig. 2.28 Quadrature encoder

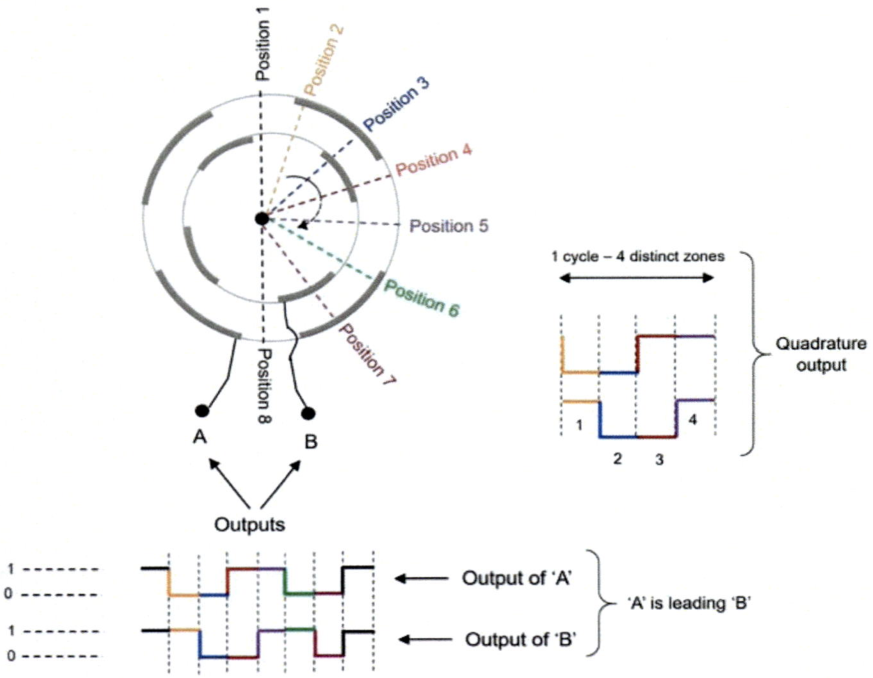

Fig. 2.29 Output of quadrature encoder

This encoder gives rise to two output signals, which are 'A' and 'B', with four zones in each cycle. In the picture, the movement of the contact ring is depicted, which starts from 'Position 1' and goes on. Here, the movement is shown in the clockwise direction. At 'Position 1', both the tracks are open, thus, the output is 1 as shown in the picture (black coloured line). Moving to 'Position 2' of the rotating contact, track 'A' is closed, but 'B' is open. This makes the output of 'B' as 1 and the output of 'A' as 0 (orange coloured line). Similarly, at 'Position 3', both 'A' and 'B' are closed, thus, the output of both is 0 (blue coloured line). Then at 'Position 4', 'A' is open and 'B' is closed, thus, the output for 'A' is 1 and 0 for 'B' (red coloured line). Hence, this pattern continues to occur. It can be observed that for one cycle, there are four distinct zones, as shown in the picture. For the state '1', 'A' is off and 'B' is on, for state '2', both are off, for state '3', 'A' is on and 'B' is off, and finally for state 4, both are on (refer to right-most of Fig. 2.29). Thus, the pulses generated from 'A' and 'B' are quadrature-encoded, and it can be said that 'A' is leading 'B', i.e., they have a phase difference of 90°. Alternatively, if the movement of the contact ring is anticlockwise, then the output of 'B' will lead. Hence, the direction of the movement can be tracked by looking into the quadrature output. In the case of a linear incremental type of encoder, the frequency or the counts indicate the speed of the linear traversal, and for a rotary type, the frequency indicates the speed of the rotation of a shaft. The incremental type encoders can be found in almost all manufacturing machines to measure the change in the displacement, speed of the axis movements. These measurements are sent to the machine controller for precise tracking of the movements and execution of the manufacturing steps. One limitation of the incremental type is the non-availability of the absolute position. Thus, it is always necessary to arrive at the home or reference position to determine or initiate an absolute measurement.

On contrary, the absolute type of encoder is capable of generating an absolute measurement. It generates a unique digital output corresponding to the position. Each bit position is separately encoded by using a dedicated optoelectronic pair. Figure 2.30 depicts the schematic diagram of an absolute type encoder.

This type of encoder consists of several multiple disks, as shown in the schematic picture, and the same are coded with opaque and transparent zones. Considering the encoder depicted in Fig. 2.30, shows 16 unique positions starting from 0 to 15. For encoding of the positions, Gray code is employed. The Gray code is beneficial as compared to the standard binary code as it results in the change of one state while moving from one position to the other. For instance, in the encoder shown here, take the 'Position 0', according to Gray code, it is encoded as 0 for track 1

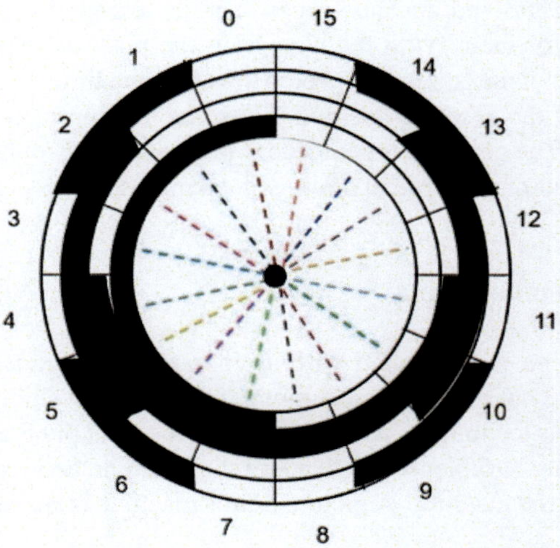

Position	Gray code
0	0111
1	0110
2	0100
3	0101
4	0001
5	0000
6	0010
7	0011
8	1011
9	1010
10	1000
11	1001
12	1101
13	1100
14	1110
15	1111

Fig. 2.30 Schematic view of an absolute encoder and corresponding Gray code

and 1 for the rest of the three tracks. Track 1 corresponds to the inner-most track. 1 corresponds to the transparent zone, which means that it will allow the light to pass through, and 0 corresponds to the opaque zone, which will not allow the light to pass through it. Moving to the next position i.e., 'Position 1', it is encoded as 0 for track 1, 1 for the next two tracks, and finally 0 for the last track. In these two positions, it can be observed that there is only a change of one state i.e., from 0111 to 0110. This is particularly useful in applications of automation as it considers only the change of one state. Otherwise, if two or three states will change simultaneously, then, there may occur a time lag. This is the limitation of standard binary coding. For reference, the Gray code for each position is mentioned in Fig. 2.30. Absolute encoders are particularly useful in applications where the absolute position is required to be determined. Since it has a unique indicator for each position, it never loses the data even after a power failure and does not require a reference position.

2.5.3.3 Summary of the Velocity and Motion Sensors

The discussion held in this section describes the sensors for measuring the velocity and motion. For this purpose, the instruments employed are tacho-generator and encoder. The tacho-generator is capable of measuring the speed but cannot provide information about changes in the position. An encoder is capable of measuring both position and speed. The encoder can be used for both linear and rotary measurements and are titled as a linear and rotary encoder, respectively. Both of these encoders are further categorized into incremental and absolute types. The incremental type generates the counts that are used for identifying the change in the position. The number of counts is sent to a device that measures the speed. However, one limitation of this type is the non-availability of the absolute position. Thus, the measurements are based on some reference point. The absolute type encoder provides an absolute position by rendering a unique digital output corresponding to each position.

2.5.3.4 Real-Life Example in Manufacturing

This section presents a CNC machine instrumented with an incremental encoder. This machine is utilized for carrying out a welding technique, friction stir welding (FSW) *(Eta Technology, WS004)*. This technique involves a rotating tool that plunges into the workpieces to be joined. The workpieces are fixed on the machine bed and the tool is mounted on the spindle of the machine. A picture of this machine is shown in Fig. 2.31.

The machine bed shown in the picture holds the materials to be welded in a suitable fixture. During the welding process, the tool plunges into the materials giving rise to frictional heat, and after some time, the bed moves linearly along the axis as shown in the picture. The frictional heat deforms the materials plastically. The combined action of the rotating tool and the translating bed stirs the material from one side to the other. To track the linear movement of the machine bed during the welding, an

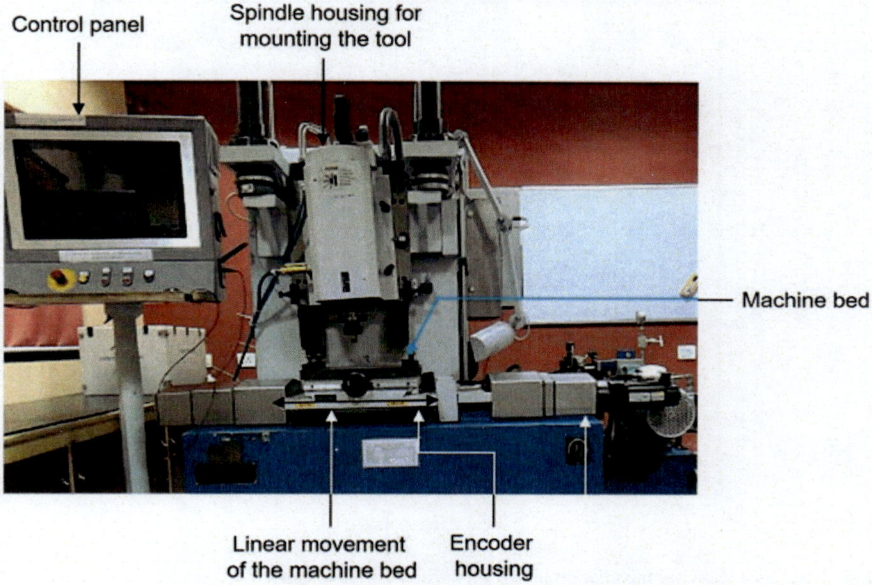

Fig. 2.31 Picture of a CNC machine used for FSW

incremental encoder is engaged beneath the bed. This encoder senses the position of the bed and the signal is sent to the control panel. The output of this encoder is a record of the change in the velocity during the welding process. Figure 2.32 shows the plot of the velocity of the bed during a welding operation. The change is shown with respect to the time. In the plot, 'C' to 'D' marks the region of welding, 'A' refers to the start of tool plunging which ends at 'B'. The velocity in 'A' to 'B' is nil because there is no movement of the bed during tool plunging. The movement starts after 'B' when the plunging of the tool followed by the dwelling period is over. It finally drops as the movement of the bed is stopped. With this plot, one can easily note the movement of the bed and the operation at any point in time.

The above discussion shows the importance of the position of the machine bed. Let us observe the importance of the velocity of the machine bed. For this purpose, Fig. 2.33 shows only the welding signal from the above data i.e., the region shown from 'C' to 'D'. The instructed velocity to the machine was 200 mm/min, which is depicted in a dotted line in Fig. 2.33. The fluctuation in the actual velocity throughout the welding process can be observed from this plot. A user can utilize this data to monitor the process. For instance, if the weld input parameters represent a good weld, then the fluctuations will be less as compared to an improper weld. Besides, the data can also be studied for monitoring the health of the motor, which is responsible for the movement of the machine bed.

This FSW machine is also engaged with a rotary encoder of incremental type to acquire the speed of the spindle. However, as the spindle is completely enclosed, it is

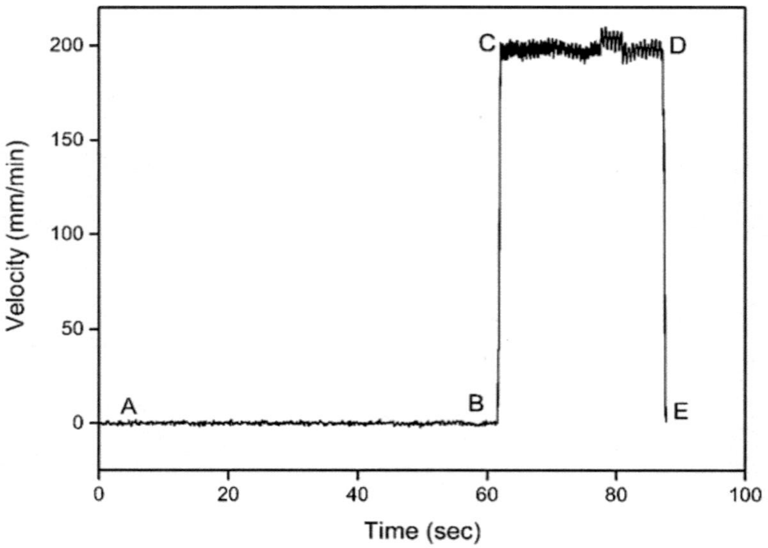

Fig. 2.32 Plot showing the change in velocity with respect to time

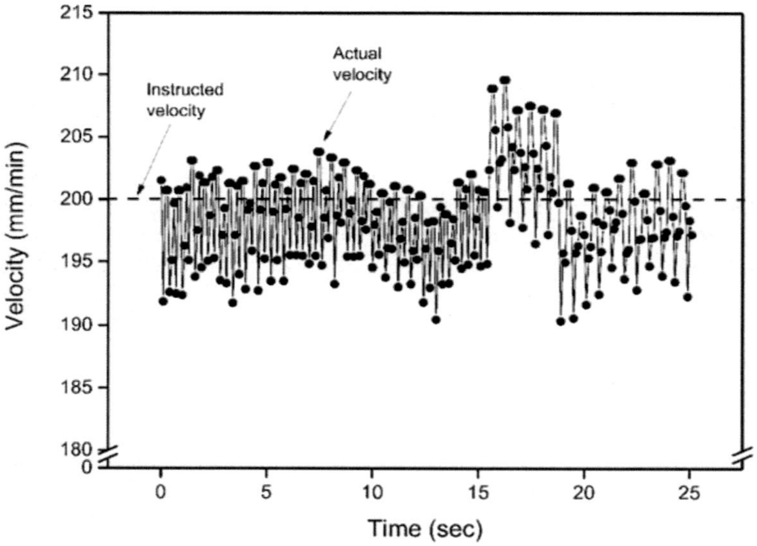

Fig. 2.33 Fluctuation in velocity during welding acquired using a linear encoder of incremental type

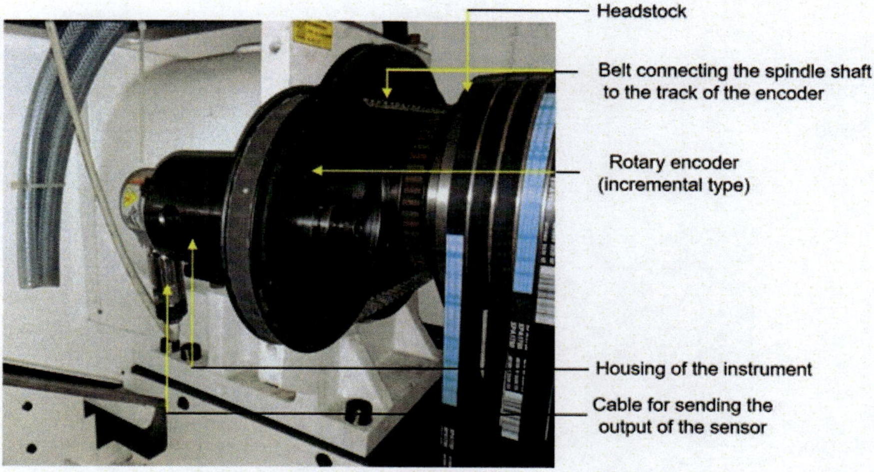

Fig. 2.34 Picture showing the spindle of a CNC machine instrumented with a rotary encoder of incremental type

difficult to get a picture of the instrument. To present the readers with a more practical view, the example of the CNC turning centre is considered again *(Ace Micromatic, LT-20 XL)*. In the turning centre, measuring the spindle rotation is one of the crucial requirements. For this purpose, the spindle is instrumented with a rotary encoder of incremental type, and a picture of this is shown in Fig. 2.34. A belt arrangement can be observed connecting the spindle shaft with the track of the encoder. As the spindle rotates, the track moves. The housing of the instrument holds the optoelectronic pair and other associated accessories. The pulses recorded during the movement of the spindle are sent to the control panel via the cable shown in the picture (Fig. 2.34). In this way, the machine keeps a track of the deviation between the instructed velocity and the actual velocity. CNC machines also need to change the speed of the spindle in real-time based on the operation being performed. For example, lower speeds are usually preferred when working with jobs of larger diameters, while the speed is eventually increased as the diameter reduces. This ensures the life of the cutting tool. To perform these operations, the encoder continuously senses the rotation of the spindle and gives measurements to the control panel. It should be noted that the position of the spindle is not important here, thus, the incremental type is suitable for this purpose.

However, consider the tool carrying station i.e., turret in the CNC machine. The station in this particular machine has provision for mounting 7 different tools. The user has the flexibility to select any tool out of the 7 depending upon the type of operation to be executed. In this case, the absolute position is important to determine than just the change in the position. Thus, the CNC machine is instrumented with an absolute encoder for tracking the position of the station, as shown in Fig. 2.35.

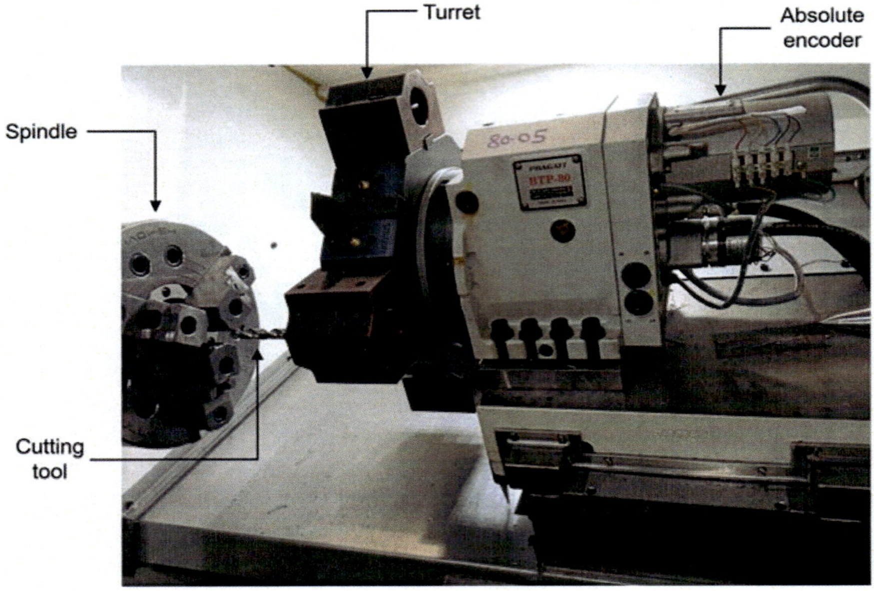

Fig. 2.35 Use of absolute type encoder in a turret of a CNC machine

Accordingly, the signal is sent to the controller by the instrument regarding its absolute position. A proximity sensor is also instrumented inside the housing for determining the clamped position of the turret. The proximity sensor also sends the signal to the controller regarding the clamping position.

2.5.4 Force

Force is a physical quantity and it is of utter importance in some manufacturing processes such as forming, machining, and a few welding techniques. In case of forming, say in sheet metal operation, the load is applied on to the punch, which deforms the material to make a change in its shape and size. Similarly, in machining, the cutting tool interacts with the workpiece resulting in the generation of the force. In welding techniques such as friction welding and FSW, there is an interaction between the two metal pieces that give rise to forces. Hence, sensing this quantity can help in monitoring these processes. The force acting in a process is generally determined by the measurement of the displacement of some typical material. The following are examples of sensors that can be used to measure force.

2.5.4.1 Strain Gauge Load Cell

A load cell is a transducer that converts the force into a measurable electrical quantity [4]. The force can be in the form of tension or compression. The load cell measures the force by employing a strain gauge as the sensing element, thus, it is referred to as a strain gauge load cell. The operating principle of the strain gauge is already discussed in this chapter (Sect. 2.5.1.3). A further dependency of the sensing element (i.e., the output) on the force will be shown through the discussion.

This sensor consists of a metallic body usually made of alloy steel that contains the strain gauge elements. These strain gauges are identical and operate on elastic properties. During tension, the force will be pulling, thus, the length of the element will increase and the area will decrease, opposite occurs during the compression force. Hence, the resistance of the element will increase in the case of tension and decrease in the case of compression as the resistance has a direct dependency on the length. These elements are connected to a Wheatstone Bridge, which has been discussed earlier to measure the change in the resistance (Sect. 2.5.13). The following expressions illustrate the dependency of the output voltage of the Bridge on the applied force (Eq. 2.19).

$$
\begin{aligned}
&\tfrac{\Delta R}{R} = G\varepsilon && \text{From Eq. 2.8}\\
&V_{out} = \tfrac{V_{in}\Delta R}{4R} && \text{From Eq. 2.13}\\
&\Rightarrow V_{out} = \tfrac{V_{in}}{4} G \tfrac{\Delta l}{l} && \text{Using Eq. 2.7}\\
&\Rightarrow V_{out} = \tfrac{V_{in}}{4} G\varepsilon && \\
&\Rightarrow V_{out} = \tfrac{V_{in}}{4} G \tfrac{\sigma}{E} && \text{Using } \textit{Hooke's law}\\
& && \sigma = E\varepsilon\\
&V_{out} = \tfrac{V_{in}}{4} G \tfrac{F}{EA} && \text{Using } \sigma = \tfrac{F}{A}
\end{aligned}
$$

$$
V_{out} \propto F \tag{2.19}
$$

where σ is the stress, E is Young's modulus, F is the force, and A is the cross-sectional area.

2.5.4.2 Piezoelectric Sensor

There are certain materials blessed with the ability to generate electric charges when subjected to stress, or vibration. Such materials are referred to as piezoelectric materials, and this natural phenomenon is termed the piezoelectric effect. To understand it more deeply, let us go through some basic concepts to finally comprehend the piezoelectric effect. It all begins with the class of materials, which are referred to as dielectric, ferroelectric, and piezoelectric [5]. A common term to note in these materials is 'electric', which gives a hint that this has something to do with electricity. These classes are distinguished based on the response of the materials to an externally applied electric field. A solid, when kept in an externally applied electric field, starts responding to the field by dynamically changing the position of the nuclei and electrons, resulting in the generation of a dipole. This phenomenon is termed electric polarization. This phenomenon is known to occur in three configurations namely electronic, ionic, and dipolar or orientation. In the first case i.e., electronic polarization, electrons are displaced against the atomic nucleus while in the second case i.e., ionic polarization, the positive ions are displaced against the negative ones. The dipolar or orientation polarization is only found in materials that have permanent dipole moments.

The dielectric material is electrically insulating and exhibits an electric dipole structure. This structure indicates a demarcation between positive and negatively charged entities at a molecular or atomic level. The ferroelectric materials are a class of dielectric materials that exhibit spontaneous polarization when kept in an external electric field. These materials experience a phase transition from high to low temperature (paraelectric to ferroelectric) at Curie temperature. In case you are wondering what paraelectric materials are, it refers to the materials, which generate dielectric polarization in the presence of an electric field and loses the same when the field is removed. Finally, the piezoelectric materials are a class of dielectric materials, which can be polarized both by an externally applied electric field and also by the application of mechanical stress. This property is termed the piezoelectricity, and is also known as pressure electricity. Thus, the piezoelectric effect is the generation of electric charge as a result of force exerted on a piezoelectric material. Some of the examples of piezoelectric materials are ceramics, quartz crystal, Rochelle salt, etc. The advantage of this natural phenomenon is utilized in sensing the forces or displacement during a process. The most common material being used is quartz. This sensor converts the mechanical stress into an electrical quantity. Nevertheless, there are certain synthetic materials as well such as ammonium diphosphate that exhibits a piezoelectric behaviour.

The discussions for the working of this sensor can be referred to in Fig. 2.36a, b. The picture shows a piezoelectric material that is being subjected to a force. Before application of any external force, the material remains in an electrically neutral state. This is shown by the molecular model, which is depicted in Fig. 2.36a. With the application of the force, the material gets deformed, which results in the separation of positive and negative charges, as shown in Fig. 2.36b. This results in the formation of dipoles. The opposite charges cancel each other while the fixed charges appear on the surface. Hence, the material becomes polarized and it results in the generation of an electric field. This effect is the piezoelectric effect [4].

This happens because the electrons inside the material get excited and jump to the next higher level creating a positive charge and a negative charge in the location which they left. The existence of these charges generates an electrical potential that is equivalent to the force applied. Figure 2.37a depicts a picture of the quartz material, a piezoelectric material under unloaded condition. Quartz consists of one part of oxygen and two parts of silicon. With the application of load, the crystal deforms and takes a shape as depicted in Fig. 2.37b. The charge develops because of the application of force, which separates the positive and negative charges.

Fig. 2.36 Picture of molecular model: **a** absence of external force and **b** presence of external load

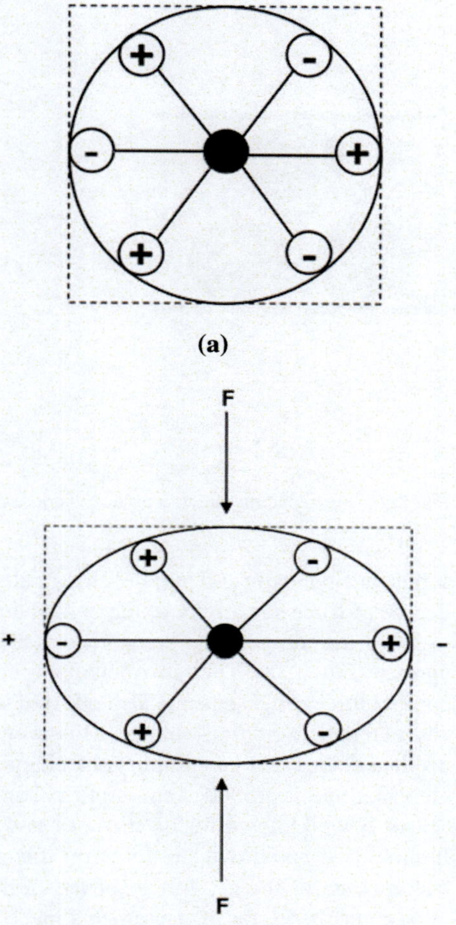

The aforementioned is the principle of the longitudinal effect, which is during the case of tension/compression force. The developed charge does not depend upon the size or shape of the crystal disks i.e., the piezoelectric element. Equation 2.20 represents the relationship between the charge developed and the force applied.

$$Q = kFn \tag{2.20}$$

In the above equation, Q indicates the charge developed on the surface of the piezoelectric material, F is the amount of force applied, k is the piezoelectric coefficient, and n is the number of crystals. The only way to increase the charge on the surface is to increase the number of crystal disks. These disks are connected in

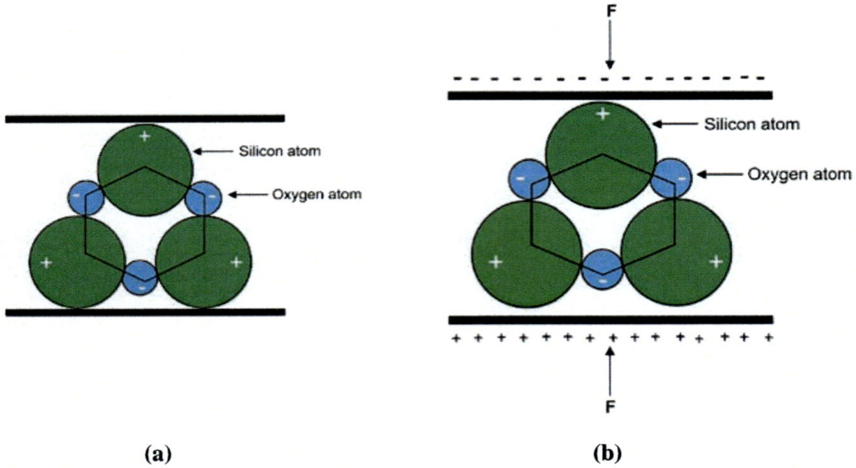

Fig. 2.37 Piezoelectric effect: **a** unloaded condition and **b** loaded condition

series mechanically and parallel electrically. The piezoelectric constant indicates the degree of force sensitivity of the crystal and also the direction of the force.

Thus, the charge that is being generated has a direct dependency on the force being applied (Fig. 2.38). The rate of charge produced depends upon the rate of change of applied force. This sensor is also referred to as an active sensor as it does not require any external excitation voltage. However, the charge generated is very small and needs a device that can amplify the charge. This device is referred to as the charge amplifier and it produces an output voltage big enough to be able to measure. The sensor is well known for its stiffness as dynamic nature for measurement purpose. It should be noted that the force on this sensor needs to be dynamic. If the force becomes static, the net voltage of the entire material will be null, which makes this sensor unsuitable for measurements involving static force. Another property of this

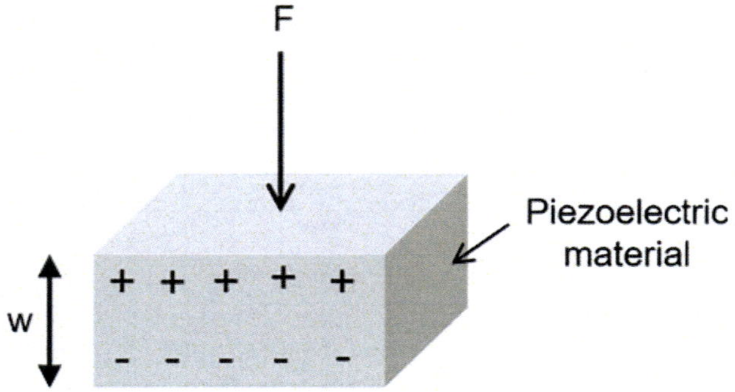

Fig. 2.38 Piezoelectric sensor for force measurements

sensor is that it also acts as an "inverse transducer". This means that if the material is subjected to some voltage, it will generate an equivalent force.

The following illustrates the dependency of the output voltage generated in the transducer on the applied force. The charge generated in the crystal depends upon the force applied to the crystal. This is represented in the expression Eq. 2.21.

$$\begin{aligned} Q &\propto F \\ \Rightarrow Q &= pF \end{aligned} \tag{2.21}$$

where Q is the charge, F is the force, p is the proportionality constant that is referred to as the "charge sensitivity". Considering the piezoelectric sensor to be a capacitor as it can store the charge, the following expression (Eq. 2.22) can be written as shown below.

$$\begin{aligned} Q &\propto V \\ \Rightarrow Q &= CV \\ \Rightarrow V &= \tfrac{Q}{C} \end{aligned} \tag{2.22}$$

$$\begin{aligned} &\Rightarrow V = \tfrac{pF}{C} \qquad \text{(Using Eq. 2.21)} \\ &\Rightarrow V = \tfrac{pFd}{\varepsilon_r \varepsilon_0 A} \qquad \text{(Using Eq. 2.14} \\ &\Rightarrow \tfrac{V}{d} = \tfrac{p}{\varepsilon_r \varepsilon_0} \tfrac{F}{A} \\ &\Rightarrow \tfrac{V}{d} = g\sigma \qquad \text{(Using } \sigma = \tfrac{F}{A}\text{)} \end{aligned}$$

$$\Rightarrow V = g\sigma d \quad \left(\text{Using } g = \frac{p}{\varepsilon_r \varepsilon_0}\right) \tag{2.23}$$

where, g is the voltage sensitivity and is expressed as the ratio of electric field intensity and the pressure (Eq. 2.23), and C is the capacitance.

2.5.4.3 Summary of the force sensors

In this particular section, two types of force sensors were discussed, strain gauge and piezoelectric. As discussed, strain gauge type load cell employs strain gauge elements that deform under the application of force. It either stretches or compresses depending upon the type of applied load. This deformation is proportional to the applied force. The deformation results in the change of the electrical resistance and hence, when it is connected in a measuring Wheatstone Bridge circuit, an equivalent voltage signal is generated. For this purpose, the strain gauge sensor is passive and requires an excitation voltage for generating the output voltage. On the other hand, a piezoelectric-based force sensor employs a piezoelectric material that can generate an electric charge under the application of force. The applied load leads to a charge shift at the molecular level resulting in the generation of an electrical charge at the surface of the piezoelectric material. This charge is converted to a voltage signal with the help of a charge amplifier. Coming to the application aspects, the piezoelectric sensor shows drift in the measurement, which is more predominant when the applied force is static and measurement is to be taken over a longer period. It has been observed that the drift in the measurement is negligible when high dynamic loads are present and measurement is taken for short intervals. This implies that a piezoelectric-based sensor is more suitable for dynamic applications, the ones involving highly transient events. Strain gauge type is more suitable for heavy-duty industrial applications requiring both static and dynamic measurements as it does not show any drift in the measurement. In a piezoelectric sensor, the charge is generated due to the change in the molecular level; the deformation to the sensing element is very less. This enables the sensor to offer high stiffness and thus, it has a high resonant frequency. This is why the piezoelectric sensor comes suitable for dynamic measurements. This also makes the piezoelectric sensor suitable to measure a wide range of forces. Contrary to this, the resonant frequency of a strain gauge type sensor depends on the nominal force of the sensor itself. The transducers designed for smaller force applications employ softer strain gauge element and hence, have low frequencies. Similarly, for the case of larger forces, there will be higher frequencies. This indicates that for fast measurements of small forces, a piezoelectric sensor is suitable and for fast measurement of large forces, a strain gauge type will be suitable. So, different strain gauge sensors will be required based upon the applications. Further, the temperature of the environment in which the sensor will work also needs to be considered. The piezoelectric sensor has a low-temperature sensitivity, which particularly allows it to be operated at higher temperatures with calibration performed at room temperature. However, the electrical resistance is strongly dependent on the temperature of a conductor, and thus, a strain gauge needs calibration at the measuring temperature. This also indicates that a strain gauge sensor may produce incorrect measurements when operated at a higher temperature for a longer duration. Finally, the piezoelectric sensors are compact and hence, require smaller space. So, applications with space constraints may need to employ this sort of sensor. Typical examples of the application of strain gauge type sensor include measuring the load of the vehicles on weighbridges for identifying overloaded vehicles, in structural applications for monitoring the load upon them, etc. The piezoelectric sensors are highly recommended for dynamic measurements such as for monitoring of the manufacturing processes as they involve transient events.

2.5.4.4 Real-Life Example in Manufacturing

For illustrating the application of the strain gauge type load cell for force measurement, let us consider a manufacturing process, deep drawing. It is a sheet metal forming process, in which, a metal blank is radially drawn into a forming die by the mechanical action of a punch. A compression testing machine is utilized to serve the operation. The picture of the machine is shown in Fig. 2.39. For drawing the metal

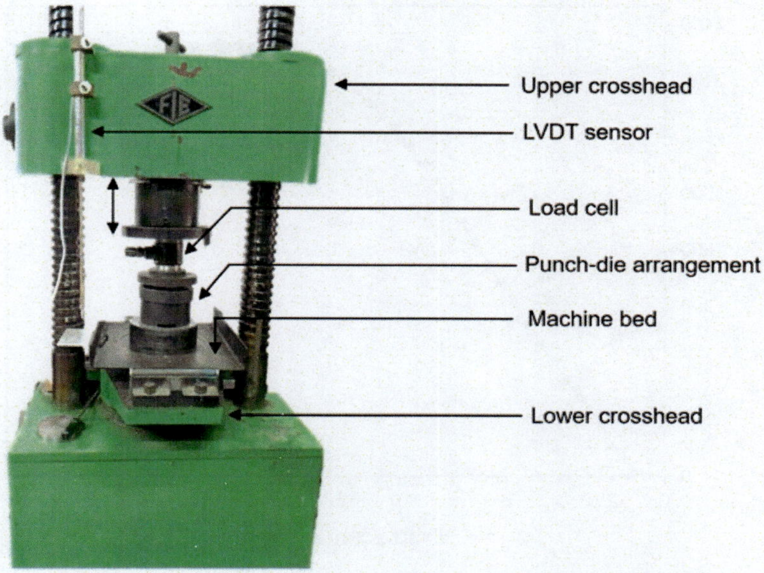

Fig. 2.39 Application of a strain gauge type load cell in a manufacturing machine

blank, it is greased and kept over a suitable die. It is referred to as deep because the depth of the part which will be obtained after the drawing process is higher than that of its diameter. The compression testing machine is driven by hydraulic power. During the drawing process, the compressive load is applied to the punch, which transfers it to the metal block. A strain gauge type sensor was employed, which, in this case, measured the load applied to the punch. This sensor is kept over the punch. It is further connected to a data acquisition card for acquiring the signal. Besides, an LVDT sensor can also be observed in the picture that monitors the displacement of the blank with the application of the load. The schematic for data acquisition from the strain gauge type force sensor is shown in Fig. 2.40.

The obtained load–displacement plot is shown in Fig. 2.41. With the help of the data acquired through the sensors, the operator will be able to track the drawing process. By studying the signal and correlating it with the quality of the drawn

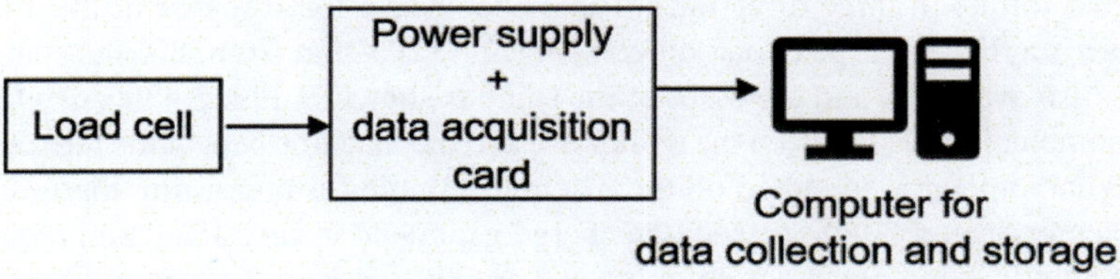

Fig. 2.40 Schematic diagram for data acquisition from the force sensor

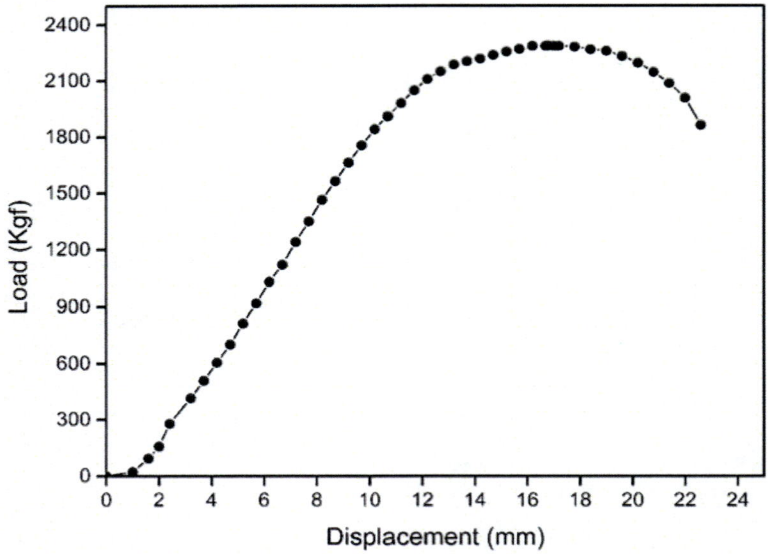

Fig. 2.41 Data captured using a strain gauge type force sensor

product, the threshold values of load required to obtain the desired quality can be determined.

For illustrating the use of a piezoelectric sensor for force measurement, let us consider a vertical milling machine *(BFW, VF 3.5)* fitted with one such sensor. A picture of this machine is shown in Fig. 2.42 and it is modified to be utilized for FSW. The sensor shown here is also referred to as a rotating type dynamometer *(Kistler, 9142B1111)* and a picture of this sensor is shown in Fig. 2.43a. This sensor is employed for force and torque measurements. Dynamometers are either single-component or can be multi-component. Here component refers to the ability of the sensor to capture the number of forces and moment. A single component can capture force in only one direction, and usually, it is the axial force acting in the vertical direction. Else, the configuration may be 3-component, 4-component, or maybe 6-component as well. The 3-component sensor can acquire forces in three directions, the 4-component sensor can acquire the same, besides, it can also acquire the torque or moment in the vertical direction, and the 6-component sensor can acquire both force and torque in three directions. The piezoelectric force sensor utilized in this example is a rotating-type dynamometer (Fig. 2.43a). Apart from this, the stationary type is also available, and a picture of the same is shown in Fig. 2.43b. A stationary dynamometer is fixed between the workpiece and the machine bed, while the rotating-type dynamometer is mounted on the machine spindle for measuring the force and torque components. As discussed, the charge generated is very small and requires a charge amplifier for signal conditioning and amplifying it, and a picture of the same is shown in Fig. 2.43c.

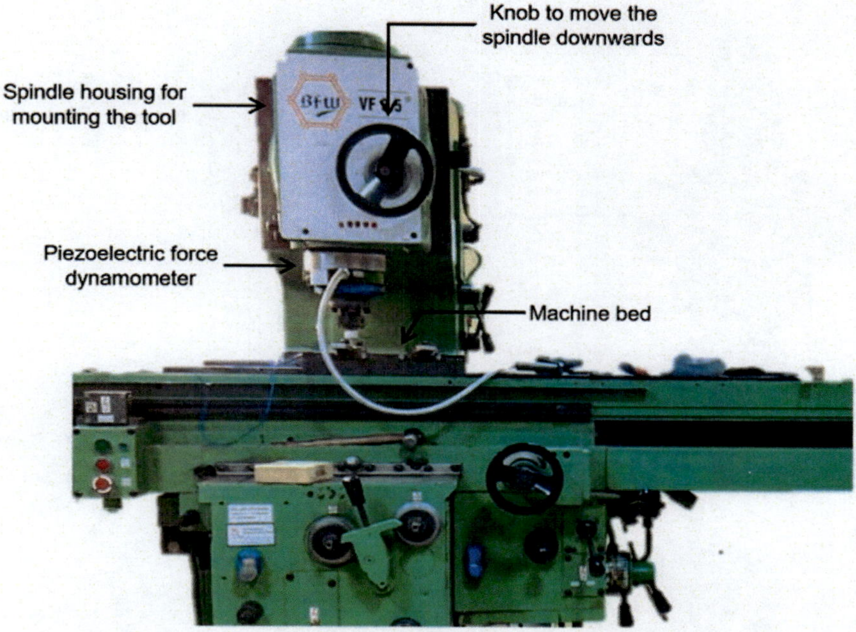

Fig. 2.42 Picture of a vertical milling machine fitted with a piezoelectric type force sensor

One end of the dynamometer is mounted on the machine spindle while the other holds the tool to be used for welding. The materials are fixed on the machine bed. As discussed in the earlier section (Sect. 2.5.3.4), FSW involves a rotating tool that plunges into the materials to be joined. The plunging of the tool gives rise to frictional heat, and the materials to be joined in turn oppose this motion, which gives rise to a force. To acquire the force, the piezoelectric force sensor is utilized here. This sensor is a 4-component dynamometer, which means, it can measure forces in three directions i.e., in X, Y, and Z directions, and torque in Z-direction. In the case of the stationary dynamometer, the sensor is directly connected to the charge amplifier via a connecting cable. However, in a rotating type dynamometer, the sensor sends the data to the charge amplifier unit through telemetry, which is then tapped by using a data acquisition card. A computer is used for collecting the signal from the data acquisition device in real-time. A schematic diagram for data acquisition from the piezoelectric sensor is shown in Fig. 2.44.

Figure 2.45 depicts the plot of an acquired force data in the FSW process with respect to time. The data acquired from the signal conditioning unit is in terms of volts. A multiplication factor is used to convert the same into an equivalent quantity in kN. This factor is provided by the manufacturer of the sensor. From the plot, it

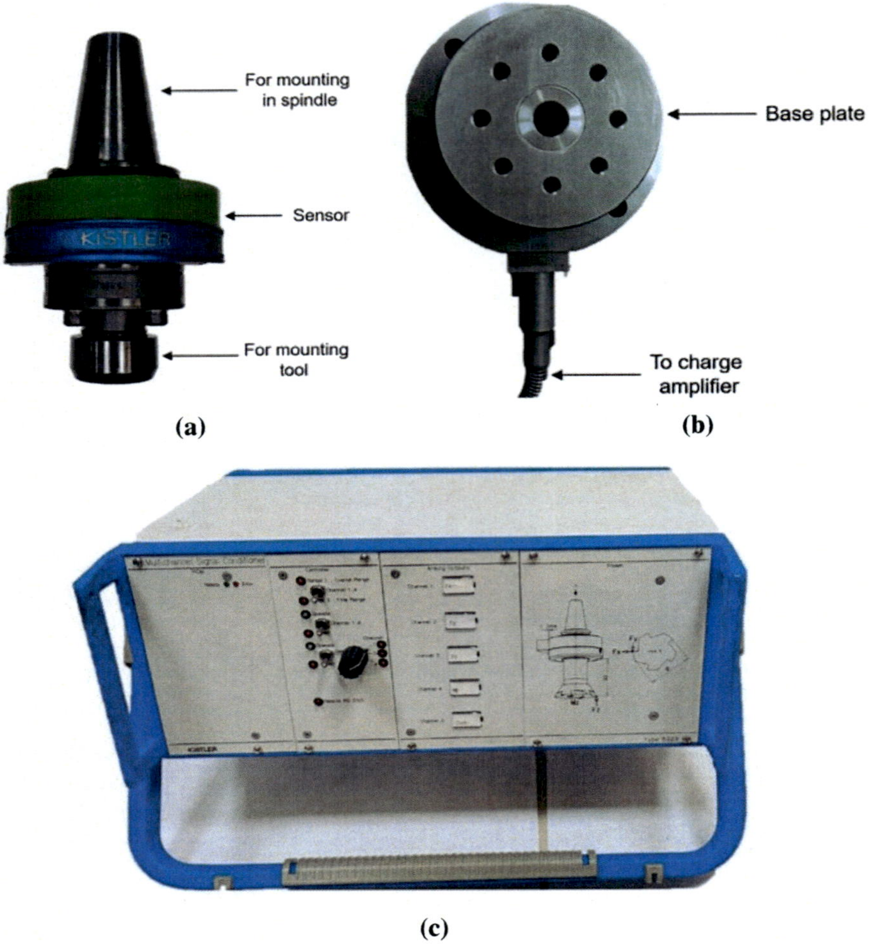

Fig. 2.43 Dynamometer: **a** rotating type, **b** stationary type, and **c** charge amplifier

can be seen how the force rises with time, which accounts for the plunging of the tool inside the workpieces.

Initially, the tool is in the air and that is why the force at the beginning is almost nil. As the tool makes a contact with the workpiece, which is demarcated as the "start of tool plunging" in the plot, the force rises and reaches a peak value. The shoulder of the tool touches next resulting in further rise of the force. This is demarcated as "shoulder contact" and "peak of shoulder plunging". The plunging event can be marked from the first two peaks that occur at the time instance of 20 and 40 s, respectively. This

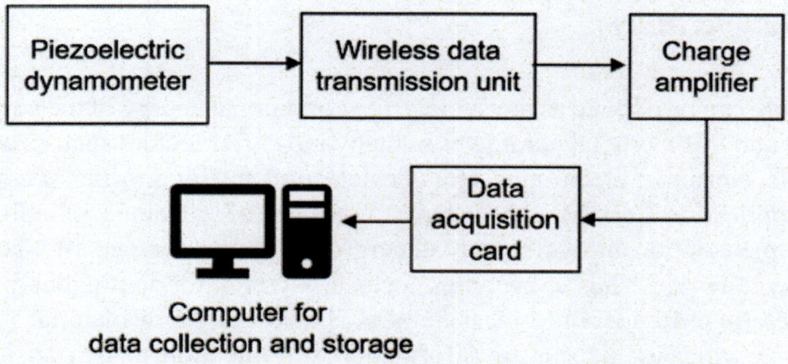

Fig. 2.44 Schematic diagram for data acquisition from the piezoelectric force sensor

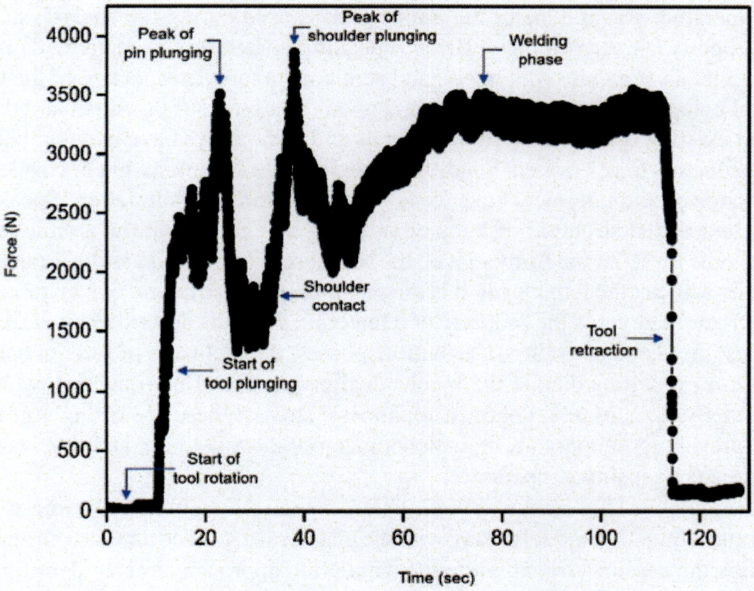

Fig. 2.45 Data collected from a piezoelectric force sensor

is followed by a stable region that denotes the welding phase, and finally, the force decreases indicating the retrieval of the tool from the workpieces. As the welding depends upon the frictional heat that is generated because of the interaction of the tool and the workpieces, if the contact between them would be insufficient, then this force signature will vary. Therefore, it can be utilized for monitoring the welding process.

2.5.5 *Temperature*

Temperature can be defined as the average heat or thermal energy of the particles in a substance and it is a crucial parameter in many industrial sectors such as aerospace, automobile, consumer electronics, etc. The detection and monitoring of temperature in many application areas could be taken as a measure of preventive reliability. There may be some component, which may undergo substantial damage after some point of a temperature rise. Thus, it becomes necessary to monitor continuously to ensure a safe operation. Temperature measurement is also crucial in material processing applications, which include plastic deformation, solidification, phase transformation, grain growth, etc. [6]. For instance, during the fabrication, or machining process, the temperature of the workpiece and the tool is monitored and measured continuously to get control over the manufacturing operation by making real-time decisions from the temperature signal. One of the popular terms used during the analysis of materials' property is microstructure. Briefly speaking, materials can be divided into four namely metals, ceramics, polymers, and semiconductors. Irrespective of the class a material belongs to, its properties depend upon the atomic arrangement and the type of bond existing between the atoms. Metals and their alloys have metallic bonding, semiconductors have covalent bonding, ceramics have a combination of covalent and ionic bonding, and polymers have weak secondary forces of attraction. Speaking in the context of the structure of a material, it can be either be at the atomic level or microscopic level. At the former level, the structure is understood as the arrangement of atoms, while at the latter level, it is known as the microstructure of a material. This microstructure governs the property of a material, which is described by grains, size of grains, the shape of grains, distribution, phases, the inclusion of foreign particles, etc. These properties dictate the mechanical properties of the material, which are tensile strength, hardness, ductility, toughness, etc. All these are strongly governed by the prevailing temperature in which a material is worked on, and this marks the importance of sensing temperature.

In the previous chapter, it was highlighted that the digital twin will comprise data-driven and simulation models along with the historical data. The data-driven approach considers the sensors' signal while the simulation approach aims at identifying the insights based on the physics of a process. Temperature is one such physical parameter that could be obtained by simulating a thermal analysis of a process. The obtained results will give a holistic view of the evolution of a process or a phenomenon. The following section discusses the types of sensors that are commonly used for measuring temperature.

2.5.5.1 Thermistor

The term "thermistor" is composed of two words, the first is thermal and the second being resistor. The thermistor is a temperature-sensitive device that exhibits a change in the resistance as a response to the change in the temperature [7]. It is made from a mixture of metal oxides that are semiconductors and processed into various forms such as beads, rods, etc. All resistors have some degree of dependency on the temperature, which is described by a parameter called the temperature coefficient. Based on this, there are two types of thermistors, negative temperature coefficient (NTC) and positive temperature coefficient (PTC).

The NTC thermistor has a negative temperature coefficient, i.e., the resistance decreases with an increase in the temperature. It exhibits a non-linear behaviour. It is commonly used as a temperature sensor and current-limiting device. This sensor is used in the temperature range of ≃55–200 °C, and the relationship between temperature and resistance is given in Eq. 2.24.

$$R_T = R_0 e^{\beta\left(\frac{1}{T}-\frac{1}{T_0}\right)}$$

$$\Rightarrow \beta = \frac{\text{In}\left(\frac{R_T}{R_0}\right)}{\left(\frac{1}{T}-\frac{1}{T_0}\right)} \tag{2.24}$$

where R_T is the resistance at temperature T, R_0 is the resistance at temperature T_0, and β is a constant that depends upon the characteristics of the material. This equation is referred to as the Beta formula. Commercially available thermistors do specify the resistance value at room temperature. From the above equation, it can be observed that the higher the value of β, the higher would be the variation in resistance for the same rise in the temperature. This will help in having a higher sensitivity and accuracy of the sensor. Further, the temperature coefficient is given in Eq. 2.25.

$$\alpha_T = \frac{1}{R_T}\frac{dR_T}{dT} = -\frac{\beta}{T^2} \tag{2.25}$$

where α_T is the temperature coefficient. From the above equation, the temperature coefficient can be observed to be negative and this is the reason it is termed as negative temperature coefficient. Using this equation and the obtained resistance values from an NTC sensor, the temperature can be obtained. NTC thermistor is available in various configurations namely bead type, disk type or metallized, and glass encapsulated, as shown in Fig. 2.46. Each of these configurations is having its own set of applications. The bead-type is fabricated from platinum alloy or lead alloy and is sintered into the ceramic body. The diameter of the bead type configuration probably ranges from 0.075 to 5 mm. Because of this small diameter, it is having a fast response time and low heat dissipation. The disk type is fabricated from oxide powders sintered into a round die and it ranges from 0.25 to 25 mm diameter. Because of a larger size as compared to the bead type, its response time is slower than the bead type. These thermistors have a temperature range of up to 150 °C. Finally, the glass encapsulated type is fabricated from a metal oxide encapsulated in a glass, ensuring a hermetic sealing. This means that the sealing prevents the passage of air,

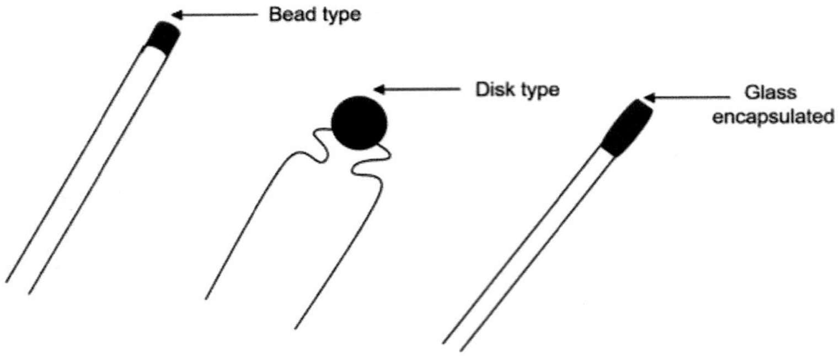

Fig. 2.46 NTC thermistors

oxygen, moisture, etc. If you are wondering, what benefit does it render, then it is the elimination of error in measurement i.e., the resistance readings, which occur with penetration of moisture. The glass encapsulated type also offer a higher temperature measuring range, usually up to 200 °C.

Owing to the aforementioned characteristics, NTC thermistor find applications in surge protection, temperature monitoring, and so on. The first application owes to its high value of resistance, as the power is switched on. Certain equipment when is switched on, draw a large amount of current, which may damage the equipment. NTC thermistor is employed in such equipment, which has a significantly high value of resistance as it is switched on, limiting the flow of current. The second application is temperature monitoring, where they could be employed for continuous monitoring of temperature.

PTC thermistor on the other hand has a positive temperature coefficient, i.e., the resistance increases with an increase in the temperature [7]. It is popular because of the linear behaviour and this makes it find application in the electrical circuits. They are mainly of two types, silistors and switching type PTC thermistor. The silistor is manufactured using silicon as the semiconductor and it exhibits a linear temperature-resistance characteristic. Usually, the relationship remains linear up to 150 °C, and applications are selected accordingly. The switching type PTC thermistor is manufactured from polycrystalline ceramic materials and is made semi-conductive by adding dopants. It exhibits a negative temperature coefficient at the least resistance value, and this leads towards a positive one when moving towards the transition temperature. The transition temperature is also referred to as the switch temperature, and it is the one from which the resistance value starts increasing rapidly with a temperature rise. Applications of PTC thermistor include over-current protection, time delay, motor starting, and so on. Speaking for the first application example, the PTC thermistor is useful for limiting the current. With a larger current flowing through equipment, for which it is not designed to handle, the thermistor resistance will increase, and once it reaches the transition temperature, the resistance will rise sharply, which will limit the current. The second application example is the time delay, which is usually required for certain electrical appliances. The thermistor, in this case, benefits in switching between high and low power requirements by maintaining a time delay, ensuring the life of the appliance. The third application is in the motor starting, where a PTC thermistor helps in switching the start-up winding as the motor is switched on, and later with the rise in temperature and resistance, the normal operation can be resumed.

2.5.5.2 Resistance Temperature Detector

The resistance temperature detector (RTD) is also known by the name resistance thermometer and it measures the temperature by correlating the change in resistance of the RTD element as a response to the change in temperature [7]. The RTD element refers to a metal, and the working principle of this sensor is the fact that as the temperature across a metal increases, there is a change in its resistance. So, to take the measurements by this sensor, an electric current is passed through it, which is located in proximity to the job whose temperature is to be measured. As the temperature of the job changes, there occurs a change in the resistance of the RTD element, which is captured by using an instrument. The instrument employed for measuring the resistance is a Wheatstone Bridge. Usually, the resistance value of an RTD element is known for some temperature values, which is utilized to correlate for unknown values of temperature. The sensor may consist of a wire of metals like platinum, nickel, copper, etc., wrapped around a ceramic or glass core and has a positive temperature coefficient. This is because, in metals, the resistance to the flow of electricity increases with the temperature rise. Thus, the resistance of the RTD element increases. A schematic picture of this sensor is shown in Fig. 2.47.

Out of the aforementioned metals, platinum is widely employed as the RTD element owing to its characteristics such as the near-linear relationship of resistance with the temperature, higher temperature coefficient enabling sufficient change in

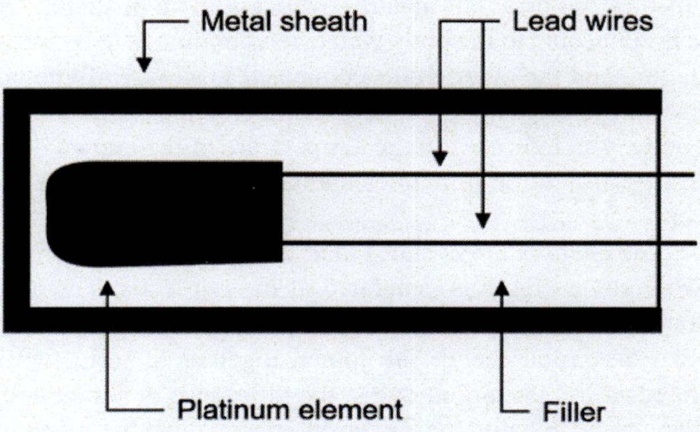

Fig. 2.47 RTD element

resistance with a temperature change, it offers a stable resistance–temperature characteristic, chemically inert, etc. It comes with a temperature range of $\simeq$100 to 600 °C. Let us now compare the RTD element with a thermistor and visualize the differences. The first is the element that they are made of; the RTD element is manufactured from pure metals such as platinum, copper, nickel, etc., while the thermistor is manufactured from semiconductors or polymers. Second is the range of measurement, which is high in the case of an RTD as it may go up to 600 °C, while it is ~200 °C in the case of a thermistor. Thus, as compared to the thermistor, the RTD has a higher temperature range, and is highly stable and give responses over a long period. However, the thermistor has a higher sensitivity than that of the RTDs owing to their smaller size and quick response.

2.5.5.3 Thermocouple

The thermocouple is another type of sensor that also measures the temperature [7]. The term thermocouple can be divided into two i.e., thermo and couple. The term thermo signifies thermal and it indicates the measurement of the temperature. The other term couple refers to a combination of two and in the case of a thermocouple, it indicates two dissimilar metals. The reason for dissimilar metals is that different metals behave differently to temperature change. This is the underlying principle of a thermocouple. As a metal wire is subjected to temperature, it will develop a voltage difference. This voltage generation is not the same for all metals. Different metals will develop different voltages. Thus, when two different metals are connected and one end of them is subjected to a change in temperature, the other end of the two develop a voltage difference, which can be measured to infer the temperature on the other end. Instead, if two similar wires are connected, then the voltage developed by the two would be the same. Thus, there would be no voltage difference, which will make it impossible to figure out the temperature at the other end. A thermocouple consists of two wires made from different metals covered with an insulator. One end of the thermocouple is connected to the body whose temperature is to be measured, known as the hot junction, and the other end is connected to a body of known temperature called to be the reference junction. Thus, the device measures the temperature of the unknown body with reference to the temperature of the known body. The sensor takes into consideration three principles namely, *Seebeck effect*, *Peltier effect,* and *Thomson effect*.

The first i.e., the *Seebeck effect* states that when two dissimilar metals are joined together at two ends, an EMF is generated at the two ends. The amount of EMF generated differs for different combinations of the metals being joined. *Peltier effect* states when two dissimilar metals are joined together to form two junctions, an EMF is generated within the circuit due to the difference in temperatures of the two junctions of the circuit. Finally, the *Thomson effect* states that when two dissimilar metals are joined together forming two junctions, a potential exists within the circuit due to the temperature gradient along the entire length of the conductors. Owing to the combination of dissimilar materials, the thermocouples are further categorized.

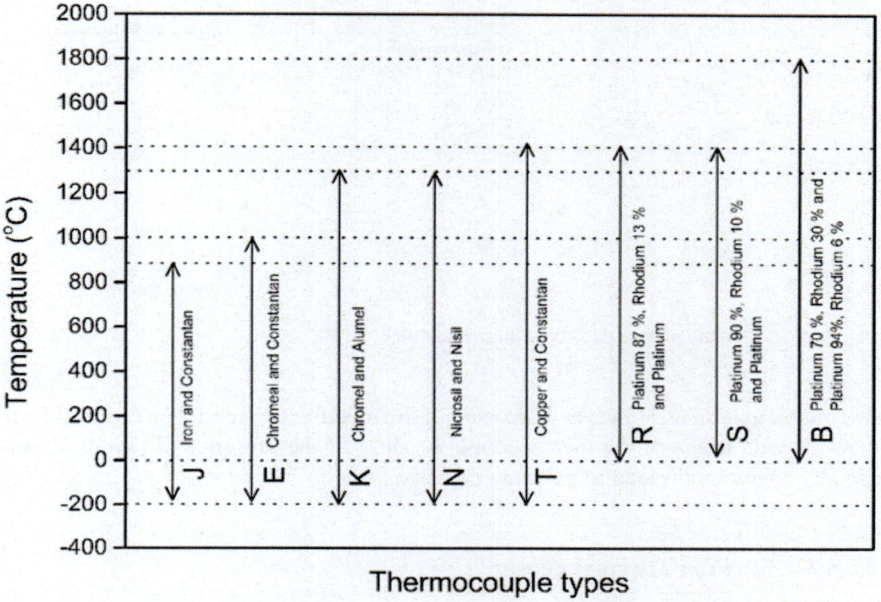

Fig. 2.48 Types of thermocouples, measuring range, and composition

Figure 2.48 shows the categories of the thermocouples along with different ranges and compositions. The types are J, E, K, N, T, R, S, and B. They differ in temperature range, accuracy, dead time, sensitivity, etc. Thus, the selection of a thermocouple would require knowledge of the range that needs to be measured, the smallest change in temperature to be measured, etc. The advantage of thermocouples includes their simple design and easy fixing on the target whose temperature is to be measured.

Out of the aforementioned thermocouple types, K-type is the most general-purpose one as it offers a wide temperature range, thus, serving a large number of applications. However, with application demanding higher temperature measurements, T, R, S, and B-types are employed. A schematic picture representing a K-type thermocouple is shown in Fig. 2.49. Almost, all thermocouple look alike, however, are differentiated by the help of colour coding. As shown here (Fig. 2.49), the K-type thermocouple is a yelow coloured one with two dissimilar wires, Chromel is yellow coloured and Alumel is ed coloured. One end of the thermocouple is shorted and it is fixed on the job whose temperature is to be measured. The other end goes into the data acquisition card for acquiring the temperature signal. Thus, all types of thermocouples are having a specific colour coding. This colour coding may also differ country wise, and the one shown here (Fig. 2.49) is of the US standard.

As compared to thermistor and RTD, the thermocouple is preferred for the wide range of temperature measurement. Besides, it is an active sensor as it does not require any external excitation as required in the case of thermistor and RTD. This also eliminates the occurrence of self-heating in thermocouples, which is found in

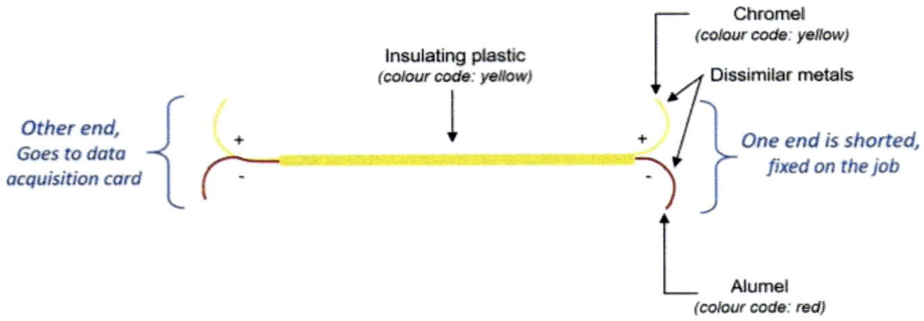

Fig. 2.49 Schematic picture of K-type thermocouple

thermistor and RTD. However, corrosion of the metallic wire could be one disadvantage for long usage of the thermocouple because of the presence of two dissimilar metals, which might lead to galvanic corrosion.

2.5.5.4 Infrared Thermal Sensor

Considering the electromagnetic spectrum, it has variants with highest to lowest wavelengths i.e., the radio waves to gamma rays. The visible light also occupies a certain range and the human eye can detect the visible light. However, bodies emit radiations that are also beyond the visible range. Thus, a human eye cannot detect it. The infrared thermal sensor is one such instrument that measures the temperature by sensing the infrared radiations [8]. This sensor senses heat being radiated by an object. Every object that is above the absolute zero temperature emits heat. This emitted heat is referred to as infrared radiation, which is above the visible light range. The infrared thermal sensor is also referred to as a thermal imager and it looks like a normal optical camera, however, the difference is that the normal optical camera takes the visible light, while the thermal camera targets the infrared range which has a higher wavelength. This can be visualized from the electromagnetic spectrum, which is depicted in Fig. 2.50. The visible range is from 400 to 700 nm in which, the normal optical cameras work, while it is the next higher wavelength range for the thermal camera starting from 700 nm to 1 mm.

The thermal imaging camera consists of an optical lens, an infrared detector that is the thermal sensor, and associated electronics. The optics present in the camera focuses the energy on the detector, which in turn, generates an electrical signal proportional to the radiation. It is non-invasive i.e., it does not need to make physical contact with the target of which the temperature is to be measured.

To compare with the thermocouple, which was discussed in the previous section, let us consider the following example. The objective is to record the change in the temperature of a metallic plate that will be subjected to the drilling operation. To measure the temperature across a hole, a few thermocouples will be required to fix it across the holes. This is particularly time-consuming and at times it may not

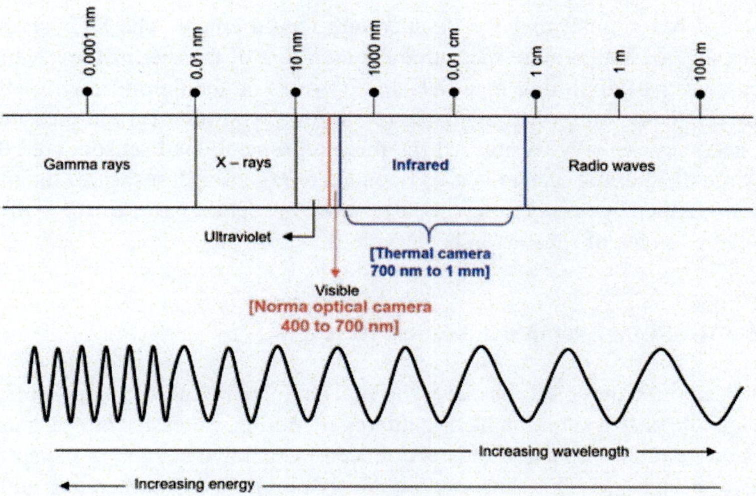

Fig. 2.50 Electromagnetic spectrum showing the visible and infrared ranges

be recommended since fixing the thermocouple may require drilling smaller holes depending upon the diameter of the thermocouple wire. Instead, a thermal imager can be utilized to record the temperature non-invasively. However, it is to be noted that a thermal imager will only be able to record the temperature at the surface, and it cannot measure the temperature beneath the surface or the internal temperature. This is where a thermocouple would be essential for taking the temperature measurements. Other advantages of thermal imager include the ability to capture the signal with respect to a point, a line or an area. The major disadvantage of this sensor is the line of sight required by the sensor for measurement. In case of interruption, disturbances are created in the measurement. Another disadvantage is the limited distance from the targeted area for the fixation of this sensor. The thermal cameras are available in different measurement ranges, resolution, etc., and again one needs to identify the particular configuration based on the application to deal with.

2.5.5.5 Summary of the Temperature Sensors

In this particular section, four types of sensors were discussed for measuring the temperature. They are thermistor, RTD, thermocouple, and thermal imager. In the case of thermistor and RTD, the temperature is measured by sensing the change in the resistance of the sensing element, voltage or EMF is measured in the case of the thermocouple, and infrared radiation in the thermal imager. The thermistor and RTD are passive sensors as they require an external excitation for operation. However, the thermocouple is an active sensor. The thermistors are of two types, NTC and PTC, and are well suited for constant temperature monitoring applications.

However, it has a small range of temperature measurement. The RTD sensor has a wider range of temperature measurement than that of the thermistors. A thermocouple consists of dissimilar materials and operates on the thermoelectric effect to measure the temperature. It is available in different types that vary in measurement range, accuracy, sensitivity, etc. All the three aforementioned sensors are contact-type, while the thermal imager is a non-contact type sensor. It measures the infrared radiations emitted by a body to detect the temperature. Thus, based on the application requirement, a type of sensor can be chosen.

2.5.5.6 Real-Life Example in Manufacturing

In CNC machines, the spindle is one of the vital components and the inappropriate use of this spindle or in a situation having improper cooling, excessive heat is generated in the spindle motor. To avoid the occurrence of excessive heating in the spindle, a thermistor is engaged. This sensor continuously monitors the temperature and when the temperature rises beyond a certain threshold, an alarm is raised. In this way, the machine is protected from the problem that might occur because of over-heating. The use of thermistor, in this example, can be felt from the health monitoring perspective.

For another example, an application in the field of additive manufacturing is considered here. Directed energy deposition (a type of additive manufacturing) also referred to as 3D printing is used to build objects/parts by depositing material layer-by-layer. It uses data generated through computer-aided designing (CAD) files or 3D scanners to replicate the design and come up with the 3D part. The deposited material can be metal/plastic/concrete. During the process of deposition, the quality of the product being manufactured depends upon the melting and solidification of the material. This marks the importance of monitoring temperature during the build. The prevailing temperature on the substrate will provide clues on the quality of the melt pool. Hence, by monitoring the temperature signal, proper control techniques can be applied to ensure the product with required mechanical and metallurgical properties. Figure 2.51 depicts an additive manufacturing set-up. A K-type thermocouple is fixed on the side edge of the workpiece. For acquiring the temperature data, the thermocouple's other end is connected to a data acquisition card that is further connected to a computer, as shown schematically in Fig. 2.52.

Figure 2.53 depicts the temperature data collected during the process. Sudden peaks can be observed in the data at regular intervals, which are followed by a decreasing trend. This is because the thermocouple is fixed in one location. The laser head, shown in the experimental set-up (Fig. 2.51) moves ahead while depositing the material on the workpiece surface. Highest temperature could be recorded when the laser head is near the thermocouple, and it decreases as the laser head moves away. From this, a map of the temperature for the molten pool can be known for one particular location i.e., the one where the thermocouple is fixed. Thus, to get information about other locations in the workpiece, one needs to fix multiple thermocouples, which makes it a tedious task. Also, the obtained data needs to be processed through various techniques for determining and minimizing the noise content, which

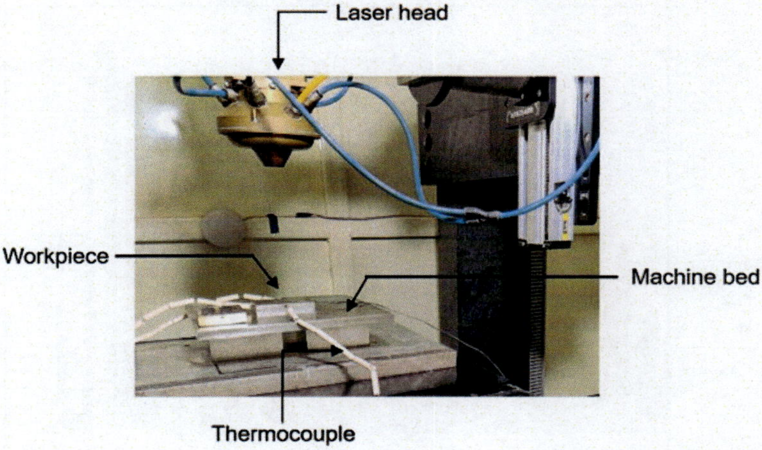

Fig. 2.51 Use of thermocouple in a manufacturing process

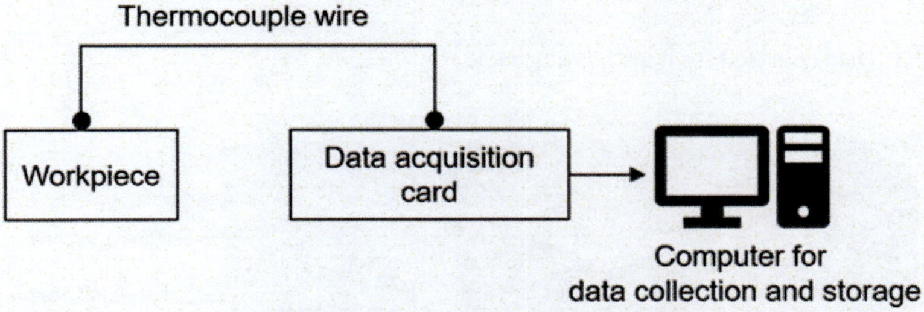

Fig. 2.52 Schematic diagram for data acquisition from the thermocouple

is required for precise monitoring of the temperature in the molten pool to acquire information on metal depositions.

Let us now shift to the use of a thermal imager for measuring the temperature. For this, an application in the friction-based sintering process is considered. Sintering is a process of compacting and forming a solid mass of material from metal powders by the application of heat or pressure. The process is performed without melting the metal powder. One of the innovative methods developed and patented in this field is the friction-based sintering process that is being employed for the fabrication of copper foam by using a pin-less friction stir tool [9]. A picture of the experimental set-up is shown in Fig. 2.54 that consists of an FSW machine *(Biss ITW, SP 02–05-10)*, a pin-less tool mounted on to the machine spindle, a fixture arrangement, and the designed set up contained in the fixture.

The custom-designed set-up fabricated for the friction-based sintering has four parts: a top sheet, a middle block, a stopper, and a bottom plate, as shown in Fig. 2.55. The middle block has a cavity to contain the metal powder and the size of the cavity

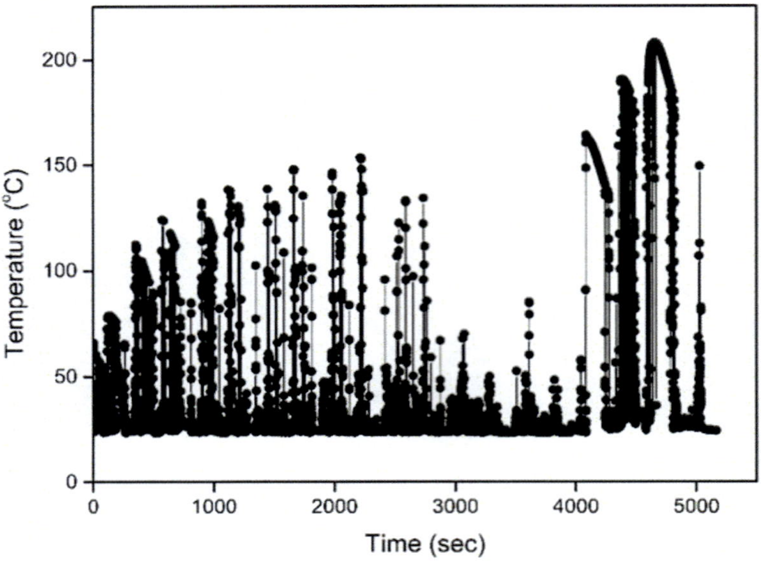

Fig. 2.53 Data collected from the thermocouple

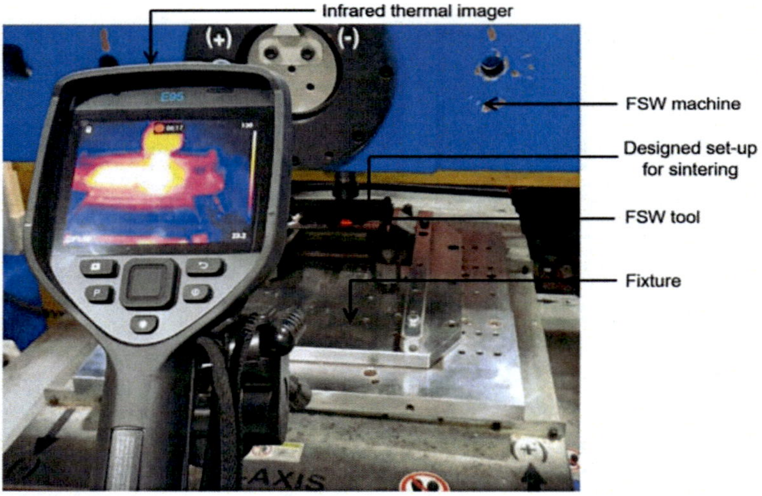

Fig. 2.54 Picture of an infrared thermal imager employed during a friction-based sintering process

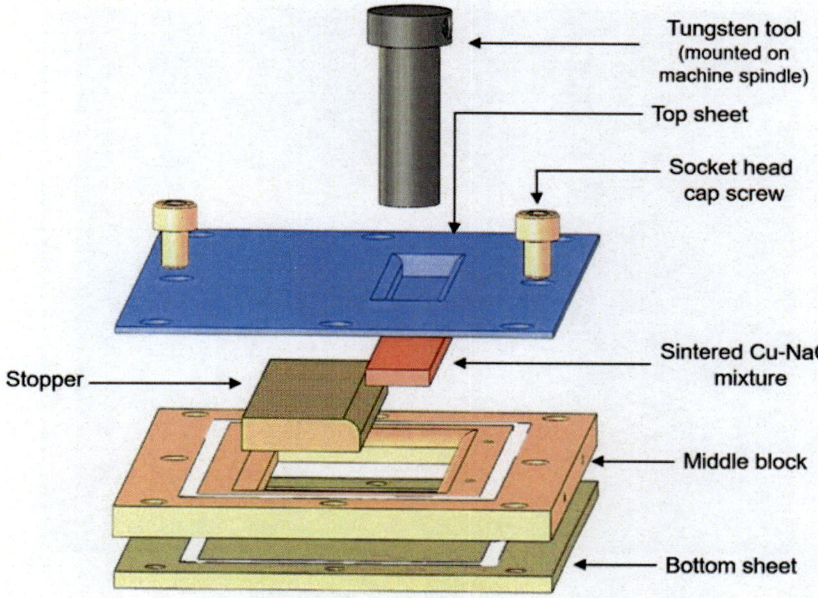

Fig. 2.55 Picture of the custom-designed set up for friction-based sintering

is controlled by the stopper. A high-speed rotating tool is plunged against the top sheet, which gives rise to frictional heat as it rubs over the top plate under axial load. The tool is also allowed to plunge inside, which further increases the axial load. As a result of this, the top plate gets deformed and the mixture kept beneath the top plate gets compacted. Thus, the process results in the solid-state sintering of the mixture.

Because of the association of heat and pressure, the measurement of temperature is of utmost importance during this process. An infrared thermal imager is utilized as shown in Fig. 2.54 that was pointed towards the experimental set-up for measuring the temperature *(FLIR, E-95)*. The output of this sensor is a thermogram that is depicted in Fig. 2.56.

The output of the thermal imager is a view of the thermal distribution graphically that is referred to as a thermogram. By observing the thermogram, one could determine the temperature of a particular point of interest and the entire zone as well. Darker colour schemes represent lower temperature values while lighter ones are used for higher temperature values. For instance, in the thermogram shown in Fig. 2.56, the temperature at the tip of the tool is 820.7 °C, which is represented by a lighter colour while the surrounding is shown in a darker colour. From this output, the advantage of the thermal imager can be felt, which is the complete view of the temperature prevailing on the target by using one instrument. In contrast, several numbers of thermocouples will be required in this case to get temperature measurements. Also, fixing the thermocouple at the tool-tip in this particular case is another problem as the rotating and plunging tool will damage the thermocouple wire.

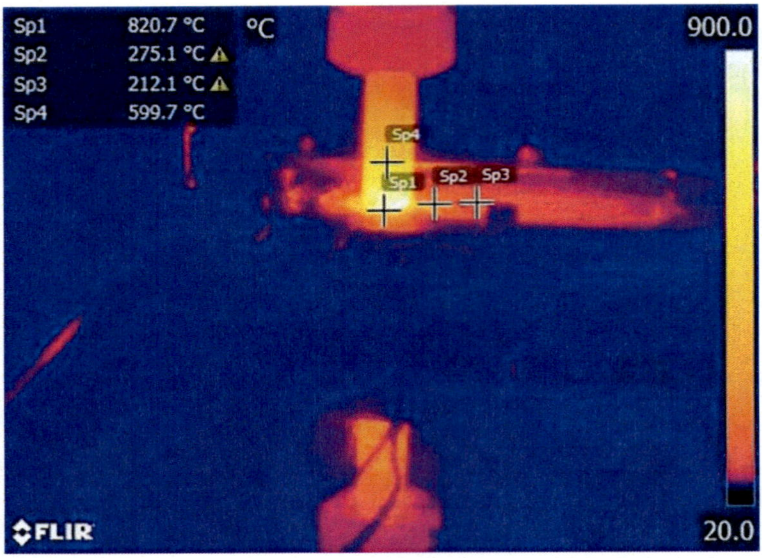

Fig. 2.56 Thermogram obtained as output from the thermal imager

2.5.6 Vibration Sensor

Vibration forms an important physical parameter in monitoring the health of a machine. The machine tools are associated with some vibrations as a part of their normal operation. The machine vibration can take different forms, it may be over smaller distances or may cover a larger one, it may be faster or slower, or it may or may not generate sound and heat. In most cases, the vibration is undesirable as it leads to damage to the machine. Thus, a suitable sensor is installed that can sense the vibration in the machines. At times, the vibration in a machine is caused because of nearby machines and rotating components. These vibrations may propagate through the machine foundations and are called independent vibration. The dependent vibrations are caused because of the process itself. The major causes of vibration in machines are acknowledged because of repeating forces, looseness, or resonance. Considering a machine, repeating forces are generated because of the imbalance or misalignment in the rotating components. The imbalance results because of variation in bolt sizes, air cavities in cast parts, broken and corroded parts, etc. The misalignment is a result of improper assembly, thermal expansion, distortion, etc. Similarly, loose parts in a machine can cause it to vibrate. It occurs in both rotating and non-rotating types of machinery. It may be because of clearances in the bearing, bolts, mismatched parts, etc. Vibration is described by two descriptors namely amplitude and frequency. The amplitude describes the severity of the vibration, while the rate of oscillation of the vibration is described by frequency. The following section explains the types of vibration sensors.

2.5.6.1 Piezoelectric-Based Accelerometer

In this list, the accelerometers are the first. They measure the acceleration of a motion [10]. The accelerometers have piezoelectric materials for sensing the vibration. To visualize the working principle, a schematic of this type of sensor is depicted in Fig. 2.57. It consists of a piezoelectric material, which generates an electrical charge when it is subjected to any external force. Because of vibration, the force is generated. This force is sensed by the piezoelectric material and it generates an equivalent voltage signal. This phenomenon is termed the piezoelectric effect, which has already been discussed in the previous section (Sect. 2.5.4.2). The piezoelectric crystal is coupled with a mass, as shown in the picture (Fig. 2.57), and it forms a mass-spring system, with a single degree of freedom.

Let us define a few basic terms to help visualize it in a better way. In the previous section, the force was discussed, which is a physical quantity, and when it is applied to an object, causes it to move in the direction of the applied force. Mass is another physical quantity that indicates the amount of matter in an object. Finally, acceleration is also a physical quantity, which is defined as the rate of change of velocity of an object in the same line of an unbalanced force. If the force is removed, the velocity will be nil and there will not be any acceleration. These three quantities i.e., force, mass, and acceleration are linked to each other, and they are defined by *Newton's second law of motion*, which is expressed in Eq. 2.26.

$$F = m \times a \tag{2.26}$$

where F is the applied force, m is the mass, and a is the acceleration. Thus, as a machine element, on which a vibration sensor is engaged, starts vibrating or it is in motion, the force will be generated, which will cause squeezing of the mass. This,

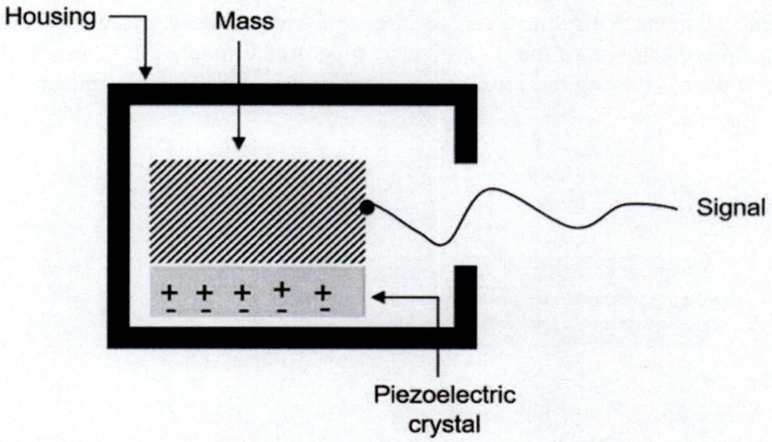

Fig. 2.57 Schematic picture of a piezoelectric-type accelerometer

in turn, will squeeze the piezoelectric material, which will generate an equivalent electrical charge. Since the mass is constant, the electrical charge developed is directly proportional to the acceleration. This is the underlying principle of the operation of accelerometers. Further, electronics are required to be engaged with the sensor for acquiring the charge developed by the piezoelectric crystal.

2.5.6.2 Capacitive Accelerometer

The second is the capacitive type, which employs the change in capacitance of two capacitor plates to measure the vibration. A schematic diagram to visualize the working principle of this type of sensor is depicted in Fig. 2.58.

It consists of two use capacitor plates, 'Plate 1' and 'Plate 2', as shown in the picture (Fig. 2.58). These two plates are fixed and they sandwich a diaphragm, which is the sensing element. With this arrangement, two capacitors are formed, each having a fixed capacitor plate and the diaphragm is shared by the two. The diaphragm consists of a mass as shown in the picture coupled with a spring system. With the application of vibration or motion, the spring-mass system experiences the force. This in turn causes a deflection of the spring-mass system, which can be expressed as represented in Eq. 2.27.

$$X = \frac{F}{K} \tag{2.27}$$

where X is the deflection of the spring-mass system caused due to the applied force, F indicates the applied force, and K is the stiffness of the spring. Because of this deflection, the capacitance of the circuit varies. The change in capacitance i.e., in the case of a parallel plate capacitor is already discussed in this chapter (Sect. 2.5.2.2). The change in capacitance is captured by the electronic circuit, which outputs an equivalent voltage signal. This sensor enables the measurement of both static and dynamic events. In comparison to the piezoelectric-type accelerometer, the capacitive-type is useful for monitoring large structures as they can acquire measurements across a

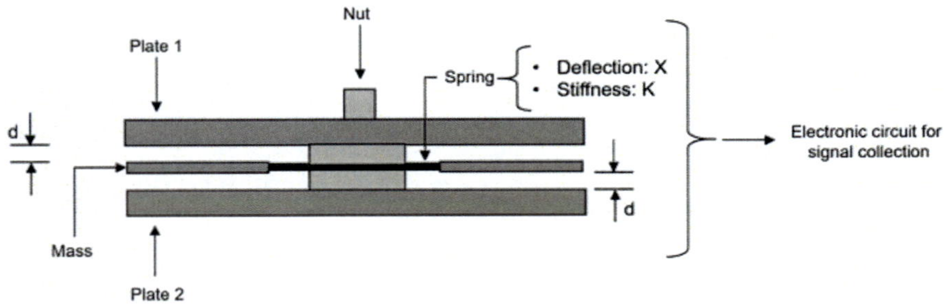

Fig. 2.58 Schematic picture of a capacitive-type accelerometer

wide frequency range. The piezoelectric type is not beneficial for static applications as the charge generation demands dynamic loading condition.

2.5.6.3 Laser-Based Non-contact Vibrometer

The aforementioned instruments, piezoelectric- and capacitive-type, are contact-type sensors. This means that they need to be in physical contact with the target object. One disadvantage is the weight of these sensors, which is added to the weight of the target object. A laser-based vibrometer has been developed to deal with this problem. This sensor utilizes a laser beam to measure the vibration in the object, hence, is non-invasive. The vibrometer employs the *Doppler effect* to acquire the vibration velocities [11]. The *Doppler effect* also called *Doppler shift* is the change in the frequency of a wave with respect to an observer, who is moving relative to the source of the wave. Let us define a few important terms to proceed further with the details of the vibrometer. These terms are wavelength, frequency, and velocity, and all three are related to waves. Wavelength is defined as the distance between two successive crests of a wave, frequency is the number of waves passing a given point per unit of time, and velocity of the wave indicates the speed of the wave, which is the distance covered per unit time. These three terms are associated via a relationship as expressed in Eq. 2.28.

$$\lambda = \frac{\vartheta}{f} \tag{2.28}$$

where λ is the wavelength of the wave, ϑ is the velocity, and f is the frequency of the wave. The said *Doppler effect* revolves around these terms. To set an example, imagine a person standing on a railway track observing the approach of a train towards him. The train appears to be slower when it is at a distance from the observing person, and it appears to be faster as it crosses by the person. The slower/faster action indicates a shift in the frequency and explains the *Doppler effect*. The expression of the *Doppler effect* is given in Eq. 2.29.

$$f\prime = \frac{\vartheta + \vartheta_o}{\vartheta + \vartheta_s} f \tag{2.29}$$

where f indicates the actual frequency, ϑ, ϑ_0, and ϑ_s denote the velocities of a sound wave, observer, and source, respectively, and f' is the observed frequency. The laser vibrometer works on the principle of laser interferometry. The vibrometer includes a laser emitting source, beam splitters, mirror, detectors, and associated electronics as its major components. A schematic diagram is depicted in Fig. 2.59 to visualize the principle employed in vibrometry.

The source emits a coherent laser beam on to the object under test, which passes through a beam splitter. This is shown as 'Beam splitter 1' in the picture. A beam

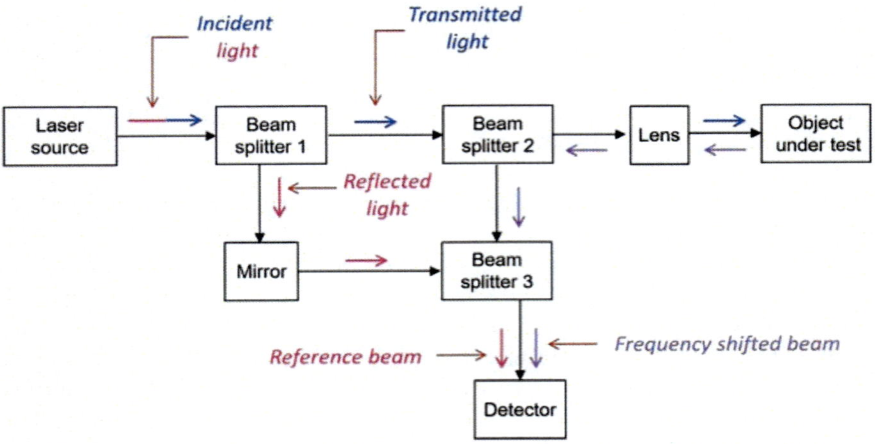

Fig. 2.59 Schematic picture of laser interferometry principle

splitter is a device employed to split an incident beam into two, which are termed as transmitted beam and reflected beam, both having equal intensities. The transmitted beam heads towards the object under the test, while the reflected beam travels towards another beam splitter ('Beam splitter 3') via a mirror. The reflected beam is termed the reference beam. This is termed so because it was the initial incident light, which was emitted by the source. Thus, it contains information such as wavelength and frequency. As the transmitted beam hits the object, it is reflected, and it passes through another beam splitter ('Beam splitter 2'). This splitter reflects a part of the light towards the third beam splitter ('Beam splitter 3'). At 'Beam splitter 3', the light beam, which is reflected from the object and the reference beam interferes, and this is why it is termed laser interferometry. Because the object is vibrating, it brings a frequency shift in the initially transmitted beam, and this becomes the frequency-shifted beam. The frequency-shifted beam and the reference beam finally goes into a detector, which detects the amount of frequency shift and determines the velocity of the object. In other words, because of the vibrating object, there is a path difference between the two beams (reference and the one reflected from the object). This path difference results in a phase lag that varies with the velocity of the vibrating object. Thus, the shift is proportional to the velocity of the target object.

To fetch more benefits out of the non-invasive sensing capability of the laser vibrometer, it is mounted on a robot end-effector. A picture of this sort of system is depicted in Fig. 2.60. The system includes three laser vibrometers *(Polytec, PSV 500 3DH)* for sensing the vibration of an object in three directions and these sensors are mounted and integrated on a robot *(Kuka, KR 90 R3700)* having a payload of 90 kg and a reach of ~2.5 m. further, the robot is mounted on a linear track *(Kuka, KL 4000)* of 5 m in length, which increases the reach of the robot.

With this set-up, the robot can move around test objects and the vibrometer can scan the surface to obtain the vibration measurements. If the vibrometer is fixed on

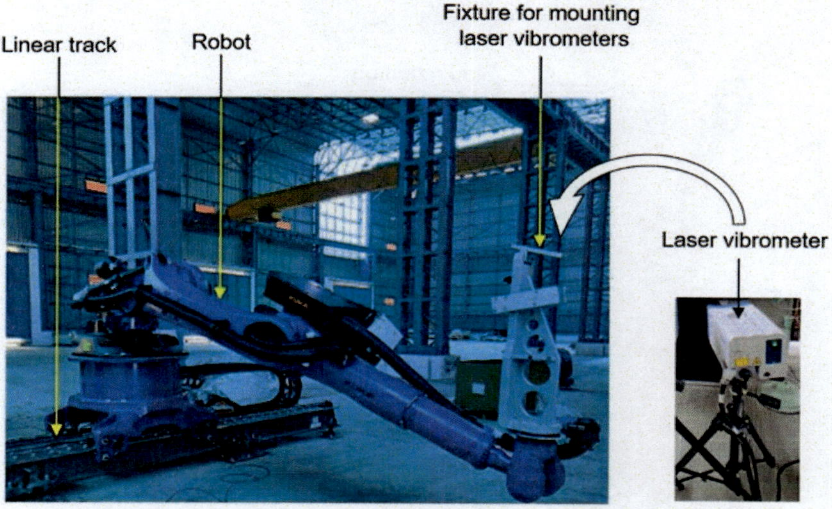

Fig. 2.60 Picture of a laser vibrometry system with a robot and a linear track

a tripod stand (as shown on the right side of Fig. 2.60), it needs to be moved every time for measuring a new surface. In this case, the robot provides higher flexibility.

2.5.6.4 Summary of the Vibration Sensors

To summarize, the discussion in this section elaborated on the sensors for measuring vibration. Two general sensors utilized for this purpose are the accelerometers and the capacitive type. Both these sensors are contact type, i.e., they are mounted on the component being measured using a suitable technique. Among these two, the accelerometer is the most preferred one owing to its versatility and durability. Besides this, there exist a few industrial applications that require non-contact means of sensing. The laser vibrometer is one such instrument that measures the vibrations in a structure through the laser system. It is advantageous over the accelerometer sensor as it does not add any extra weight to the component being measured. However, it is expensive and requires precise handling as well.

2.5.6.5 Real-Life Example in Manufacturing

As an example of the application of sensing vibration, this section illustrates a motor fitted with a vibration sensor *(IFM, VTV122)*. A picture of the set-up is shown in Fig. 2.61. The set-up shows a hydraulic power pack with a 3-phase induction motor. The objective here is to determine a fault in the bearings of the motor. Bearing is one of the critical components of a motor that helps in supporting the rotor and

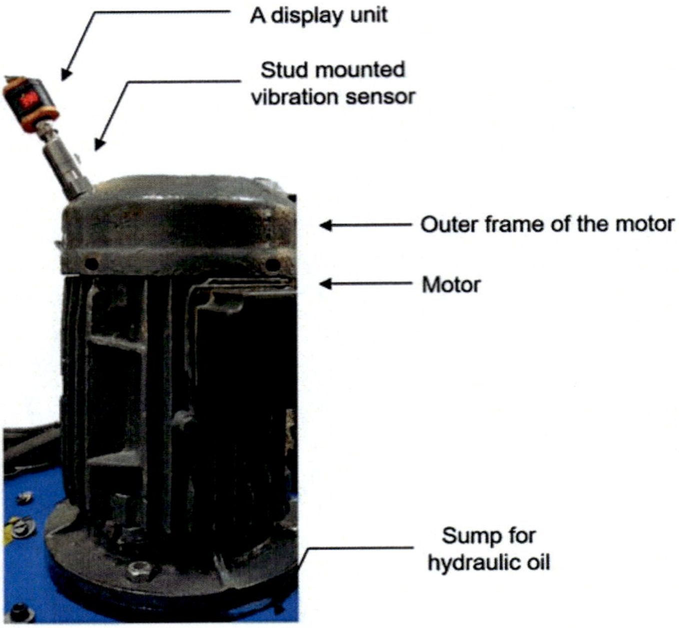

Fig. 2.61 Picture of a vibration sensor fitted on to a motor

transferring the load from the shaft to the motor frame. However, with time, the condition of the bearings in the motor degrades as it suffers wear and tear. Thus, it is of utmost importance to determine the condition of the bearings in the motor in real-time, which otherwise may result in machine failure.

To capture the vibration, an accelerometer is used, and it is mounted on the motor frame by a stud. Mounting via studs is the most preferred method because it ensures the rigidity and proper mounting of the sensor on the target. The sensor is connected to a data acquisition device which collects the signal and stores it in a computer. The schematic of the sensor connection and data acquisition is shown in Fig. 2.62.

Figure 2.63 shows a plot of the vibration measurement collected from the motor. The output is in terms of voltage. This data corresponds to a healthy bearing that was newly installed in the motor. A peak at the beginning of the plot can be observed, which indicates the start of the motor. As the rotation begins, there will be a torque requirement, thus, a force will be acting. Once it reaches the rated rotational speed, the vibration level comes down and thereby, the signature can be seen to be stable. The signature can only be stable if the machine components are healthy, i.e., the vibration is negligible. A portion of this region is shown in a smaller plot in the same figure to highlight the vibrations occurring in the motor. With the degradation in the condition of the bearing, there would be high fluctuations in the signature. Thus, for fault analysis, one needs to replace the healthy bearing with a faulty one and may observe the change in the amplitude of the signal. This is because the amplitude may

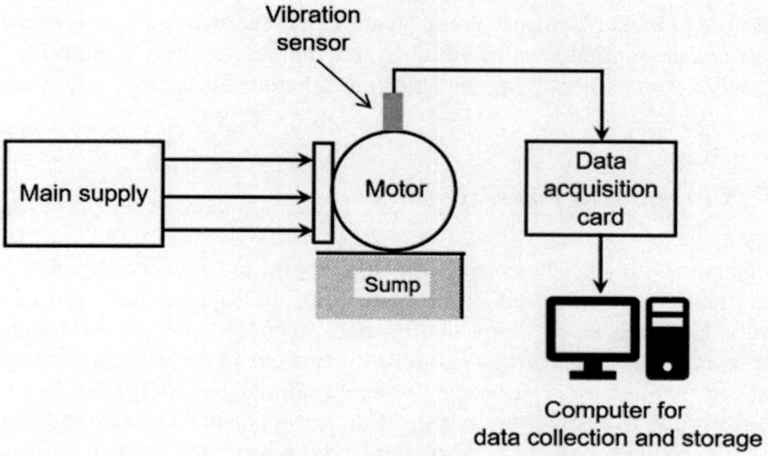

Fig. 2.62 Schematic for data acquisition from the vibration sensor

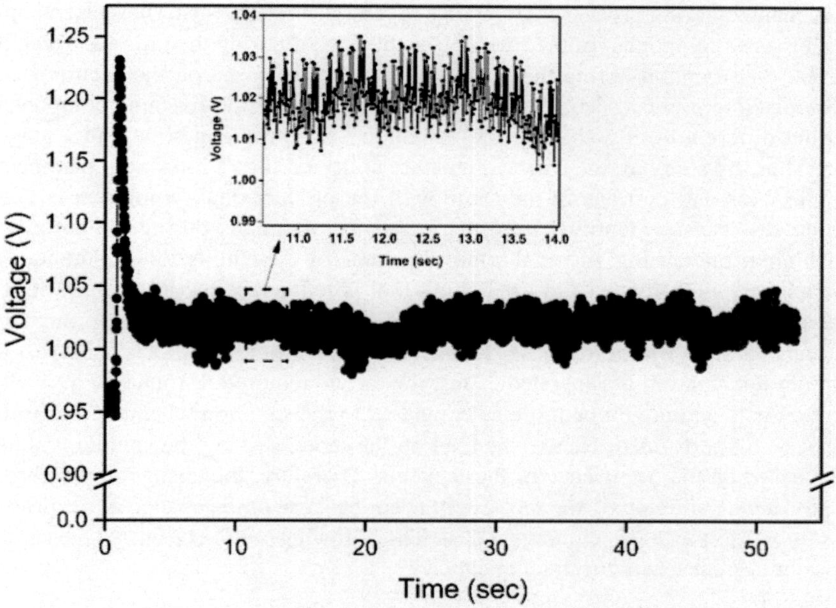

Fig. 2.63 Plot of the vibration data with respect to time

rise because of broken/damaged components. Once a knowledge base is established that certifies the normal level of vibration, continuous vibration monitoring of the structure can help in identifying the faults much before a catastrophic failure.

2.5.7 Current and Power Sensors

Every manufacturing machine involves motors for the movements. Depending upon the load and type of the work to be performed, the capacity and type of motor may vary. However, if any motor malfunctions, then the operation of the machine will get hampered. Considering negligible fluctuations in the incoming voltage, the current consumption by a component during the normal operation will take a certain value. However, with time, the health of the components degrades and thus, the current consumption would also vary. Hence, by sensing the current, the health of machine components such as the motor can be determined. Current sensing is also advantageous for process monitoring.

A manufacturing process may involve several discrete events. These events may trigger some component of the machine as they occur during the process. Thus, the current consumption during the process will also vary based on the occurrence of those discrete events. Also, the current consumption by the machine components will be different during an abnormal situation i.e., faulty condition and a normal one. Thus, by studying the current signature collected for various input parameters of a process and correlating the same with the product quality, inferences about the process can be identified from the signal. For example, during the process of machining, say turning, material removal is done by the cutting tool. In the turning operation, the job rotates and the cutting tool is fed across the job to turn out the material from the job. The current consumption during the sole job movement and cutting operation will be differing. This is because as the cutting tool is fed across the rotating job, friction is generated. Thus, the torque required to rotate the job when the tool is in action will be more as compared to the situation when the machining is not being performed. Hence, the load on the motor driving the spindle will also vary based on the occurrence of these events. Therefore, the change in the current or power consumption by the motor will occur because of the change in the process conditions during the machining process. The following section discusses the various types of current sensing technologies.

2.5.7.1 Direct Sensing—Sensors Based on Ohm's Law

Let us discuss the basic principle of current sensing. Whenever a current is flowing in a wire or a circuit, two things occur simultaneously, one is the voltage drop and the other is the generation of a magnetic field surrounding the conductor that is carrying the current. Based on these two things, a current sensor is designed and is categorized into two namely direct sensing and indirect sensing techniques. This

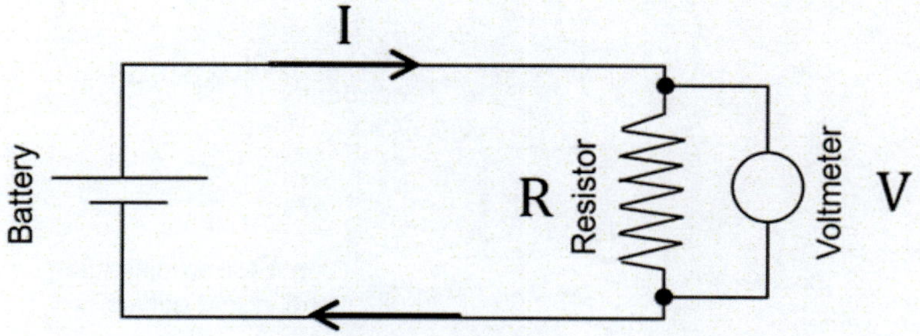

Fig. 2.64 Schematic picture depicting Ohm's law

section will cover the direct sensing techniques. The direct sensing technique is based on *Ohm's law* that is schematically represented in Fig. 2.64. This law states that a current through a conductor between two points is directly proportional to the voltage across the two points. The equation for *Ohm's law* is shown in Eq. 2.30.

$$I = \frac{V}{R} \tag{2.30}$$

where I denotes the current flowing through the circuit, R is the resistance of the resistor, and V is the voltage. This also implies that the current flowing in a circuit is inversely proportional to the resistance offered and directly proportional to the voltage.

In the case of direct sensing techniques, sensors employ this law to measure the current flowing through a conductor. The first example is a shunt resistor, which is employed in the circuit, in which, the current is to be measured. The voltage drop across the shunt resistor is used as a proportional measure of the current flow. A schematic of the same is shown in Fig. 2.65, which depicts a current-carrying conductor and a shunt resistor connected in the conducting path. Because of the presence of this resistor, the voltage drop across the resistor could be measured using *Ohm's law*. This is performed by employing the necessary circuit to collect the voltage signal and the current flowing in the circuit is determined.

A shunt resistor can be used both for AC and DC measurements. As a shunt resistor is connected in the conducting path, the resistance of the shunt resistor is very low, usually a few milliohms. This ensures a low value of voltage drop. A few advantages of employing a shunt resistor for current measurement include its ability to make a precise measurement of the current flowing through the conductor, faster response enabling detection of faults quickly, and low cost. However, as the shunt resistor is employed in the conducting path, it results in power loss, and thus, it is restricted to the low current applications only. Besides using this shunt resistor, the other method is to utilize the intrinsic resistance of the conducting element in the circuit. A copper trace as a low-value resistor is employed in this approach. This is a better approach than the use of a shunt resistor as the power losses are significantly

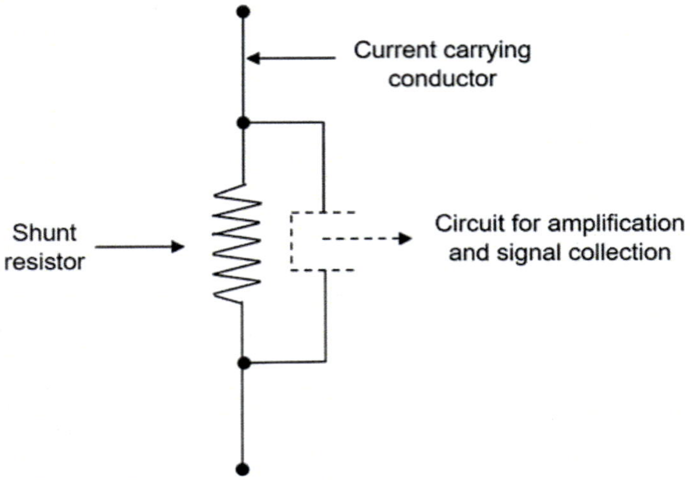

Fig. 2.65 Schematic picture of shunt resistor for current measurements

reduced. However, an amplifier with a high gain is required for measuring the output voltage as the voltage drop across the trace is very small. Hence, in a nutshell, it can be mentioned that while *Ohm's law* is the simplest way of measuring the current flowing in a circuit, the major drawback is the electrical connection to be established between the current to be measured and the sensing circuit.

2.5.7.2 Indirect Sensing—Sensors Based on Faraday's Law

The indirect sensing technique is based on *Faraday's law* that explains how a magnetic field would interact with an electrical circuit to produce an EMF. It is also known as electromagnetic induction and the expression for a coil with N number of turns is shown in Eq. 2.31.

$$\text{EMF} = -N\frac{d\varphi_B}{dt} \tag{2.31}$$

where EMF is the electromotive force and φ_B is the magnetic flux. This law indicates that when a conductor is exposed to a changing magnetic field, a current is induced in the conductor. The *Lenz law* of electromagnetic induction specifies the direction of this induced current. It says that the direction of the induced current is opposite to the initial changing magnetic field that produced it. Therefore, the representative equation consists of a negative sign. The induced EMF is directly proportional to the rate of change of the magnetic field. Based on this law, instruments named the Rogowski coil is used to measure the current. In the following paragraph, the Rogowski coil is discussed.

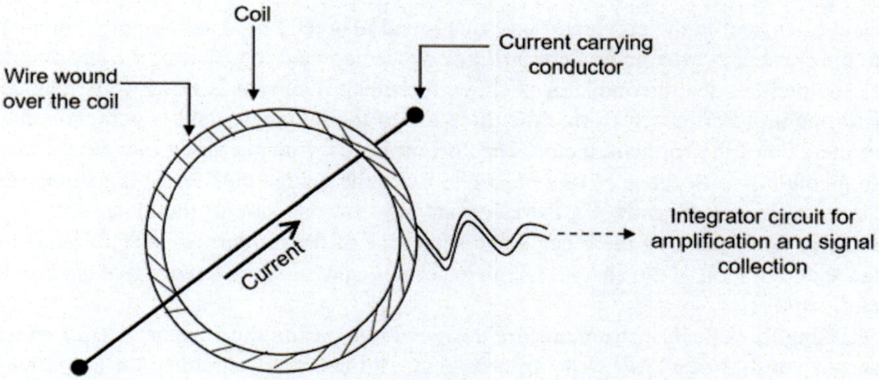

Fig. 2.66 Schematic picture of Rogowski coil for current measurements

The Rogowski coil is an open-loop current transformer that employs an open-air core with a low inductance to measure the current flowing through a conductor. The sensor has been named after the scientist *Walter Rogowski* and a schematic picture of this sensor is depicted in Fig. 2.66. It works on the principle of electromagnetic induction that is discussed above. The construction of the sensor comprises a coil that is wounded with wire from one end to the other. This coil surrounds the current-carrying conductor for which, the current is to be measured. As the conductor is carrying current and is introduced inside the Rogowski coil, an EMF is induced in the coil based on *Faraday's law* of electromagnetic induction. The EMF induced in the coil is given in Eq. 2.32.

$$\text{EMF} = M\frac{dI}{dt} \tag{2.32}$$

where M is the mutual inductance and $\frac{dI}{dt}$ represents the rate of change of current. Let us define this new term, inductance and then mutual inductance. Inductance can be referred to as a property of a conductor to induce EMF in another conductor. Thus, when two time-varying current-carrying conductors are placed close to each other, the magnetic flux varies owing to the varying current, which results in an induced EMF in each of them. This is termed mutual inductance and therefore, the Rogowski coil is selected such that it has a low value of inductance. Because of the low inductance, amplification is required to obtain the signal. The Rogowski coil is connected to an integrator circuit, which integrates the voltage induced in the coil to measure the current. Owing to the working principle of this sensor, it can only measure AC.

Other instruments exploit the measurement of the magnetic field because *Faraday's law* does not deal with the static magnetic fields. One such principle is the *Hall effect*, which was invented by the scientist, *Edwin Hall*. This law indicates the generation of a voltage difference in a current-carrying conductor transverse to an electric current in the conductor and an applied magnetic field acting perpendicular to the current. A schematic picture (Fig. 2.67) is depicted to visualize the statement. In the picture, a semiconductor is shown in which a current is flowing. A magnet is brought closer to it such that the direction of the magnetic field is perpendicular to the axis of the semiconductor. The current is also flowing along that axis of the semiconductor. Because of this magnetic field, the charge carriers of the semiconductor experience a drift i.e., the electrons move to one side of the plate while the positive holes move to the other. This is because of the *Lorentz force*. This leads to the accumulation of the charge, which generates a potential difference that is termed *Hall voltage*.

Using this effect, instruments are designed to measure the current. A *Hall effect* sensor consists of a *Hall element*, a magnet, and necessary circuitry for amplification and measurement of the signal. A schematic picture is shown in Fig. 2.68 to understand the construction of a *Hall effect* sensor. In the picture, a current-carrying conductor is shown for which the current is to be measured. The *Hall effect* sensor surrounds this conductor, which consists of a magnetic core (blue coloured) and a *Hall element* (purple coloured). The *Hall element* is placed perpendicular to the direction of the magnetic field in the sensor. As the *Hall element* is introduced to the current-carrying conductor, drift in its charge carrier occurs, which leads to the generation of a voltage difference i.e., the *Hall voltage*.

The voltage induced because of the magnetic field of the current-carrying conductor is proportional to the current flowing in that conductor. However, the induced voltage is of a very small quantity, thus, this instrument requires a high gain amplifier for data acquisition. This sensor can be utilized for both AC and DC measurements.

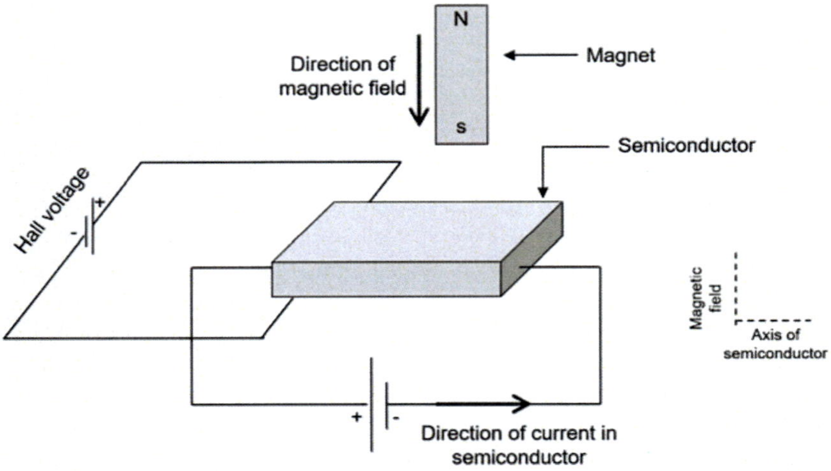

Fig. 2.67 Schematic picture of *Hall effect*

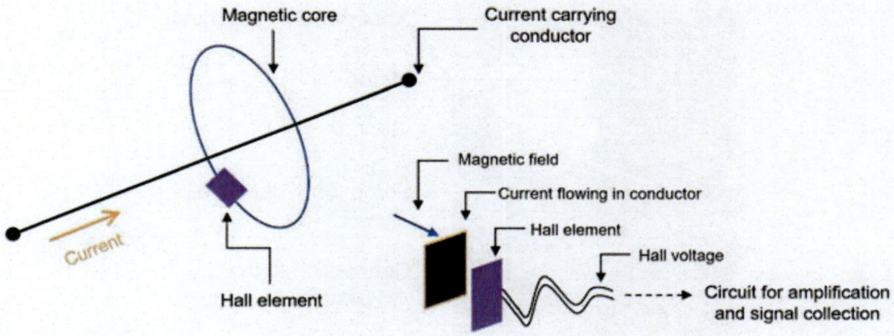

Fig. 2.68 Schematic picture for construction of a Hall effect sensor

2.5.7.3 Summary of the Current and Power Sensors

To summarize, the discussions in this section elaborated on the sensors for measuring current. The sensors are categorized into two namely direct sensing and indirect sensing. The direct sensing method utilizes *Ohm's law* for measuring the current. The examples under this method include the use of shunt resistor and trace resistance sensing. The shunt resistor is directly connected in the line of the current-carrying conductor, which leads to a voltage drop across the resistor and the current is measured. It poses a problem because of high power losses and alternatively, the trace sensing method is employed. This reduces power loses but requires high-gain amplifiers for getting the output signal. The indirect sensing method utilizes *Faraday's law*, which is utilized by a sensor such as Rogowski coil. The Rogowski coil is used in measuring high-frequency current measurements. It also requires an integrator circuit for amplification of the output signal. Because *Faraday's law* is not capable of sensing the current generating a static magnetic field, the magnetic sensors are utilized for this purpose. *Hall effect* current sensor comes under this category that works on the principle of *Hall voltage* generation. They are beneficial because they provide electrical isolation, high bandwidth, and good accuracy.

2.5.7.4 Real-Life Example in Manufacturing

As an example, the motor and the experimental set-up that has been mentioned during the discussion on vibration sensor (Sect. 2.5.6.5) is again utilized in this section to present the application of a power sensor for capturing the variation in the signature of the power consumption over time. Figure 2.69 shows the experimental set-up that consists of the motor connected to a hydraulic line and a power sensor *(Montronix, PS-200-NG)* integrated with the supply line of the motor. This is a *Hall effect* sensor having three separate sensors (observe the yellow coloured cables in Fig. 2.69). Three cables are for the three phases of the induction motor. The objective of integrating the power sensor is to check the health of the motor. As the performance of the

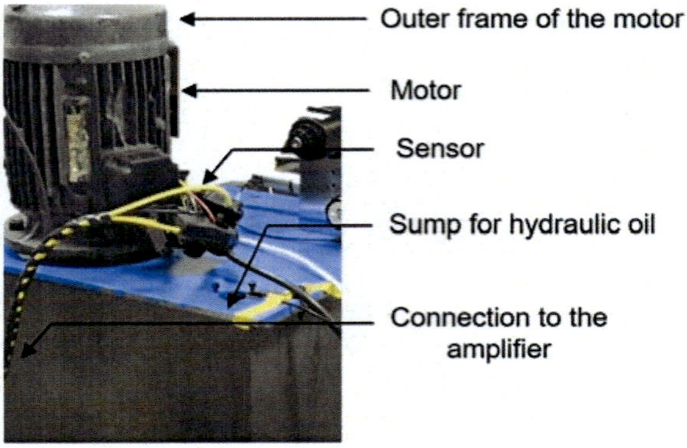

Fig. 2.69 Power sensor connected in the motor

motor degrades with time, the power consumption varies. The signature of power consumption by the motor is one of the physical parameters that could be utilized as an indication of its health. A schematic diagram for acquiring the data from the power sensor integrated into this arrangement is shown in Fig. 2.70.

As mentioned, the three *Hall effect* sensors are used for three phases of the motor and are shown in blue colour designated as I_1, I_2, and I_3. From the corresponding lines, three more wires are connected to capture the voltage and these are represented in orange colour designated as V_1, V_2, and V_3, respectively. All these go into an amplifier, which amplifies the signal and is powered by a DC supply. A data acquisition card acquires the amplified signal, which is then collected in a computer. Figure 2.71 shows a plot depicting the variation in power consumption by the motor over time. One could observe that there is no significant variation in the amplitude

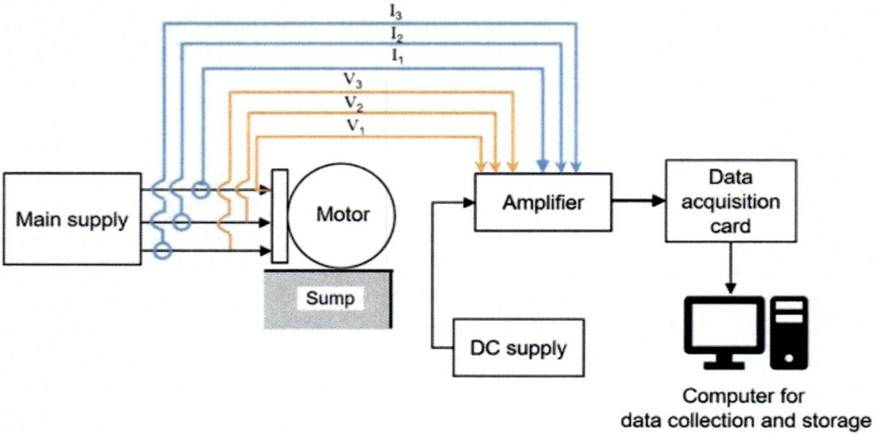

Fig. 2.70 Schematic of the data acquisition from the power sensor

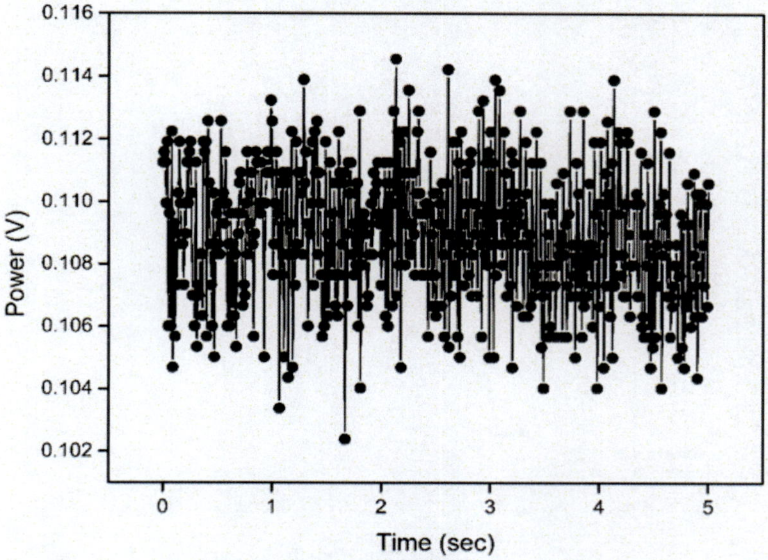

Fig. 2.71 Plot of power consumption by the motor

of the signal over time. This is because the plot corresponds to a healthy motor. With degradation, the power consumption is expected to vary and the same could then be harnessed to identify the differences from the healthy signal.

As a second example, a study on the usefulness of the current signal is shown for detecting tool wear during a drilling process. In this study, the current signal of the spindle motor is acquired by using a *Hall effect* current sensor *(Tektronix, A621)*. The current signal acquired during the process was studied for correlating with the tool wear. The root mean square (RMS) value of the current was found to be displaying a good correlation with the increasing tool wear [12]. A plot of this feature is shown in Fig. 2.72. In this way, the current signal can be studied to infer characteristics of the machining process.

2.5.8 Pressure Sensor

The manufacturing machines employ energizing systems to operate their components. These systems may be pneumatic, hydraulic, etc. In either case, fluid is involved, the air in case of pneumatic, and oil in case of hydraulic. To drive and control the movement of the components, it is important to control the energy to be provided by the air or the oil. This energy can be referred to as the pressure, which means, the force to be exerted per unit area, as expressed in Eq. 2.33.

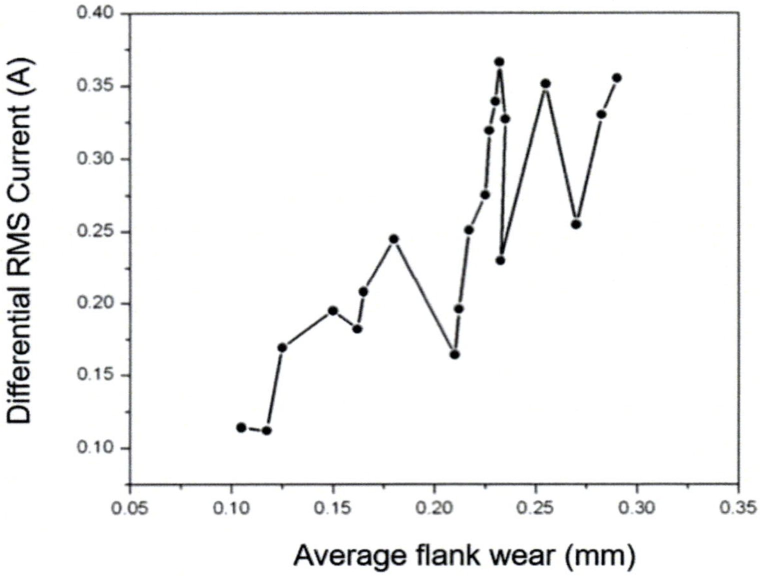

Fig. 2.72 Plot of RMS current for detection of tool wear during the drilling process

$$P = \frac{F}{A} \tag{2.33}$$

where P denotes pressure, F is the force, and A is the area. Pressure is an extremely important quantity and an instrument termed a pressure sensor is utilized to determine the pressure. It is utilized for both process and health monitoring applications. Measurement of pressure is usually carried out in three configurations namely absolute, gauge, and differential [13]. During absolute pressure measurement, one end of the sensor is exposed to the target of which the pressure is to be measured, while the other end is a known pressure corresponding to absolute vacuum. A gauge pressure sensor measures the pressure of a target corresponding to the atmospheric pressure. Finally, a differential pressure sensor measures the difference in pressure experienced at two points. In the following sections, various types of pressure sensors are discussed.

2.5.8.1 Piezoelectric Type

The principle of piezoelectricity has already been discussed in this chapter (Sect. 2.5.4.2), which is the generation of electric charge in a piezoelectric crystal when it is subjected to an external force. This principle is also employed for sensing pressure. A schematic picture for a piezoelectric-type pressure sensor is shown in Fig. 2.73.

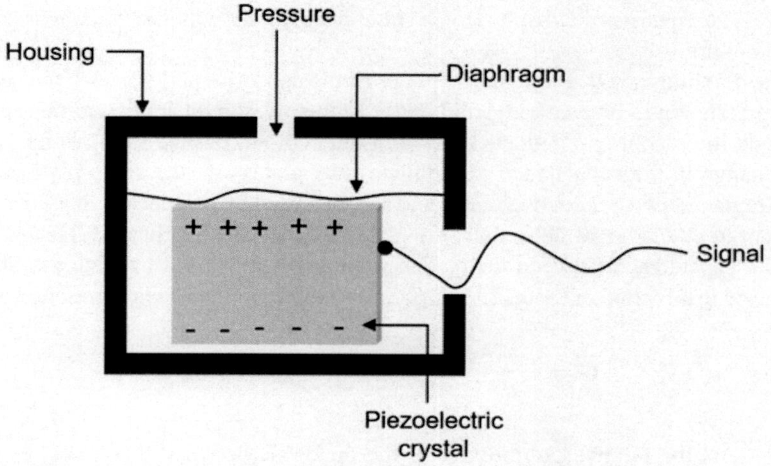

Fig. 2.73 Schematic picture of a piezoelectric-type pressure sensor

Here also, the sensor consists of a piezoelectric crystal that is exposed to the pressure acting on a target. A diaphragm is present inside the housing, which is a flexible membrane and it deflects due to applied pressure. The amount of deflection of the diaphragm is directly proportional to the applied pressure. Upon application of the pressure, the diaphragm deflects, and this, in turn, leads to the elastic deformation of the piezoelectric crystal resulting in the generation of an equivalent electric charge. This charge is transferred to a charge amplifier for amplification and signal collection. A piezoelectric-type sensor is an active one as it does not require any external excitation voltage. However, the charge generated is very small and needs a device that can amplify the charge, like in the case of a dynamometer and an accelerometer. This device is referred to as the charge amplifier and it produces an output voltage big enough to be able to measure.

The piezoelectric sensor does not have any moving parts so it is robust i.e., it is well known for its stiffness. Since the deformation caused to the piezoelectric crystal is significantly small, it offers a wide range of pressure measurement. It also can withstand high temperature, which makes it suitable for measurement for pressure in high-temperature applications such as jet engines, power plants, etc. However, it should be noted that the pressure on this sensor needs to be dynamic, then only, an equivalent electric potential will be generated. If the pressure becomes static, the net voltage will be null, which is a limitation of this sensor.

2.5.8.2 Capacitive Type

The capacitive-type pressure sensor employs the principle of measuring the change in capacitance for measuring the pressure. A schematic of this sensor is shown in

Fig. 2.74. It consists of a diaphragm, a static plate, along with the housing as shown in the picture.

The diaphragm acts as one of the plates of the capacitor and it is also the sensing element (shown in blue colour, solid line). As the pressure is applied to this sensing element, the diaphragm deflects (shown in blue colour, dashed line), which results in a change in the capacitance of the plate. The plates i.e., the static plate and the diaphragm are two conductive plates and are separated by a dielectric medium. They are kept in an electrical field; thus, the plates are electrically charged. Hence, there exists a capacitance between them. This expression for this capacitance is already discussed in this chapter (Sect. 2.5.2.2), and as a reference has been mentioned below.

$$C = \frac{\varepsilon_r \varepsilon_0 A}{d} \qquad \text{From Eq. 2.14}$$

where ε_r is the relative permittivity of the dielectric between the plates, ε_0 is the permittivity of free space, A is the overlap area between the plates, d is the separation between the plates, and C is the capacitance. From the expression, it can be observed that if the separation distance varies, the capacitance would also vary. Other terms in the expression are constants. Thus, the value of capacitance is inversely proportional to the distance by which the capacitors or the plates are separated. As the distance is dependent on the pressure that the diaphragm would experience, the change in capacitance would produce an equivalent output. For charging the plates, an external excitation voltage is supplied to the sensor which leads to the generation of electrostatic forces, thus, it is a passive sensor.

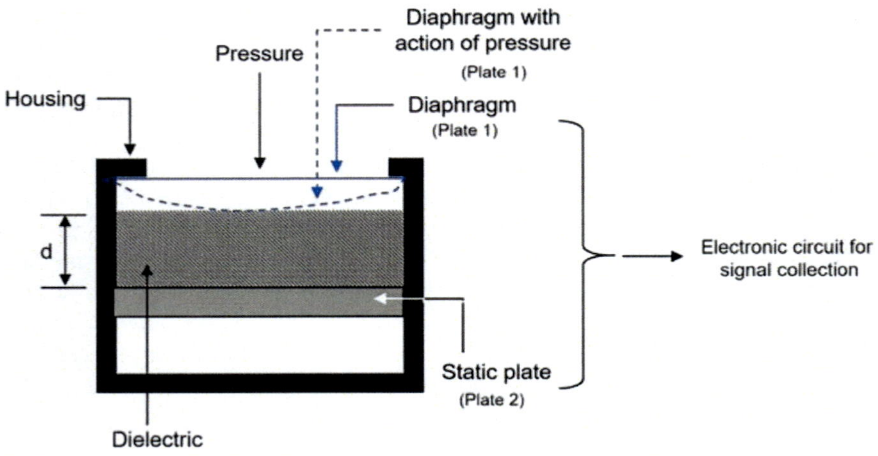

Fig. 2.74 Schematic picture of a capacitive-type pressure sensor

2.5.8.3 Strain Gauge Type

This sensor employs a strain gauge element to measure the pressure. The construction of this sensor is like the ones shown for piezoelectric and capacitive type sensors, just that, in this type, it is the strain gauge element that is acting as the sensing element in place of a piezoelectric crystal and a capacitor, respectively. The strain gauge element is attached to the diaphragm, so, it either stretches or compresses as per the applied pressure. The working principle of the strain gauge element has already been elaborated in this chapter (Sects. 2.5.1.3 and 2.5.4.1). As the strain gauge element stretches or compresses, there is a change in the resistance of the element. This is following the electrical resistivity equation, which states that the resistance to the flow of current by a conductor is directly proportional to its length and inversely proportional to the area. Hence, as the pressure would be applied to the element, it can produce an equivalent output. To acquire the output, the element is connected to a Wheatstone Bridge, which measures the change in the resistance. Thus, an external excitation is required to energize the strain gauge element, hence, it is also a passive sensor. However, unlike the piezoelectric-type sensor, the strain gauge type is suitable for both static as well as dynamic pressure conditions. But the size of the sensor may vary depending upon the range of pressure measurement to be performed.

2.5.8.4 Summary of the Pressure Sensors

In this section, the importance of sensing pressure is presented for process and health monitoring applications. The instruments for sensing pressure are based on the following principles, piezoelectric, capacitive, and strain gauge element. The piezoelectric type instrument utilizes piezoelectric crystals, which generate an electric charge upon deformation. This electric charge produced by the sensor is equivalent to the pressure applied to it. In the capacitive type, a diaphragm is employed which acts as one of the plates of the capacitor. As the pressure is applied, the diaphragm moves, and the capacitance of the capacitor changes. The change in capacitance is equivalent to the pressure experienced by the diaphragm. Lastly, the strain gauge element senses the pressure because of the change in the resistance of a strain gauge element. As the resistance changes due to the application of pressure, an equivalent voltage is generated in the Wheatstone Bridge. The piezoelectric type instrument is highly sensitive with a high-frequency response but it is applicable for dynamic loading condition only. The strain gauge type instrument on the other hand is suitable for both static as well as dynamic conditions but is less sensitive than that of the piezoelectric type. Both capacitive and strain gauge type instruments are passive as they require an excitation voltage. The piezoelectric type is an active sensor.

2.5.8.5 Real-Life Example in Manufacturing

In this section, the use of a pressure sensor is discussed from the health monitoring perspective. The pressure sensor *(IFM, PN2271)* is connected in a hydraulic power pack, as shown in Fig. 2.75. This hydraulic power pack is the one, which is also shown in Figs. 2.61 and 2.69 during the discussion on the use of accelerometer and power sensor. Here the objective is to measure the inlet pressure. During the operation of the hydraulic power pack, the pressure of the fluid flowing in the line is to be fixed. The solenoid valve connected in the hydraulic power pack, as shown in the picture regulates the flow of the fluid. The pressure sensor is connected at the inlet point of the solenoid valve to measure the inlet pressure. As the inlet pressure is fixed through a regulator, the objective is to ensure that the pressure remains fixed. This is achieved by using the pressure sensor which is shown in Fig. 2.75. A schematic diagram for acquiring the data from the pressure sensor integrated into this arrangement is shown in Fig. 2.76

As the pressure sensor, in this example, is for the sake of ensuring the inlet pressure of the fluid to the solenoid valve, the pressure value is almost constant. This was directly verified from the display unit fixed on the pressure sensor, as shown in Fig. 2.75. The data acquisition rate, in this example, is only 1 Hz. This is so because the pressure is one such quantity that does not change very fast and hence, it is captured at lower frequencies. However, an important consideration would be to engage pressure sensors in the inlet and outlet lines that are going up to the actuator. This is because, if there occurs any problem in the system, a drop in the pressure

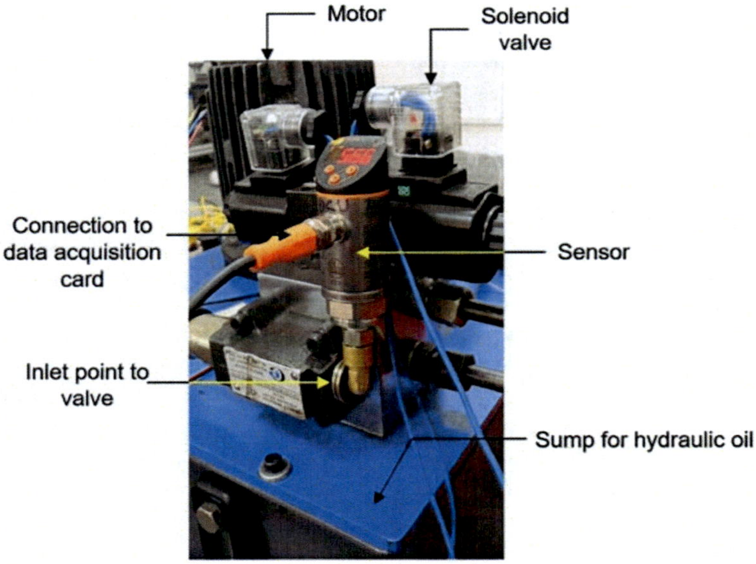

Fig. 2.75 Use of pressure sensor for health monitoring

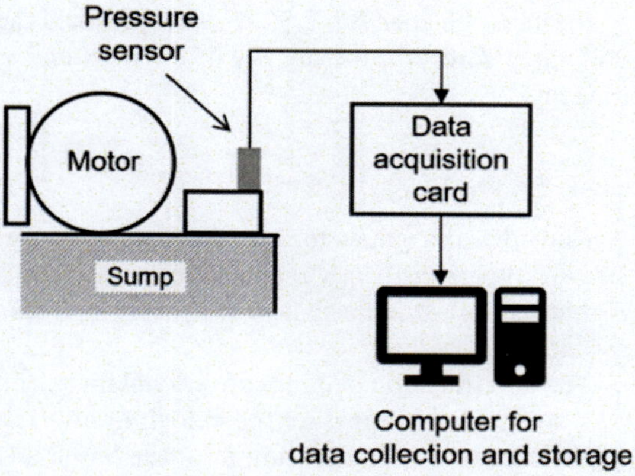

Fig. 2.76 Schematic of the data acquisition from the pressure sensor

could happen, which would be prevalent at the inlet and outlet lines. In that case, the health of the manufacturing system can be monitored.

2.5.9 Flow Rate Sensor

In addition to the measurement of the pressure of a fluid in the case of the energizing system, the flow rate is another physical quantity that is crucial in the process and health monitoring applications. Considering the chemical process industry, which involves mixing various types of liquids, where each of them will be of a certain quantity and will be added at a certain rate, instruments would be required to measure and control the rate of flow of the fluids. In other cases, flow rate measurement will be useful for health monitoring purpose. In process industries, such as oil refineries, there are pipelines termed as instrument air-line that carries air as the fluid for end applications. Here, the objective is to ensure the rate of flow of the air as required for the application. This will, in turn, ensure the health of the components included in the end application.

Two things are to be taken care of while selecting a proper flow rate sensor i.e., to understand what is required to be measured, the volume of the fluid or the mass. Considering Eq. 2.34, F_v denotes the rate of flow of a fluid, A is the area of cross-section of the pipe in which the fluid is flowing, and V is the average velocity of the fluid.

$$F_v = A \times V \tag{2.34}$$

Consider unit values for Eq. 2.34, i.e., let the velocity be 1 m/s, cross-sectional area be 1 m^2, then the flow rate will be 1 m^3/s. This is referred to as the volumetric flow rate as it provides the volume of the fluid flowing per second.

Further, consider Eq. 2.35, F_m also denotes the flow rate, but is called the mass flow rate. Here, ρ is the density of the fluid under the test and F_v is the volumetric flow rate.

$$F_m = \rho \times F_v \tag{2.35}$$

Consider unit values for Eq. 2.35 i.e., let the density of the fluid under the test be 1 kg/m^3, volumetric flow rate be 1 m^3/s, then the mass flow rate will be 1 kg/s. This is referred to as the mass flow rate as it provides the mass of the fluid flowing per second.

The aforementioned things are crucial for selecting a type of sensor. Besides these two, some sensors measure the velocity only. The following sentences discuss the types of sensors for measuring flow rate based on these design considerations.

2.5.9.1 Positive Displacement Sensor

This sensor measures the volume of the fluid passing through it. It consists of a mechanical arrangement such that a known volume of the fluid is entrapped each time and this fluid interacts with rotating components. Once this fluid interacts, the components rotate, and as the components complete one full revolution, the known volume can pass through it. The number of rotations is counted and it is directly proportional to the volume of fluid passing per unit time. Here, the rotational velocity is directly proportional to the flow rate. The examples under this category include oscillating piston, oval gear, nutating disc, vane, rotor, etc. Depending upon the fluid characteristics, a configuration of this type is selected.

2.5.9.2 Mass Flow Sensors

This sensor measures the mass of the fluid passing through it. The most common design of this sensor includes a hot-wire type configuration. The wire is heated by electric current and a temperature sensor is utilized to measure the temperature of the wire. As the fluid flow starts, convective heat transfer occurs, which results in a temperature drop across the wire. The electrical resistance of the wire decreases, which, in turn, increases the current flow through the wire to maintain the temperature. The amount of current drawn provides a measure of the mass of the fluid flowing through the sensor. Another design considers the differential temperature as a measurement of the mass flow rate. Coriolis flowmeter is another example that comes under the category of the mass flow sensor. The principle of measurement is that the fluid flowing through an oscillating tube exerts inertia, causing the tube to twist, and the degree of twisting is proportional to the mass flow rate of the fluid. The Coriolis flowmeter consists of a tube that is energized by a fixed vibration such that the resultant phase shift is null. The fluid under the test is allowed to pass through this tube which causes a change in the vibration of the tube because of the mass flow momentum. Thus, there is a phase shift that is proportional to the mass of the fluid flowing through the tube. It should be noted that the hot-wire design configuration is dependent upon the characteristics of the fluid, however, the Coriolis flowmeter measures the real mass flow. Besides, the mass of a material is not affected by temperature and pressure, whereas the volume does. Hence, the target application needs to be properly understood for selecting a type of flow rate sensor.

2.5.9.3 Velocity Measuring Flow Sensor

This type of instrument measures the velocity of the fluid passing through it. The examples under this category are mechanical *(Hall effect)*, electromagnetic *(Faraday's law)*, and ultrasonic *(Doppler effect)* types.

The first example i.e., the mechanical type flow meter includes a wheel that is mounted on the path of the fluid flow. A schematic picture of this sensor is shown in Fig. 2.77. From the figure, a wheel can be observed that is installed in the pipe section, where the flow of the fluid is to be measured.

As the fluid flows through the pipe, this wheel rotates. On one of the blades of this wheel, a magnet is attached such that the magnetic field is perpendicular to the current flowing in the *Hall element*. As the blade with the magnet approaches the *Hall element*, there is a drift in the charge carriers of the *Hall element*, and a potential difference is generated. This is the *Hall voltage* and this has been discussed in details in a previous section (Sect. 2.5.7.2). The rotation of the wheel is detected by using a magnetic *Hall effect* sensor that generates an equivalent output signal. Equivalent because the rotation of the wheel will be directly proportional to the velocity of the fluid in the pipe. The illustration on the extreme right of the figure (Fig. 2.77) shows the pulses that are recorded as the wheel rotates. The frequency of the pulses is proportional to the rotation of the turbine and the rotation is proportional to the velocity of the fluid. This type of sensor is advantageous owing to its simplistic

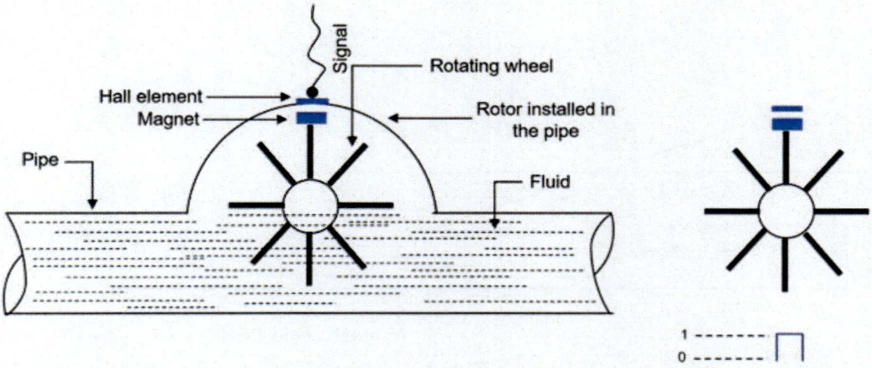

Fig. 2.77 Schematic picture of a velocity measuring flow sensor (mechanical type)

design. The challenge will arise if dirt or foreign particles get stuck to the rotating wheels, or the continuous interaction of the fluid with the rotating component results in wear. In such cases, maintenance of the instrument will be an additional task.

The second example is the electromagnetic type, which is based on the principle of electromagnetic induction. The principle of electromagnetic induction is already discussed in this chapter (Sect. 2.5.7.2). It explains how a magnetic field would interact with an electrical circuit to produce an EMF. A schematic diagram of the electromagnetic type flow sensor is shown in Fig. 2.78. The construction of the sensor consists of an electromagnetic coil and necessary electronics circuit to capture the induced EMF.

The electromagnetic coil in this instrument generates the magnetic field around the fluid passing through it. As the fluid passes by, the magnetic field varies and a voltage is induced in the electrodes present in the instrument. This voltage is proportional to the rate of flow of the fluid in the pipe. The same can be expressed as shown in Eq. 2.36.

$$\text{EMF} = B \times d_i \times V \tag{2.36}$$

where B is the magnetic field strength, d_i is the internal diameter of the pipe through which the fluid is flowing, V is the velocity of the fluid, and EMF is the induced voltage. From this expression, it can be observed that the induced voltage is directly proportional to the velocity of the fluid and the magnetic field. Compared to the mechanical type, as it does not involve any rotating components, it is advantageous. However, it is limited only to conductive fluids. Also, it is important to observe that the fluid under test should not form any coatings on the surface of the electrodes. Contaminated fluids can also pose a problem while using this type of sensor.

The third example is the ultrasonic type of flowmeter which emits ultrasonic energy into the fluid under test to measure the flow velocity. A schematic diagram

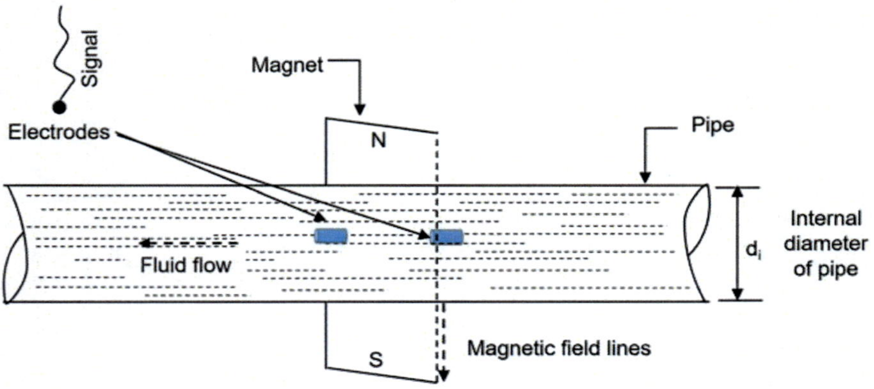

Fig. 2.78 Schematic picture of a velocity measuring flow sensor (electromagnetic type)

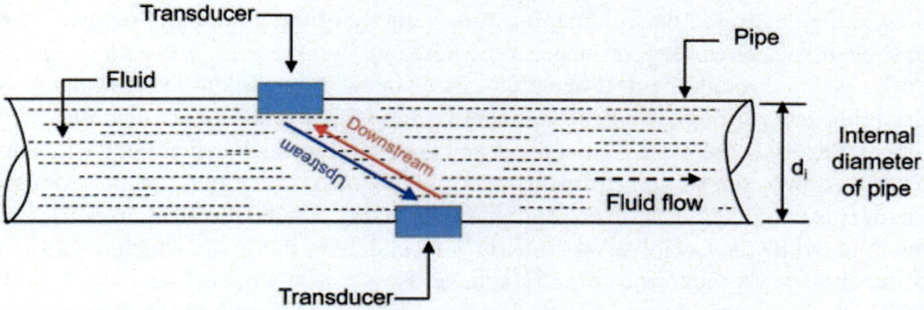

Fig. 2.79 Schematic picture of a velocity measuring flow sensor (ultrasonic type)

for understanding the construction and principle of this type of sensor is shown in Fig. 2.79. The instrument consists of two transducers, one on the upstream and the other on the downstream side, as shown in the picture (Fig. 2.79).

The working principle is the *Doppler effect*, which has already been discussed in this chapter during the discussion on vibrometer (Sect. 2.5.6.3). In this flow sensor, either of the transducers forms a transmitter and a receiver. They emit ultrasonic waves, which is received at the other end after reflection. In absence of fluid in the pipe, there will be no change in the frequency of the wave. But as the fluid flows through the pipe, a frequency shift occurs that is termed as the *Doppler effect*, and it is proportional to the rate of flow of the fluid. Observing the schematic picture (Fig. 2.79), the time required by an ultrasonic wave to move towards the upstream side will be more as compared to the one approaching the downstream side. This is because in the former case, the wave will move against the direction of the fluid flow. So, it will experience some resistance by the fluid. The difference in the time to travel across the ends i.e., upstream and downstream will be proportional to the velocity of the fluid passing through the pipe. The expression for the same is shown in Eq. 2.37.

$$V \propto \Delta t \tag{2.37}$$

where V is the velocity of the fluid and Δt is the difference in the time of travel for upstream and downstream. This instrument is applicable for both conductive and non-conductive fluids. It can also be used as a non-invasive type of instrument. The limitation of this sensor comes with certain fluids that do not allow the ultrasonic waves to pass through them.

2.5.9.4 Summary of the Flow Rate Sensors

This section discussed the importance of measuring the flow rate of the fluids for process and health monitoring applications. The instruments designed for measuring flow either determine the volume, mass, or velocity of the fluid under the test. The positive displacement sensor inspects the flow rate by measuring the volume of the fluid passing through them. It operates a mechanical arrangement, which rotates as the fluid passes through it and the number of rotations in a unit time are recorded. The other category is the mass flow sensors and the examples included are hot-wire and Coriolis flowmeter. Both instruments measure the mass of the fluid flowing through them. However, the hot-wire configuration depends upon the physical properties of the fluid, while the Coriolis flowmeter is independent of the same, which makes it a better choice over the former. Finally, some sensors measure the velocity of the fluid. Under this category, the examples are mechanical, electromagnetic, and ultrasonic types. The mechanical ones consist of a wheel that rotates as the fluid passes through it. It employs a *Hall effect* type magnetic sensor to detect the rotations per unit time and generate an equivalent output voltage. The sensor is not suitable for fluids with contaminants as the foreign particles can adhere to the blades of the wheel and damage them. The electromagnetic type works on the principle of *Faraday's law* of electromagnetic induction and is suitable for conductive fluids. The ultrasonic one measures the velocity via the *Doppler effect* and is suitable for both conductive and non-conductive fluids.

2.5.9.5 Real-Life Example in Manufacturing

In this section, the use of a flow rate sensor is discussed from the health monitoring perspective. The FSW machine *(Eta Technology, WS004)* depicted in Fig. 2.31 is considered. This machine is energized by a hydraulic system. It controls the movement of the spindle axis by using two double-acting cylinders. The objective here is to investigate the quality of the hydraulic oil returning to the tank. For this purpose, a flow rate sensor, a turbidity sensor, and a K-type thermocouple are engaged in the return line, as shown in Fig. 2.80. The turbidity and flow rate sensors shown in the picture are Arduino-based. In this section, only the flow rate sensor *(YF-S201)* is discussed and the turbidity sensor is discussed in the next section. Also, a discussion on the Arduino board could be found in the next chapter (Chap. 3). A schematic diagram depicting the data acquisition from these sensors is shown in Fig. 2.81. This flow rate sensor comes under the category of velocity measuring sensor and it is a mechanical type. It has a wheel that rotates as the fluid passes through it and a *Hall effect* sensor that generates an equivalent voltage.

The sensors were connected to an Arduino board that was connected to a computer for data collection and storage. As mentioned, these sensors are engaged to check the quality of the oil. With time, the quality of the oil will degrade because of the wear and tear occurring inside the machine elements. This degradation will give rise to high temperature in the oil or may include foreign particles, and the flow rate will be hampered. Thus, if the sensors are installed in fresh conditions i.e., new oil, then one can determine the condition of the oil continuously, if recorded. This will ensure the machine is always up. Figure 2.82 shows the data recorded from the flow rate sensor. Sharp peaks can be observed in the data that corresponds to the movement of

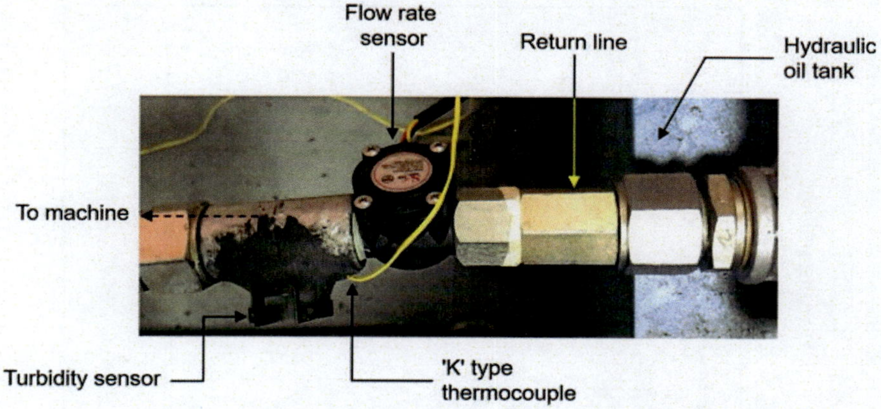

Fig. 2.80 Picture depicting a flow rate sensor engaged in a machine

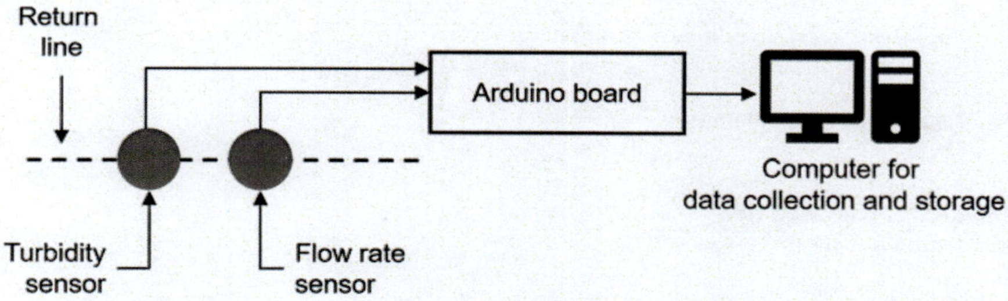

Fig. 2.81 Schematic diagram for data acquisition

the spindle axis. The first indicates the downward movement of the spindle and the second one is for the retraction of the spindle. During the welding phase, the flow rate almost remains constant. This is an indication of the healthy condition of the oil. To identify the degradation, constant monitoring of the same will be required.

2.5.9.6 Turbidity Sensor

Turbidity, simply indicates, how clear fluid is. The importance of monitoring the oil quality has been discussed in the preceding section, in which turbidity measurement is one important feature. This will prevent damage to the components. Figure 2.83 shows a schematic diagram of a turbidity sensor. It consists of two legs, one contains the transmitter and the other contains the receiver. The body contains the associated electronics. It employs optical sensing technology. The legs are dipped inside the pipeline containing the fluid i.e., the oil under the test, as shown in Fig. 2.84. The transmitter emits optical light, which is received by the receiver at the other end, and

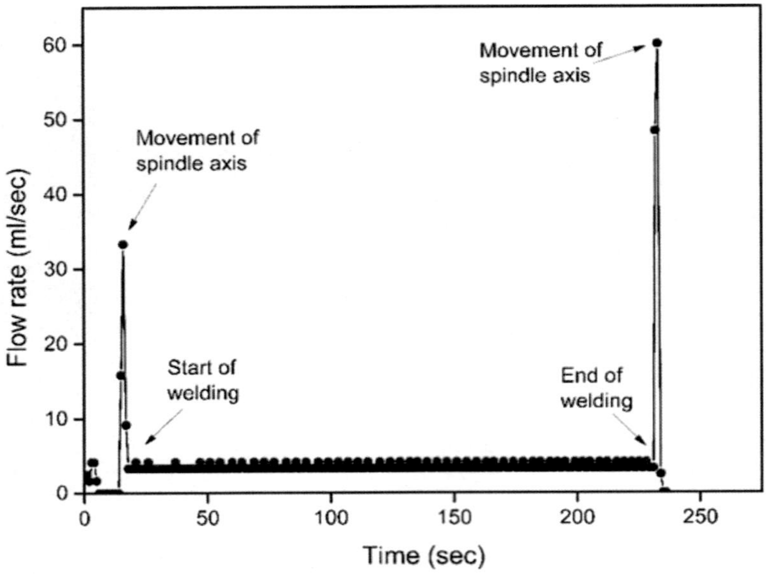

Fig. 2.82 Variation in oil flow rate

Fig. 2.83 A schematic view of the turbidity sensor

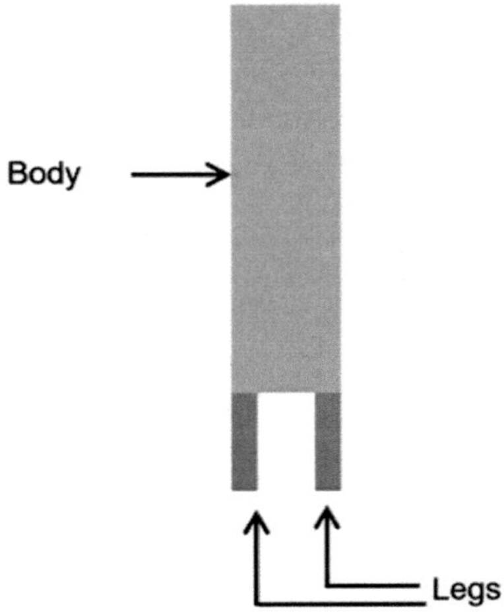

the receiver generates an equivalent electrical signal. During the ideal situation, when there is no movement of the fluid, all the emitted light is received by the receiver. As the oil flow starts, depending upon the condition of the oil i.e., the inclusion of foreign particles, the amount of light that the receiver should receive will vary. If the oil is in fresh condition, it will be totally clear. Thus, the signal amplitude will be high. As the oil degrades, the amount of light that will be received at the receiver end

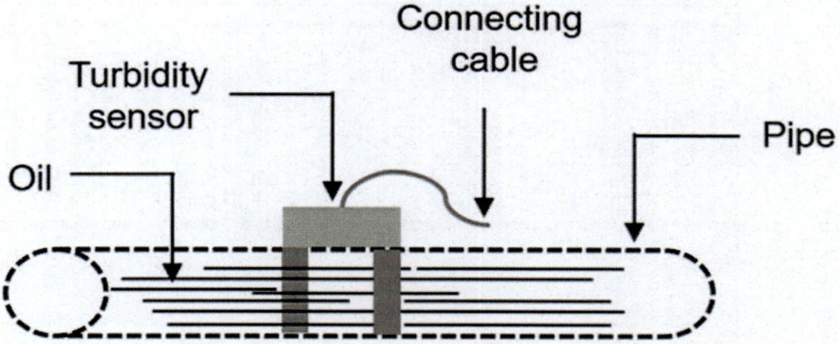

Fig. 2.84 Turbidity sensor fitted in a pipe

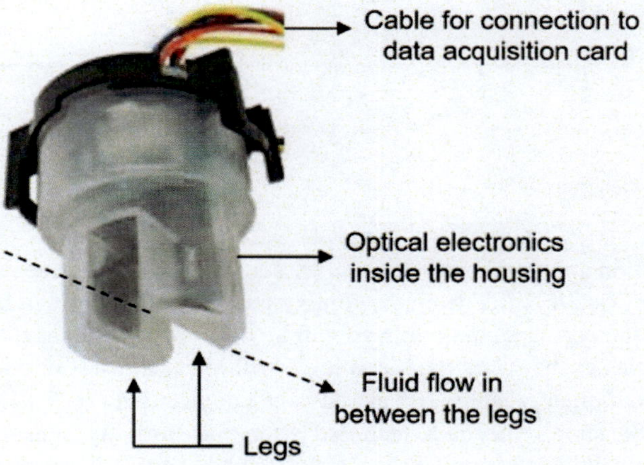

Fig. 2.85 A turbidity sensor

will decrease, so, the signal amplitude will decrease. This will form an indication of the degrading oil quality. It can also be an inference for the increasing wear in the mechanical components of the machine for which the monitoring is being performed.

Real-Life Example in Manufacturing

Referring to Figs. 2.80 and 2.81 that are depicted in the preceding section, the engagement of a turbidity sensor in a manufacturing machine is already shown. The employed turbidity sensor is Arduino-based *(SKU)*, and a picture of the sensor is shown in Fig. 2.85. This sensor is having two legs that go inside the pipeline. The head of the sensor contains the associated electronics for acquiring the signal. This is an optical type of instrument i.e., it sends light, which is received by a receiver at the other end. The sensor can produce data in the range of 0–5 V, where 0 indicates the

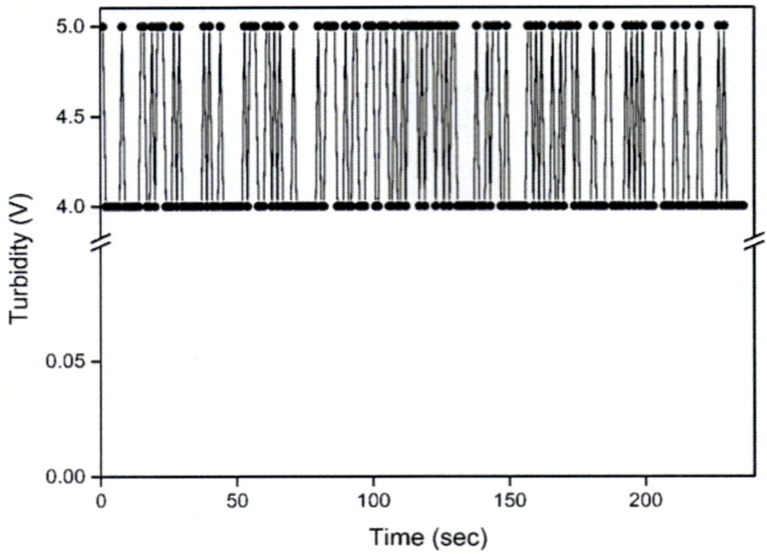

Fig. 2.86 Variation in turbidity of oil

dirtiest condition and 5 indicates the healthy condition. This is so because as the oil will be healthy i.e., it is free from the foreign particles, the light will be completely received and the corresponding voltage will be high. Otherwise, as the inclusion of foreign particles in the oil starts, the voltage will lower down. For data acquisition from this sensor, the procedure has already been depicted in Fig. 2.81.

Figure 2.86 shows the data recorded from the turbidity sensor. As already mentioned, the oil which is under the test is fresh and new. A corresponding voltage between 4 and 5 V can be observed from the data. Thus, it corresponds to a high amplitude signal. This is because the oil is fresh and it does not contain any foreign particle. However, with time, the oil quality will degrade, and the engaged turbidity sensor will pick up that.

2.6 A Guide on Sensors for Digital Twin

Figure 2.87 gives the total picture of the sensors for process and health monitoring that are discussed in the preceding sections.

The selection of a sensor will primarily depend upon the application domain. It may happen that for capturing a phenomenon, two or more types of sensors can be employed. For instance, in the event of identifying breakage of the cutting tool during machining, current, force, vibration, or a temperature sensor can be employed. Here, the first aim should be to choose the best amongst them. To identify the best sensor

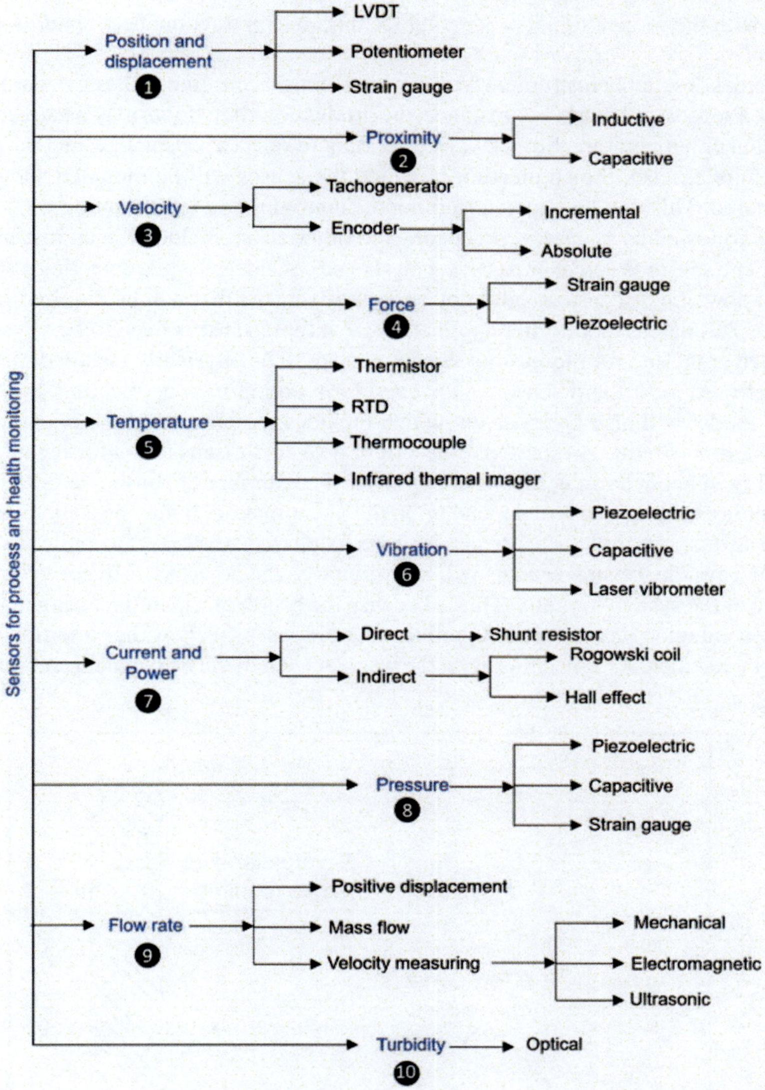

Fig. 2.87 List of sensors for process and health monitoring

for a phenomenon, signal processing will be required to harness features from the data. The correlation of the features with the process is observed and accordingly, the sensors can be ranked. Another important consideration is cost. Not all sensors are cheap, some are very expensive, and even they may require a large sampling rate for data acquisition. Manufacturing always aims to reduce the cost without compromising with quality. Thus, consideration of the cost factor will be important along with the proper signal processing technique to extract the best features out of a signal.

Besides, a suitable instrument type can be selected according to the environment in which, a sensor will work. For instance, the production shop floors may be affected by vibration or temperature. So, it will be important to select a sensor such that the effect is negligible. Also, the application in which the sensor will be mounted should be considered. This will help in determining the configuration for mounting the sensor. This is followed by accuracy, precision, and other errors included in an instrument. Whether a sensor is active or passive, can also be considered while selecting a sensor.

Till now, the discussions have covered many sensors that will be required for the process and health monitoring applications. All these sensors need to be physically installed at the site for monitoring. Since they are to be physically installed, they are also referred to as 'hard sensor'. However, for a digital twin application, the simulation model will also be involved, which includes the physics of the phenomenon for which it is built. A qualitative description of this discussion with regards to a digital twin application is shown in Fig. 2.88. The number of sensors to be selected for monitoring purpose will depend upon the phenomenon, it may be a single sensor and at a time, multiple sensors can also be employed. However, to build a digital twin of a system, several sensors will be required as the objective is to get a complete picture of the system digitally. This is because the health of a manufacturing machine will depend upon its age, ongoing process conditions, as well as the condition of the nearby machines. As a machine ages, the wear and tear in the mechanical components

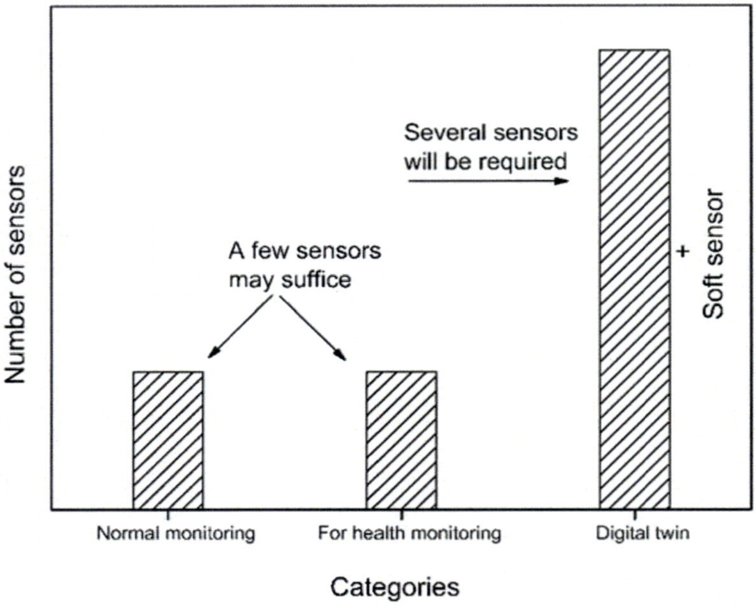

Fig. 2.88 A qualitative description of the digital twin in the context of sensors

increases which gives rise to heat and sound. If the parameters are not optimized, the machine will not run smoothly and may give rise to unnecessary heat and vibration. Finally, the nearby machines will also be influential because of the foundation being shared, which allows the vibration to be induced in machines. Hence, to take care of these factors, several sensors will be required to develop a digital twin. Also, the inferences from the simulation will be utilized in the digital twin, as discussed in the previous chapter. As mentioned, a physical sensor is also referred to as a hard sensor and the simulation is a soft sensor. The digital twin will be a mixture of information from the hard and soft sensors.

2.7 Performance Indices of a Sensor

By this time, the chapter has described the importance of sensors in manufacturing, both for monitoring the process as well as the health of the machine. The next thing is to identify a type of sensor required for monitoring a phenomenon occurring in a process or health of a critical component in a machine. This selection is based upon the application and is the first step towards the development of a digital twin model for that application. However, sensors are further characterized by certain features that define how well a sensor is behaving and matching its target. For instance, a very simple example is a 'multimeter', which is also known as 'volt-ohm-milliammeter'. It is a general-purpose instrument employed for measuring voltage, resistance, current, etc. A picture of this instrument is shown in Fig. 2.89.

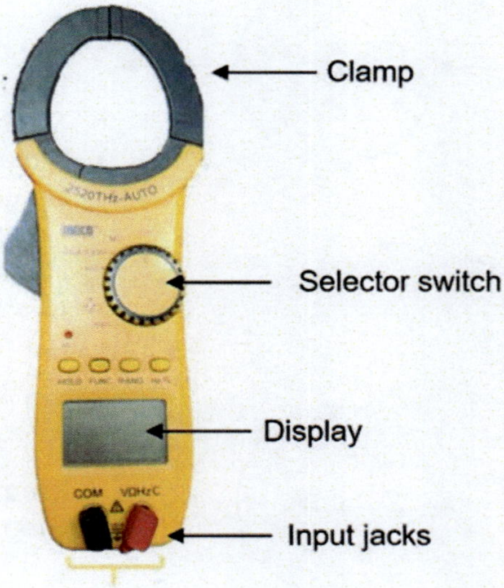

Fig. 2.89 Picture of a multimeter

This instrument can be selected based on how much voltage or current or resistance can be measured, AC or DC, the smallest change in the value that can be recorded, etc. All these can be listed as its features and depending upon the application, one can be selected. Thus, it is crucial to understand these features/characteristics of the sensors.

The characteristics can be categorized into two namely static and dynamic [14]. The former category refers to those features, which define the performance of a sensor in ambient conditions. The latter category is associated with the response of a sensor to the changes in the target (i.e., the one being measured) with time. These features are referred to as the performance indices and they help in determining the uncertainty associated with the outputs of a sensor. The features under each category have been depicted in Fig. 2.90, and discussed in the following section.

"*Range*" and "*span*" are the two primary features in the case of selecting a sensor for a purpose. The range of a sensor defines the ends within which the input values can lie. The span is the difference between the maximum and the minimum input values. For example, let us consider a sensor having a specified range of 0–100 units. This indicates that the minimum and maximum values that the sensor can read are 0 and 100 units, respectively, and the span is 100 units (i.e., 100−0).

The next feature is the "*linearity error*" which is defined as the deviation of the output of the sensor from a straight line. It is always expected from a sensor to produce an output that must be linear to the input. However, in a real-life scenario, this is not so. Thus, a deviation might be seen from the actual linear line. This deviation is referred to as the linearity error. Consider the plot shown in Fig. 2.91 that shows

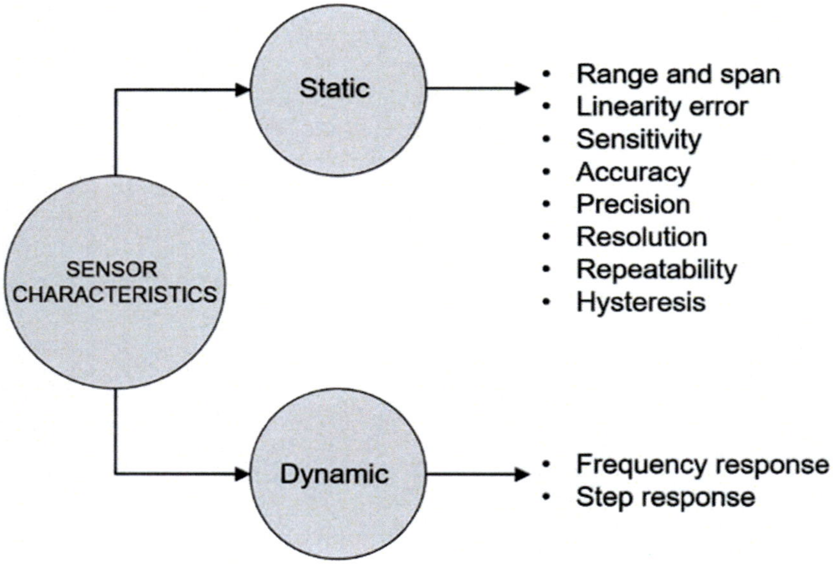

Fig. 2.90 Characteristics of a sensor

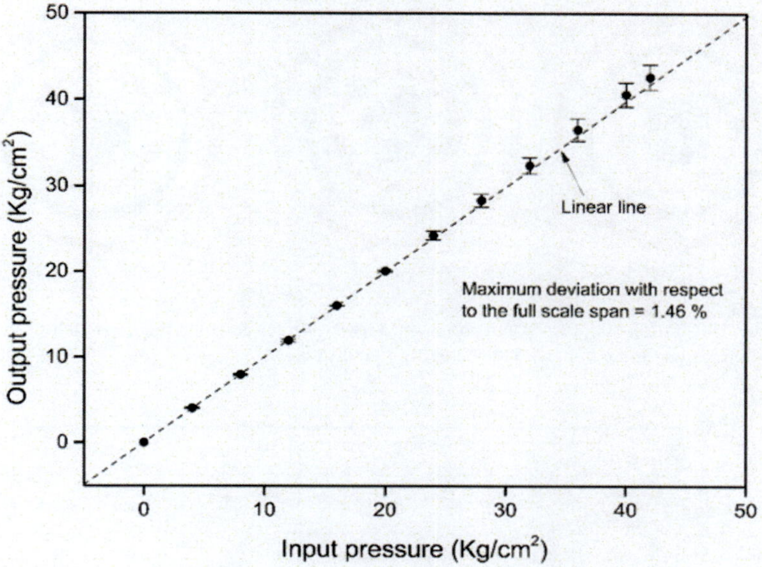

Fig. 2.91 Picture showing the linearity error in a sensor's output

the data corresponding to a pressure sensor. The input values and the corresponding output values are shown in the plot. The deviation could be observed and maximum it is 1.46% of the full-scale span.

Next is the "*sensitivity*" parameter that accounts for the effect produced by the external environment on the measuring system. For example, the pressure sensor that is considered here has a sensitivity of 0.0023%/°C. This indicates that the output produced by the sensor will be affected by 0.0023%/°C change in the temperature.

"*Accuracy*" and "*precision*" are two more static features. Accuracy is defined as the amount of uncertainty or veracity in measurement with respect to a standard or reference value. It is expressed in percentage of the full-scale span. Precision refers to the degree of reproducibility of the measurements. Consider Fig. 2.92, in which, repeated measurements are compared to dots that are shown in the figure. Figure 2.92a illustrates the case of low accuracy and high precision.

The accuracy describes the closeness of the dots to the bullseye at the centre. Dots that are closer to the bullseye are considered more accurate. Thus, the closer a system's measurements are to the accepted value, the more accurate the system is considered to be. As the dots shown in Fig. 2.92a are away from the centre (target), it has low accuracy. However, the dots (measurements) are closer to each other, which resembles the precision is high. Similarly, Fig. 2.92b illustrates the case of high accuracy and low precision, and Fig. 2.92c shows high accuracy and high precision.

Let us consider the plot shown in Fig. 2.93, the data of which is acquired from the same pressure sensor. The data was collected in three iterations for the same input conditions. The three iterations are designated as 'Acquisition 1', 'Acquisition 2',

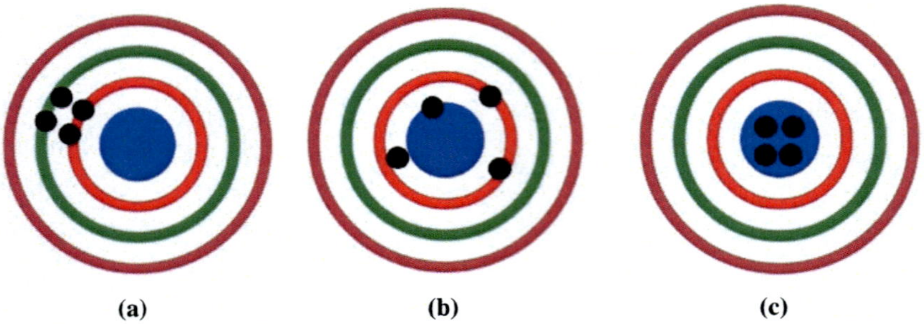

Fig. 2.92 Illustration of accuracy and precision: **a** low accuracy and high precision, **b** high accuracy and low precision, and **c** high accuracy and high precision

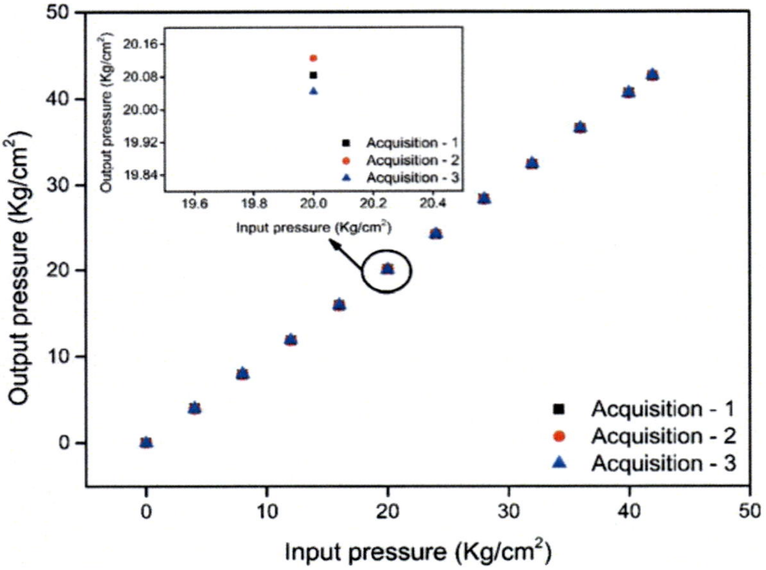

Fig. 2.93 Picture showing a sensor's output (accuracy and precision)

and Acquisition 3. The obtained output values can be observed to be very close to the actual input values, indicating the sensor to be an accurate one. Far the values from the standard linear line (i.e., the true values), lower will be the accuracy of that sensor. Precision is known by observing the outputs, which can be defined as how close the measurements are for the same given set of inputs. For a better view, a portion of the plot is zoomed and a smaller plot is shown, where the deviations in the outputs obtained in the three iterations for the same input condition can be observed. The observations are close to each other, which indicates the instrument is precise as well.

The term accuracy also considers a few other things such as the linearity error and "*zero error*". The linearity error has already been discussed which means the deviation of an output value from the linear line. The zero error resembles the values produced as output for zero input. For instance, in the plot shown in Fig. 2.93, the outputs produced for zero input are 0.032, 0.025, and 0.031 kg/cm^2, respectively in three iterations. The uncertainty in the measurement because of the zero error can be given as shown in Eq. 2.38. For the shown example, the uncertainty, UC, in the measurement for the zero error is 0.0184 kg/cm^2.

$$\text{UC(zero error)} = \frac{\text{Maximum zero error}}{\sqrt{\text{number of iterations}}} \tag{2.38}$$

The precision can also be interpreted by another term that is referred to as "*repeatability*". It is defined as the ability of a measurement system to produce the same output when taken again and again for the same input and under the same conditions. Another terminology is the "*stability*" of a sensor. A measurement system will be said to be stable if it can give the same output when employed to measure a constant input over a longer period. A specific term called "*drift*" is used to express the change in output that will occur over time.

"*Resolution*" of a sensor refers to the smallest change in the input value that can be detected and produce an observable change in the output. For example, let us consider two voltmeters, both with a range of 0–10 V. One of the voltmeters has the smallest division to be 0.1 V, and the other one has the smallest division to be 0.2 V. During actual measurement, if the input voltage would be 5.1 V, then the first voltmeter will show the output as 5.1 V. However, the second will show the output to be 5 V. This is so because the smallest change that can be measured with the second voltmeter is 0.2 V. Thus, a change of 0.1 V in the input cannot be captured by this voltmeter. Thus, the resolution of the first voltmeter will be 0.1 V and that of the second will be 0.2 V.

"*Hysteresis error*" refers to that characteristic of a sensor because of which, it is unable to repeat the measurements faithfully in the opposite direction of operation. Consider Fig. 2.94 that depicts the data acquired from the same pressure sensor in forward and reverse directions. A small portion of this plot is zoomed-in that depicts the difference in the data captured for the two directions.

Besides all these, there is a range of input values for which there is no output by the sensor. The "*dead time*" refers to this time length from the start of application of input until the output begins to respond. Finally, the frequency response is a measure of how quickly a sensor responds to a change in the physical quantity. The response of a sensor to a physical change can be defined in two approaches, response time and flat frequency response. Response time is defined as the time required by a sensor signal to change from 0 to ~50% of the full-scale range of that sensor. Flat frequency response corresponds to the maximum frequency value that a sensor can respond to without any distortion in the signal.

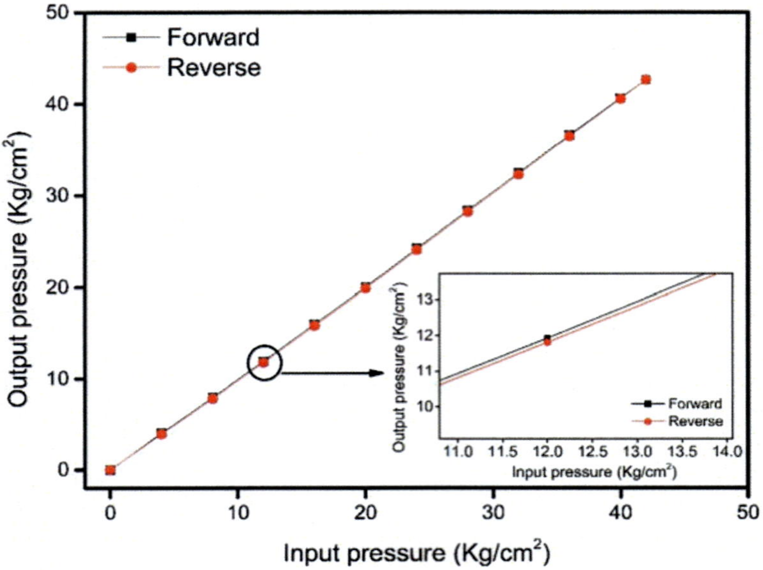

Fig. 2.94 Picture showing a sensor's output (hysteresis)

2.8 Summary

This chapter discussed the importance of sensors and transducers in industrial automation. The role of sensors in manufacturing is also discussed. A comprehensive guide has been outlined describing various sensors for process and health monitoring applications. Real-life examples in manufacturing are also discussed. In the context of building a digital twin, several sensors will be required to be engaged with the physical machine. The digital twin will consist of information from the soft sensors in addition to the data that will be captured from the hard sensors. Later, the chapter discussed the characteristics of a sensor, which lists the deciding factors for selecting a sensor. In a nutshell, this chapter provided the idea of selecting a sensor based upon an application and engaging it with necessary electronics for acquiring the data. The next step is to harness information from the data, which could be fulfilled by applying the signal processing techniques.

2.9 Questions for Practice

1. *List the sensing principle of various types of sensors.*
2. *Define piezoelectric effect and deduce its relation with force.*
3. *Define inductance and capacitance.*

4. *What is the difference between incremental and absolute encoder? Which type of encoder should be chosen for sensing a tool station?*
5. *What is Hall effect?*
6. *What is Doppler effect?*
7. *What is mutual inductance?*

References

1. Nyce DS (2016) Position sensors. Wiley, Hoboken, NJ, USA
2. Clergeot H, Placko D, Detriche JM (1988) Electrical proximity sensors. Sensors and sensory systems for advanced Robots. Springer, Berlin Heidelberg, Berlin, Heidelberg, pp 295–308
3. Ruocco SR (1988) Robot sensors and transducers. Springer, Netherlands, Dordrecht
4. Ştefănescu DM (2020) Handbook of force transducers. Springer International Publishing, Cham
5. Dahiya RS, Valle M (2013) Robotic tactile sensing. Springer, Netherlands, Dordrecht
6. Clemens H, Mayer S, Scheu C (2017) Microstructure and properties of engineering materials. Neutrons and synchrotron radiation in engineering materials science. Wiley-VCH Verlag GmbH & Co. KGaA, Weinheim, Germany, pp 1–20
7. Huynh T (2015) Fundamentals of thermal sensors. In: Jha CM (ed) Thermal sensors. Springer, New York, New York, NY, pp 5–42
8. Vollmer M, Möllmann K-P (2017) Infrared thermal imaging. Wiley-VCH Verlag GmbH & Co, KGaA, Weinheim, Germany
9. Sharma VM, Racherla V, Pal SK (2019) Synthesis of open-cell copper foam using friction sintering. Int J Adv Manuf Technol 103:3163–3174. https://doi.org/10.1007/s00170-019-03562-z
10. Nandi A, Ahmed H (2019) Condition monitoring with vibration signals. Wiley
11. Kroschel K (2020) Laser doppler vibrometry for non-contact diagnostics. Springer International Publishing, Cham
12. Patra K, Pal SK, Bhattacharyya K (2007) Artificial neural network based prediction of drill flank wear from motor current signals. Appl Soft Comput 7:929–935. https://doi.org/10.1016/j.asoc.2006.06.001
13. Fraden J (2010) Pressure sensors. Handbook of modern sensors. Springer, New York, New York, NY, pp 375–397
14. McGrath MJ, Scanaill CN (2013) Sensing and sensor fundamentals. sensor technologies. Apress, Berkeley, CA, pp 15–50

Chapter 3
Signal Processing for Digital Twin

3.1 Overview

The previous chapter discussed the importance and role of sensors in manufacturing. With the data in hand, one needs to process it for extracting meaningful information. The process of information extraction from the signal is termed signal processing, which is discussed in details in this chapter. A few case studies in manufacturing are also discussed. In the context of a digital twin, real-time information extraction from signals is vital and necessary.

3.2 Introduction to Signal

A signal can be thought of as a piece of information about the behaviour of a system and is the output of a sensor. The way stethoscope is a classic example of a sensor, the output of the stethoscope can be a classic example of a signal. It is a sound that the doctor receives through the stethoscope when diagnosing a patient. Figure 3.1 is a schematic representation of sensor-to-signal. It is the output of sensors that are engaged with manufacturing machines.

3.2.1 Meaning of the Term 'Signal'

In the case of machining, the cutting tool removes material from the workpiece. This gives rise to forces, temperature, etc. Based on the prevailing machining conditions, physical parameters i.e., force or temperature will vary. Thus, force or temperature data will provide clues about the machining conditions. In simple terms, the force or temperature data can signal the occurrence of events in the machining process. The

S. K. Pal et al., *Digital Twin – Fundamental Concepts to Applications in Advanced Manufacturing*, Springer Series in Advanced Manufacturing,
https://doi.org/10.1007/978-3-030-81815-9_3

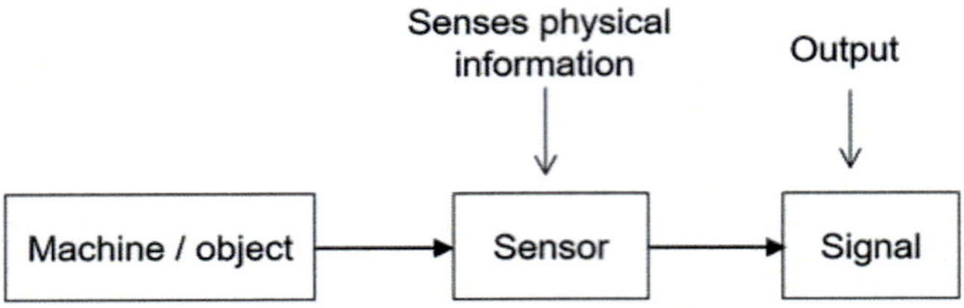

Fig. 3.1 Schematic diagram of the sensor to signal

events could be the initiation of wear in the cutting tool, or a dent in the surface of the cutting tool, or maybe the complete breakage of the cutting tool. Thus, force or temperature data are termed signals. Signals are typical to a process from which they are acquired. For instance, the examples shown in the previous chapter depict several signals collected in different processes. Their nature depends upon the process.

Let us consider the process of friction stir welding (FSW) in which force is one of the essential parameters as discussed in the previous chapter. A plot of the force signal is shown in Fig. 3.2 in which the occurrence of events in the welding process is depicted. This includes the start of tool rotation, tool plunging, shoulder contact, welding phase, and tool retraction [1, 2]. Schematic pictures of these events are also included in the plot for better understanding. The following paragraph explains these events briefly.

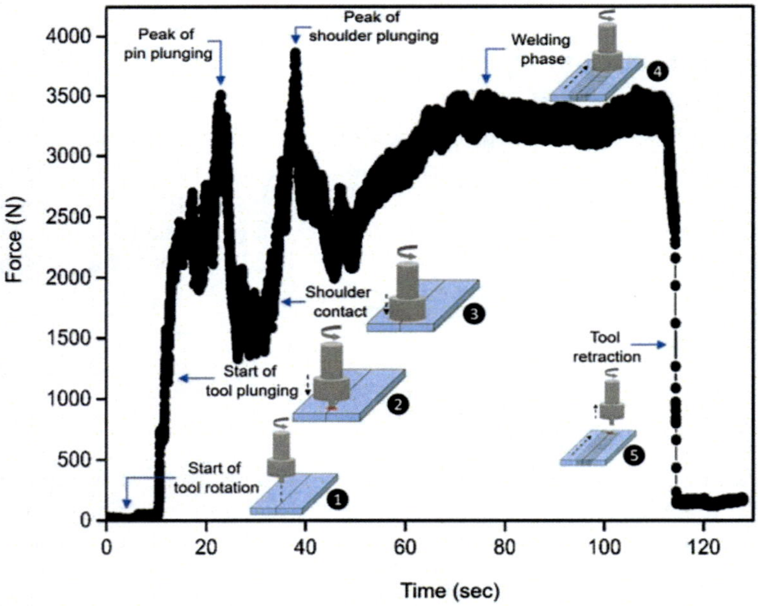

Fig. 3.2 A force plot acquired in FSW process

The tool used for welding in FSW is mounted on to the machine spindle and is one of the input parameters of welding [3]. The welding process begins with the rotation of the tool (denoted as 1 in Fig. 3.2). During this event, the force can be observed to be nil. This tool has two features, pin and shoulder. The pin is the protruding part and the shoulder is the base surrounding the pin. The length of the pin is a little lesser than that of the thickness of the materials to be welded and its diameter depends on the type of materials and their thickness. The shoulder diameter is 3–4 times to that of pin. As the tool approaches the materials to be joined, the pin touches first and thus, the force rises (denoted as 2 in Fig. 3.2). As the pin continues to plunge, a peak value is reached and then the force drops as the material surrounding the pin soften. The softening occurs because of the frictional heat and plastic deformation caused by the action of the tool. This is followed by contact of the shoulder that makes the force rise again (denoted as 3 in Fig. 3.2). The tool dwells here for a small period to ensure the deformation of the metals to be joined. The machine bed traverses next with a linear velocity. This is the second parameter in this welding process. Small fluctuations in force can be observed from the welding phase (denoted as 4 in the Fig. 3.2). This is because of the continuous engagement of the tool with the base materials. Finally, at the end of welding, the tool retracts out and the force drops (denoted as 5 in Fig. 3.2).

Hence, the force plot signals the occurrence of events of the FSW process. So, the essence is that the welding process can be tracked following this plot or signal. The next important feature is to investigate the amplitude as depicted in the vertical axis. The obvious question is: does it always remains the same. If not, what is the reason behind its variation? Of course, the amplitude is not the same, and in this case, it depends upon other conditions included in the welding process. This means the signature or the behaviour will vary for different conditions, and this is the underlying fact due to which the signals are acquired and processed in manufacturing processes.

Another example to cite here is the signature of the motor power consumption which is depicted in Fig. 3.3. The experimental setup consists of an induction motor mounted on a hydraulic sump. The power derived from the motor is used to actuate a double-acting cylinder. The direction of actuation is controlled by using a solenoid valve. A power sensor is engaged to capture the signature of power consumption. During the idle stage i.e., when the motor is not active, the power consumption is nil (denoted as 1 in Fig. 3.3). As the motor starts, the power rises (denoted as 2 in Fig. 3.3). This is because of the initial torque that is induced in the motor to reach its desired speed. Thereafter, the power remains stable (denoted as 3 in Fig. 3.3).

For more information, the stable region is zoomed and a plot of the same is shown in Fig. 3.4. A pattern can be observed from the plot like continuous rise and fall of the power consumption. Intuitively, it indicates the load on the motor during the actuation of the cylinder. As the actuation begins i.e., forward and backward motion, a load acts on the motor, which increases the power consumption. Thereby, as the load is removed, the power consumption falls. As the cylinder is continuously actuating, the pattern is repeating itself. This is also validated by using a stopwatch measuring the time taken during the actuation period. An important point to note here is the condition of the mechanical components that are involved in this study. At present,

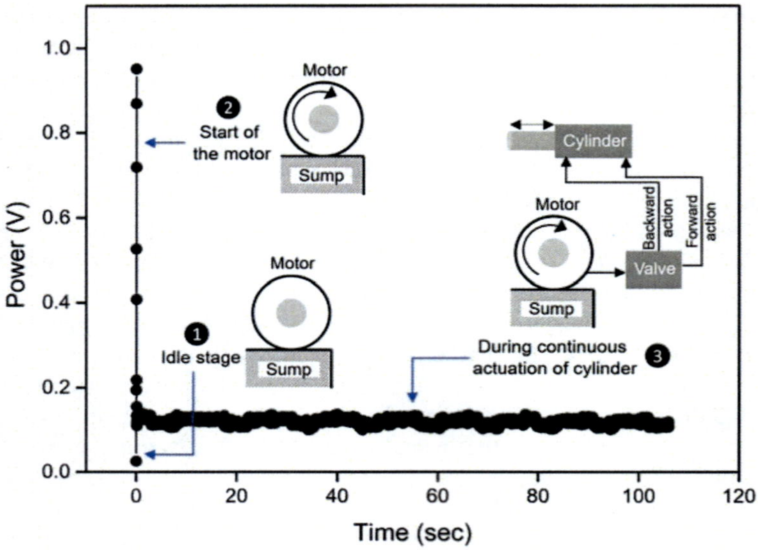

Fig. 3.3 Signature of motor power consumption

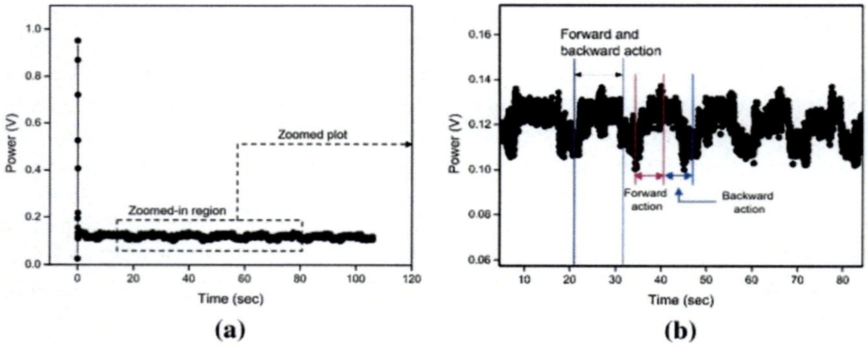

Fig. 3.4 Power signature: **a** normal plot and **b** zoomed-in plot

they are healthy. With time, their life will be affected and the signature of power consumption can signify the changes occurring in their lives. For instance, if the motor condition is poor because of some faults in the rotor or bearings, the power consumption will be different.

3.3 Signal as Indirect Means of Monitoring

The monitoring of manufacturing processes can be grouped into two techniques, direct and indirect, as shown schematically in Fig. 3.5.

The former refers to the actual measurement. Considering the machining process, where the objective is to monitor the quality of the cutting tool, the direct monitoring technique will require the cutting tool to be removed from the machine and test it under a microscope to determine the wear. Similarly, in the case of welding, the quality of the welded sample can be determined by testing the sample in a universal testing machine. The observation here is the involvement of offline assessment of the product quality. At the time, the tests are destructive, i.e., a portion of the entire sample is to be cut for the test. Thus, the direct monitoring approach is time-consuming as well as it is capital intensive.

In contradiction to the above, the indirect monitoring approach refers to the sensing of physical parameters to assess the product quality [4]. These are the signals which are acquired in-process during the manufacturing operations. It should be noted that the results obtained from the direct monitoring approach are more accurate as they involve machines or instruments for testing. In the case of the indirect approach, information needs to be extracted from the signal, thus, it is always less accurate than

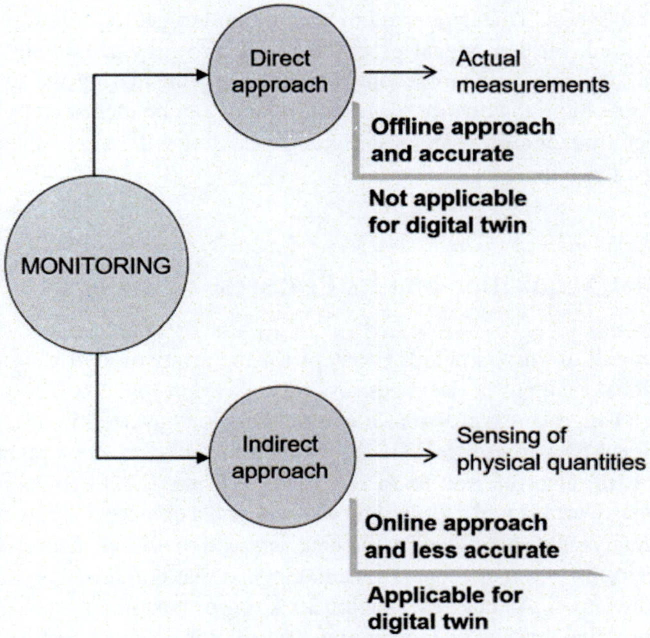

Fig. 3.5 Schematic diagram of monitoring of a manufacturing process

the direct monitoring approach. However, the biggest advantage of indirect monitoring comes from the acquisition of the signals in-process during the manufacturing operations, which makes it more suitable and economic for industrial applications. The objective of a digital twin is to provide real-time information about the physical device. Thus, the digital twin along with the sensing devices form a part of the indirect monitoring approach.

3.4 Importance of Signal Processing in Digital Twin

Signals carry a lot of information about the process to which they belong. But not all the information needs to be visible from the raw signal. To correlate the information contained in the signal with the quality of the product or certain phenomenon, analysis of the signal is required. This analysis is termed signal processing. This representation of the signal to time is termed as time-series and are referred to as a 1-D signal. This chapter will deal with the 1-D signal only. The other type is the image, which is referred to as a 2-D signal, and it will be discussed in the next chapter.

The manufacturing machines involve a lot of electronic components which generate electromagnetic radiation. This affects the sensing element and corrupts the signal. Also, the vibrations coming from the nearby machines are responsible for corrupting the signal. This is referred to as noise and to get the true signal, filtering is often required. Further, signal processing will also help in ranking the sensors suitable for capturing a phenomenon. For instance, considering the example of a machining operation, the quality of the cutting tool can be identified by processing force, temperature, or current signals. Signal processing will help in determining the best amongst these three.

3.5 Signal Acquisition and Its Features

This section will discuss in details the steps for the acquisition of a signal. A basic representational diagram of data acquisition is shown in Fig. 3.6. The sensor is for sensing the change in the physical quantity. This is converted into an equivalent electrical signal by a transducer. This signal is transferred to a computer via a data acquisition card, also referred to as DAQ [5, 6]. Thus, DAQ acts as an interface between a real-world signal captured by a sensor and a computer where the signal is stored and analysed. In this section, the data acquisition will be discussed in detail.

DAQ performs two major operations namely, signal conditioning and analogue to digital conversion. As outlined previously, a sensor contains the sensing element which captures the physical phenomenon for which it is intended, and it generates an equivalent output signal. The source signals are continuous and are termed analogue signals. Figure 3.7 shows a continuous signal. On the horizontal axis is the time and

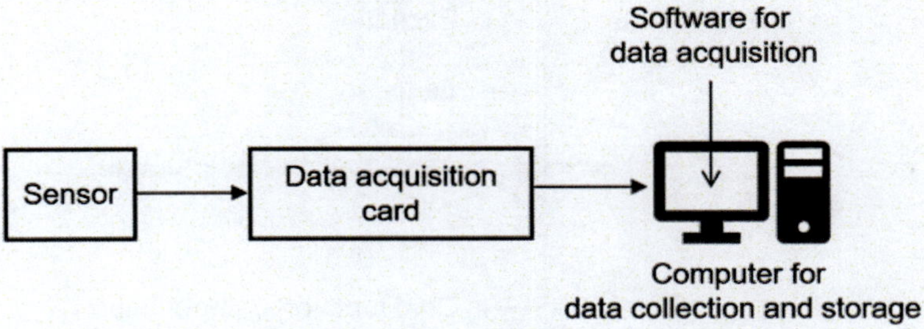

Fig. 3.6 Schematic diagram for data acquisition

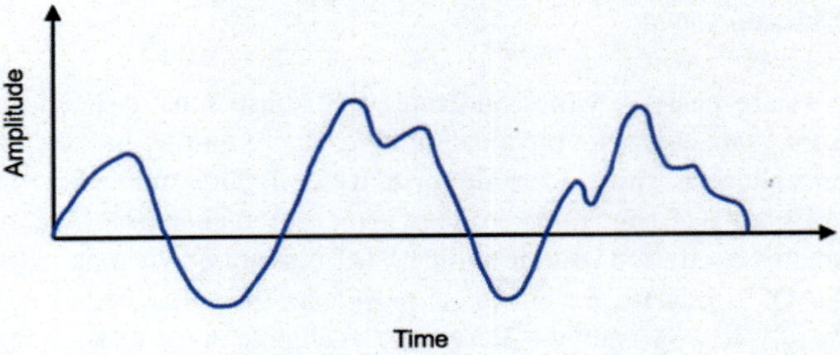

Fig. 3.7 Plot of a continuous signal

on the vertical axis is the amplitude of the signal. The signal is continuous means that the value of the signal exists at each instance of time.

The source signal is passed through a "signal conditioning" unit that refers to the pre-processing of a signal. It is one of the vital elements in the signal acquisition method as it processes and prepares the signal to match the hardware device. The following section discusses the operations performed by a signal conditioning unit.

3.5.1 Signal Conditioning

Figure 3.8 lists the operations performed by a signal conditioner that includes excitation, isolation, amplification, attenuation, filtering, cold-junction compensation, etc. The signal conditioning unit is located close to the sensor to reduce the corruption of the signal by external factors. The following paragraph explains each of these operations.

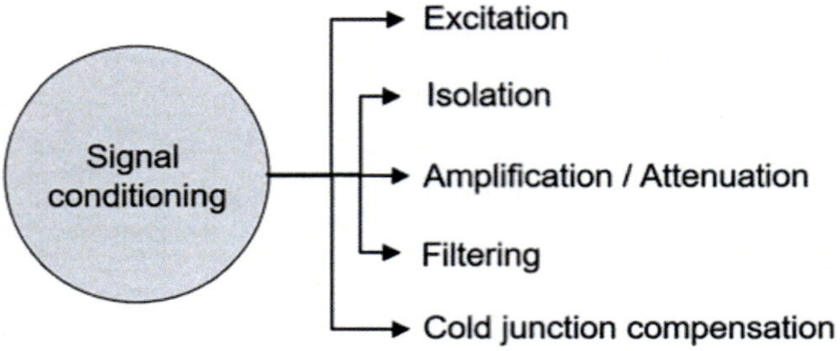

Fig. 3.8 Operations performed by a signal conditioner

3.5.1.1 Excitation

A few sensors are passive, which indicates that these sensors need an excitation voltage/current to measure the physical change. For example, in the case of LVDT, an excitation voltage is supplied to the primary coil. This induces a voltage in the secondary coils and a change in the voltage is measured because of the movement of the core. Here, the excitation voltage to the LVDT is supplied via a signal conditioning unit of the DAQ. Similarly, in a strain gauge sensor, the resistance changes with the change in the physical parameters. To record the change in the resistance, an external voltage is supplied across the transducer circuitry. However, for an active sensor, such as piezoelectric-based, excitation is not required. Likewise, for temperature measurement, thermistor, RTD, and thermocouple are widely used sensors. However, excitation will be required for the first two sensors while the thermocouple does not need it.

3.5.1.2 Isolation

The signal conditioning unit also looks after the safety of the system and isolation is one such feature that ensures the same. A measured signal may be a high-voltage spike is a response to a change in the system. This high-voltage spike may damage the DAQ system and hence, it is necessary to eliminate the same.

3.5.1.3 Amplification

Amplification indicates increasing the voltage level of the measurement. The amplitude of the output generated by the sensor is usually very low and thus requires amplification. This is also required to make the amplitude of the measurement compatible with the hardware device. The operation is shown schematically in Fig. 3.9. This is fulfilled by employing an amplifier that applies a gain to the signal before sending it to the next electronic system. For instance, the output voltage generated by a

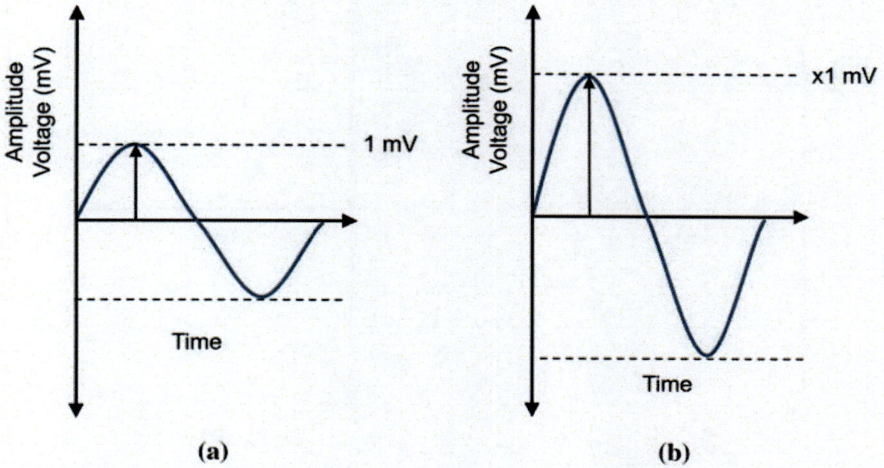

Fig. 3.9 Amplification of a signal

piezoelectric-based sensor is very small and thus, it requires amplification so that it is readable by the DAQ system.

3.5.1.4 Attenuation

Attenuation means to lessen the amount of something. In signal conditioning, attenuation refers to the lessening of the amplitude of the signal. It is the opposite of amplification. DAQ cards work in a range i.e., +5 to −5 V or maybe +10 to −10 V. If the measured signal is beyond the range of the DAQ, then the required processing of the signal could not be performed. Thus, the voltage level of the signal is decreased. This is fulfilled by employing a voltage divider circuit. This circuit consists of two resistors in parallel and when an input voltage is applied to the circuit, it generates an output voltage, which is a fraction of the input voltage. Recap Eq. 2.6 and Fig. 2.8 in Sect. 2.5.1.2 of Chap. 2, where a voltage divider circuit was discussed for the potentiometer. The process of attenuation is shown schematically in Fig. 3.10. The raw signal is having amplitude beyond the range of DAQ (i.e., +10 to −10 V), thus, it is lessened, as shown in the right-hand-side image.

3.5.1.5 Filtering

Signal conditioning also demands the removal of unwanted components from a signal. This process is referred to as filtering. The operation is shown schematically in Fig. 3.11. The plot on the left can be observed having many fluctuations. This can be attributed to the inclusion of noise in the source signal which may occur because

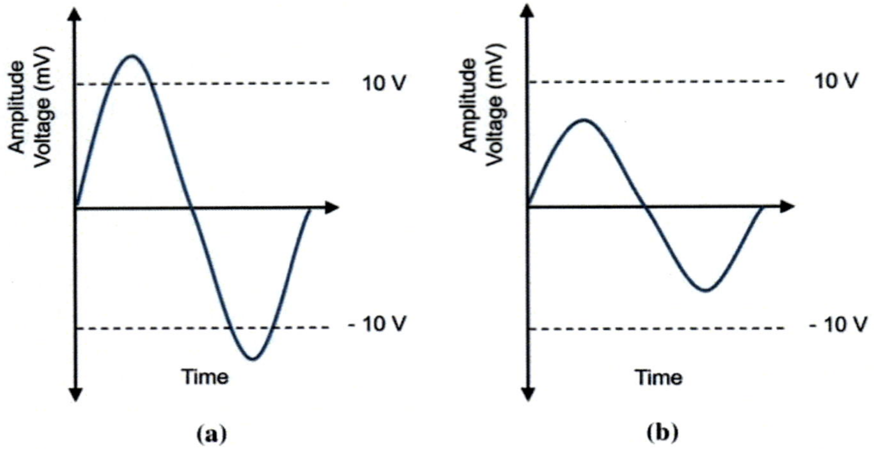

Fig. 3.10 Attenuation of a signal

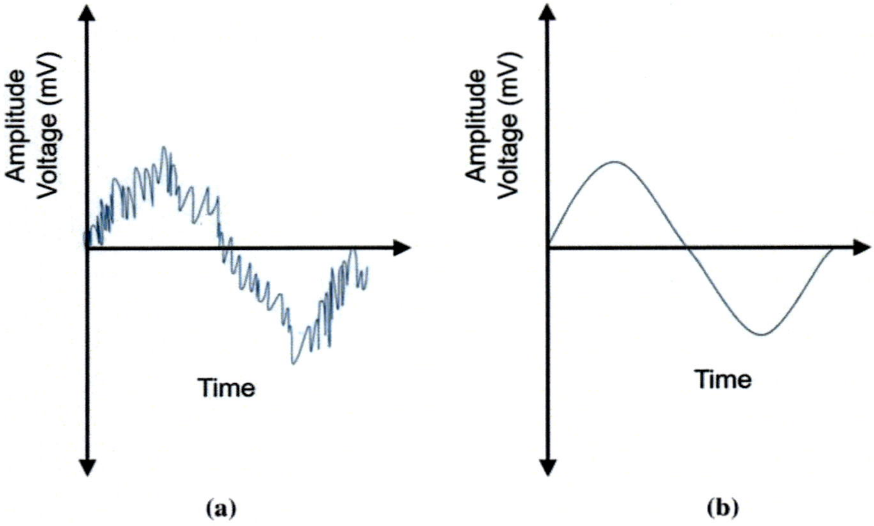

Fig. 3.11 Filtering of a signal

of several reasons. In such a scenario, a suitable filtering technique is applied on the raw signal that filters out the unwanted components and brings out the true signal, as shown on the right side of Fig. 3.11. More details about the filtering techniques are discussed in a later section (Sect. 3.8).

3.5.1.6 Cold-Junction Compensation

The "cold-junction compensation" is required during the thermocouple measurement [7]. The thermocouple consists of two dissimilar metals that generate a voltage difference across the hot and cold junctions. The voltage difference is because of the presence of free electrons which diffuse at different rates depending upon the temperature in the conductor. The electrons present at the hot junction diffuse more rapidly because of higher energy as compared to the electrons present at the cold junction. The hot junction refers to the end of the thermocouple which is embedded into the object whose temperature is to be measured. The other end forms the cold junction. Owing to the difference in the energy amongst the free electrons in a conductor, charge is built, and thus, a potential difference is generated. As the thermocouple is made up of two metals, the voltage difference is proportional to the temperature gradient. This indicates that the thermocouple measures the temperature difference rather than the absolute temperature. Thus, to determine the absolute temperature, it becomes essential to determine the temperature of the other end of the thermocouple. To do so, the cold junction is connected across the terminal of the measuring device which is of another metal and the process is referred to as cold junction compensation. A schematic of this is depicted in Fig. 3.12 wherein 'A' is the hot junction and 'B' and 'C' form the cold junction. The thermocouple is made up of 'Metal 1' and 'Metal 2'.

Classically, junctions 'B' and 'C' were immersed in an ice-bath, which gave it the name cold-junction. Practically, they are connected to a third metal 'Metal 3' as shown in the picture, which is usually copper. A block of copper surrounds 'B' and 'C' junctions so that they remain thermally stable. This is important because unless the temperature of the cold junction is known, the temperature of the hot junction cannot be determined. For measuring the temperature of the cold junction, RTD, thermistor, etc. are utilised. Thus, it is compensated by applying a proper correction.

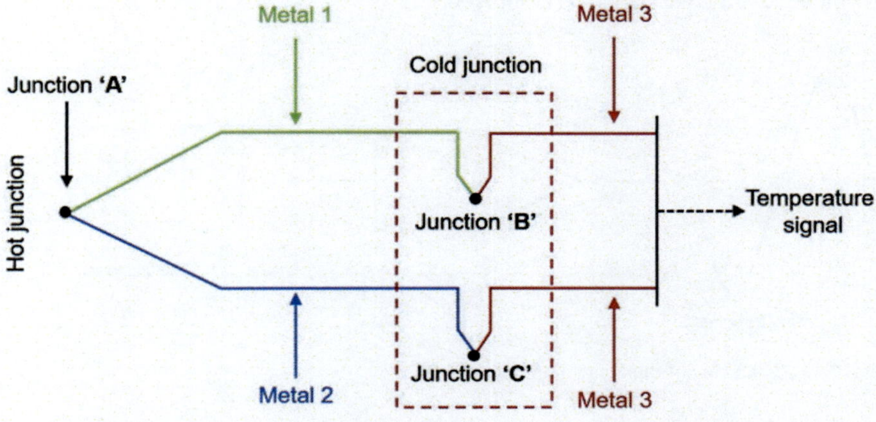

Fig. 3.12 Schematic for cold junction compensation

This processed signal is then received by the hardware device for analogue to digital conversion. This device is operated and controlled by software.

3.5.2 Analogue-To-Digital Converter

The analogue-to-digital converter also referred to as ADC takes the conditioned signal as the input and digitizes the same to obtain a digital signal as shown in Fig. 3.13 [8]. The conversion makes the signal readable on a computer. This is the purpose of having an ADC and it is thus found in all DAQ systems.

But the importance of introducing ADCs here is to highlight the importance of resolution. The resolution of an ADC is the number of bits it uses to digitize an input signal. Earlier, the ADCs came in an 8-bit configuration. Let us try visualizing the importance of this matter. For an n-bit ADC, the number of discrete levels that can be produced is 2^n. For instance, in the case of an 8-bit ADC, the number of discrete levels will be 256. The least significant bit that can be detected by the ADC is the reciprocal of the number of discrete levels as expressed in Eq. 3.1.

$$LS = \frac{1}{2^n} \tag{3.1}$$

where LS denotes the least significant bit of an ADC and n denotes the resolution of the ADC. Thus, for the 8-bit ADC, the least significant bit will be 3.90. Thus, considering the case of a 1 V signal (±0.5 V), the performance of different digitizers is listed in Table 3.1.

From the above table, it can be observed that as the resolution increases, the smallest bit can be measured. For the case of manufacturing applications, where signals are dynamic, mostly 24-bit ADCs are used. However, in some applications, 16-bit or 12-bit ADCs are also employed.

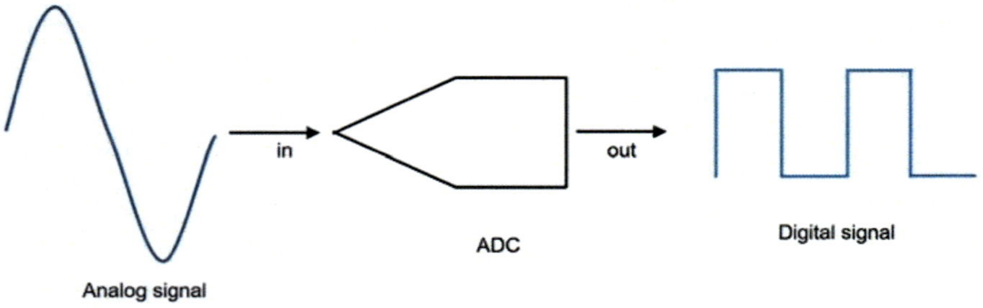

Fig. 3.13 Schematic of input and output to an ADC

Table 3.1 Least significant bit of digitizers

Sl. no	Resolution	Dynamic range	Least significant bit
1	8-bit	1:256	3.90 mV
2	10-bit	1:1024	0.976 mV
3	12-bit	1:4096	0.244 mV
4	14-bit	1:16,384	61 μV
5	16-bit	1:65.536	15 μV
6	24-bit	1:1,67,77,216	0.059 μV

3.5.3 The Sampling Rate of a Signal

Two fundamental questions in the acquisition of a signal are: (a) how many and (b) how often? [9]. The real-world signals are analogue. However, they are required to be converted to digital form to process them on a computer. The digital signals are discrete in time and amplitude. To convert the analogue signal to a digital one, the process followed is termed sampling. It is defined as the process of measuring the instantaneous values of continuous signals in a discrete form, which results in samples. The discretization of the signals is performed by fixing the period or frequency which is termed as sampling period or sampling frequency, respectively, as shown in Eq. 3.2.

$$F_s = \frac{1}{T_s} \tag{3.2}$$

where F_s refers to the sampling frequency and T_s refers to the sampling period. The sampling frequency denotes the number of samples to be collected per second and is also called the sampling rate. The sampling period is the reciprocal of the sampling frequency. The concept of sampling is important and critical in cases where the analogue signal is required to be reconstructed from the digital signal. For this, a criterion called "*Nyquist rate*" is followed. "*Nyquist rate*" states that for a band-limited signal with W being the highest frequency component, to reconstruct the original signal effectively, the sampling rate should be at least twice the highest frequency component, as shown in Eq. 3.3.

$$F_s \geq 2W \tag{3.3}$$

where, F_s refers to the sampling frequency and W refers to the highest frequency component.

Based on the "*Nyquist rate*", the Nyquist sampling theorem has been proposed which mentions the minimum number of samples required for a band-limited signal. The theorem states that a signal can be exactly reproduced if it is sampled at a rate that is at least twice the maximum frequency component present in that signal. The

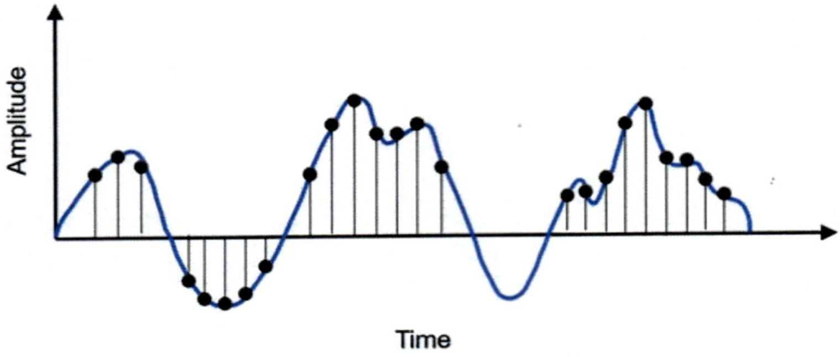

Fig. 3.14 Plot of a discrete signal

continuous signals are discretized which makes them readable on a computer. The continuous signal as shown in Fig. 3.7 is sampled at regular intervals. Figure 3.14 shows the discretization of a continuous signal, which means that the value of the signal exists at regular intervals of time.

The following is an example to visualize the importance of sampling rate while acquiring a signal. For this, the experimental set-up consisting the induction motor mounted on the hydraulic sump is considered for which the signature of power consumption is explained in the previous sections. This time, a vibration sensor i.e., an accelerometer is fitted on the motor for acquiring the vibration signature. As discussed in Chap. 2, the vibration signature is of particular interest for evaluating the health of certain components. In this case, it is the health of the motor. A picture of the experimental set-up is shown in Fig. 3.15.

The rated speed of the induction motor employed in this experimental set-up is 1480 rpm. To determine the speed of the motor, a tachometer was utilized. Thus, the fundamental frequency will be around 25 Hz i.e., 1480/60 rps. As per the *Nyquist rate*, the sampling rate should be at least twice the highest frequency component. For the time now, the fundamental frequency is known, which is 25 Hz. Thus, the sampling rate must be at least 50 Hz. To test the same, data has been captured at different sampling rates i.e., (a) 12 Hz, (b) 48 Hz, (c) 120 Hz, and (d) 1000 Hz. These four rates are chosen intentionally to visualize the difference amongst them. The time-domain signal acquired at each of the aforementioned sampling rates is shown in Fig. 3.16.

To visualize the importance of *Nyquist rate*, the time domain signal is transformed into the frequency domain using Fourier transform. This transformation depicts the frequency associated with a signal. Fourier transform is explained in details in a later section. However, at this point of time, let us understand a few details about the frequency information. The term frequency refers to the number of occurrences of an event per unit of time. Thus, in signal processing, the frequency domain could help in determining cyclic events in a signal, which is not found in a time domain signal. For the above shown signals that are acquired at different sampling rates, the

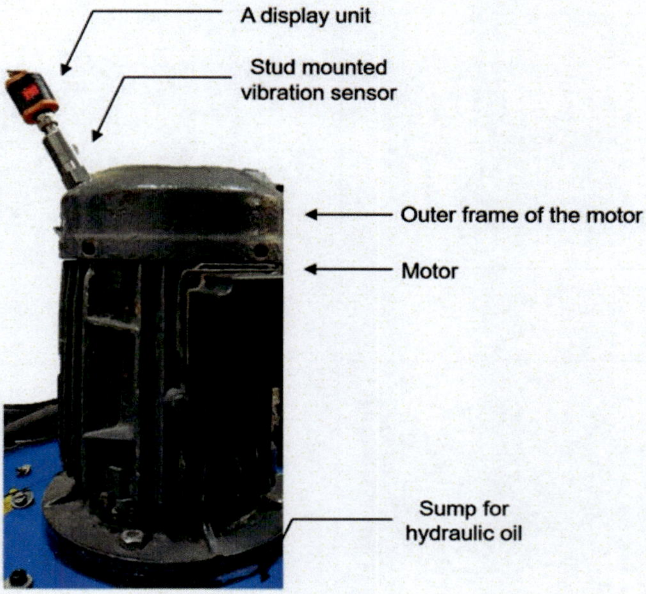

Fig. 3.15 Picture of an experimental set-up: vibration sensor fitted on to a motor

amplitude of the signal is known over the time they have been acquired. In simple terms, events versus time.

Here comes an important information about the signals. The experimental set-up utilised in this work consists of an induction motor, and one of the bearings of this motor is a faulty one. Let us not bother about the type and intensity of fault at this moment. Rather, the target is to capture the variation from the signals. The frequency information of these signals is shown in Fig. 3.17.

From the frequency information, it can be observed that the signal acquired with a sampling rate of 12 and 48 Hz are insufficient to characterize the cyclic events occurring in the signal. Peaks in the signal, if any, mostly corresponds to noise. The signal acquired with 120 Hz which is above the *Nyquist rate* shows two peaks, one close to 30 Hz and the other one at 50 Hz. As the motor considered in the experimental set-up runs at 1480 rpm, the fundamental frequency is around 25, thus, 50 Hz is one of the harmonics of the fundamental frequency. However, the signal acquired with a rate of 1000 Hz, which is at a very high rate, not only shows the fundamental frequencies of the signal, but also has a few peaks at higher frequencies. These frequencies may correspond to the behaviour of the faulty bearing in the motor. Hence, it is crucial to collect signal at a rate much higher than that of the *Nyquist rate* so as to capture the minute details of a phenomenon. This is more important for manufacturing processes as the fault frequencies are different. To build a digital twin, where the objective may be to capture the behaviour of a phenomenon of a process or diagnose the health of

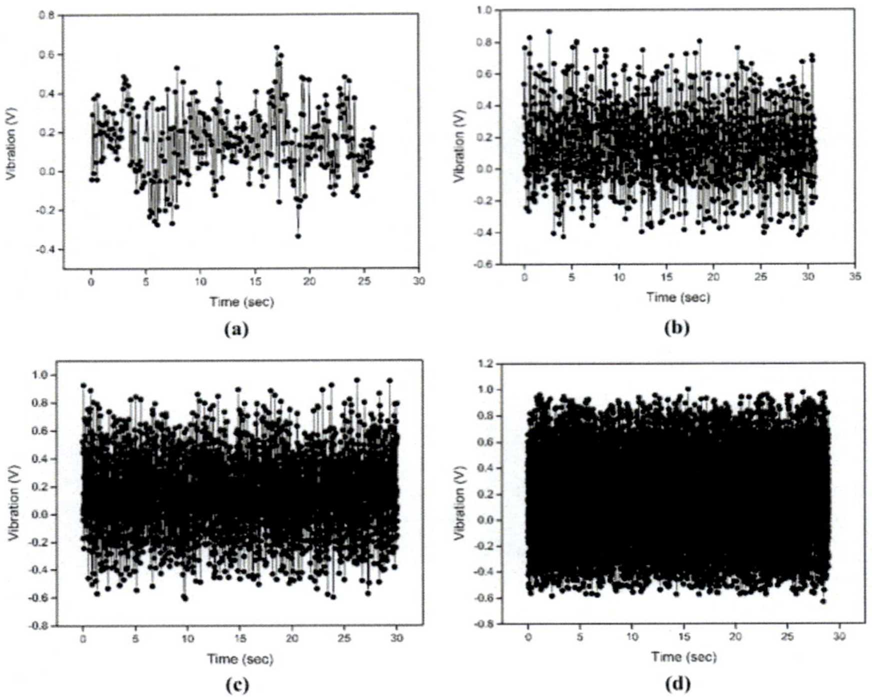

Fig. 3.16 Time-domain signal acquired at different sampling rates: **a** 12 Hz, **b** 48 Hz, **c** 120 Hz, and **d** 1000 Hz

a machine, the sampling of the signals is to given importance because it is the only one upon which the entire analysis will depend.

3.5.4 Signal Acquisition

For acquiring a signal, a data acquisition system is utilised. They help in measuring the signal and transferring them to a computer for further processing. The operations which are discussed in the preceding sections such as signal conditioning, analogue to digital converter, and sampling are featured by a DAQ system. It is a hardware to which a sensor is connected and it enables acquisition of the data sensed by that sensor. A DAQ is configured by connecting it to a computer.

Figure 3.18 shows the picture of a DAQ system *(National Instruments, 6210)*. This card has a 16-bit resolution and 24 channels. The channels are where a sensor is connected with the card. On the other side, this card is connected to a computer through a USB support. A software is required for connecting to the card. For this DAQ, the LABVIEW software is utilised. A number of manufacturers are providing

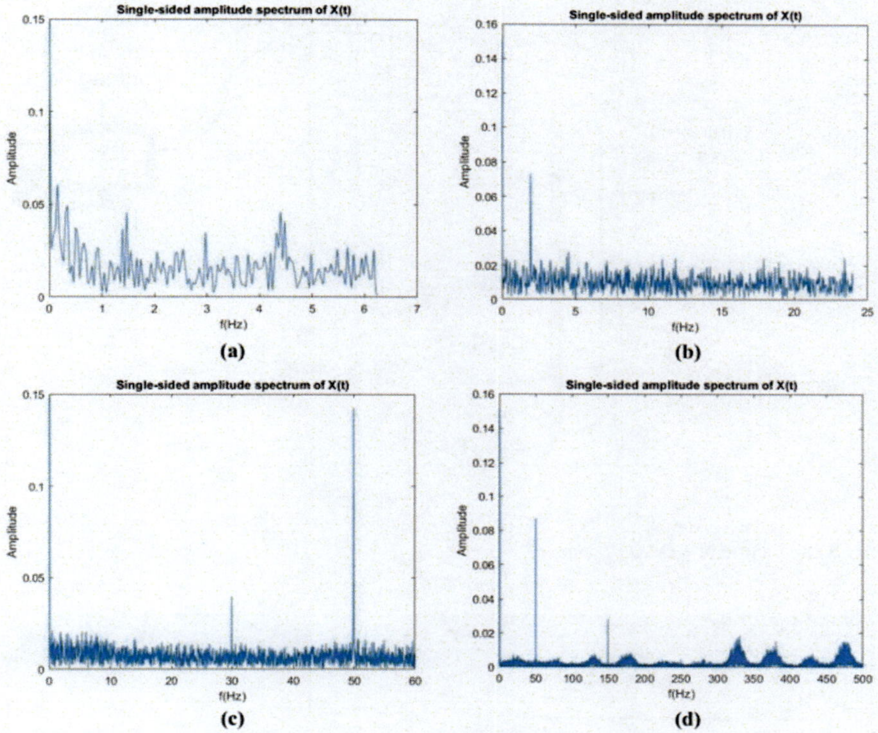

Fig. 3.17 Frequency information of the time domain signal acquired at different sampling rates: **a** 12 Hz, **b** 48 Hz, **c** 120 Hz, and **d** 1000 Hz

DAQ systems with their customized software. As an example, the LABVIEW software is briefly outlined for connecting to the shown DAQ system.

For sampling the signal and collecting it to a local file in the computer, a virtual instrument, also referred to as VI can be built. The first step is to configure the data acquisition card and the signal to be collected in the software. Figure 3.19 shows the circuit diagram of the VI built in the LABVIEW. The data acquisition card is configured, in which, the type of signal is selected first. Figure 3.19 shows the screenshots of this step, where an "Analog input" is selected. Following this, the physical parameter is selected, for example: voltage, temperature, strain, current, etc. Once the type of sensor is selected, other details such as sampling rate, number of samples to be collected, range of the signal, etc. can be configured. After setting this environment, the VI can be run for acquiring the signal. The signal is sampled as per the defined sampling rate and is available with respect to the time. This signal is termed as the raw signal or the time-series. Raw means that it is the direct output of a sensor and time-series means the output is acquired with respect to the time.

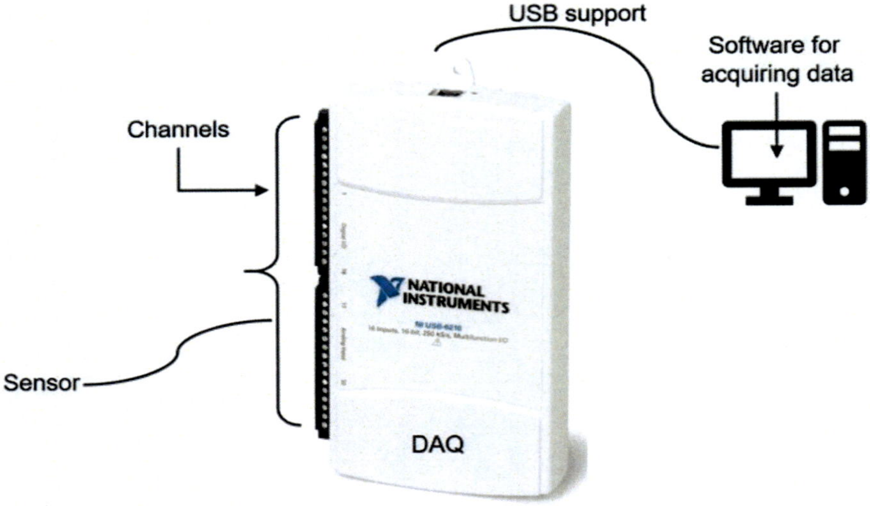

Fig. 3.18 Picture of a DAQ system

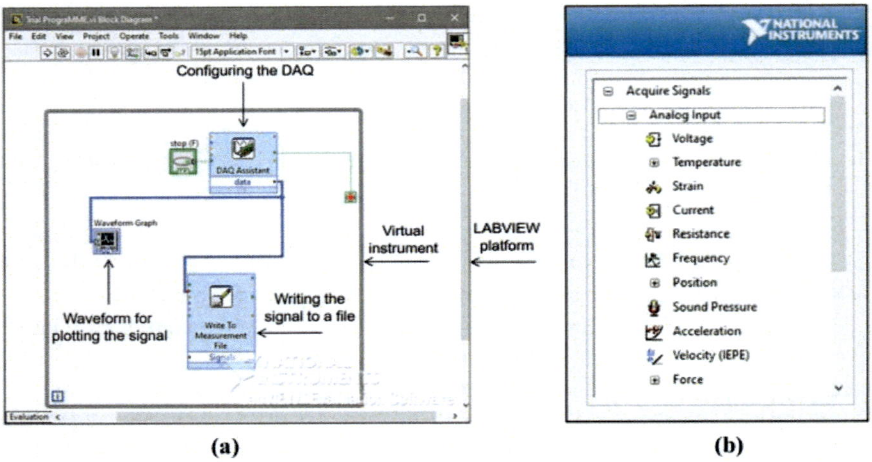

Fig. 3.19 Screenshots of configuring the data acquisition device and signal to be collected in the software

Another thing to note here is the difference between the DAQ system for measuring temperature and others. When it comes to measuring temperature, sensor such as thermocouples and RTDs require a DAQ system. The system designed for this purpose include features such as open-thermocouple detection, cold-junction compensation, along with other features such as isolation, common mode voltage, etc.

3.6 Arduino Microcontroller

Arduino refers to a development platform that is being leveraged for building IoT projects. It consists of a programming board and a software [10]. The board consists of digital and analogue pins, a microcontroller, an interface for connecting to the device to a computer, etc. The software titled "Integrated Development Environment" (IDE) is used to write the program, which are uploaded to the board via a suitable interface. Arduino is very popular because of the simplistic design and easy-to-connect interface. The Arduino board can interact with a variety of devices, thus, makes it suitable for the Internet of Things (IoT) projects. Different boards have been released and made available and amongst them, *Arduino UNO* is the most common and widely used. A picture of this board is shown in Fig. 3.20. The components of the board are also shown in the picture.

To make the board interact with the objects, sensors are used that are compatible to the Arduino board and are referred to as Arduino-based sensors. Two such sensors have already been introduced in Chap. 2 (Sects. 2.5.9 and 2.5.10), those are for measuring flow rate and turbidity. In this section, the application of one more such sensor is shown. It is an ultrasonic sensor *(HC-SR04)* that uses sonar signal to determine the distance of object non-invasively. A picture of this sensor is shown in Fig. 3.21. It consists of a transmitter and a receiver unit. It can sense objects approximately in the range of 2–400 cm.

This sensor was utilized in a prototype-vehicle to sense the obstacle in front of it. The objective is to avoid the obstacle, if any. The experimental set-up is shown in Fig. 3.22 that consists of a four-wheeler vehicle equipped with DC motors, a servo

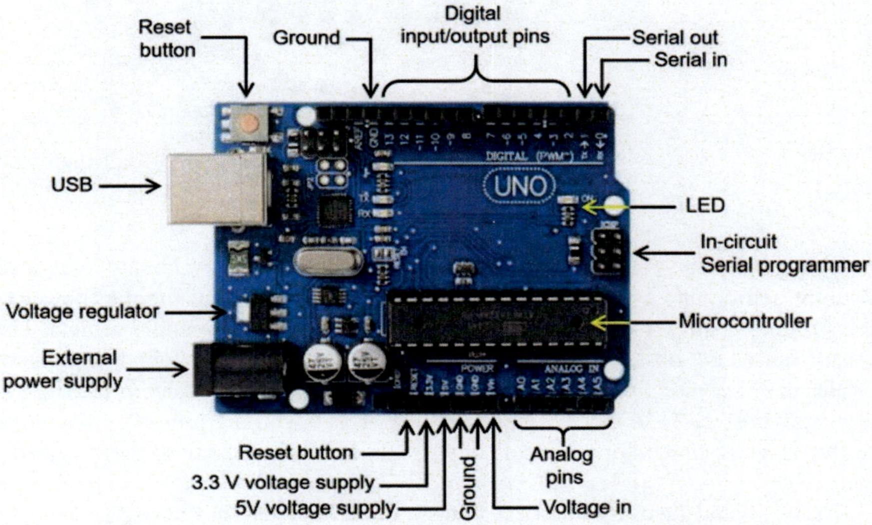

Fig. 3.20 Picture of an Arduino UNO board

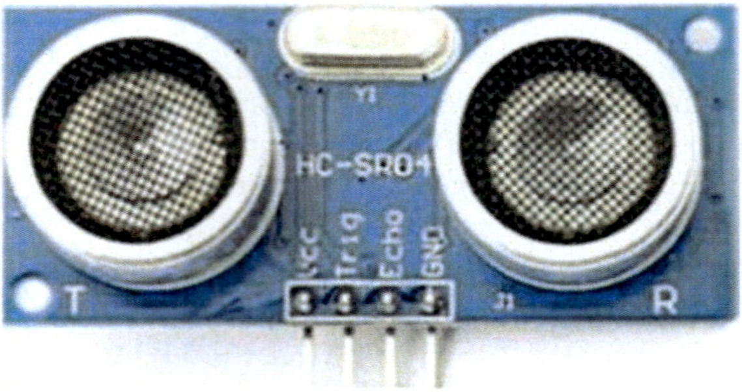

Fig. 3.21 Arduino-based ultrasonic sensor

Fig. 3.22 Application of an ultrasonic sensor

motor on which the ultrasonic sensor is mounted, an *Arduino UNO* board integrated with the sensor, and a power supply unit. The power supply unit supplies power to the Arduino board for switching the four DC motors. This powers the vehicle. The sensor senses the obstacle present in its way. If any obstacle is found, the sensor rotates in either direction, i.e., left or right, to determine whether any of the sides is free from obstacle, and thereby makes a move. It is a working project and is under further investigation on optimization of the path, and time taken in decision making, etc.

Today, several such Arduino-based sensors are available. They have gained popularity because of their low-cost and easy-to-use features. They are the best solution for building prototypes, or coming up with innovations in IoT. Figure 3.23 depicts

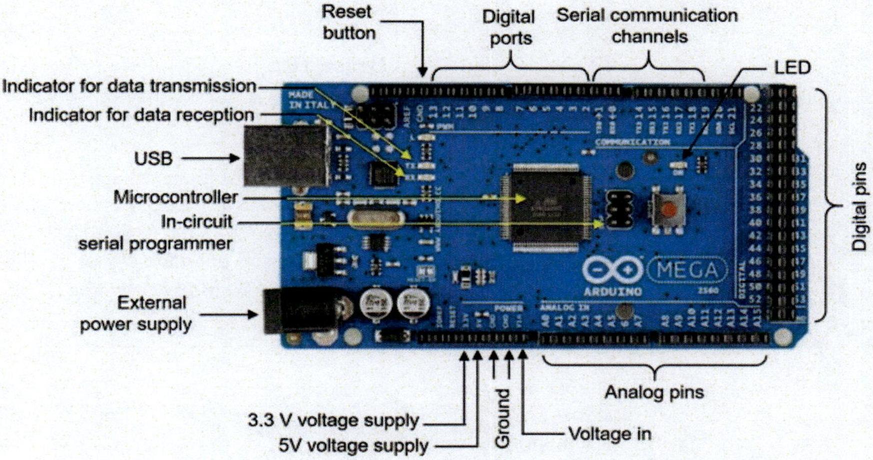

Fig. 3.23 Picture of an Arduino Mega board

the picture of *Arduino Mega*, which is one of the most advanced boards. The board has 16 analog pins and 54 digital pins, a total of 70 pins for connecting the devices. It consists of a powerful microcontroller that supports multiple serial tasks.

However, it is vital to answer how relevant and capable are the Arduino-based sensors and Arduino microcontrollers for the industrial-grade applications. Most of the Arduino-based sensors are bare i.e., the sensing element is exposed to dust, temperature, moisture, etc. These are a few critical factors that affect the working of the sensor. In an industry, these are common to the shop floors, and thus, the sensors and data acquisition boards need to be capable of working against these severe conditions. The industrial-grade equipment such as PLC, sensors, and other computing devices are so manufactured that they intend to work under severe conditions. In addition, an industrial programmable logic controller (PLC) is so designed that it is capable to deliver an equivalent voltage for actuation of the industrial-grade actuators. This is one major reason behind the low-cost of the Arduino boards and compatible sensors as compared to the normal sensors. However, they are an excellent choice for building IoT prototypes.

3.7 Input/Output Module and PLC for Industrial Applications

In the previous chapter, the term 'instrumentation' was introduced, which is referred to as the art of measuring and controlling via instruments. A schematic representation of the same is depicted in Fig. 3.24, which consist of three components namely sensors, controllers, and actuators required for industrial automation. Sensor is the

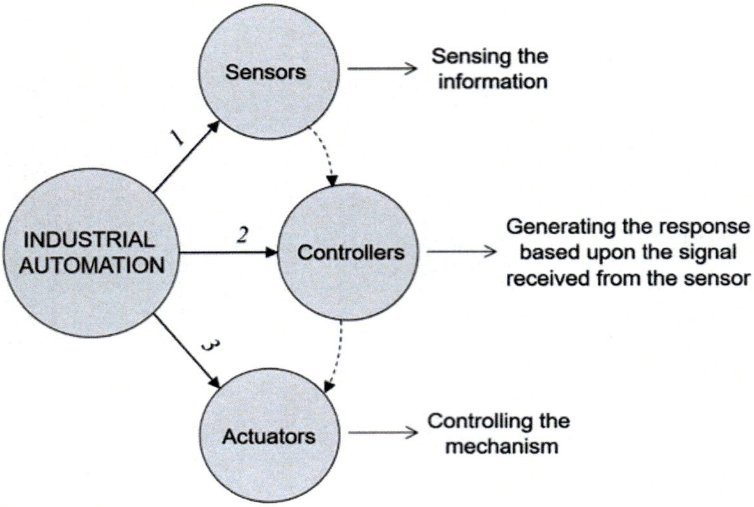

Fig. 3.24 Components of industrial automation

one that senses changes in a physical quantity. It is accompanied with a transducer that converts the information sensed by a sensor into an electrical quantity. On the other hand, an actuator is one that coverts a control signal into an action. In between a sensor and an actuator, a controller is present that receives the information from a sensor, processes it, and sends an output signal to an actuator. In this section, controllers will be discussed.

In an industrial automation, a PLC is a widely employed electronic device. In the previous section, discussion was held on a microcontroller, which is also a controller, however, it is a small computing device with less core. It is a general-purpose electronic device that can be utilized for building small prototypes. PLC is a high-level microcontroller and is specially designed for industrial applications. It has a sturdy design making itself applicable to harsh industrial environment. This is the reason that makes microcontrollers be employed for embedded system and PLC for the industrial automation.

PLC consists of three major components namely processor, input/output (I/O) module, and power supply unit as shown schematically in Fig. 3.25.

The processor is the central processing unit of a PLC that calls, executes, and stores the user-defined programs. Next is an I/O module that resembles a device through which a PLC communicates with a physical system via sensors and actuators. Most of the industrial tasks are coupled with PLC-based automation solution. A PLC operates in three distinct steps as shown in Fig. 3.26. The first step includes 'input' wherein a PLC reads data from the input module. In the second step 'program', a PLC executes the user-defined program and finally in the third step 'output', the output devices are energized or de-energized as specified in the program. For example, in a process industry, pressure is one of the crucial parameters that is monitored continuously.

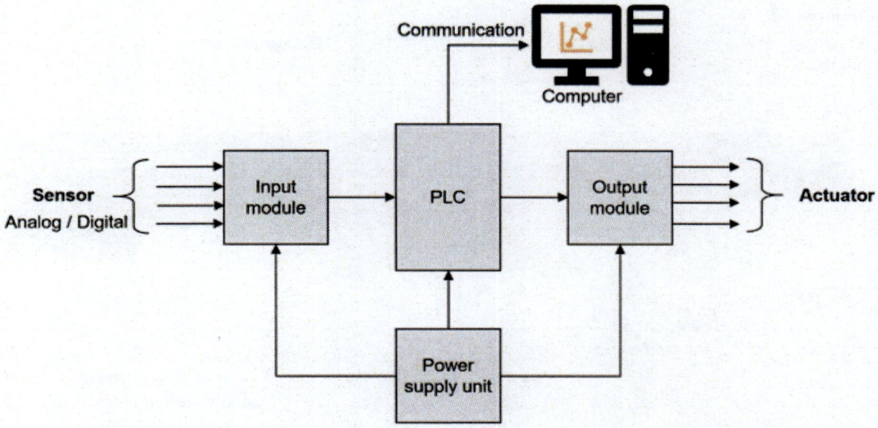

Fig. 3.25 Schematic picture of PLC and I/O module

Fig. 3.26 Steps for execution of a PLC

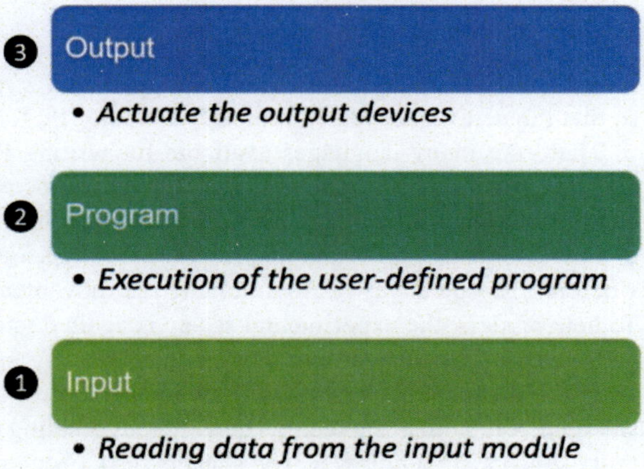

A pressure sensor collects data which is sent to a PLC via an I/O module. In this case, the PLC is programmed with a threshold value of the pressure. A threshold value, in this example, indicates a critical amount of pressure above which if the pressure rises, it might be improper for the process. Thus, the objective is to limit the pressure within the identified threshold value. Thus, during the process, if the pressure rises above this threshold value, the PLC generates a control signal that actuates a mechanism to open and release the pressure in real time. Another example could be triggering a chain of activities in a sequence, which is also achieved via programming a PLC. PLC is advantageous because it offers flexibility in terms of programming, fast in operation, reliable device, etc. PLCs originated sometime in 1968 for controlling the production activities on the shop floor. With the advent of

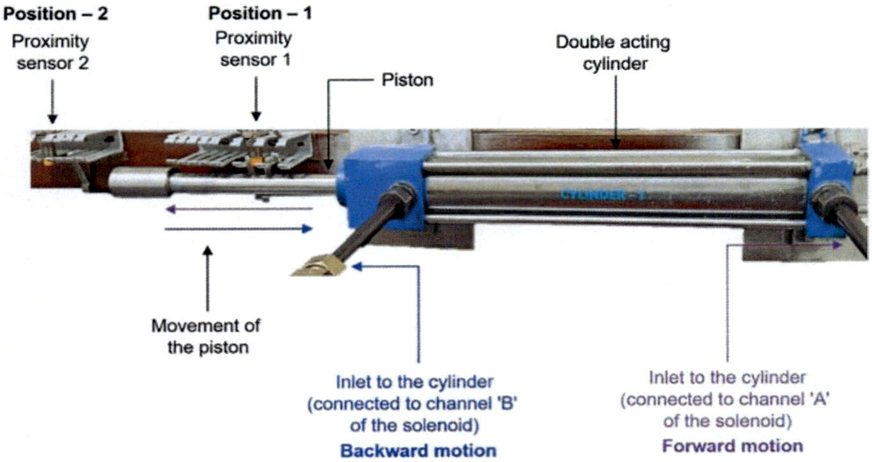

Fig. 3.27 Picture illustrating the experimental set-up for automatic actuation of a double-acting cylinder

Industry 4.0, the modern-day PLC is equipped with various communication protocols so that the data can accessed remotely enabling IIoT.

There are many languages available for writing PLC programs such as ladder diagram, function block diagram, structured text, instruction list, sequential function chart, etc. For a practical demonstration, an example is discussed to present the usage of a PLC. Consider the example presented in Chap. 2 on the use of proximity sensor, wherein, the objective was to automate the movement of a double-acting cylinder. To help readers, the experimental set-up is again depicted in Fig. 3.27.

The set up consists of a double-acting cylinder, a solenoid coil, and a hydraulic system. To switch between these movements, a user must continuously operate the solenoid coil, which is a tedious task. For automating the same, two inductive proximity sensors are employed. The sensors have been engaged as shown in the experimental set-up. The proximity sensor at 'Position 1' detects the presence of piston and then sends a signal to the solenoid coil which actuates the cylinder to move in the forward direction. Once the piston reaches the end, i.e., 'Position 2', it is detected by a second proximity sensor. The control signal sent to the solenoid coil stops the forward motion of the cylinder and actuates it to move in the backward direction. Thus, the process is automated.

This automation is enabled by using a software named "AUTOMGEN". It is a tool employed for PLC programming and designing control algorithms. Here, a ladder program is written for PLC programming. The diagram of the same is shown in Fig. 3.28. Here, once the step function is on, it checks for the position of the cylinder by checking the proximity sensors and accordingly, the signal is sent to the solenoid valve.

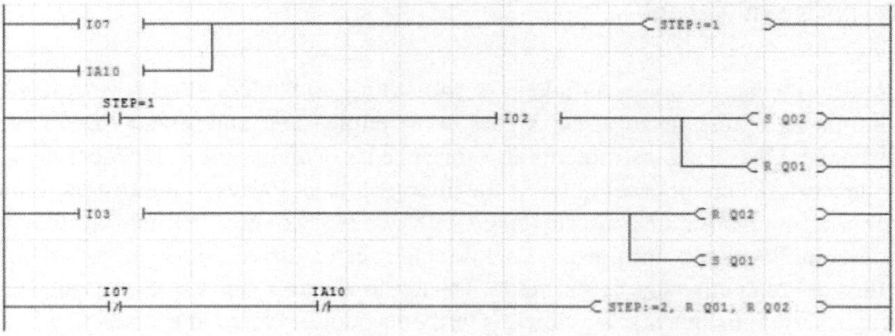

I I : NO function (Normally open) I/I : NC function (Normally closed) ⊂ ⊃ : output coil in a row

Fig. 3.28 Circuit diagram made in the AUTOMGEN software

In the program, I07 and IA10 represent two switches for starting an operation. These are physical switches. Step 1 and 2 define sequential programming. They are for starting and termination of the program. I02 and I03 are the signals that signify the switching of a proximity sensor. Here, I02 corresponds to the proximity sensor 1 that is fixed at 'Position 1', and I03 corresponds to the proximity sensor 2 that is fixed at 'Position 2'. Q01 and Q02 are the signals for switching the solenoid coil, wherein, Q01 corresponds to channel 'B' of the solenoid and Q02 corresponds to channel 'A' of the solenoid. Finally, the letters S and R written in the program indicate set and reset, respectively. Let us understand the execution of this PLC program.

As mentioned, it is a ladder program. This means that the execution of the program follows a ladder structure. As I07 is switched on, 'STEP: = 1' function is the output (row 1). In the second row, as 'STEP: = 1' function is on, it checks whether I02 is switched on. I02 will be switched on only if the piston of the cylinder is at 'Position 1' where the proximity sensor 1 is fixed. If it is the case, then Q02 is fed with a feedback signal (row 2). This feedback is a 24 V signal to switch the channel 'A' of the solenoid coil. Thus, the cylinder moves in the forward direction. As the cylinder moves in forward direction and reaches proximity sensor 2, which is fixed at 'Position 2', I03 will be switched on. This triggers a feedback signal to Q01 i.e., channel 'B' of the solenoid coil. As this program continues, the cylinder is continuously actuated. Finally, if I07 is switched off, which can be observed in the last line of the program in the shown picture, 'STEP: = 2' function is activated that resets both Q01 and Q02. Reset means sending a 0 V signal.

3.8 Noise in Signal

Generally, the term 'noise' is taken as something unwanted. It usually refers to the humming sound in an audio signal or to unpleasant/loud sound that creates disturbance. Measuring instruments also suffer because of this noise. This section will provide a brief description of noise of the instruments and ways to remove the same. The measuring instruments are attributed with two types of error namely, systematic and random. Because of the presence of systematic error, the measured value is always less than or greater than the actual value. The random error refers to the unwanted and irregular variations in a signal, which is termed as noise. Noise in a signal can be of two natures, low-frequency and high-frequency. The low-frequency noise is usually because of the variations in the object under the test. For example, during a welding process, if the sound signature is acquired, then it may get corrupted with low-frequency noise due to the variations in the materials (i.e., composition, thickness, geometry, etc.) being welded. It can also be due to the variation in deposition of the filler material on the surface of the weld joint. On the other hand, the high-frequency noise is usually the electrical noise coming from the external sources. It causes spikes in the true signal because of issues concerning electrical circuitry and electronic components involved in the circuit. The high-frequency noise could also because of the instrument and changes in the natural signal. Other factors such as over-amplification, uncontrolled gain, etc. could also cause occurrence of high-frequency noise in a signal. A signal acquired from a sensor is termed as the raw signal and it consists of both true signal and noise. This raw signal is then processed via the signal processing techniques to remove the noise and derive the true signal. The practice of removing unwanted components from a signal is referred to as 'filtering' or 'denoising'. It is always a challenging task to identify the presence of noise in a signal. Thus, noise is always estimated out of a signal. It cannot be completely removed but it can be reduced. The below-mentioned section discusses various types of filtering techniques that are employed in signal processing.

3.8.1 Median Filter

The term 'median' refers to the middle of a sorted dataset. In median filtering technique, each data point is replaced by the median value of the considered dataset. The median filter is a non-linear filtering technique utilized for removing the additive noise of impulsive nature. An important parameter in this technique is the width of the filter. To illustrate this filter, consider the signal shown in Fig. 3.29a, which is the force signal obtained during FSW process. This signal was collected by using a piezoelectric dynamometer with a sampling rate of 100 Hz. Out of the entire signal, only the welding part has been considered for the analysis. This is so because it represents the actual welding, while other parts of the signal represent the plunging,

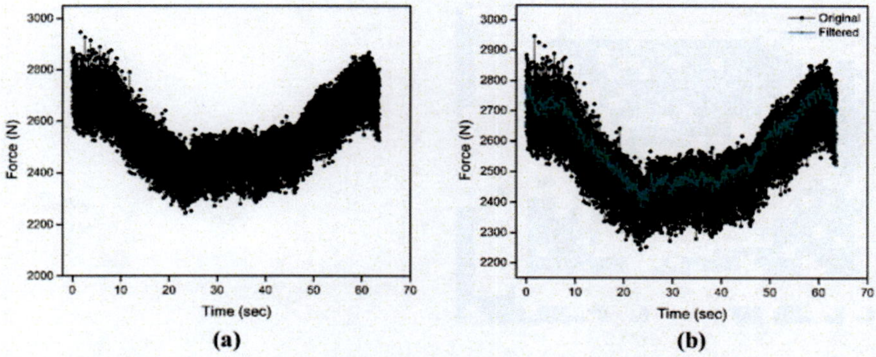

Fig. 3.29 **a** Original signal and **b** filtered signal

Table 3.2 Sample dataset

$x(n)$	1	2	3	4	5	6	7	8	9	10	11	12	13
y	2725	2814	2831	2828	2767	2660	2648	2784	2813	2884	2872	2884	2631

dwelling, and retraction of the tool. For filtering, let us select a window size of 13, just to visualize the idea.

For the above signal, let us consider the first few points, as listed in Table 3.2.

Let us try out the calculation for one data point. Say for, $x(1)$, the median value will be calculated based on the following: $x(n-1)$, $x(n-2)$, $x(n-3)$, $x(n-4)$, $x(n-5)$, $x(n-6)$, $x(n)$, $x(n+1)$, $x(n+2)$, $x(n+3)$, $x(n+4)$, $x(n+5)$, and $x(n+6)$. 13 samples are considered for $x(1)$. These values for $x(1)$ are: 0, 0, 0, 0, 0, 0, 2725, 2814, 2831, 2828, 2767, 2660, and 2648. The sorted order of this will be: 0, 0, 0, 0, 0, 0, 2648, 2660, 2725, 2767, 2814, 2828, and 2831. Here, the median value is 2648. This median value will replace the original value. In this way, for each value, the median value is calculated as per the window considered, and original value are replaced.

Further, let us visualize the same filtering technique with one another example. The force signal considered above is that of a defect-free weld, which was identified after the welding was done. The welded sample did not have any surface defect. Let us consider a bad weld and its corresponding force signal. The signal of a defective weld is chosen intentionally to illustrate the usefulness of filtering a signal. The raw force signal acquired for the welded sample is depicted in Fig. 3.30a. Recap the force signals acquired in the FSW process that are shown in Chap. 2, Sect. 3.2 of this chapter, and Fig. 3.29a. A clear difference in the plot could be observed between those signals and the one considered here (Fig. 3.30a). High fluctuations can be seen in the force signal. This is because of the improper combination of the input parameters selected for welding that has failed to fuse the material well, and the same is reflected in the force signal.

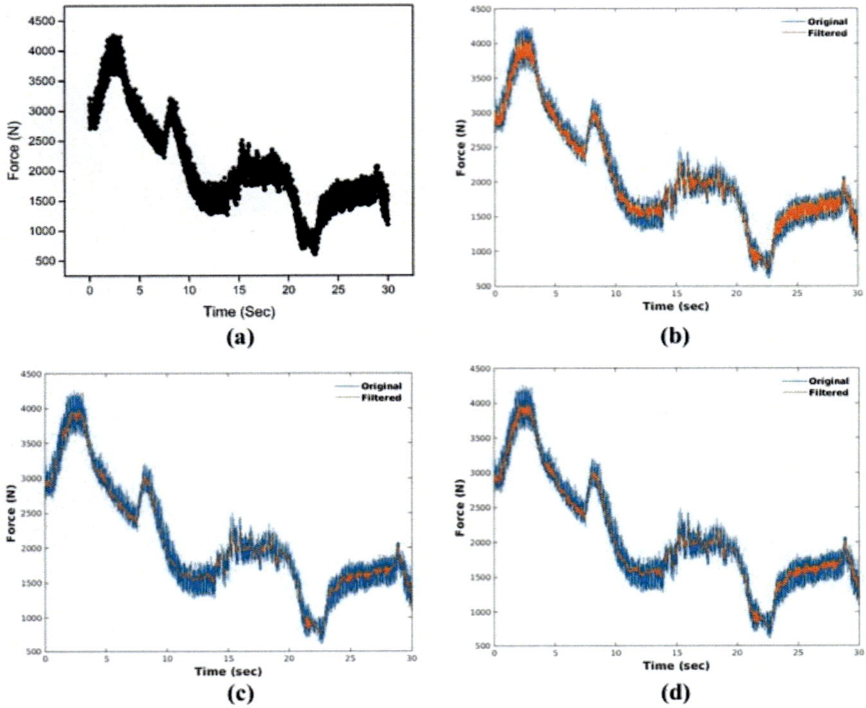

Fig. 3.30 Picture of: **a** Original signal and **b–d** filtered signal with order = 5, 9, and 13, respectively

Figures 3.30b–d show the filtered signals. For this, different window sizes of the median filter are considered. First is an order of 5, second is an order of 9, and third with an order of 13. Variations in the signal are more prominent in case of the filtered ones, which is important for manufacturing. This shows that the feature of the true signal is getting hidden by the addition of noise. Thus, filtering the signal is important. With order of 5, i.e., window size = 5, variations in a few portions can be observed, which were not prominent in case of the original raw signal.

Median filter is a non-linear filter. It is taking the neighbourhood data samples, calculating the median value, and utilizes the same to replace a sample of a signal. From the filtered signals shown in Fig. 3.30b–d, it can be observed that with increasing order of the median filter, the filtering effect also increases. Let us list a few important things. As a median filter employs neighbourhood data samples and calculates the median value, it does not introduce any new data point. It utilizes the existing samples in a signal for smoothing. Thus, it is preferred for removal of impulse type of noise or to remove sudden fluctuations of a signal. Though the filtering effect is increasing with the order of the filter, it loses a few details. Besides these, an advantage of employing median filter is that it preserves the edge information of a signal. Figure 3.31 shows the filtered signal and a picture of the welded sample for better visualization of the

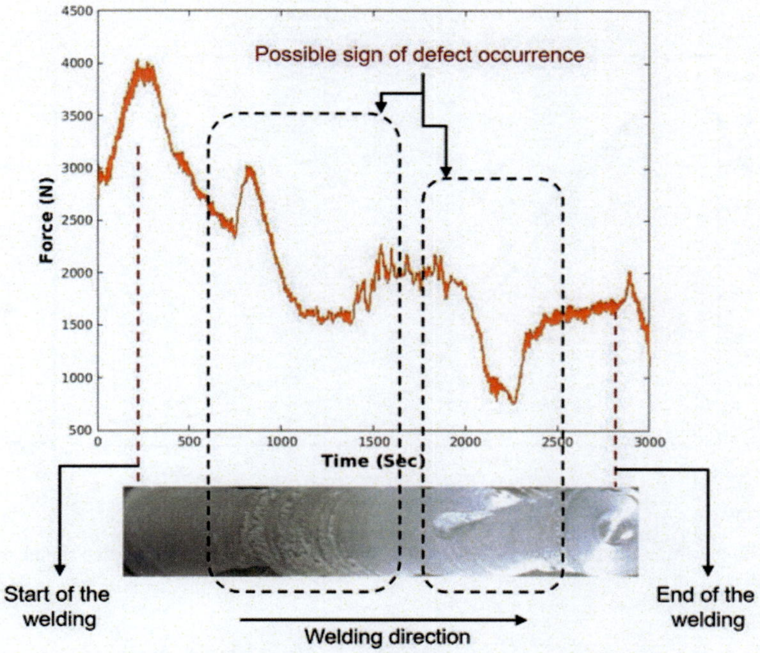

Fig. 3.31 Mapping of the filtered signal with physical sample

signal. The start and end positions of the welded sample have been marked in the figure. A few portions of the welded sample can be seen to be having defects. The corresponding portions of the signal are having sudden fluctuations which are the possible signs of defect occurrence. Of course, filtering is not enough. Further signal processing would be required to map the defects with the fluctuations in the signal and validate it. With filtering done, unwanted components have been removed.

For comparing different orders considered for the median filtering, certain performance measuring features can be selected such as, signal to noise ratio (SNR), root mean square error (RMSE), correlation coefficient (CC), etc. A higher value of SNR depicts the efficiency of a filtering function in denoising a signal while a lower value of RMSE supports the same and CC measures how well the denoised signal follows the original signal; so, a larger value is desirable. The way to computation of these features is shown in Eqs. 3.4–3.6. With these features, the optimum filtering function for a noisy signal can be obtained.

$$\mathrm{SNR} = 10\log_{10}\frac{\sum_{i=1}^{N} x^2(i)}{\sum_{i=1}^{N} |x(i) - x_{\mathrm{den}}(i)|^2} \tag{3.4}$$

$$\mathrm{RMSE} = \sqrt{\frac{1}{N}\left\{\sum_{i=1}^{N} [x(i) - x_{\mathrm{den}}(i)]^2\right\}} \tag{3.5}$$

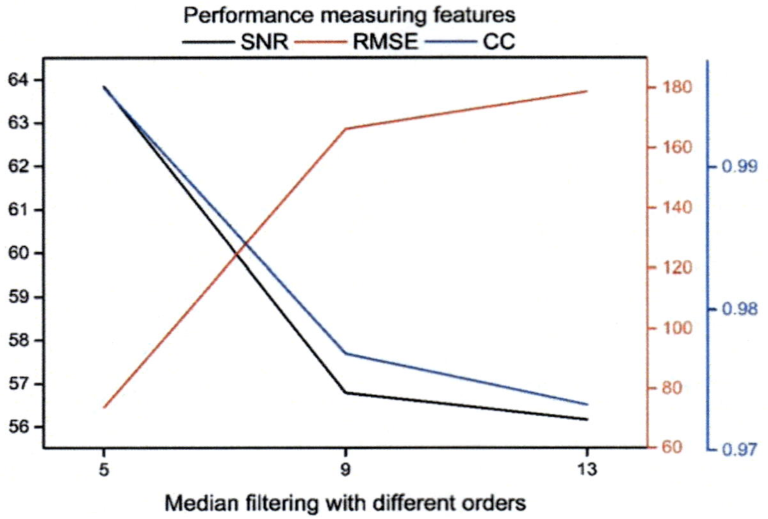

Fig. 3.32 Performance for different orders of median filter applied on to the force signal shown in Fig. 3.30a

$$r = \frac{\sum_{i=1}^{N}(x_i - \overline{x})(x_{\text{den}} - \overline{x_{\text{den}}})}{\sqrt{\left\{\sum_{i=1}^{N}(x_i - \overline{x})^2 \sum_{i=1}^{N}(x_{\text{den}} - \overline{x_{\text{den}}})^2\right\}}} \tag{3.6}$$

where SNR denotes signal to noise ratio, x is the original signal, x_{den} is the denoised signal, RMSE denotes root mean square error, r is the correlation coefficient, and N is the number of samples of the signal. For the orders of median filtering considered for the force signal, the performance measuring features are evaluated and the results are depicted in Fig. 3.32. From the plot, it could be observed that with order = 5, a higher value of SNR and CC is obtained with lower RMSE. Thus, it could be considered as the optimum for this signal.

3.8.2 *Gaussian Filter*

Gaussian filter is a low-pass filter, which means it can be utilized for removing the high-frequency components. The Gaussian distribution is represented in Eq. 3.7 where, σ is the standard deviation of the distribution.

$$G(x) = \frac{1}{\sqrt{2\pi}\sigma} e^{-\frac{x^2}{\sigma^2}} \tag{3.7}$$

The nature of the distribution is so that it has a zero-mean value. The Gaussian filter is a linear low-pass filter in the frequency domain, which convolves a Gaussian function over a signal. It is effective in removing the Gaussian noise. The degree of smoothing is controlled by the standard deviation (σ). It can be observed that the nature of the signal is dependent on the value of σ, i.e., higher the value of σ, smoother will be the distribution of the filter. Thus, the extent of filtering a signal will depend upon the choice of σ. Figure 3.33 shows the Gaussian distributions for various values of σ such as 10, 20, and 30. The variations amongst the distributions can be observed from the plots i.e., increase in the flatness, which eliminates the high-frequency components from a signal.

To illustrate the filtering action of the Gaussian filter, the same was applied to the force signal, which has been considered during the illustration of median filter (depicted in Fig. 3.30a). Figure 3.34 shows the filtered signals obtained after applying Gaussian filter. The effect of increasing σ can be observed from these plots.

Upon comparing these plots with those obtained after median filtering, it can be observed that the edges of the signal are still steep, which occurs because of the inclusion of a median value of the considered window in the case of median filtering. However, in the case of Gaussian filter, the signal is smoother. This is due to the nature of the Gaussian distribution where the weights are reducing radially. For the considered signal and corresponding Gaussian distributions, the performance measuring features are evaluated and the results are depicted in Fig. 3.35. From the plot, it could be observed that with $\sigma = 30$, a higher value of SNR and CC is obtained with lower RMSE. Thus, it could be considered as the optimum for this signal.

3.8.3 Wavelet-Based Filtering

Another popular method utilized for denoising is referred to as the transform based, where the signal is transformed into frequency domain and a threshold is applied on the low frequency components to create a sparse matrix [11]. Generally, the high frequency components in a signal represent certain abrupt changes or abnormalities which are of major importance. Hence, even after applying threshold on the redundant low frequency components, the portions of signals which are of major importance are retained. Out of the many techniques involved in the transform-based method, denoising using wavelets is one of the popular ones owing to its robustness in analyzing non-stationary signals. Wavelets can be defined as waves which exists for a limited duration of time, have an irregular shape, and a zero-mean value [12]. The wavelet analysis generates approximation and detail coefficients, and by nature, the detailed coefficients are sparse, which is the primary reason for its huge range of applications in the field of compressive sensing [13]. In this section, explanation will be confined to the use of denoising only. Detailed discussion on wavelet transform will be done in a later section.

The process of denoising by using wavelets involves three distinct steps namely, decomposition, alteration of the wavelet coefficients, and reconstruction as depicted

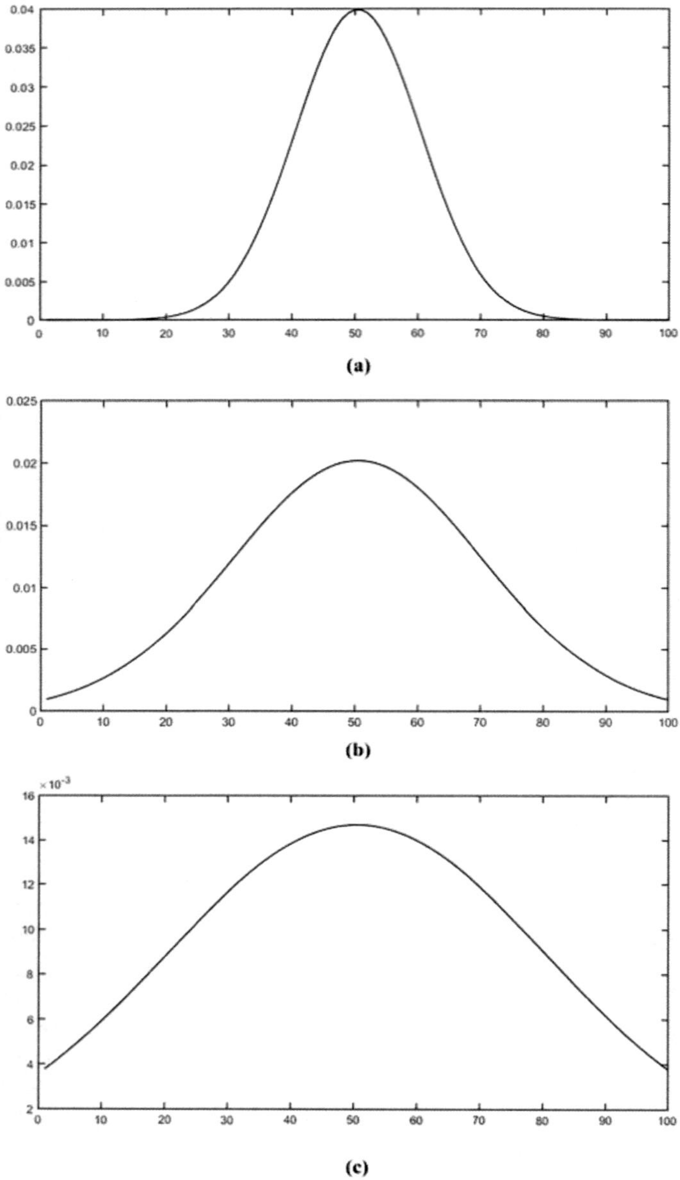

Fig. 3.33 Gaussian distribution for different values of σ: **a** 10, **b** 20, and **c** 30

Fig. 3.34 Gaussian distribution for different values of σ: **a** 10, **b** 20, and **c** 30

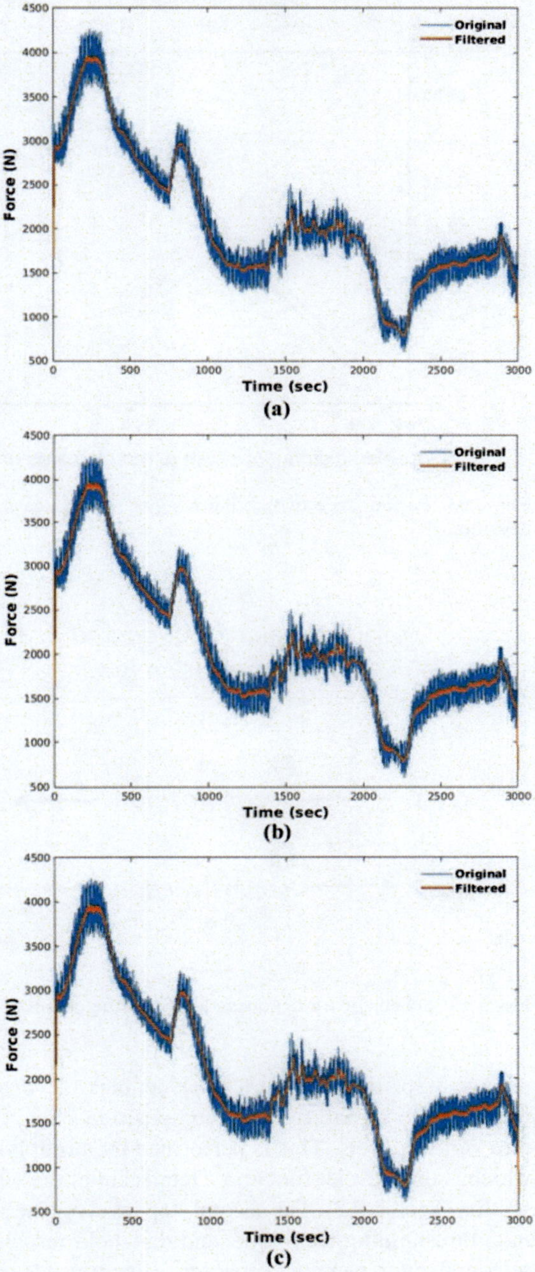

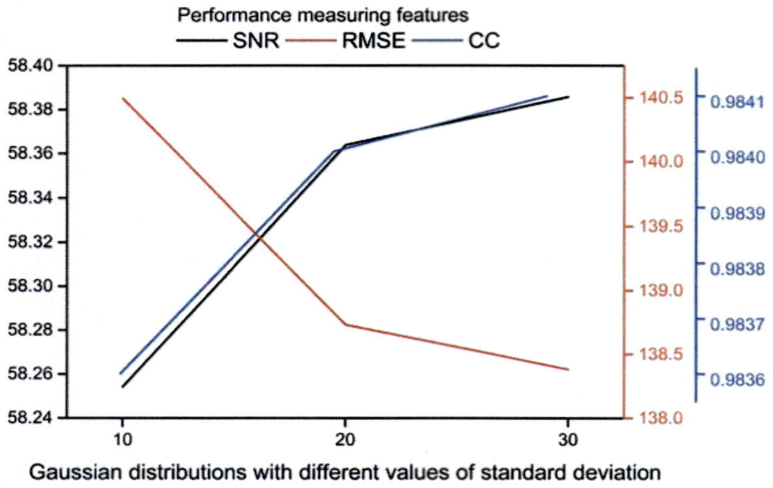

Fig. 3.35 Performance of filtering a signal with Gaussian distributions with different standard deviations

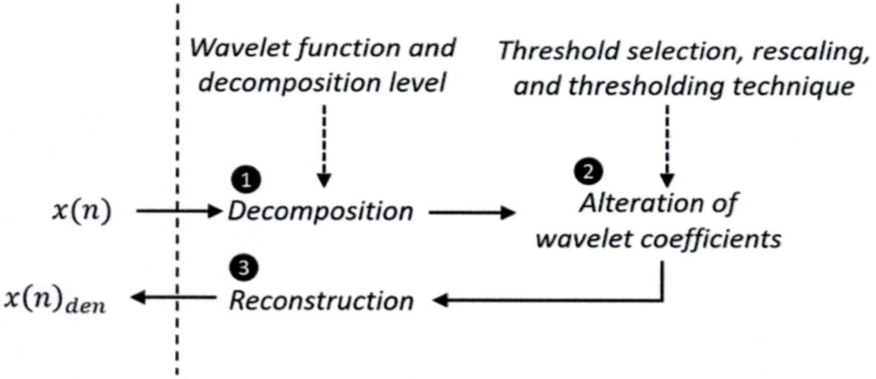

Fig. 3.36 Schematic for denoising a signal using wavelet

in Fig. 3.36. In the figure, $x(n)$ denotes the original signal and $x(n)_{\text{den}}$ denotes the denoised signal. In the decomposition stage, the original signal is decomposed into various levels. This is performed by multiplying the signal with a basis vector, which is the wavelet function. Details about the wavelets will be discussed in a later section (Sect. 3.9.3). The second step involves the selection of the threshold selection rule, thresholding technique, and threshold rescaling. There are different threshold selection rules namely, *Sqtwolog, Rigrsure, Heursure, and Minimaxi.* These rules make use of the statistical regression of the noisy coefficients over time to obtain a non-parametric estimation of the reconstructed signal without noise. *Sqtwolog* method has a fixed threshold value and *Rigrsure* uses the Steins' unbaised estimate of risk and determines the threshold value. *Heursure* is a combination of the first and the second methods, and *Minimaxi* used a fixed form of threshold value to give minimax performance for mean square error. The thresholding techniques can either be soft or hard. Finally, reconstruction is performed.

To illustrate this filtering, the force signal shown in Fig. 3.30a is considered which has also been considered for illustrating the use of median and Gaussian filtering techniques. Here, for the sake of example, three different wavelet functions are considered namely *Daubechies*, *Symlet*, and *Coiflet* to showcase the effect on filtering of the signal. For this purpose, the other factors considered were the *Heursure, soft thresholding, multi-level scaling,* and two levels of decomposition for each wavelet function. Figure 3.37 show the filter in action, wherein, (a) is for filtering with *Daubechies*, (b) is for *Symlet*, and (c) is for *Coiflet*.

Wavelet-based denoising is critical because of the parameters involved. One important parameter is the selection of the mother wavelet function. This is so because the wavelet function is the one that will approximate the original signal and capture the transient characteristics hidden in the signal. In addition, it also affects the frequency spectrum. Similarly, as shown in the schematic diagram for wavelet-based denoising, other factors are to be chosen carefully. Let us gather a few observations on the obtained signals. After applying median and Gaussian filters, the filtered signals were smoother, however, the ones obtained after applying wavelets have large fluctuations. This is so because in wavelet denoising technique, the wavelets coefficients are obtained by decomposing the original signal employing a wavelet function. A wavelet function localizes a signal in both time and frequency domains. From Fig. 3.37, it could be observed the information at the edges is well retained as compared to the median and Gaussian filtering techniques. Also, the wavelet filtering technique is effective in capturing the transient phenomenon in the signal. This is where the wavelets are more useful. Things such as transient events, time and frequency domains, and wavelet functions will be clearer in the discussions to be held in the following section (Sect. 3.9).

For the wavelet functions considered in this example, the performance features are evaluated and the results are depicted in Fig. 3.38. From the plot, amongst all the wavelet functions, with *Coiflet*, SNR is high, RMSE is low, and CC is also high. Thus, in this case, *Coiflet* wavelet function can be the optimum for the considered application. However, it should be noted that this is not a generalized scenario and can differ for varying applications. Thus, analysis is required to be done. One reason for the better results obtained with *Coiflet* wavelet function could be attributed to its symmetry property, whereas *Daubechies* is asymmetric and *Symlet* is near-symmetric.

In the filtering section, three types of filters were discussed, median, Gaussian, and wavelet-based. For each of the techniques, a real-life signal has been taken, i.e., the force signal captured in FSW process. The same signal was filtered with the discussed techniques. Further, to compare amongst these techniques, let us select the optimum ones from the individual techniques, which are: order = 5 in case of median filtering, $\sigma = 30$ for Gaussian filtering, and *Coiflet* as the base wavelet function for

Fig. 3.37 Denoised signals: **a** *Daubechies*, **b** *Symlet*, and **c** *Coiflet* wavelet functions

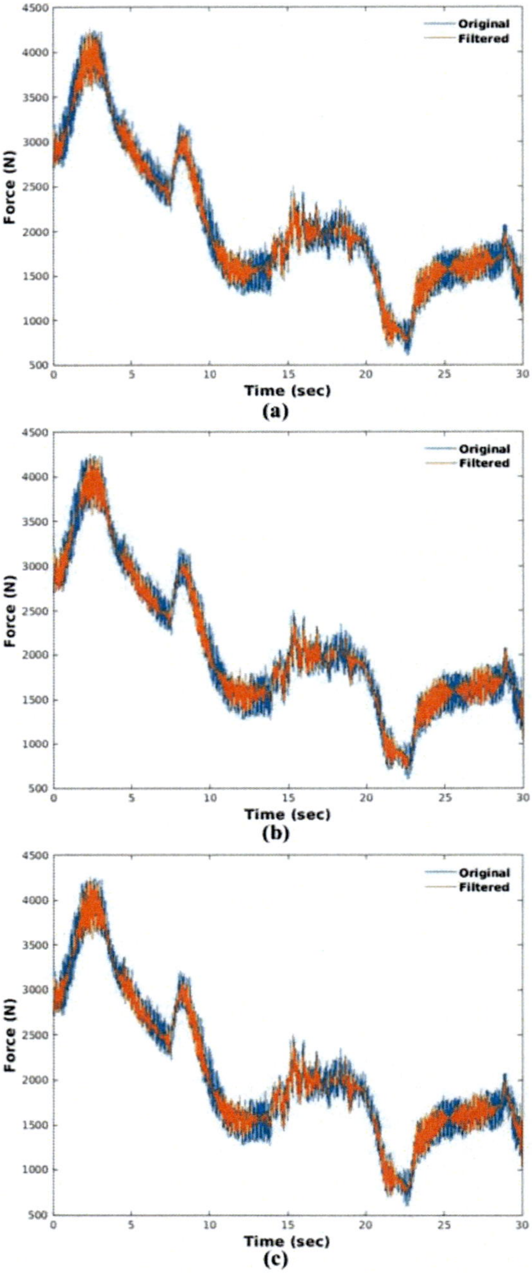

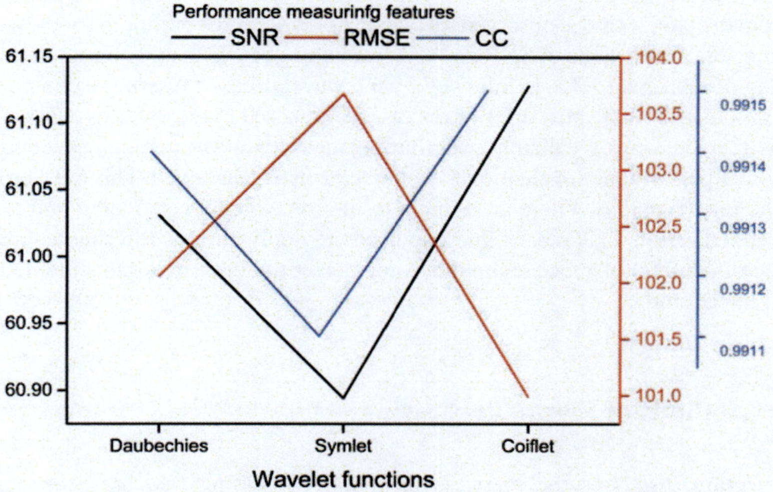

Fig. 3.38 Comparative study of different wavelet functions used in denoising signal

wavelet-based filtering. Here, the comparison takes the three performance evaluation measures and the time of execution, which is crucial for the real-time application and creation of the digital twin. The results obtained are depicted in Fig. 3.39.

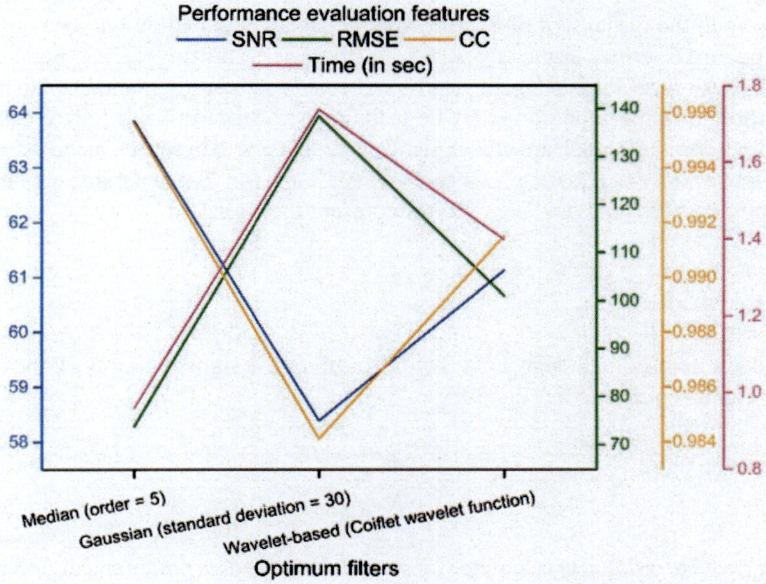

Fig. 3.39 Comparative study of considered filtering techniques for the example force signal

From the plot, it can be observed that the features are most optimum for the median filtering, i.e., a high value of SNR and CC, low value of RMSE, and minimum time for filtering of the signal. This is followed by the wavelet-based filtering technique, and Gaussian distribution show inferior results as compared to the other two techniques. This is a typical way of realising a few filtering methods and determining the optimum one. Other performance evaluation measures can also be involved. This must be noted that the results obtained here are typical to the force signal considered and is not a generalized result. This means that one needs to apply various filtering techniques, evaluate the effect of those techniques, and select the optimum one suitable to the signal under study.

3.9 Methods of Signal Processing

This section discusses the methods of signal processing that are time domain, frequency domain, and time–frequency domain analyses. For each of the method, the advantages and limitations are also discussed. At the end of each section, case studies are demonstrated in details for getting insights on the use of that method.

3.9.1 Time Domain Analysis

Referring to the discussion of Sect. 3.5 on the signal acquisition, the data obtained from the card is termed as the raw signal or time-series. This time-series signal is also termed as the time domain signal. This is so because the representation of this signal is performed in the time domain. The techniques that are utilized for the analysis of a time-domain signal are the statistical parameters. These are mean, standard deviation, root mean square, skewness, kurtosis, etc. [14]. These parameters become the features of a signal, and are used to represent the signal.

3.9.1.1 Mean

Consider a time-domain signal, $x(t)$ with length of the signal being N, $\overline{x}$ the mean, which can be computed as (Eq. 3.8):

$$\overline{x} = \frac{1}{N}\sum_{i=0}^{N-1} x_i \tag{3.8}$$

The mean of a signal is calculated by summing all the data points and dividing it by the total number of data points. It is a measure of the central value of a signal. It also referred to as a measure of the location of a signal. Often the term DC in electronics refers to the direct current and it is the mean value. AC refers to the alternating current and it depicts how a signal fluctuates around the mean value. As the real-life signals are random, a more generalized parameter named standard deviation is used.

3.9.1.2 Standard Deviation

The standard deviation of a signal can be computed using Eq. 3.9.

$$\sigma = \sqrt{\frac{1}{N-1} \sum\nolimits_{i=0}^{N-1} (x_i - \overline{x})^2} \tag{3.9}$$

where σ represents the standard deviation and $\overline{x}$ is termed as sample mean. In this equation, each sample of the signal is being subtracted from the mean value of that signal. This means the deviation of each sample is being calculated from the mean value. Thus, the average deviation is determined by summing the individual deviations and dividing the output by the total number of data points. This output will be the average deviation. In case of signals, the deviation in power is important rather than the deviation in the amplitude. Thus, the interest lies in calculating the power which occurs because of the deviation from the mean value. To achieve this, the squared term is there in the equation. If the square root will be removed from the equation, it will square the term in the left-hand-side of the equation, and it is then referred to as "variance", which indicates the power of the fluctuation. Otherwise, the square root compensates for the presence of the squared term and is referred to as standard deviation. It is a measure of the variability in the signal.

Another thing to note is the denominator in the equation i.e., $(N-1)$. The point is that the features such as standard deviation and sample variance attempt at estimating a signal, and thus, they underestimate the actual parameters of the signal. In other words, every signal is having some statistical noise that gives rise to error. This error is given as depicted in Eq. 3.10.

$$\text{Err} = \frac{\sigma}{N^{1/2}} \tag{3.10}$$

where Err is the error and σ is the standard deviation of the signal. If the value of N is large, then the error will be small. But if N is small, the error will be large. This indicates the importance of data and states that it is hard to comment on the process with fewer samples. Ideally, Eq. 3.11 should have been employed for determining the standard deviation.

$$\sigma = \sqrt{\frac{1}{N} \sum\nolimits_{i=0}^{N-1} (x_i - \mu)^2} \tag{3.11}$$

where μ is the mean of the signal. It should be ideally employed in place of sample mean ($\overline{x}$) as shown in Eq. 3.8. Since $\overline{x}$ is the sample mean, thus, the value would be close to μ. Hence, comparing Eqs. 3.9 and 3.11, the output will be larger in Eq. 3.11 than Eq. 3.9. This indicates that the standard deviation, which is being referred in Eq. 3.9 underestimates the true standard deviation of the signal. To compensate for this error, N is replaced by $N-1$, to make the standard deviation a little bigger. This correction is termed as Bessel's correction. Returning to the discussion, the intention is to determine the standard deviation in the process. To do so, the sample mean needs to be determined first.

3.9.1.3 Root Mean Square

The root mean square (RMS) value of a signal can be computed using Eq. 3.12.

$$\mathrm{RMS} = \sqrt{\frac{1}{N}\sum\nolimits_{i=0}^{N-1}(x_i)^2} \tag{3.12}$$

Here RMS represents the root mean square. This parameter also quantifies the energy of a signal. As discussed in the preceding section, σ measures the deviation of the signal from the mean value. Thus, it only measures the AC portion of the signal. The RMS value quantify the energy both for AC and DC portions. Consider Eq. 3.9, if the DC portion becomes zero, the equation is like Eq. 3.12.

3.9.1.4 Skewness

The skewness (SKEW) of a signal can be computed using Eq. 3.13.

$$\mathrm{SKEW} = \frac{1}{N}\sum_{i=1}^{N}\frac{(x_i-\overline{x})^3}{\sigma^3} \tag{3.13}$$

SKEW refers to the skewness of a signal. After referring to the location and variability measures i.e., mean and the standard deviation, the symmetry of the signal is studied next. For this purpose, the parameter skewness is calculated. It shows whether a signal is skewed towards right or left. One can visualize the distribution of a signal around the centre-point and identify the symmetry in the signal.

3.9.1.5 Kurtosis

The kurtosis (KURT) of a signal can be computed using Eq. 3.14.

$$\mathrm{KURT} = \frac{1}{N}\sum_{i=1}^{N}\frac{(x_i - \overline{x})^4}{\sigma^4} \tag{3.14}$$

KURT refers to the kurtosis of a signal. It is also a statistical parameter that indicates the sharpness of a signal. Often it is referred to as the peakedness and it is the fourth moment of a signal about its mean to the fourth power of its standard deviation, as shown in Eq. 3.15.

$$\mathrm{KURT} = \frac{\mu_4}{\sigma^4} \tag{3.15}$$

where μ_4 denotes the fourth moment and σ is the variance. The kurtosis value of a normal distribution is 3. A normal distribution is one that has a symmetrical distribution of the data resembling a bell shape. Equation 3.16 shows how the kurtosis value for a normal distribution is 3. This is also referred to as the "excess kurtosis", it is the actual kurtosis of a normal distribution subtracted from 3. This makes the kurtosis of a normal distribution be 0.

$$\mathrm{KURT} = \frac{3\sigma^4}{\sigma^4} = 3 \quad \left(\because \mu_4 = 3\sigma^4\right) \tag{3.16}$$

The following paragraph presents a case-study for the application of kurtosis as a parameter for process evaluation.

3.9.1.6 Real-Life Examples in Manufacturing

The following paragraph presents case-study for the application of time-domain parameters for process evaluation.

Let us consider the case of FSW process in which the axial force signal is collected as a measure for process evaluation. Figure 3.40 depicts the force plot for different values of tool rotational speed, i.e., 600, 1000, 1400, and 2600 rpm. At the beginning of the plot, a sharp rise in the force amplitude can be observed. This depicts the start of welding and the amplitude is high because of the interaction of the welding tool with the cold material. Rotational speed is one of the welding parameters upon which the amount of frictional heat availability depends. Other parameters, in this case, were kept constant. It should be noted that the force variation during the welding phase is only shown here.

To utilize the acquired force signal to evaluate the process, time-domain analysis is performed, and mean of those signals is extracted. Figure 3.41 shows the mean value of the four force signals depicted in Fig. 3.40. The mean value can be observed to be the highest for the force signal collected during welding with 600 rpm, followed by 1000, 1400, and 2600 rpm. As the rotational speed increases, the interaction time between the tool and the material increases. This means the frictional heat is more during welding with 2600 rpm and it decreases as the speed goes down to 600 rpm.

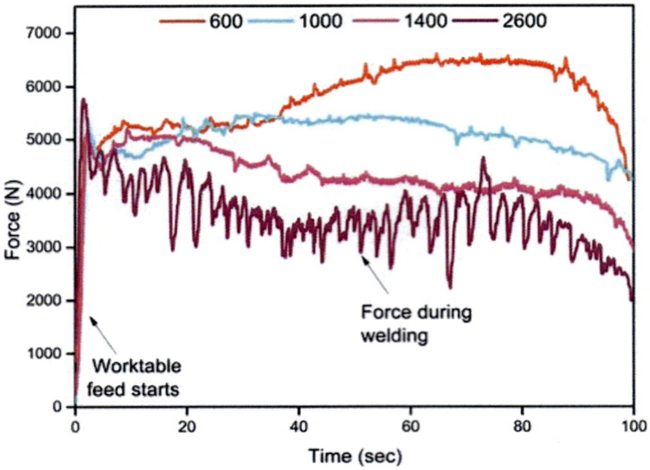

Fig. 3.40 Force signal during welding for different values of tool rotational speed

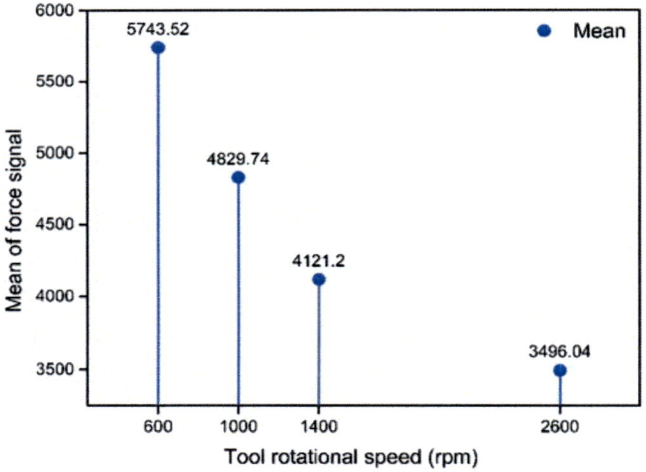

Fig. 3.41 Mean value of force signals

Thus, mean here is an indication of the power of the signal. The obtained mean values can be utilized for making decisions.

A case of a gas metal arc welding (GMAW) technique is considered, where the objective is to monitor the welding defect [15]. The signal being utilized for the same is the arc sound because it has been found useful in providing information related to the molten metal, arc column, and the metal transfer modes [16–19]. During arc welding, arcing occurs and the phenomenon has a typical sound. This is followed by the metal transfer which also generates sound. For this purpose, the arc sound signal is collected during the welding. The welded sample along with the sound signal is shown in Fig. 3.42 [15]. Kurtosis of the sound signal is extracted as a feature and the kurtosis map is also shown in the figure. The weld bead surface image is divided into four regions, in which, the "region 3" consists of a blow-hole defect. The corresponding signal is also divided into regions and the differences can be observed

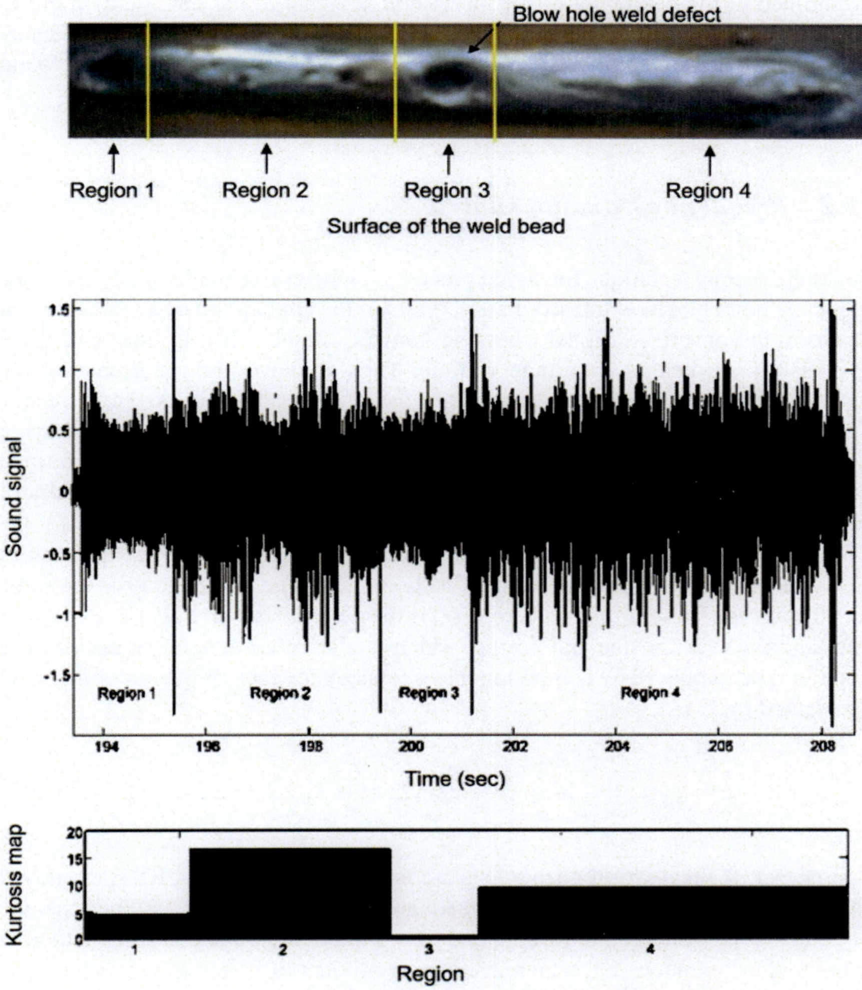

Fig. 3.42 Mapping of sound signature and kurtosis map with weld image [15]. Reprinted from Journal of Material Processing Technology, Volume 210, Issue 10, Kamal Pal, Sandip Bhattacharya, Surjya K. Pal, Investigation on arc sound and metal transfer modes for on-line monitoring in pulsed gas metal arc welding, 1397–1410, 2010, with permission from Elsevier

from the kurtosis map. The kurtosis value is significantly less for the region with the defect as compared to other regions of the weld bead. Thus, kurtosis of the sound signal is an effective indicator of the blow-hole defect.

3.9.1.7 Limitation of Time Domain Analysis

As outlined in the beginning of the discussion, a signal in the time domain represents the occurrence of events with respect to time only. For identification of many criticalities in manufacturing, the frequency information is vital, which is not found in the time domain signal. This requires processing signals in other domains.

3.9.2 Frequency Domain Analysis

This is the second technique for signal processing wherein a signal is analyzed in the frequency domain. Here, transformation refers to the application of a mathematical transform to convert the signal from one domain to other [20]. In time domain, a signal is represented as amplitude vs time. Thus, it shows the occurrence of the events in a process with respect to the time. However, many-a-times, the frequency of those events occurring in a process is of higher importance. But the time domain signal does not contain the information about the frequency. Thus, it is transformed via a technique termed as Fourier transform (FT) to obtain the frequency content of a signal. The obtained result is termed as frequency spectrum.

The Fourier series is employed for the periodic phenomena and is expressed as the weighted sum of sines and cosine functions [21]. The expression for the same can be observed from Eq. 3.17 where, $f(t)$ is the periodic function, $\frac{1}{2}a_0$ is a constant term, and the rest are sine and cosine, which is also referred to as sinusoids. The function is decomposed by computing the expression for each frequency component represented by k.

$$f(t) = \frac{1}{2}a_0 + \sum_{k=1}^{\infty}(a_k \sin 2\pi kt + b_k \cos 2\pi kt) \tag{3.17}$$

However, the real-world phenomena are not periodic, and thus, FT is employed which refers to an integral taken over the whole domain. It is thus assumed that the function is represented on the whole real axis. Equation 3.18 depicts the method to compute the continuous FT, which is the inner product of a time series with a basis function as shown in the equation.

$$\begin{aligned} X(F) &= < x, e^{i2\pi Ft} > \\ \Rightarrow X(F) &= \int_{-\infty}^{\infty} x(t)e^{-i2\pi Ft}dt \end{aligned} \quad (3.18)$$

where $x(t)$ is the time domain signal, $X(F)$ computes the FT of the signal by decomposing it to complex exponential functions of various frequencies, t indicates time, and F indicates the frequency of the signal. The inner product computes the similarity between $x(t)$ and another function ($e^{i2\pi Ft}$). So, the inner product will result in a higher value if the degree of similarity between the two functions is high, otherwise, it will result in a low value. FT is built upon the Fourier series which decomposes a function by using the sinusoids as the basis functions. As discussed, in the case of FT, the sinusoids are the basis function. Thus, the frequency content will be higher if the signal matches the frequencies of the sinusoids and will be lesser if it does not match. Since, most of the signals are digitally sampled instead of being continuous, an adapted version of FT is employed, which is termed as the Discrete FT (DFT). Equation 3.19 depicts the method to compute the DFT [22]. The frequency spectrum is the representation of the signal in the frequency domain, and it provides information on the phase shift related to the sinusoids and their frequencies of the signal under consideration. In simple terms, the time-domain signal is multiplied with the basis vectors for transformation. In real-world scenario, a signal is collected in discrete time intervals. The discrete signal is represented as x_k, and the FT for such a signal can be computed as shown in Eq. 3.19.

$$X_k = \frac{1}{N}\sum_{n=0}^{N-1} x_n e^{\frac{-j2\pi kn}{N}} \quad (3.19)$$

where, x_n is the time domain signal with n being the sample number, k being the wave number, and N indicates the total number of samples in the signal. Comparing this expression with the one computing continuous FT, here also the frequency components in a signal are evaluated, however, instead of running the integral over the entire real domain, the same is being determined at discrete points, i.e., from $n = 0$ to $N - 1$. Thus, the expression determines the frequency bins (k) for each sample number (n). Also, the term F in case of the continuous FT is replaced here by $\frac{k}{N}$ and (n) corresponds to t. The total number of samples in a signal is equal to the time period of a signal divided by the sampling rate. For instance, if a signal is sampled at 0.1 s (Δt) for a time period of 10 (T) seconds, then the total number of samples in the signal will be 100 as per Eq. 3.20.

$$N = \frac{T}{\Delta t} \quad (3.20)$$

Let us expand the expression of discrete FT for a better understanding of the frequency spectrum. The same is represented in Eq. 3.21 where, $\frac{2\pi kn}{N}$ is considered as b_n.

$$X_k = \frac{1}{N}\left(x_0 e^{-b_0 j} + x_1 e^{-b_1 j} + x_2 e^{-b_2 j} + \cdots + x_{N-1} e^{-b_{N-1} j}\right) \tag{3.21}$$

To further expand this above equation, the Euler's formula is utilised, which establishes the fundamental relationship between a complex exponential function and trigonometric functions. The formula is shown in Eq. 3.22 for a real number x.

$$e^{jx} = \cos x + j\sin x \tag{3.22}$$

Accordingly, Eq. 3.22, can be re-written as follows (Eq. 3.23):

$$X_k = \frac{1}{N}\begin{pmatrix} x_0[\cos(-b_0) + j\sin(-b_0)] + x_1[\cos(-b_1) + j\sin(-b_1)] \\ +x_2[\cos(-b_2) + j\sin(-b_2)] + \ldots + x_{N-1}[\cos(-b_{N-1}) \\ +j\sin(-b_{N-1})] \end{pmatrix} \tag{3.23}$$

After simplifying the above expression, the outcome is a complex number as represented in Eq. 3.24.

$$X_k = A_k + jB_k \tag{3.24}$$

Let us now visualize the importance of this complex expression. The real and imaginary components obtained by computing the DFT is pictorially depicted in Fig. 3.43 by drawing them in a complex plane. The magnitude and phase angle can be obtained as shown in the expressions in the figure.

3.9.2.1 Interpretation of the Frequency Spectrum

Figure 3.44 shows the plot of a force signal acquired during FSW process. In the previous chapter, FSW process has been introduced and also the importance of force behaviour in the welding process is highlighted. The welding parameters were: rotational speed of the tool was 1120 rpm and welding speed was 80 mm/min. The data was acquired at a sampling rate of 100 Hz. From this plot, the events occurring during the welding such as the start of the plunging of the tool, contact of pin and shoulder and the start of the welding phase can be observed. This is the characteristic of the time domain signal. However, the frequency content of the signal is unknown.

To transform the above signal into the frequency domain, fast Fourier transform (FFT) is employed. FFT is a faster way to compute the DFT of a time series. Figure 3.45 shows the FFT plot of this signal. For the aforementioned time series data, the fundamental frequency can be determined by dividing 1120 (rotational speed in rpm) with 60 (to determine the rate per second), which is ~19 Hz. In the FFT plot,

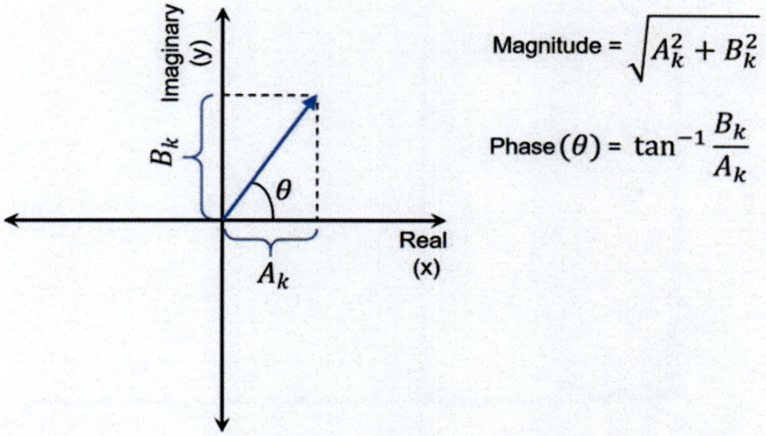

Fig. 3.43 Representation of real and imaginary components obtained from DFT

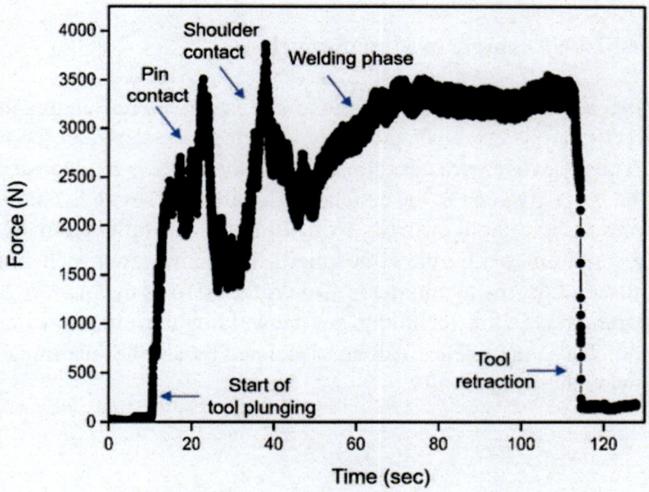

Fig. 3.44 Plot of a raw force signal versus time

a peak can be observed at 19 Hz, which indicates the fundamental frequency of this signal. Other peaks could also be observed which indicates the harmonics of the fundamental frequency. This is how the FFT plot can be interpreted and can be utilized to look for frequencies other than the fundamental one and its harmonics along with their amplitudes.

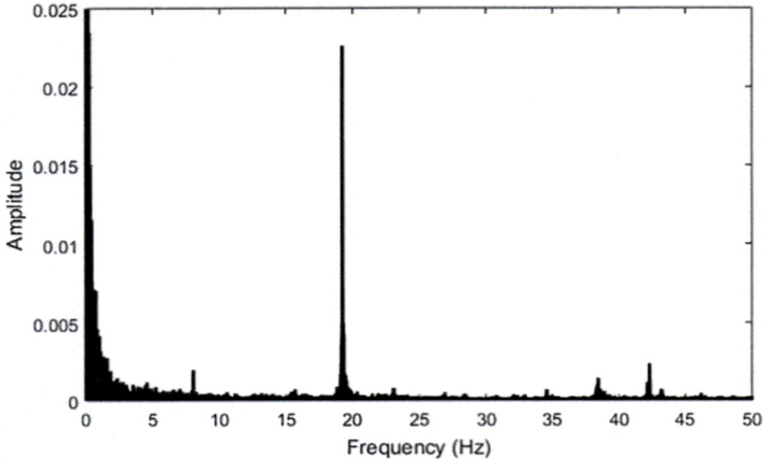

Fig. 3.45 FFT plot of the force signal

3.9.2.2 Real-Life Example in Manufacturing

As an example, a study on the GMAW technique is considered. In this joining technique, the electric arc is established between an electrode wire and the metals to be joined [15]. The electrode wire gets consumed in the process by depositing the filler material in the gap between the base metals. The electrode wire is continuously fed into the process using a spool of wire. A shielding gas is employed to protect the arc from getting contaminated by the atmosphere. The inert gases such as helium and argon are utilised, thus, the technique is often referred to as metal inert gas welding. The major parameters in this technique are the welding current, arc voltage, and the welding speed. These parameters can be associated by another parameter termed as the arc energy, which is given in Eq. 3.25 [15].

$$Q = \frac{IE}{v} \tag{3.25}$$

where, I stands for the welding current, E denotes the arc voltage, v represents the welding speed, and Q refers to the arc energy.

Often, a variant of GMAW called the pulsed-GMAW (P-GMAW) is employed to obtain superior welded joints while joining of thin sheet metals. The parameters included are peak current, background current, pulse frequency, and pulse duration in addition to the aforementioned welding parameters. Because of the pulsing nature, the overall heat input to the weld decreases. In contrast to the continuous GMAW technique, the peak current is supplied in intervals as required for melting of the base metals. The background current helps in maintaining the stability of the arc during the welding and is thus chosen accordingly. With these settings, the heat does not

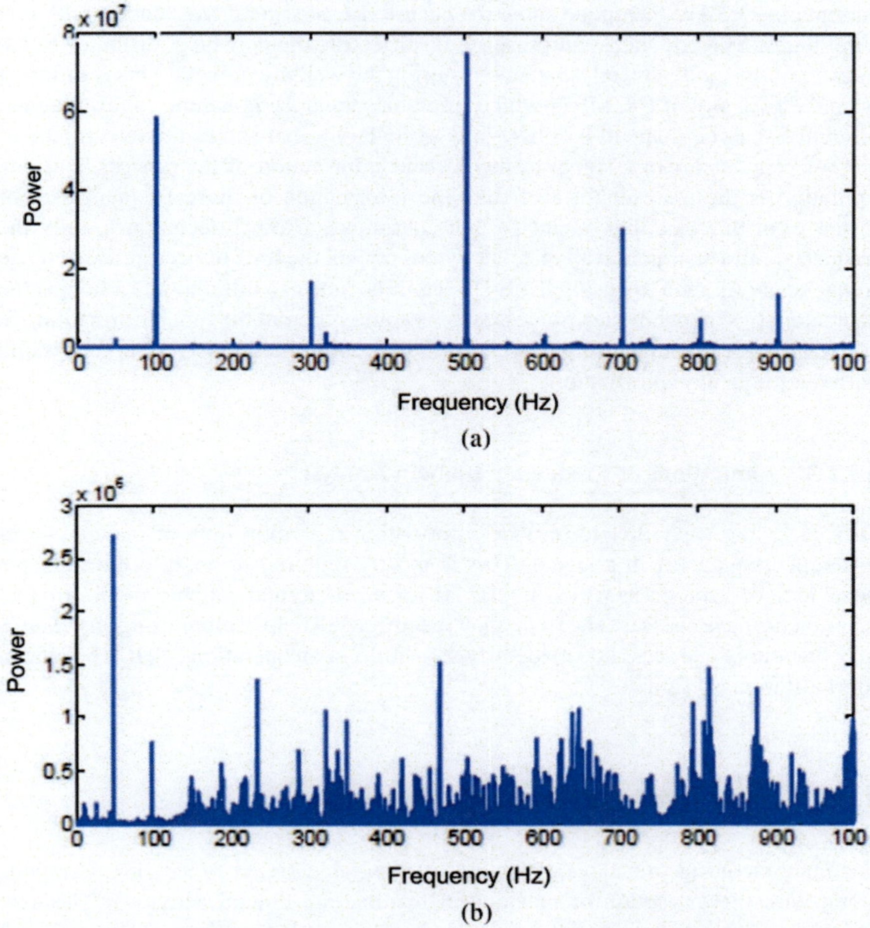

Fig. 3.46 FFT plots of sound signal acquired during: **a** pulsed GMAW and **b** continuous GMAW

concentrate in the weld zone and dissipation occurs. This eliminates problems such as formation of coarser grains, wider heat affected zone, and distortion.

Here, the characteristic of P-GMAW process is demonstrated by applying FFT to the sound signal that is captured during the welding process. The importance of sound as a signal in welding process has already been discussed. For better visualization, the FFT of the sound signal captured in P-GMAW technique is compared with the one of continuous GMAW technique. The FFT plots are shown in Fig. 3.46a, b, respectively. From the plots, the frequencies present in the sound signals captured during the two techniques can be visualized. It can be observed that the amplitude (power) of the frequencies in case of the P-GMAW technique are more than that of the amplitude in case of continuous GMAW process. Beside this, the FFT plot of continuous GMAW technique show the occurrence of several frequencies with low amplitudes. This indicates the presence of different sounds which attributes to the transfer of the molten metal, spatter, etc. during the welding process. This is different from the FFT plot of P-GMAW which resembles the pulsing nature. In this manner, inferences can be mapped by observing an FFT plot that shows the occurrence of various frequencies in a signal to the domain information of the process. But one limitation is the unavailability of the time information of those frequencies. For instance, in this example of pulsed and continuous GMAW techniques, only the frequency information is available, but it does not tell the time of occurrence of those frequencies. As such, by using the FFT technique, one can differentiate amongst the characteristics of techniques/processes, but cannot pinpoint the time information. To emphasize, for manufacturing processes, the time information is very crucial along with the frequency information.

3.9.2.3 Limitations of Frequency Domain Analysis

DFT lacks the capability to provide information regarding time of occurrence of different frequencies in a signal. This is mostly required to analyse the transient behaviour of a process, which is crucial for accurate and reliable monitoring of manufacturing processes. The frequency spectrum will help in monitoring any change in behaviour of a process and to identify safe limits of an operation [2]. It is beneficial for the stationary signals.

3.9.3 Time-Frequncy Domain Analysis

For manufacturing processes, selection of a suitable signal processing technique along with the extraction of meaningful features are crucial activities. The time domain analysis technique allows to study the signal through various statistical features, however, it does not provide information about the spectral components. This requires applying mathematical transform on the time domain signal to obtain the spectral information, which is performed by applying FT. One thing that must be noted is that the FT analyses the signal during its entire time span and identifies frequency components as if they were present during the entire signal [23]. Thus, it does not provide information about the time distribution of the signal. Obviously, this is not the case in most of the applications.

Often the signals acquired during the manufacturing process are non-stationary in nature [24]. This means the signal consists of frequency components that change over the time. This happens because of sudden changes in the process which may sometimes indicate the occurrence of faults in the process. For such signals, the time–frequency based analysis technique is useful. Several time–frequency based analysis techniques do exist, however, in this chapter, two popular and widely employed techniques namely short-time Fourier transform and wavelet transform are covered. The details of these two techniques are explained in the following paragraphs.

3.9.3.1 Short-Time Fourier Transform

The short-time FT (STFT) is a technique employed to identify the temporal distribution of a signal along with the frequency information. In this technique, the DFT is calculated by applying a sliding window of constant width [25]. As the window slides over the entire signal, it helps in identifying both spectral and temporal information contained in a signal. The evaluation process can be observed pictorially in Fig. 3.47. The time series signal is represented as $x(t)$, and $g(t)$ represents the window function. This function slides over the signal which is centred at time τ. For each specific time interval τ, the DFT is computed for the segment of the signal within the window, and then the window slides to next time interval and again the DFT is computed for the next segment. In this manner, DFT for each segment of the signal is computed and a 2D time–frequency representation of the signal is obtained.

In case of STFT, the inner product can be expressed as shown in Eq. 3.26, and the same can further be expanded, as shown in Eqs. 3.27 and 3.28 [25].

$$\mathrm{STFT}(\tau, F) = < x, g_{\tau,f} > \tag{3.26}$$

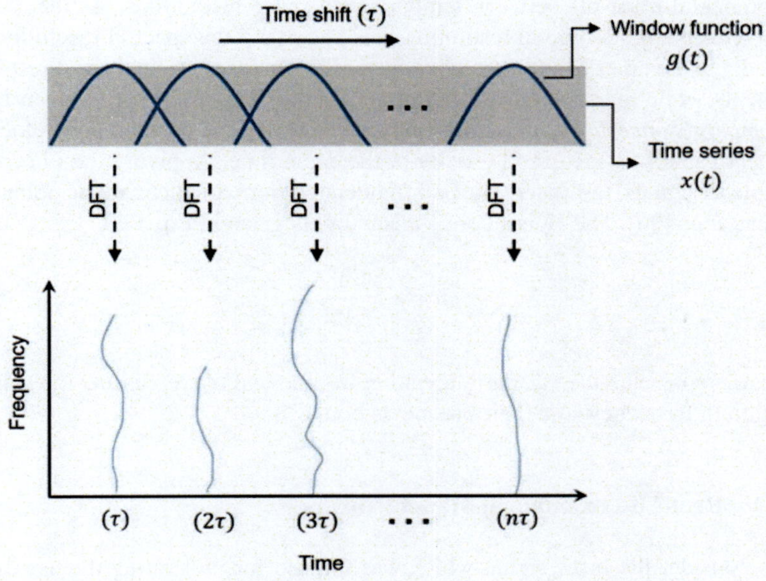

Fig. 3.47 Illustration of STFT of a signal

$$\Rightarrow \text{STFT}(\tau, F) = \int_{-\infty}^{\infty} x(t) g_{\tau,f}^{*}(t) dt$$

$$\Rightarrow \text{STFT}(\tau, F) = \int_{-\infty}^{\infty} x(t) e^{-i2\pi f t} g(t - \tau) dt \tag{3.27}$$

$$X_{\tau,k} = \sum_{n=0}^{N-1} x_n w_{n-\tau} e^{\frac{-j2\pi kn}{N}} \tag{3.28}$$

where g is the windowing function, τ is the time variable of the spectrogram. It can be observed that the above expressions also compute the similarity between a signal and the window function. Thus, STFT technique can be seen as a proper balance in between the time-domain and frequency-domain techniques for signal processing. This is because the time-domain approach has no information on the frequencies present in a signal and the frequency-domain approach lacks the time information. STFT helps in providing both the information.

An important parameter in STFT analysis is the window function upon which the spectrum is dependent. The window functions include Hann, Hamming, Gaussian, etc. The parameters to be chosen in a window function are the length and the overlapping. These two parameters decide the temporal and space resolution of the spectrum. But a question here is the accuracy of the time and frequency information and possibility of deriving them simultaneously. The answer is no. This is because there is a fundamental trade-off between temporal and space resolutions. As the window size is reduced, the temporal resolution gets better and the spectral resolution gets affected. On the other hand, if the window size is increased, the temporal resolution detriments while the spectral resolution gets better. This trade-off is given by the "*Heisenberg's uncertainty principle*", which states that the position and velocity of an object cannot be measured exactly at the same time. Applying the uncertainty principle to signals, the exact time and frequency information cannot be obtained at the same time [26]. The Heisenberg's inequality is given in Eq. 3.29.

$$\sigma_t \sigma_f \geq \frac{1}{4\pi} \tag{3.29}$$

where, σ_t represents the standard deviation in time and σ_f represents the standard deviation in frequency, and their product is bounded.

3.9.3.2 Real-Life Example in Manufacturing

Let us consider the force signal which was utilised for illustrating the use of Fast Fourier Transform for converting a time domain signal into the frequency domain.

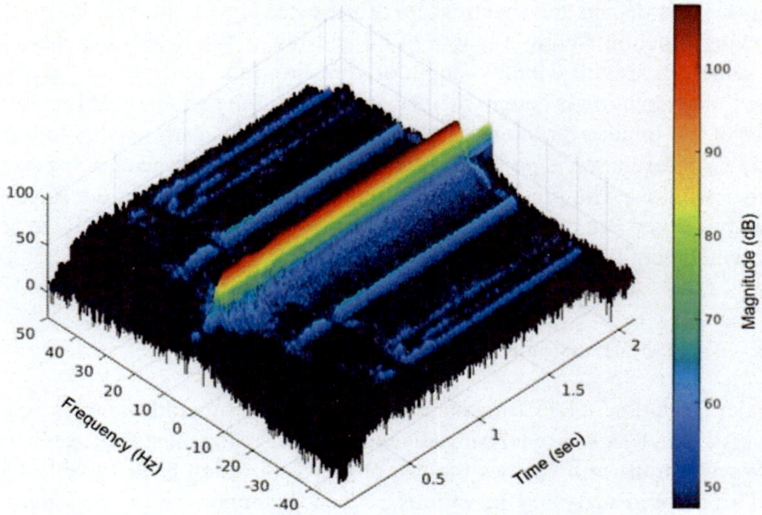

Fig. 3.48 STFT plot of a signal

For that signal, the fundamental frequency was determined 19 Hz (Fig. 3.45), and the same was validated from the FFT plot. A few other peaks were also observed which indicated the harmonics of the fundamental frequency. However, the time information is not available with Fourier transform and this limits the applicability of the technique in manufacturing. Thus, STFT can be applied to determine the time and frequency information, and the application of STFT on the force signal is shown in Fig. 3.48. It is a 3D plot which provides information regarding the frequencies present in a signal, the time of occurrence, and the corresponding power of the signal at those frequencies. The major peak can be found to be at 19 Hz and it is present throughout the length of the signal. However, small variations in the power of this frequency component can be observed. This is because of the variation in the rotational speed of the tool during the welding operation. Similarly, the time information could be determined for other frequency components as well.

With the above obtained information, let us discuss about the time and frequency resolutions. Time resolution is defined as how well a transform can resolve rapid variations in the time domain. Frequency resolution signifies how well the changes in frequencies of a signal can be tracked. In STFT analysis, time and frequency resolutions are directly dependent upon the width of the window considered. It should be noted that the time resolution is proportional to the length of the windowing function while the frequency resolution is proportional to the bandwidth of the windowing function. Hence, a window of shorter length will offer good time resolution and a wider window will be good for frequency resolution. Let us try to visualize this by performing an analysis on the considered force signal.

Figure 3.49a depicts the spectrogram of the force signal obtained with a *Kaiser* Windowing function having a length of 64 and Fig. 3.49b shows the spectrogram of the same signal with window length of 256. From the plots, it can be observed that the time resolution is better with shorter length of the window. This can be said because of the minute variations that are seen in the plot with regards to time, but the frequency resolution is poor. A better frequency resolution can be observed with the wider window. The presence of the frequency components in the force signal are clearly visible. But the time resolution is now poor. Thus, it is crucial to select a proper windowing function and its width.

3.9.3.3 Wavelet Transform

The major limitation of STFT technique is the fixed size of window that is employed for analysis which lacks in analysing all the frequency components present in a signal [24]. Wavelet transform enables the use of wavelets which have variable window sizes. This helps in analysing the various frequency components within a signal. The wavelet transform can be referred to as a time–frequency technique with variable sized windows. A wavelet can be interpreted similar to a waveform however it has a zero average value and limited bandwidth.

Figure 3.50a shows a sine wave and Fig. 3.50b shows a wavelet. From these pictures, it can be observed that the sinusoids are waves that sustain from $-\infty$ to ∞, i.e., they do not have limited duration. The sinusoids are widespread along their definition axis. Thus, these functions which are employed in the FT signify it as a global transform. Thus, one can obtain the information of the frequencies (scale) but not about their locations. Thus, one needs to apply a local transform for studying the frequencies and their locations, which is fulfilled by the wavelets. Of course, STFT could also be utilised for this purpose as it employs a window function, however, the fixed window function poses problem. Wavelets do have a limited bandwidth as shown in the picture. Also, the sinusoids are smooth and predictable, wavelets are irregular and asymmetric.

The wavelets are the basis functions just like the sinusoids in case of FT. During wavelet analysis, an input signal is decomposed into scaled and shifted versions of a wavelet. Just now, two new terms are introduced, one is scale and the other is shift, also known as translation. The scale refers to dilation or contraction of the wavelet and shift refers to the translation of the wavelet along the entire signal. Using these two features, a wavelet analyses an input signal, and as observed from Fig. 3.50b, wavelets are capable in fetching more information from the transient signals as they have sharp variations. The wavelet employed for an analysis is referred to as the base wavelet and is denoted as $\Psi(t)$.

The expression for computing wavelet transform of an input signal through the notion of inner product can be written as shown in Eq. 3.30.

$$W(s, \tau) = \langle x, \Psi_{s,\tau} \rangle$$

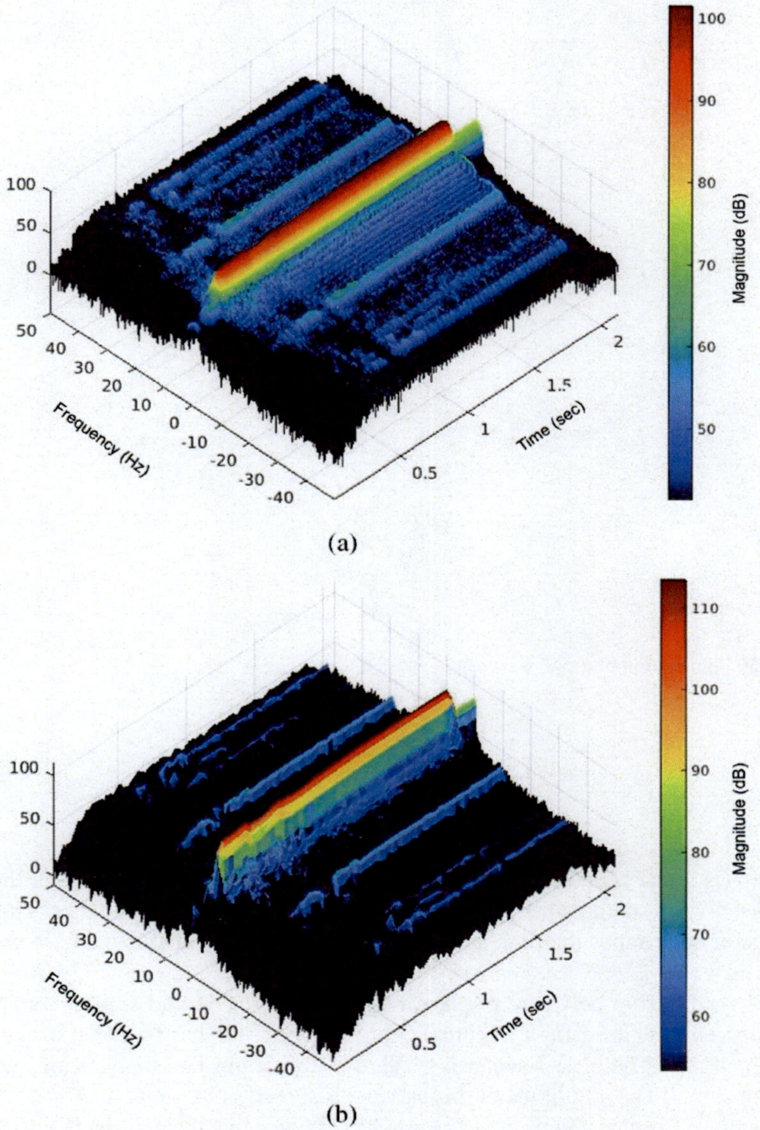

Fig. 3.49 STFT plot of force signal with window length: **a** 64 and **b** 256

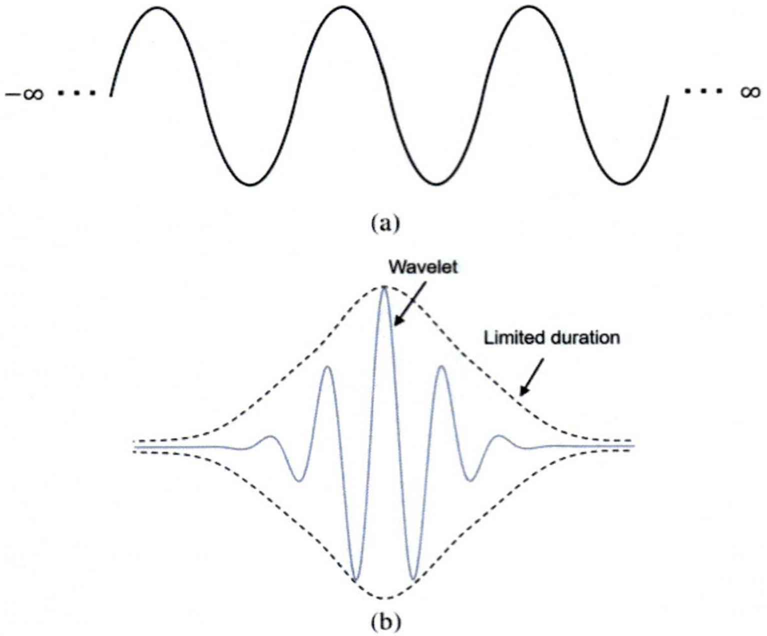

Fig. 3.50 Illustration of a sine wave and a wavelet

$$\Rightarrow W(s,\tau) = \frac{1}{\sqrt{s}} \int_{-\infty}^{\infty} x(t)\Psi^*\left(\frac{t-\tau}{s}\right)dt \tag{3.30}$$

where $x(t)$ is the signal to be analysed, $\Psi^*(t)$ is the complex conjugate of the base wavelet $\Psi(t)$, τ is the shift parameter, s represents the scale parameter. The scale parameter determines the time and frequency resolution of the scaled base wavelet, which is $\Psi\left(\frac{t-\tau}{s}\right)$.

The relationship between a base wavelet and its scaled and shifted version can be observed in Fig. 3.51a–d. Figure 3.51a shows only the base wavelet over a time-domain signal. The base wavelet is centred at the origin i.e., the start of the time-domain signal. The amplitude of this wavelet is shown as 'A' units. In Fig. 3.51b, the base wavelet is translated by $\tau = 1$. It can be observed that the wavelet is shifted, and is now centred at one unit from the origin in the horizontal direction ($\tau = 1$). The amplitude remains the same. In the third figure, the base wavelet that was originally positioned at the origin is scaled by 2 units. This makes the amplitude to reduce by $\sqrt{2}$ as per the expression of the wavelet transform. Finally, Fig. 3.51d shows both the shifted and scaled version of the base wavelet.

The expression shown in Eq. 3.30 refers to the continuous wavelet transform in which similarity is inferred in between an input signal and a scaled and shifted

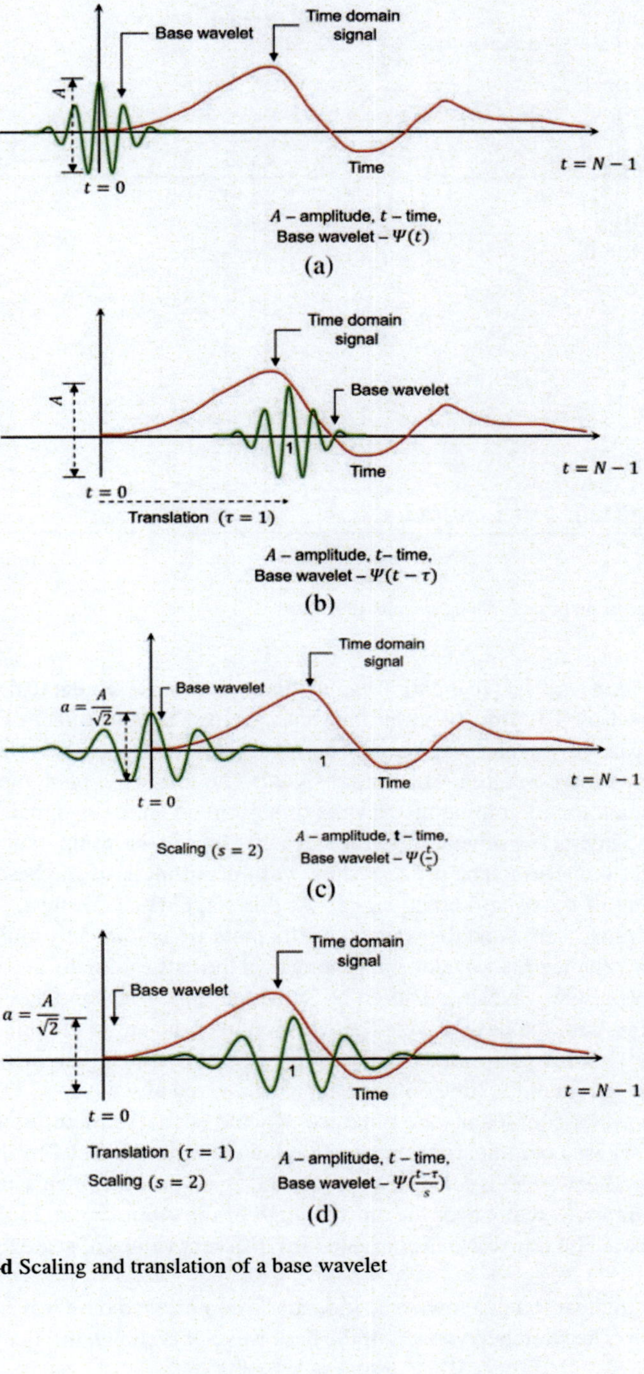

Fig. 3.51 a–d Scaling and translation of a base wavelet

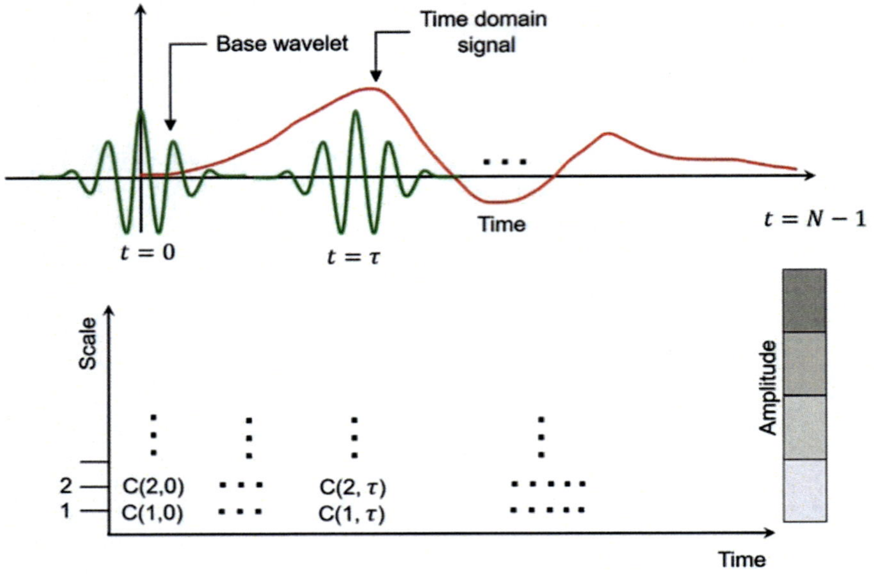

Fig. 3.52 Operation of continuous wavelet transform

version of a base wavelet. To mention again, this expression is similar to the one which was discussed for FT. However, the difference being the sinusoids in FT and base wavelet in case of wavelet transform. The result derived from a wavelet transform is a time-scale representation of an input signal. The following paragraph discusses the implementation of continuous wavelet transform on an input signal.

Figure 3.52 depicts a schematic of the steps involved in the computation of continuous wavelet transform. The process starts with positioning of the base wavelet at the beginning of the time-domain signal. At this stage, $(s = 1)$, thus, it represents the base wavelet itself. Next the wavelet coefficients are obtained by multiplying the input signal with the base wavelet, integrating over the entire domain, and multiplying by $\left({}^{1}/_{\sqrt{s}}\right)$. Next, the wavelet is shifted by some value $(t - \tau)$, and again the signal multiplication with the base wavelet, integration over the entire domain, and multiplication by $\left({}^{1}/_{\sqrt{s}}\right)$ is performed. The scale parameter is still 1. This step is repeated till the entire length of the time domain signal is covered at scale $(s = 1)$. With these steps, the wavelet coefficients are obtained at scale $(s = 1)$ and different values of translation. These coefficients are represented in the figure as C(1,0), C(1,τ), etc. In the next step, the wavelet is scaled by a value, and again the coefficients are computed by translating the wavelet over the entire length of the input signal. In this way, the continuous wavelet transform is computed for different values of scale for the signal [27].

It is now understood that low-scale indicates a compressed base wavelet as shown in Fig. 3.53a. The stretched version of the base wavelet is shown in Fig. 3.53b which

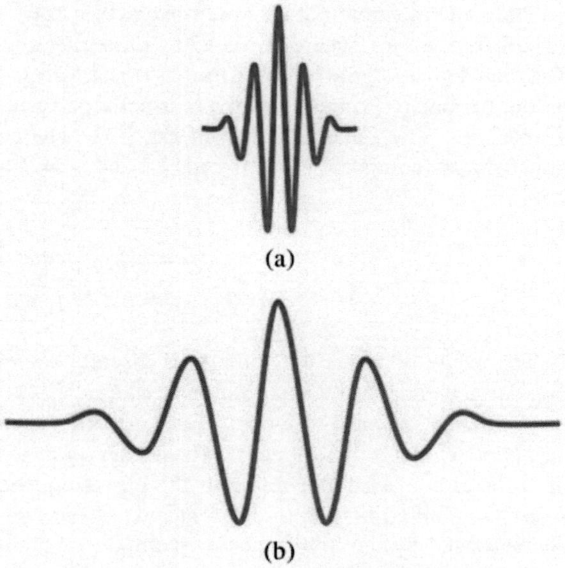

Fig. 3.53 Illustration of: **a** compressed and **b** stretched wavelet

corresponds to high-scale. By using the low-scale base-wavelet during computation, high frequency information within a signal can be derived. Similarly, the low frequency information will be derived by using the stretched version of the base wavelet. This is so because at low scales, rapidly changing information within the signal (high frequency components) can be well identified and the slow varying information (low frequency components) can be identified at high scales. The rapidly changing information is also termed as detail information, and the slow changing ones are termed as coarser information.

One thing to ponder here is the name of the technique, i.e., continuous wavelet transform, is it not applicable for discrete signals? To recapitulate, a real-world signal is analog in nature, i.e., it is continuous in both time and amplitude. However, in order to process it in a computer, it is converted to a digital form. A digital signal is discrete in time and amplitude. The continuous wavelet transform is applicable to digital signals, however, the term 'continuous' refers to the continuous change of the parameters, s and τ as they can take any value during the computation.

Because of continuous variation of the parameters s and τ in case of continuous wavelet transform, a large amount of redundant information is generated. The redundancy is not desirable in applications such as where real-time needs are very crucial or in case of compression of a signal. Since the wavelet coefficients are sparse in nature, wavelet transform finds applications in image compression. In other applications, such as signal denoising or feature extraction from a real-life signal, the redundant information may be desirable as it helps in keeping as much information as possible. In order to deal with this redundant information, discretization is performed, which results in discrete wavelet transform.

The implementation of discrete wavelet transform takes two filters with different cut-off frequencies, where one is a low-pass filter and the other is a high-pass filter. The time domain signal is passed through these filters. The discretization is performed on the parameters s and τ, which is usually performed at a logarithmic scale with a base of 2. The same is shown in Eq. 3.31. The coefficients in discrete wavelet transform are computed by taking $s_0 = 2$ and $\tau_0 = 1$, which corresponds to a dyadic grid.

$$\begin{aligned} s =& s_0^j, \tau = k s_0^j \tau_0;\ j, k \in \mathbb{Z} \\ s =& 2^{-j}, \tau = k2^{-j};\ j, k \in \mathbb{Z} \end{aligned} \tag{3.31}$$

The usage of filters for computing discrete wavelet transform comes from the popular algorithm known as "*Subband coding*" [24]. This algorithm is explained in the following sentences. Two terms that are often used in discrete wavelet transform are "approximation" and "detail". The former refers to the low frequency components of a signal and the latter refers to the high frequency components. An important point to be noted here is the significance of the two terms. In general, a signal is characterized by the low frequency components as they vary slowly and provides the actual identity of a signal. High frequency components indicate abrupt changes which means that they may signify occurrence of anomalies in the signal.

Let us consider an example for better understanding. In a machining process, for example during turning, the job is mounted on the machine spindle and it rotates while the cutting tool is fed against the job to remove material from its surface. This will give rise to force which is acting on the job because of the contact with the cutting tool. If the parameters at which the machining process is being done are optimized ones, the cutting operation will be smooth. However, if the condition is not so, there would be impulses indicating large variations in the signal. The actual force behaviour in the machining process is indicated by the low frequency component that are referred to as approximation coefficients. The abrupt changes in the signal (faults) will correspond to high frequency components that are referred to as detail coefficients.

Discrete wavelet transform yields approximation and detail coefficients of a time domain signal by analysing it through the filters. The analysis is performed for different frequency bands with different resolutions. In the filtering process for computing the coefficients, the signal is first passed through a series of low-pass and high-pass filters. Figure 3.54 illustrates this process of decomposing a signal into approximation and detail coefficients through filtering. As shown in the picture, $x[n]$ is the time-domain signal, and $g[n]$ and $h[n]$ are the impulse responses of a high-pass and a low-pass filters, respectively. The input signal has a frequency ranging from 0 to f, and as it passed through a half-band low-pass filter, the frequency content of the signal spans from 0 to $f/2$ for this case. For the high-pass filter, the frequency content of the signal will span from $f/2$ to f. The length of the signal after passing through the filters remains the same, which means that the net output (length) is

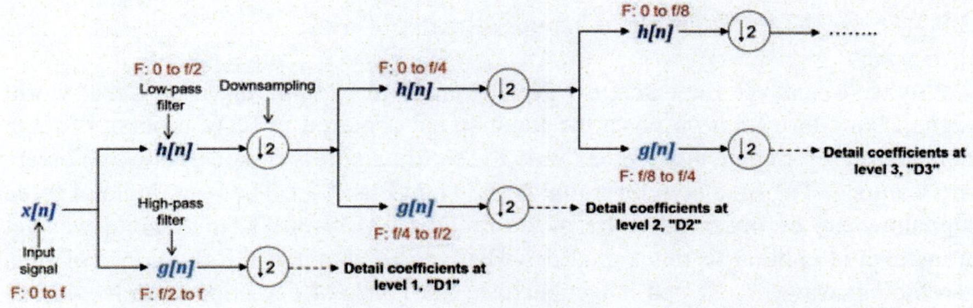

Fig. 3.54 Illustration of discrete wavelet transform

doubled that of the length of the original signal. To get rid of this, it is downsampled by 2, as shown in the figure. Thus, the length of the signal becomes half of the original length. This output obtained from the high-pass filter constitutes the detail coefficients of the original signal (denoted as "D1"), and that of from the low-pass filter constitutes the approximation coefficients. These are the information at the first level of decomposition, which is also referred to as level 1. The output of the low-pass filter is further decomposed by passing it through another set of high-pass and low-pass filters resulting in the detail and approximation coefficients at the second level of decomposition, also referred to as level 2.

Another thing to note is the time and frequency resolution during computation. In the first level, the output obtained from the high-pass filter has a length which is half of the length of the original signal after downsampling. Thus, the time resolution is decreased by a factor of 2. However, it spans only half of the frequency of the original signal i.e., from $f/2$ to f, thus, the frequency resolution is doubled. Later, in the second level, the output obtained from the high-pass filter has a length that is one-fourth of the length of the original signal. Thus, the time resolution is half and the frequency resolution are doubled that of output of the first level. Comparing to the original signal, the time resolution is 1/4th and the frequency resolution is increased by 4 times. Similarly, in the third level of decomposition, the time resolution decreases by a factor of 8 and the frequency resolution increases by 8 times as compared to the original signal. This indicates the trade-off between time and frequency content, i.e., a good time resolution can be obtained at high frequencies, and a good frequency resolution can be obtained at low frequencies. Most of the real-world signal that are encountered in manufacturing processes contain vital information in the high frequencies. The high frequency components indicate fast occurring changes in a signal which might attributed to anomalies. Also, they are transient in nature, i.e., they have varying frequency content because of abrupt changes in the process. Hence, a time–frequency approach such as discrete wavelet transform will be more useful for this purpose.

3.9.3.4 Real-Life Example in Manufacturing

Let us try to analyse the discrete wavelet transform by applying it on a real-world signal. For illustration purpose, the force signal acquired in FSW process is again considered. From the entire signal, only the welding part is considered, while others are trimmed. The time domain signal is shown in Fig. 3.55. The magnitude of force signature can be observed to be in the range of 3100–3500 N. Discrete wavelet transform is applied to this signal to visualize the obtained information. For this, the mother wavelet selected is Daubechies. Six levels of decomposition have been varied out on the time domain signal. As discussed, the analysis begins by passing the time-domain signal through a high-pass and a low-pass filter, which results in the generation of detail and approximation coefficients.

Figure 3.56 shows the plot of detail coefficients obtained by decomposing the original signal. The detail coefficients are result of passing the original signal through a high-pass filter. The original signal consists of 3000 samples and after the first level of decomposition, D1 can be observed to be having a length of 1600 samples. It is half of the span of the original signal. In this way, the detail coefficients at various levels are obtained. From the nature of the plots, it can be observed that the detail coefficients bring out the noises or the high-frequency content in the signal at different levels. This is particularly important for identifying faults in a signal. Looking at the Fig. 3.56, the first is the plot of coefficients of D1. It could be observed that many parts of the signal are close to zero while the rest are having abrupt changes. The coefficients close to zero indicate that probably there were no significant change in the original signal. To dive into more details, D2 further spans half of the above and the details are obtained. Here again, a few parts of the signal are close to zero. The

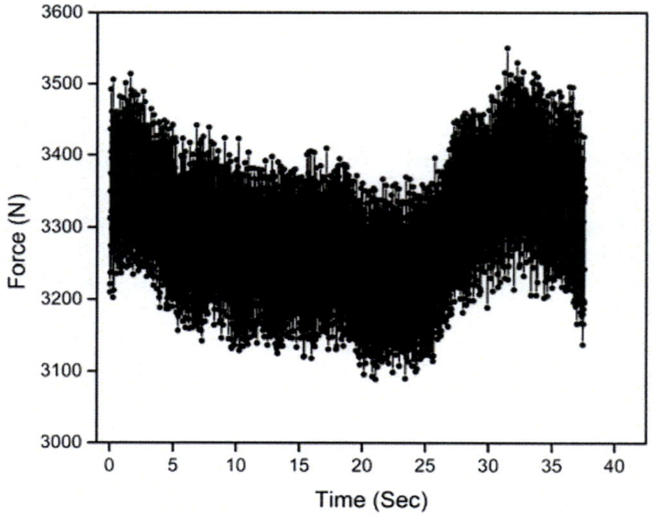

Fig. 3.55 Time domain force signal

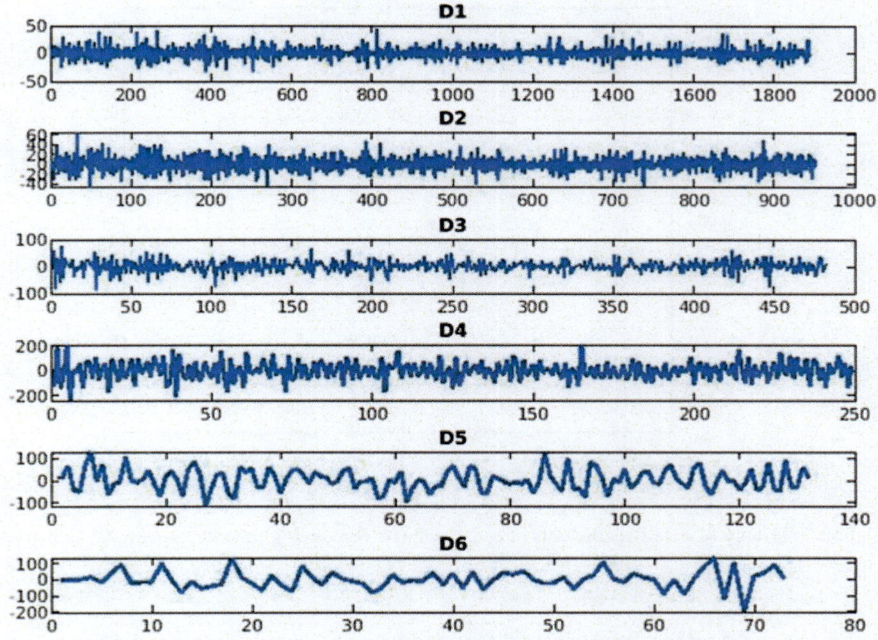

Fig. 3.56 Plot of detail coefficients: D1–D6 (Defect-free weld sample)

variations are less in this particular signal because it corresponds to a good weld quality. This means that the parameters opted to produce the weld were optimum ones and thus resulted in a good quality weld. The corresponding force signal, which is depicted in Fig. 3.55, thus has negligible fluctuation in the welding region.

Another thing to note here is the energy of the wavelet coefficients at each level. As discussed earlier, the approximation coefficients represent the behaviour of the signal which corresponds to the low frequency information. The detail coefficients capture the abrupt changes in the signal which is the high frequency information. Thus, the approximation coefficients carry maximum energy. Let us visualize the energy content of the detail coefficients obtained for the considered force signal, as depicted in Fig. 3.57. The energy content of D1 can be observed to be the highest, followed by D2, D3, and so on. Obviously, one would wonder the importance of energy content in the detail coefficients at each level. The answer is to pick the coefficients which are of higher importance and reject the ones with lower efficiency. Considering the objective of building a digital twin for a process or a machine, the crucial thing is to collect data, process it, and make meaningful prediction to control the process or the machine in real-time. This hints that the processing has to be powerful, which is not spoken here in terms of hardware, but in terms of the analytics. For this force signal case, D1 and D2 show higher energy content, which means they contain most of the information, and a part may also be with D3. Thus, others can be eliminated. Hence, the processing time would be reduced, even by a very small amount of time,

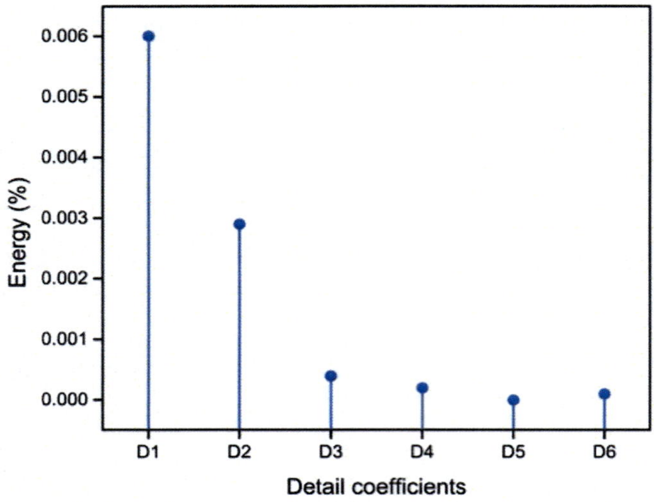

Fig. 3.57 Energy of detail coefficients at each level (corresponding time domain signal shown in Fig. 3.55)

which is crucial for the industrial applications. For a practical demonstration purpose, the original signal is decomposed up to three levels only. This is as per the energy content of the detail coefficients. The time was recorded for both the cases, i.e., for decomposing the signal into three levels and six levels. The result of the same is shown in Fig. 3.58, which shows a time saving of 482.01 ms (0.48201 s). This shows the importance of deriving meaningful information out of signal processing.

With the above information in hand, let us try to analyze a signal resembling a defective weld. The objective would be to devise a technique using which defective and defect-free welds can be identified through discrete wavelet transform. For this, the time domain signal of the defective weld is depicted in Fig. 3.59.

The signal can be seen to be having severe fluctuations throughout its length. For the analysis, the signal was decomposed by using a mother wavelet function and passing it through the low-pass and high-pass filters. The detail coefficients obtained for this case are shown in Fig. 3.60. Clear differences could be observed by comparing the detail coefficients of force signals of defective and defect-free welds. Let us look into the energy content of the detail coefficients of this signal, which is depicted in Fig. 3.61. For this signal as well, most of the energy content is found in D1, D2, and D3. Thus, others can be neglected. For comparison between the signals of defective and defect-free welded sample, let us first look more into the energy content. Figure 3.62 shows the energy content in the approximation coefficients, determined for the defective and defect-free welded samples. It can be observed that the approximation coefficients in case of defect-free accounts for 99.98%. This is because of the lesser fluctuations in the signal. However, for the defective weld, the energy content is only 99.5% and the rest lies with the detail coefficients. This is due

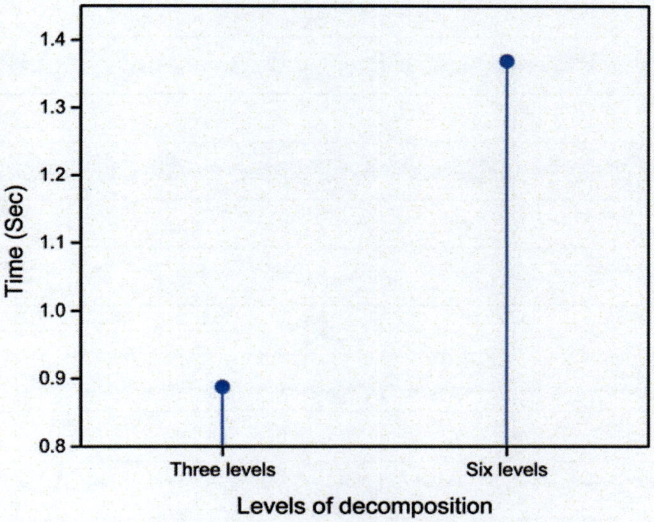

Fig. 3.58 Time elapsed in decomposing signal up to three and six levels

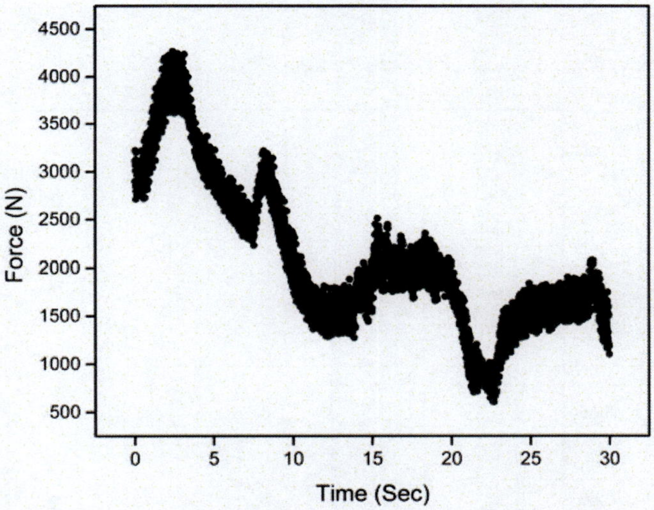

Fig. 3.59 Time domain force signal

to the higher fluctuations in the signal owing to the bad quality of the welded sample that is getting captured in the detail coefficients at various levels.

Finally, let us compare the above analysis with the actual welded sample for more insights. Till now, a few analytics are made by looking into the approximation and detail coefficients of the decomposed signals. For further analysis, a feature is

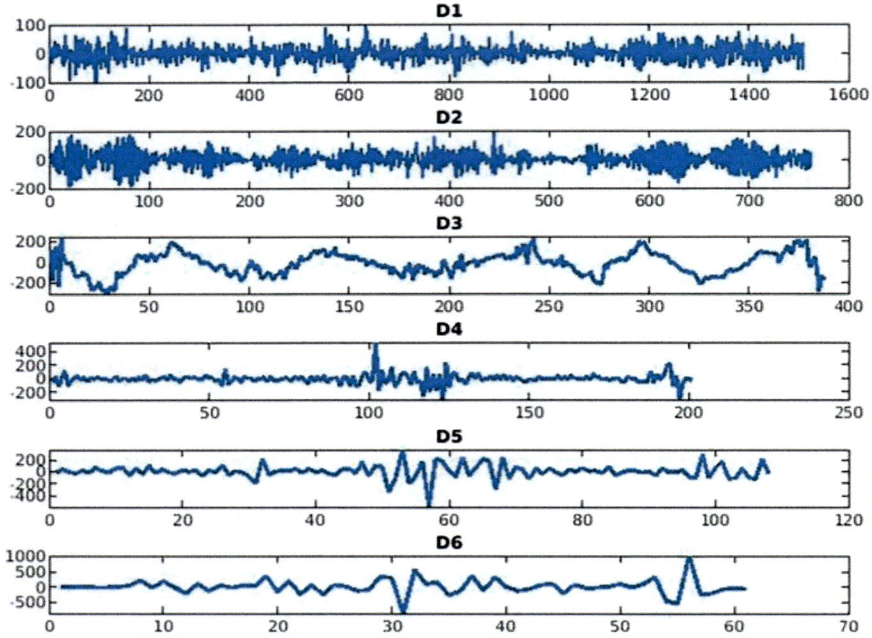

Fig. 3.60 Plot of detail coefficients: D1–D6 (Defective weld sample)

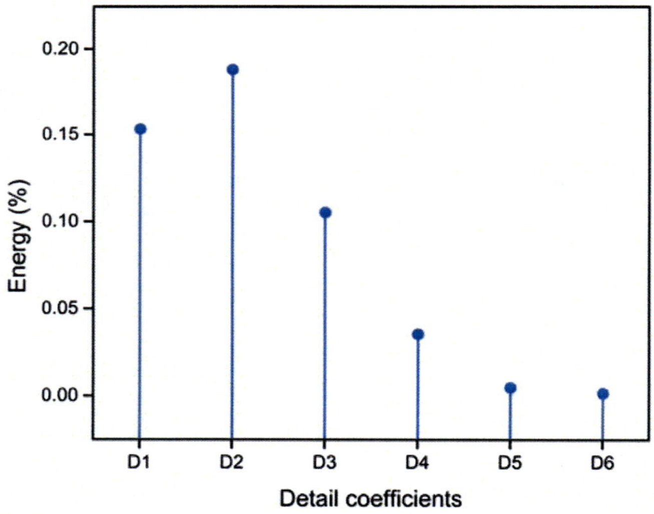

Fig. 3.61 Energy of detail coefficients at each level (corresponding time domain signal shown in Fig. 3.59)

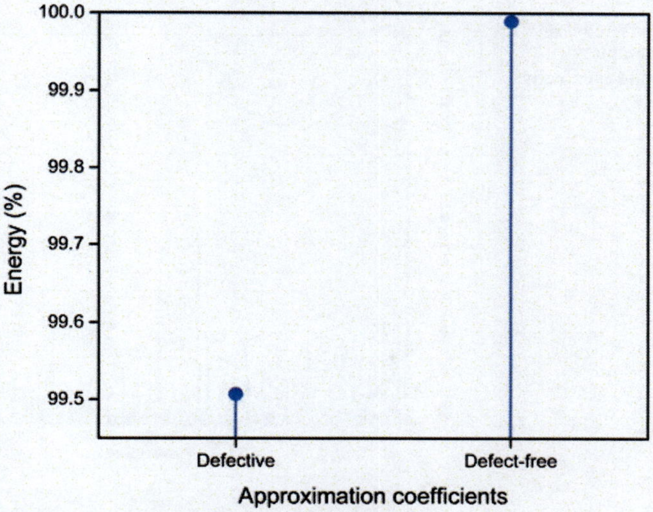

Fig. 3.62 Energy of approximation coefficients of force signal of defective and defect-free welded sample

extracted out of the detail coefficients, i.e., sum of square of errors of the detail coefficients [23, 28, 29]. Let us first understand this feature. For determining the error at each level, the coefficients obtained at one level are subtracted from the mean value of all coefficients obtained at that level. For this example, D1, D2, and D3 are considered as they contain most of the information, both for defective and defect-free welded samples. Thus, for each of these three levels, the errors were determined, squared, and were finally summed up, which forms the feature. Figure 3.63a shows the feature plot along with the defect-free physical welded sample and the same is shown for defective one in Fig. 3.63b. In case of the defect-free sample, the initial peak in the feature plot is possibly because of the start of the tool movement from plunging stage. The physical welded image can be seen to be free from defect and the feature plot has minute fluctuations, the value of which is almost within 0 to 1. The feature values vary distinctly for the defective welded sample and it shows large variations. A sharp increase in the feature plot can be observed for the region with presence of a defect, the values of which go up to 6. In order to develop this idea into a reliable scheme for weld-defect monitoring, a large number of welds could be fabricated, some of which will be defective and rest will be defect-free. For each of them, similar analysis could be performed and the values of the feature plot can be observed for the various cases. From them, a threshold value can be defined to segregate the weld into defective or defect-free based on that value. In this way, the analysis could be automated and executed in real-time.

Further to the above mapping, to ensure the analysis, the integrity of the welded samples were verified via computer tomography scanning (CT-scan) technique. Figure 3.64 shows the pictures obtained from CT-scan analysis for defective and

Fig. 3.63 Feature plot along with welded sample: **a** defect-free and **b** defective

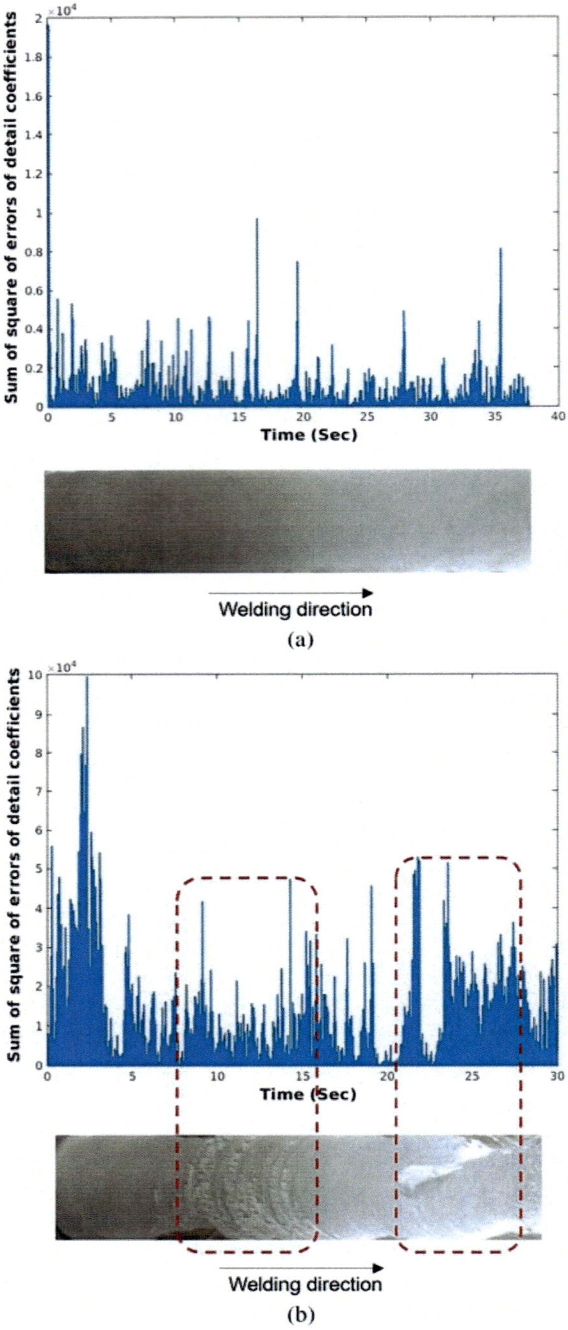

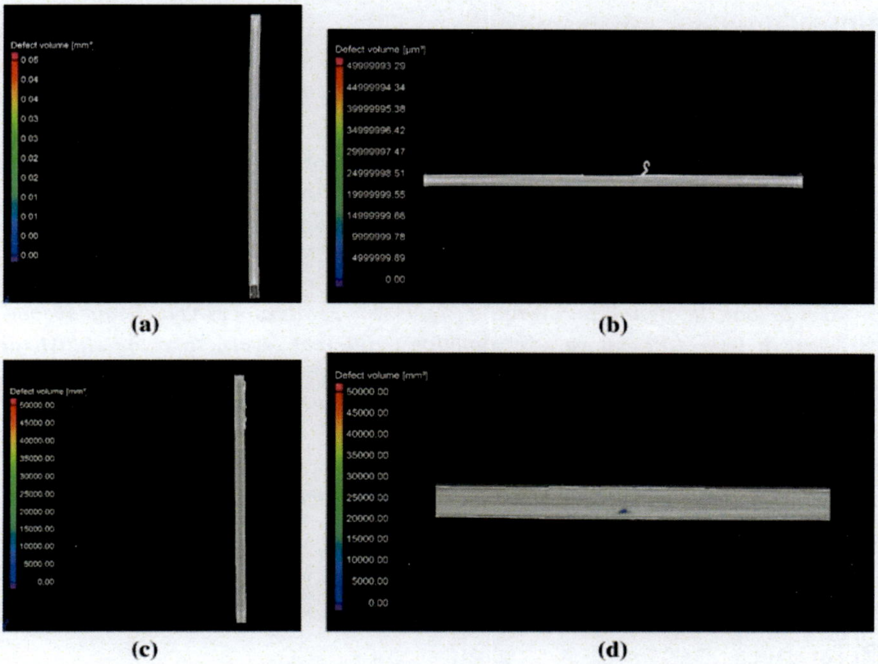

Fig. 3.64 CT-scan images of: **a** and **b** defect-free and **c** and **d** defective welded samples

defect-free welded samples in Fig. 3.64a, b, respectively. Internal defect can be observed for the said defective sample and the same is also identified from the signal as high peaks. No defect is observed for the said defect-free sample.

3.10 Summary

This chapter brought the significance of term ‘signal’ and described a few popular methods for processing of the signals. In addition, the chapter also discussed about signal acquisition, signal conditioning, and signal filtering in detail. Analysis of a few real-life case-studies in manufacturing have also been included in the chapter for practical insights. The upcoming chapter will discuss about the use of image processing in digital twin application.

3.11 Questions for Practice

1. *Take DAQs of two different resolutions, and collect signals using those two. What difference is observed in the signal?*
2. *For any manufacturing process, try operating it at two extreme conditions such that one will give a defect-free product and defective with the other. Engage two sensors relevant to the process. Implement signal processing techniques on the acquired signals* via*time and frequency domain techniques. Make a list of the statistical features and frequency information.*
3. *Try to load the signals into Wavelet Processing Toolbox of MATLAB and decompose it with different wavelet functions, different levels, take the statistical features of the decomposed signals to see the difference.*
4. *Rank the sensors used in Sl. no. 2 based on the results obtained in Sl. no. 2 and 3.*
5. *What is Heisenberg uncertainty principle? Take a signal and apply STFT with different window sizes and visualize the difference in the corresponding outputs.*
6. *Learn about different wavelet functions.*

References

1. Mishra D, Sahu SK, Mahto RP, et al (2019) Friction stir welding for joining of polymers. Springer Singapore, pp 123–162
2. Mishra D, Roy RB, Dutta S et al (2018) A review on sensor based monitoring and control of friction stir welding process and a roadmap to Industry 4.0. J Manuf Process 36:373–397. https://doi.org/10.1016/j.jmapro.2018.10.016
3. Sahu SK, Mishra D, Mahto RP et al (2018) Friction stir welding of polypropylene sheet. Eng Sci Technol an Int J 21:245–254. https://doi.org/10.1016/j.jestch.2018.03.002
4. Mishra D, Pal SK, Chakravarty D (2021) Industry 4.0 in welding, pp 253–298
5. Instruments N data acquisition card. https://www.ni.com/en-in/shop/data-acquisition.html
6. ADLINK data acquisition card. https://www.adlinktech.com/en/Data_Acquisition.aspx
7. Padmanabhan TR (2000) Process instrumentation i temperature. In: Industrial instrumentation. Springer London, pp 420–459
8. Smith SW (2003) ADC and DAC. In: Digital signal processing. Elsevier, pp 35–66
9. Nyquist's sampling theorem. https://www.cs.cf.ac.uk/Dave/Multimedia/node149.html
10. Wikipedia Arduino. https://en.wikipedia.org/wiki/Arduino#:~:text=MostArduinoboardsconsistof,memory%2Cpins%2Candfeatures.&text=Arduinomicrocontrollersarepre-programmed,theon-chipflashmemory
11. Mishra D, Shree S, Gupta A et al (2021) Weld defect localization in friction stir welding process. Weld World 65:451–461. https://doi.org/10.1007/s40194-020-01028-5
12. Mallat S (2009) A wavelet tour of signal processing
13. Gao RX, Yan R (2011) Wavelets: theory and applications for manufacturing. Wavel Theory Appl Manuf 1–224. https://doi.org/10.1007/978-1-4419-1545-0
14. Pal SK, Kharagpur T (2008) Development and validation of varioys soft computing based models for pulsed by Sukhomay Pal
15. Pal K, Bhattacharya S, Pal SK (2010) Investigation on arc sound and metal transfer modes for on-line monitoring in pulsed gas metal arc welding. J Mater Process Technol 210:1397–1410. https://doi.org/10.1016/j.jmatprotec.2010.03.029

16. Tam J (2005) Methods of characterizing gas-metal arc welding acoustics for process automation
17. Saini D, Floyd S (1998) An investigation of gas metal arc welding sound signature for on-line quality control high-speed signal data acquisition and computer-aided analysis of sound signature may reveal conditions that generate weld defects, pp 172–179
18. Pal K, Pal SK (2011) Monitoring of weld penetration using arc acoustics. Mater Manuf Process 26:684–693. https://doi.org/10.1080/10426910903496813
19. Tarn T-J, Chen S-B, Fang G (2011) Robotic welding, intelligence and automation. Springer, Berlin, Heidelberg
20. Smith SW (2003) The discrete fourier transform. In: Digital signal processing. Elsevier, pp 141–168
21. Engelberg S (2008) Digital signal processing. Springer, London
22. Smith SW (2003) The fast fourier transform. In: Digital signal processing. Elsevier, pp 225–242
23. Roy RB, Ghosh A, Bhattacharyya S, et al (2018) Weld defect identification in friction stir welding through optimized wavelet transformation of signals and validation through X-ray micro-CT scan. Int J Adv Manuf Technol 99:623–633. https://doi.org/10.1007/s00170-018-2519-3
24. Sundararajan D (2015) Discrete wavelet transform. Wiley, Singapore
25. Gröchenig K (2001) The short-time fourier transform. pp 37–58
26. Cohen L (2001) The uncertainty principle for the short-time fourier transform and wavelet transform. In: Wavelet transforms and time-frequency signal analysis. Birkhäuser Boston, Boston, MA, pp 217–232
27. Kumari S, Jain R, Kumar U, et al (2016) Defect identification in friction stir welding using continuous wavelet transform. J Intell Manuf 1–12. https://doi.org/10.1007/s10845-016-1259-1
28. Kumar U, Yadav I, Kumari S et al (2015) Defect identification in friction stir welding using discrete wavelet analysis. Adv Eng Softw 85:43–50. https://doi.org/10.1016/j.advengsoft.2015.02.001
29. Mishra D, Gupta A, Raj P et al (2020) Real time monitoring and control of friction stir welding process using multiple sensors. CIRP J Manuf Sci Technol. https://doi.org/10.1016/j.cirpj.2020.03.004

Chapter 4
Image Processing for Digital Twin

4.1 Overview of This Chapter

Improvement of product quality by reducing downtime is the ultimate aim in manufacturing. The product quality can be monitored from the process signals or images acquired at the time of manufacturing. For example, in machining, damage or wear of cutting tool causes degradation in product quality. This damage or wear can be monitored from the process signals such as force, acoustic emission, power, vibration etc. acquired during machining. The condition of the cutting tool can also be monitored from its images. In welding operation, weld quality can be monitored by observing the weld pool or bead at the time of welding. Accordingly, the process parameters can be updated to control the welding process for obtaining defect-free weld. In casting and rolling, cast and rolled surfaces can be monitored visually to find accepted or rejected part. In assembly operation, visual observation of individual component is also important. However, manual observation takes a lot of time and the process is halted at the time of measurement. Manual measurement is not possible in real-time as well. This is also not useful for adaptive correction of the process and thus, the rejection rate becomes high involving loss of money in terms of materials, energy, and human effort. To overcome this situation, manufacturing process zone, such as rolled surface in sheet or bar rolling, cast surface in continuous casting, machined surface or cutting tool in machining, weldment or weld line in welding, and manufactured components at the assembly line can be inspected in real-time by using machine vision. Machine vision system enables more flexibility, repeatability and precision in inspection since it is invasive. Machine vision comprises image acquisition, processing, and decision making. Unlike the other signal processing application, image processing allows describing an object or process zone in two dimensions (2-D) or three dimensions (3-D) for extraction of more direct information. In some of the applications, the description of the process zone can only be possible from the process zone images. For example, during welding in a circular path through a robot, the weld line should be followed for accuracy. The weld path can be tracked

S. K. Pal et al., *Digital Twin – Fundamental Concepts to Applications in Advanced Manufacturing*, Springer Series in Advanced Manufacturing,
https://doi.org/10.1007/978-3-030-81815-9_4

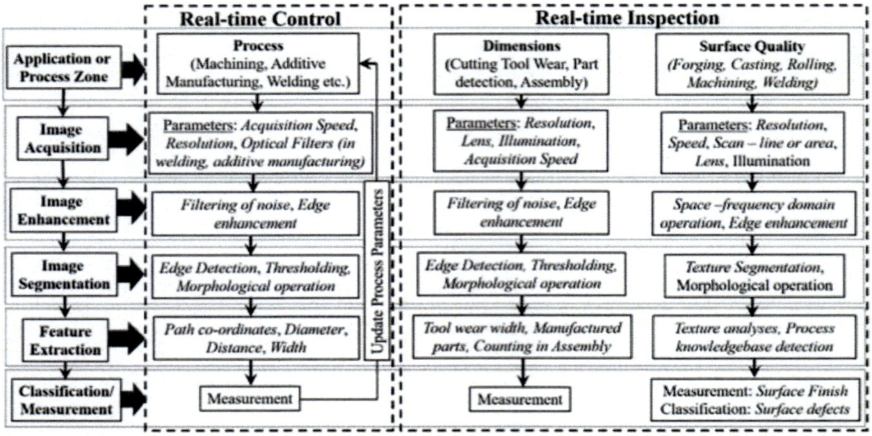

Fig. 4.1 Flow-chart for stage-wise implementation of image processing for process control and inspection

from its images by real-time path planning. Hazardous or difficult-to-access zones can also be monitored by using image processing, such as in hot forging and rolling. Advancement of computing and communication hardware make image processing more relevant for industrial applications in manufacturing, which leads to unmanned manufacturing in the era of the fourth industrial revolution.

Image processing or in a broad sense machine vision is applied mainly for two purposes in manufacturing—process monitoring for subsequent control, and real-time quality assurance. An overview of the application of machine vision in manufacturing is presented in Fig. 4.1.

Real-time application of image processing is mainly in process control and inspection, as shown in Fig. 4.1. Mainly six stages, such as, selecting process zone or application zone for image acquisition, acquisition of an image of the target zone, image enhancement, image segmentation, feature extraction, and finally decision making through automated classification or measurement, are involved for successful implementation of machine vision for both process control and inspection in manufacturing. The importance of these stages is explained briefly with examples of relevant manufacturing processes in the following sections.

4.2 Selection of Process Zone or Application Zone

Proper selection of process zone is the most important step for the application of machine vision in manufacturing. In unmanned manufacturing or inspection, automated selection of process zone or region of interest (RoI) is the most concern in terms of augmentation through artificial intelligence. Before going for the automated

RoI selection, the discussions will be about the applicability of machine vision in various manufacturing processes in real-time process control, and real-time process inspection.

4.2.1 Real-Time Process Control

Acquisition of process zone image for process information, subsequent image processing for extracting features, and decision to be made from the extracted features for updating process parameters are three major steps in process control. All these steps should be performed in real-time. Accuracy and speed of processing are two major factors for machine vision system in real-time process control. The time needed to give feedback, using image processing, should be much less as compared to the response time of the manufacturing process itself. Otherwise, many events may be missed out. Here, some process control applications in the field of machining, additive manufacturing and welding are mentioned below.

4.2.1.1 Machining

In real-time process control of conventional and non-conventional machining, machine vision was utilized. One application in machining is the compensation of a tool's position using the feedback from a machine vision system. In profile grinding, which is a conventional machining process, finish machining of the surface of a complex contour is performed. Here the ground tool position should be controlled with respect to the workpiece surface to obtain a precise surface finish. Theoretical profile of contour is used to track the relative motion between grinding tool and workpiece. Images of tool and profile are acquired to monitor the XYZ coordinates of contour path for estimating the offset between the theoretical and actual profiles. This offset value was then used to control the work table to allow average profile error within 3 μm [1]. Position compensation of tool in non-conventional machining is required to reduce machining error. In the micro-electric discharge machining (μ-EDM) process, the EDM tool is responsible for material removal by maintaining a spark gap between the tool and the workpiece.

Spark gap increases with an amount of ΔZ, as shown in Fig. 4.2, due to an increase in tool wear, which affects the material removal rate, and hence the quality of machining. Therefore, monitoring of tool position with respect to the workpiece via machine vision system was performed for compensation of the tool position to maintain the required spark gap in [2, 3]. The machine vision system is also employed in controlling the electrochemical machining (ECM) process. For instance, in an ECM application to generate miniature blind cavities, a machine vision system was used to generate feedback for controlling the process parameters. Here, an electrolyte is used for material removal where the flow profile of the electrolyte is responsible to generate the required profile of the cavities. Scanning micro electrochemical flow

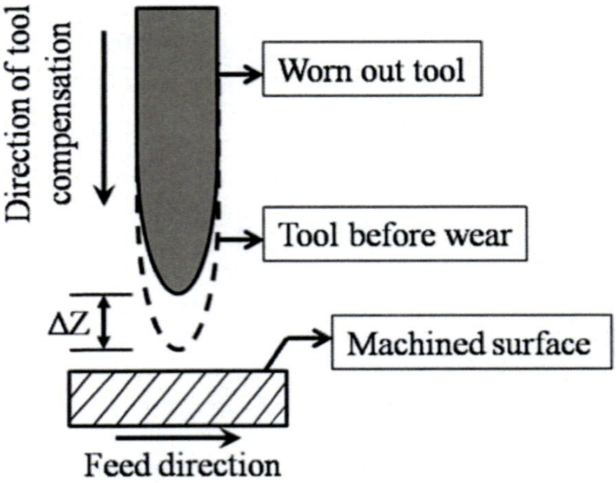

Fig. 4.2 Schematic diagram of the effect of tool wear in μ-EDM process

cell (SMEFC) process is a kind of ECM process where a vacuum is maintained for controlled flow to avoid splashing of electrolytes (Fig. 4.3).

A potential difference between the tool and workpiece is generated and the electrolyte is fed to the workpiece for material removal. A vacuum gap is maintained to control the upward flow of electrolyte to avoid splashing as shown in Fig. 4.3. The precise dimension of the machined cavity is dependent on the electrolyte droplet diameter. Thus, monitoring of electrolyte droplet diameter through image processing was performed to control current density, vacuum gap and flow rate of electrolyte [4]. Finally, cavities of 900 μm diameter were fabricated with an error of 4.75 μm only.

In another aerospace application, micro-holes in an array are needed to be machined in aero engines for efficient cooling. The precision of hole diameters and

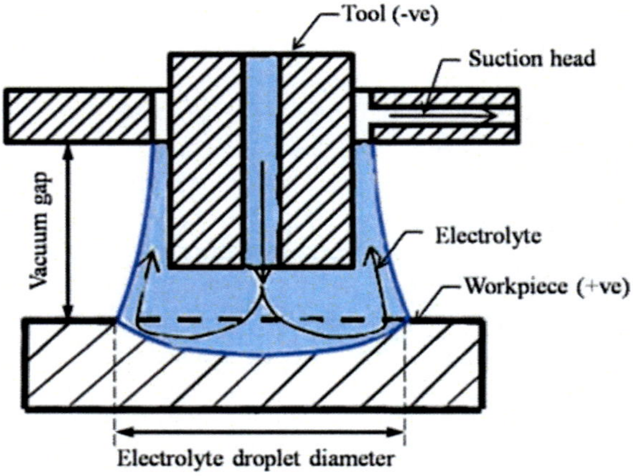

Fig. 4.3 Schematic diagram of SMEFC process

depth are major factors as the diameter of the holes ranges from 0.2 to 1.8 mm [5]. In this case, first, a laser drilling is performed followed by ECM to remove the burrs and the recast layer is produced during laser drilling. At the time of ECM, precision positioning of the tool at the hole centre is essential to obtain precision in hole dimension by avoiding tool-workpiece interaction. In this perspective, image processing was used for accurate centring of the electrode, where holes were machined with a maximum error of less than 12 μm [5].

4.2.1.2 Additive Manufacturing

Metal additive manufacturing is a process where metal powders are blown from a coaxial nozzle and molten using a laser. The metal in the molten state is deposited on a substrate material layer by layer to build geometry or to refurbish a product. In this process, dilution or the accuracy of deposition is dependent, mainly, on the laser power for a constant scan speed. The dilution of the molten pool can be monitored using a machine vision system by determining the width of the molten pool or geometry. Subsequently, the laser power can be controlled from the information of molten pool geometry [6–8].

Image of molten pool acquired using co-axial imaging i.e., coaxial with the laser head is shown in Fig. 4.4a. Width or other geometrical information of molten pool was calculated using image processing and this information was utilized for controlling the laser power. Cladded parts produced with and without controlling laser power are shown in Fig. 4.4b, c. It can be seen that the part produced without controlling the laser power from molten pool information is more non-uniform as compared to the part produced with control of the laser power. Scanning speed or relative speed between the coaxial nozzles with laser head was controlled from the feedback information obtained from an infrared camera [9]. Time–temperature history was obtained from the infrared camera, from which scan speed was controlled for controlling the microstructure of produced layer in directed-energy deposition (DED).

Fig. 4.4 **a** Molten pool image from coaxial imaging; laser cladded parts **b** without control and **c** with control of laser power with the information of molten pool geometry [8]. Reprinted by permission from Springer Nature Customer Service Centre GmbH: Springer Nature, The International Journal of Advanced Manufacturing Technology, A feedforward controller for tuning laser cladding melt pool geometry in real time, S Moralejo et al. 2016

4.2.1.3 Welding

Weld pool and seam are some process zones to be imaged for controlling parameters in various welding processes, especially fusion welding. In arc or inert gas welding, the weld pool needs to be monitored to control the process parameters such as weld current. Heat accumulation at the end of the weld is more due to heat transfer in an arc welding process if the welding current or voltage level is constant throughout the process. It is affecting the weld quality by resulting in a non-uniform width of the weld. Mitigation of this issue can be possible by controlling weld current according to the visual information of weld pool geometry [10]. A clear distinction between the quality of welds with and without control of welding current in tungsten inert gas (TIG) welding can be observed in Fig. 4.5.

Tracking of the weld seam and monitoring of weld gap plays an important role in controlling the position of weld torch during robotic welding application. An example of process zone selection in position tracking of weld line is shown in Fig. 4.6, where

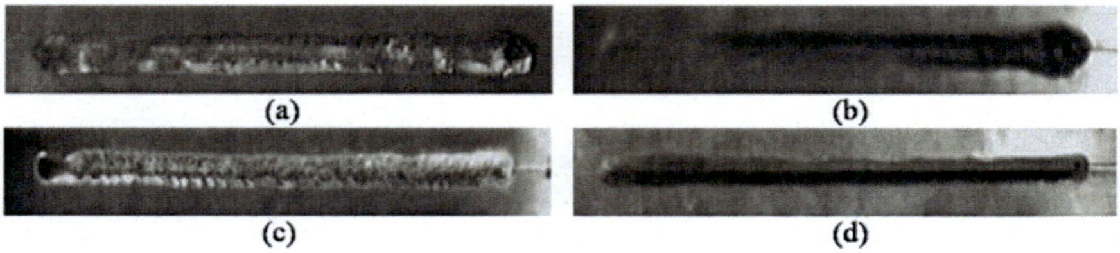

Fig. 4.5 **a** Topside and **b** backside images of welds with constant welding current and **c** topside and **b** backside images of welds with controlled welding current through monitoring of weld pool width using image processing [10]. Reprinted by permission from Springer Nature Customer Service Centre GmbH: Springer Nature, The International Journal of Advanced Manufacturing Technology, Obtaining weld pool vision information during aluminum alloy TIG welding, J. J. Wang et al. 2004

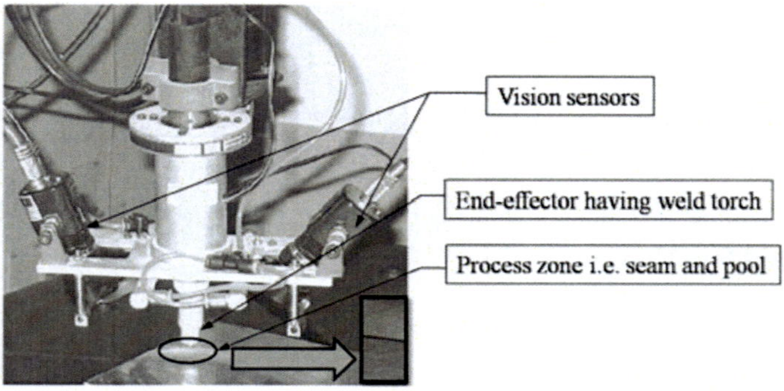

Fig. 4.6 Pre-weld tracking of weld seam in robotic gas tungsten arc welding (GTAW) process (inset: an image of weld seam) [12]. Reprinted by permission from Springer Nature Customer Service Centre GmbH: Springer Nature, Journal of Intelligent and Robotic Systems, Autonomous acquisition of seam coordinates for arc welding robot based on visual servoing, L. Zhou et al. 2006

weld seam image before gas tungsten arc welding (GTAW) was acquired through a vision system for further processing. Here two vision systems were attached with the robotic arm, one near the end-effector, and the other vision system was used to acquire weld pool image for online monitoring of weld quality [11].

Tracking of weld path, the gap between two plates to be welded are utilized for control of welding robot's motion or end effectors' position in robotic welding. Joint penetration was controlled updating process parameters through monitoring of weld pool. Laser power and scan speed can also be controlled in the laser welding process through weld pool monitoring [13]. Tracking of seam coordinates during robotic gas tungsten arc welding (GTAW) was performed by taking feedback from machine vision system for both straight and curved line with ±5 mm linear and ±5° angular error when welding speed is 20 mm/s [12]. A good quality weld is obtained in robotic welding by controlling the weld voltage or current from the feedback information of offset between weld torch centre and weld seam centreline. In one example, the centre of the weld was monitored during a GTAW process for tracking the position and for controlling the voltage to obtain a good quality weld line [14]. In another example, GTA welding of curved lines was performed by tracking the weld seam from the information of offset between weld pool centre and seam centre with ±0.5 mm accuracy [15]. This has been applied for welding of rocket flange. In pulsed metal arc gas (MAG) welding, the offset was controlled for obtaining weld with an error of less than 0.4 mm [16].

The offset between the wire centre and seam centre was identified through image processing and the motion of the robot was controlled accordingly to obtain a good quality weld along a curve, as shown in Fig. 4.7. Tracking of two welding heads of

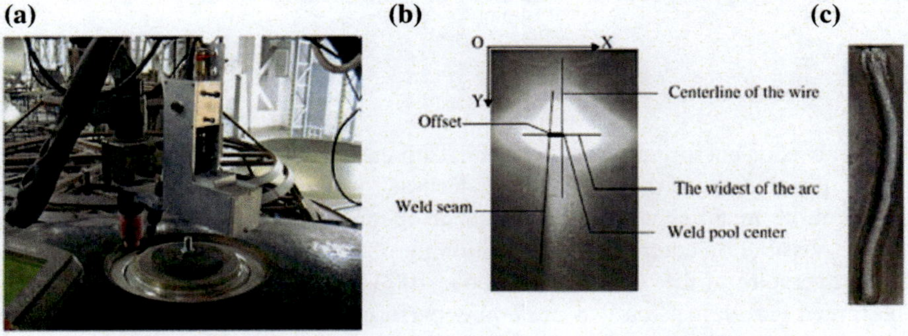

Fig. 4.7 **a** Welding by tracking the curved seam line from the feedback information of offset between weld pool and seam centres, **b** offset between weld pool centre and seam centre, **c** obtained weld after controlled tracking [15, 16]. For (**a**): Reprinted by permission from Springer Nature Customer Service Centre GmbH: Springer Nature, Journal of Intelligent and Robotic Systems, Real-time seam tracking technology of welding robot with visual sensing, Hongyuan Shen et al. 2010. For (**b**) and (**c**): Reprinted by permission from Springer Nature Customer Service Centre GmbH: Springer Nature, The International Journal of Advanced Manufacturing Technology, Passive vision based seam tracking system for pulse-MAG welding, Zhen Ye et al. 2012

robotic butt welding having welded seam of 0.2 mm gap in presence of CO_2 gas for shielding was performed using machine vision within 100 ms control time [17].

4.2.2 *Real-Time Inspection Using Machine Vision*

Real-time inspection of the manufactured surface to determine surface finish or to detect surface defects is indispensable for manufacturing industries to improve product quality, maintain goodwill, and reduce cost in batch production. Also, modification in a batch production process in the product line can be realized from the information obtained through real-time inspection. The application of machine vision in real-time inspection for various manufacturing processes is discussed in the following sub-sections.

4.2.2.1 Casting

Casting is the basic operation in conventional manufacturing. The continuous casting process, a variant, is used for producing semi-finish steel with high quality, energy-saving and low cost. However, cracks present on a cast surface become catastrophic at the time of subsequent operation like rolling. Bleed is one kind of surface defect present in continuous casting. But, automatic detection of bleed is challenging as there is minimal difference between bleed and its background due to the presence of oscillation and water spray marks. Image processing with rule-based detection technique was applied for real-time bleed detection within 6% detection error [18].

4.2.2.2 Forging

In the precision hot forging process, the final dimension of the forged part is dependent on the positioning of the upper die. Inaccuracy in the dimensions of forged part causes more material waste in subsequent operations. Improvement in precision can be possible by controlling the positioning of the upper die. On the other hand, the temperature of the workpiece is above 1000 °C. Thus, real-time inspection of the forged part dimension was preferably performed by machine vision to avoid a hazardous inspection process. [19, 20]. Light or laser strip was projected on to the hot workpiece, as shown in Fig. 4.8, and the corresponding images were acquired. Then the dimension was calculated from extracted points on the strip.

High accuracy dimension measurement (99%) can be realized through this process so that material loss in subsequent post-operation with subtractive manufacturing can be avoided [19].

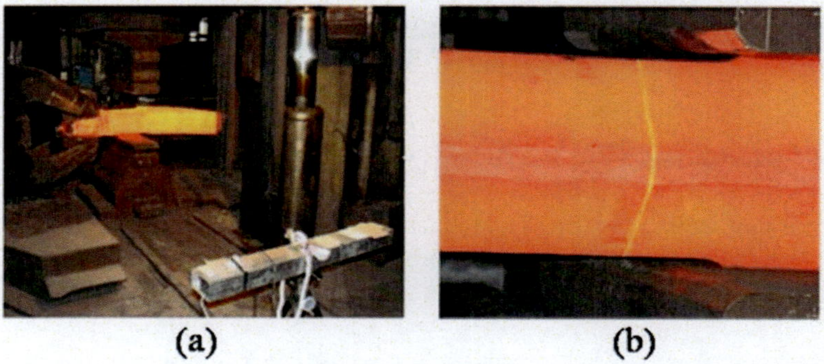

Fig. 4.8 **a** Forging job measuring set-up, **b** laser light fall on the job for measurement using laser triangulation [20]. Reprinted by permission from Springer Nature Customer Service Centre GmbH: Springer Nature, The International Journal of Advanced Manufacturing Technology, Spectral selective and difference imaging laser triangulation measurement system for on line measurement of large hot workpieces in precision open die forging, Drago Bračun et al. 2016

4.2.2.3 Rolling

In hot and cold rolling, the whole batch of production is rejected due to the presence of defects, which involves loss of energy, time, and money. Thus, real-time inspection during the rolling process is necessary to avoid this situation. Image processing plays here an important role due to its non-contact and flexible nature.

During hot rolling, the high temperature workpiece surface is only accessible for inspection from a safe distance. In this case, inspection using a machine vision system is the only alternative to avoid hazardous manual inspection. In hot rolling of bars, scales due to oxide formation, 'seam' like defects due to overfill or underfill of material, and cracks are some common but catastrophic defects. In this context, a special arrangement to acquire the image of the hot-rolled bar (over 1400 °C) was employed, which is termed as HotEye® imaging system to acquire images by avoiding high-temperature glares [21]. Finally, the classification was performed by applying the process knowledge-based classification technique.

During hot rolling of flat surface at 550–700 °C temperature having 3–18 m/s speed, images were acquired and processed for classification of various defects such as patches, crazing, pits, inclusions scales etc., with 98% accuracy [22]. Detection of surface defects during hot rolling was performed using a field-programmable gate array (FPGA), within 12.5 μs time for each image with 92% accuracy [23].

In cold rolling, patches of liquid coolants are present on the rolled surface, where defects and the patches become similar in visual appearance. Image processing plays here an important role in segmenting the defects from the liquid patches automatically within a short duration of 10 ms [24].

Surface roughness of rolls during reconditioning was determined with less than 0.5 nm resolution from the image of speckle pattern created on the surface of rolls with a computing speed of 200 Hz with the help of FPGA computing, where semiconductor logic circuits are used for computation to facilitate faster-embedded computing [25].

4.2.2.4 Machining

Monitoring of cutting tool wear and machined surface quality are two areas where image processing is used to maintain product quality and to reduce machine tool downtime.

Tool Wear Monitoring

In conventional machining such as turning, milling, drilling, etc. product quality, especially the machined surface quality and surface integrity are dependent on the condition of the cutting tool. The main merit of tool condition is tool flank wear width to decide the remaining useful life (RUL) of the cutting tool or cutting insert [26]. As per ISO 3685:1993, if the average tool flank wear width ($VB_{average}$) of uniform wear zone is higher than 300 μm in semi-finish machining, then the tool is considered as a worn tool. Majorly there are four conditions of cutting tool—unworn, semi-worn, worn, and broken. The surface finish of a machined part is dependent on these four conditions. Thus, there are two options to monitor tool flank wear using machine vision, i.e., directly observing the tool condition using microscope or camera-based imaging system or indirectly by extracting features from images of the machined surface [27, 28]. The advantages of detection of tool flank wear using image processing are in terms of non-contact, flexible, and fast detection. Also, there will be no requirement to detach the tool insert from the tool holder in machine vision-based monitoring, which prevents the misalignment of the insert in the tool holder as well, due to refitting it. This avoids the degradation of quality. Faster and timely detection of tool condition can save time and money by reducing machine tool downtime and enhancing product quality. Non-contact and flexible detection using machine vision can also be applied in multiple machines due to the advancement of edge and cloud computing for enhanced productivity with less downtime. Faster and reliable detection of tool flank wear through image processing is a major advantage towards Industry 4.0 era. It can be discussed by an example of profile milling, where a large number of cutting inserts (Fig. 4.9) are involved to machine a curved or slanted surface for obtaining a high-quality surface finish. Therefore, the condition of each insert should be monitored from time to time to achieve precision at the sub-micron level. In [29, 30], faster detection of the condition of multiple inserts (24 inserts) were performed just after the machining, only within a very short duration (3 min) without detaching any of the inserts from the profile milling head by using machine vision. In this case, a very low cost and flexible computing system (Raspberry Pi) was utilized for the detection of tool flank wear by automatic RoI selection [31].

However, in the case of direct tool condition monitoring, detection of tool flank wear (within 300 μm) needs costly lens set-up i.e., narrow field-of-view macro lens

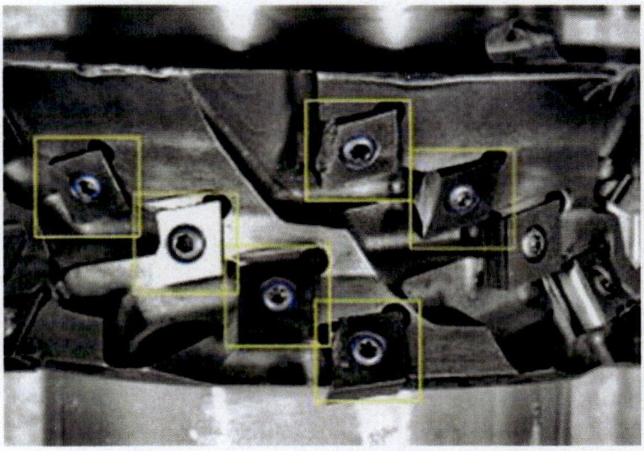

Fig. 4.9 Inserts connected to the profile milling head [30]. Reprinted by permission from Springer Nature Customer Service Centre GmbH: Springer Nature, Springer eBook, Cutting edge localisation in an edge profile milling head, Laura Fernández-Robles, George Azzopardi, Enrique Alegre et al. 2015

for precise detection. But, in [32], a more semantic image segmentation technique was applied on the images of tool flank wear acquired with a comparatively low-cost wide field-of-view lens system. Wear detection accuracy of 91% was achieved in this application.

Batch scale automated inspection of the used cutting tool can also be performed by employing the machine vision technique. For example, automated inspection of wear of micro-drill used in PCB manufacturing was performed using image processing [33]. Nowadays, the application of machine vision-based inspection has started in the cyber-physical production system (CPPS). In [34], an inspection of wear on drill bit was performed with more than 96% accuracy [34].

However, the direct tool wear measurement system has some limitations, such as the requirement of costly image acquisition set-up and the difficult-to-access process zone. Therefore, in some cases, tool wear was predicted by analyzing the machined surface image, as the acquisition of image from the machined surface is comparatively easy, and the required technique for the acquisition of image is less costly.

The difference between quality of surfaces, machined by a fresh tool insert and by a dull tool insert, respectively, can be observed in Fig. 4.10. More non-uniform image texture is visible in Fig. 4.10b, i.e., surface turned by a dull tool as compared to Fig. 4.10a, i.e., surface turned by a fresh tool. Classes of tool wear i.e., unworn, semi-worn or worn, were predicted from the machined surface images in turning and milling [35–40]. Progressive flank wear values were also predicted by processing images of the machined surfaces in turning and milling processes [41–47]. This is also required in on-machine tool detection of tool wear from the machined surface images.

In some machining process, such as in belt grinding, direct detection of the tool wear is a major challenge. Here a belt having abrasive particles is utilized to grind

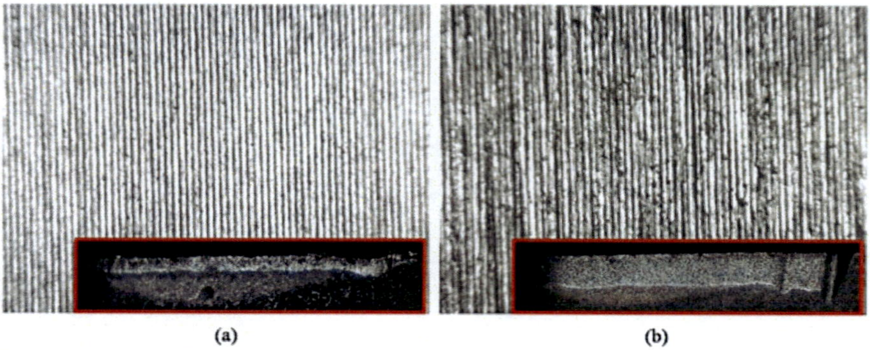

Fig. 4.10 Images of machined surface turned by **a** fresh tool with tool flank wear width ($VB_{average}$) = 75 μm, and by **b** dull tool with tool flank wear width ($VB_{average}$) = 303 μm [insets: corresponding tool flank wear images].

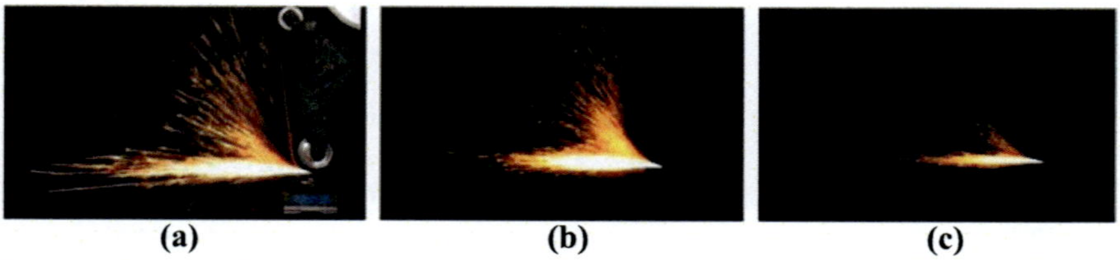

Fig. 4.11 **a** Acquired image of spark field, a pre-processed image of spark field for **b** semi-worn belt and **c** worn belt in belt grinding [48]. Reprinted by permission from Springer Nature Customer Service Centre GmbH: Springer Nature, The International Journal of Advanced Manufacturing Technology, A new in-process material removal rate monitoring approach in abrasive belt grinding, Lijuan Ren et al. 2019

the surface to obtain the required depth of grinding. In this case, each time, the belt needs to be detached for manual inspection, qualitatively. On the other hand, a spark field generates at the time of grinding, the area of which correlates with the grinding depth, as shown in Fig. 4.11. Hence, in [48], images of spark field were acquired and processed to predict the wear and grinding depth in real-time.

Surface Finish Inspection

The surface finish or surface roughness of a machined object needs to be determined to ensure product quality. In conventional surface inspection technique i.e., with a stylus-based surface profiler, 1-D attribute of surface can only be measured after stopping the machining. Thus, unmanned inspection in multiple machines is not possible. In this situation, a machine vision system can be utilized for unmanned inspection of 2-D or 3-D surface finish in multiple machines without stopping the job motion. Therefore, the required online modification in process parameters can be possible to get the desired product quality for realizing unmanned machining. Real-time inspection of surface finish can also be possible to inspect the surface finish of parts in an assembly line. Therefore, the machine vision-based surface finish inspection can also be an integral part of Industry 4.0 or digital twin, for faster and ubiquitous inspection. However, at the time of machining, the acquisition of good quality images from the machined surface, with a low-cost acquisition system, is a challenge due to image blurring. In this context, deblurring was performed on the images taken from a milled surface moving with 25.4 mm/s speed. Extracted features had a very good correlation, i.e., 0.85–0.93 R^2-value, with measured surface roughness by the conventional stylus method [49]. A low-cost webcam system was integrated into the machine tool for inspection of milled surface, where machining parameters were updated based on the extracted features from the machined surface [50]. It is highly realizable as the feed rate in machining is directly related to the machined surface image [41]. Valve seals in internal combustion engine require high precision in surface finish and geometry to reduce leakage and achieving high efficiency in performance. However, there are three surface segments with 30°, 45°, and 60° inclinations on the circular valve seal, where the circular surface with 45° inclinations is mainly responsible to seal the leakage. Surface chatter may occur in machining the valve seal due to improper process parameters or tool wear. The circularity of the valve seal surface is also another parameter to improve an engine's performance by reducing leakage. In this case, a non-contact machine vision-based inspection is indispensable to detect the geometry and surface finish of the valve seal for corrective action in machining [51]. In-situ detection of surface finish in milling and turning was attempted in [52–54]. In-situ surface finish evaluation in turning is challenging as the rotating curved surface needs to be imaged without motion blurring. But, surface finish up to 10 μm was evaluated using machine vision with ± 2 μm average accuracy [53].

In any robotic machining, localization of workpiece with respect to the robot coordinate system is necessary to utilize it in an unmanned situation. In this case, automatic identification of parts in 3-D was performed using machine vision [55]. At the time of assembly operation i.e., drilling and riveting, automated identification of the position of the holes is necessary, especially, if one wants to use robotic drilling and riveting. In this application, image processing plays a major role in the automatic identification of hole position to enhance positional accuracy. Image processing was used to identify the accurate position for robotic drilling in curved surface for aircrafts' parts [56]. Similarly, the inspection of countersinks was performed by using machine vision to achieve high precision for subsequent assembly operation in the aircraft's panel [57].

Welding

Welding is a joining technique, where two plates of the same or different materials are joined in various configurations by fusing materials in molten condition followed

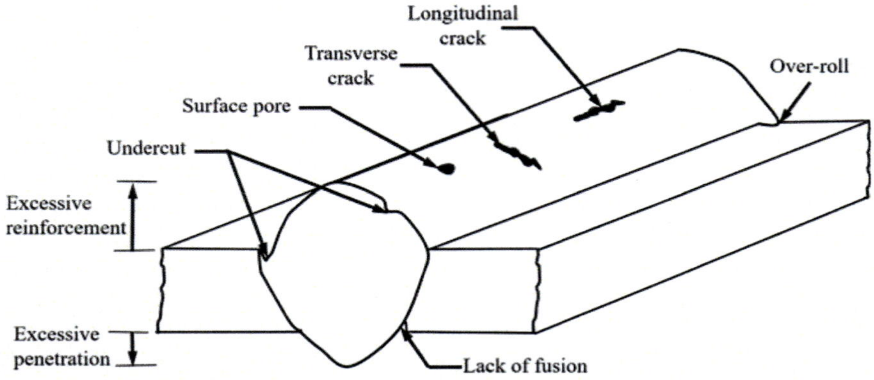

Fig. 4.12 Schematic representations of various surface defects in welding

by solidification. Welding current, traverse speed, weld gap between two plates, gap between weld plates and the torch etc. are major parameters involved, improper control of which results in defective weldment. Image processing can be applied to detect the weld defects and accordingly, action can be taken to control the required parameters. There are surface and internal defects where surface defects can be detected by any digital camera. However, X-ray image or radiographs are required to detect internal defects, which can also be processed for automatic defect detection or classification. Various surface defects that occur in welding are shown in Fig. 4.12.

Lack of fusion at the root due to improper penetration for incorrect positioning of plates, surface porosity due to entrapment of gas for the improper flow of shielding gas, undercut and excessive reinforcement and penetration due to high welding current and low traverse speed, cracks due to arc termination, internal strains and over-roll due to slow speed and improper torch angle are some of the major weld defects.

Internal as well as surface defects in welding can be detected by using radiographic images in welding. However, the detection time of the defect is high, if each portion of the weld is imaged by X-ray, manually. Thus, a sequential image processing system was developed in [58, 59], where the long pipe was imaged through X-ray in moving condition. Real-time defect (such as undercut, reinforcement etc.) detection just after the welding, without detaching the welded part from the weld fixture was performed, where 3-D measurement of weld bead has also been realized using structured light principle [60, 61]. From the 3-D image, 1-D maps were extracted in two different directions to detect the amount of reinforcement and undercut. On-line detection of defects (humping) from the thermal images of the weld pool was done during the GTAW process [62]. A weld inspection system at shell-tube heat exchanger was realized, where the inspection system was attached with the welding robot [63].

Blowhole, lack/excess of material, misalignment of the welding centre, thin/large welding joint are four major defects present in fuel injectors' weld, which need to be detected to enhance performance. For instance, a machine vision set-up was employed in [64] to acquire weld images of fuel injectors, where a colour camera or RGB camera having 4 μm pixel pitch along with LED-based diffused illumination was utilized to detect the transitions present in micron range in defects like blow holes or lack/excess of materials. A very accurate (i.e., more than 97%) defect classification and localization technique on fuel injectors was realized using a single board computer based on a graphical processing unit (GPU), through an advanced deep neural network [64]. Detection of defects on the surface of friction stir weld (FSW) was also performed using image processing techniques [65–67]. 3-D reconstruction for detection of the volume of defects in FSW was performed using the laser triangulation principle [68]. Faster recognition of defective and no-defective canisters is required to reject the defective canisters from the assembly line. In this regard, a real-time machine vision system was developed [69]. To produce a weld with a high depth to width ratio, the keyhole laser welding process is adopted, where a highly focused laser beam having high power density (>107 W/cm^2) is utilized to melt and vaporize the materials to be welded by avoiding heat conduction. Vaporization of the materials forms a cavity (i.e., keyhole) by the penetration of laser beam so that the collapse of molten walls can be avoided. Then the keyhole is traversed to form the weld between two plates to be welded. Factors such as the size of keyhole and penetration are dependent on the process parameters like laser power, scan speed, beam focus etc. Final weld quality is dependent on the size of the keyhole and penetration. Therefore, machine vision-based monitoring of keyhole was employed to predict weld quality in [70, 71]. Also, the process parameters can be controlled online by using the result of monitoring. In laser welding of thick plates, the amount of plasma plume i.e., ion-ion, ion–electron, and electron–electron interaction between metal atoms, varies with the process parameters is visible along with the spatters.

Machine Vision Metrology

Automatic identification of parts, measurement of parts' dimensions, and checking of assembly in an assembly line are the major objectives to apply machine vision in metrology. In this line, identification of electronic components along with the motor, measurement of their dimensions and checking of electronic board assembly were performed in real-time on the conveyor belt where a machine vision-based digital twin was developed [72].

In an automatic assembly line of O-rings on plastic carriers (Fig. 4.13), a vibrating hopper is used to feed the O-rings into the primary wheel where plastic carriers are placed for assembly with O-rings. The unassembled O-rings are fed to the secondary wheel and are transferred to the collection bin with the help of an air knife. The suction of O-rings to the hopper occurs after transfer through the air-knife. Assembled carriers with O-rings are transferred from the primary to the secondary wheel via transfer track 1. Then unassembled carriers are transferred from the secondary to the primary wheel via transfer track 2. In this assembly process, four types of faults i.e., jamming of O-rings in two transfer tracks, electromechanical faults in air-knife, and

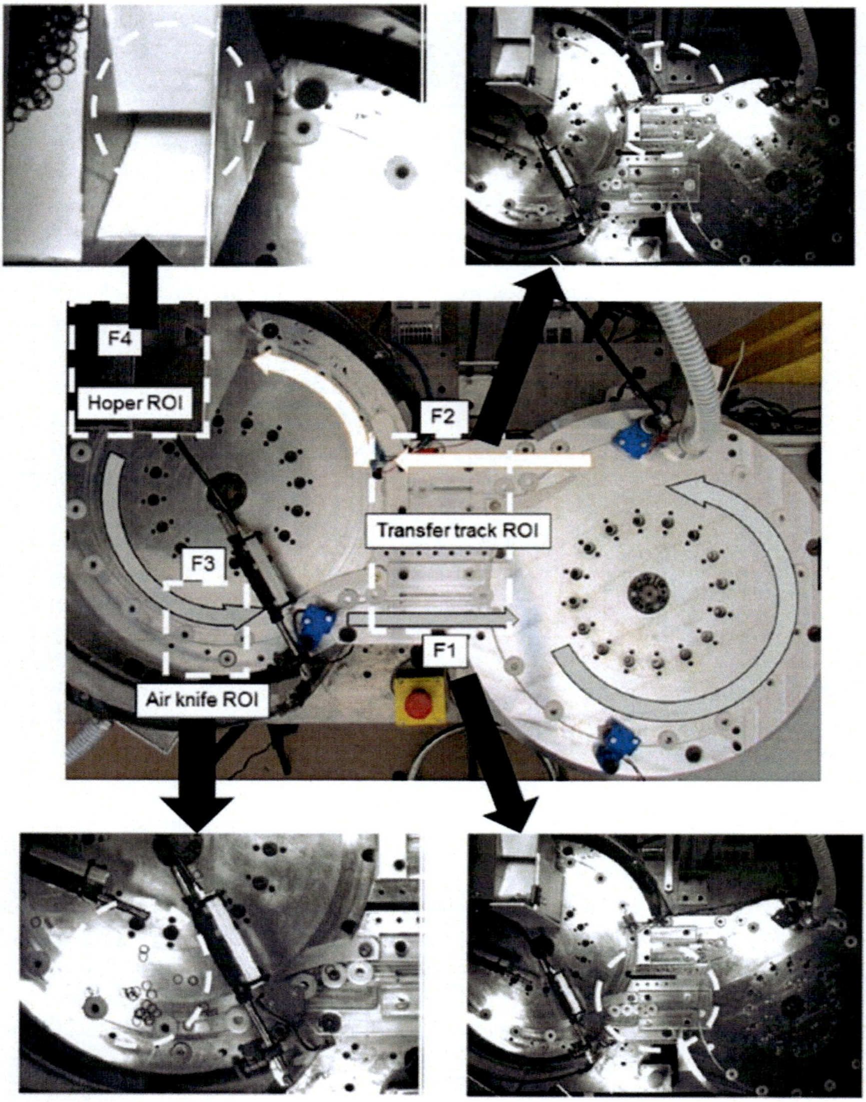

Fig. 4.13 Faults in an automated assembly line of O-rings with plastic carriers [73]. Reprinted by permission from Springer Nature Customer Service Centre GmbH: Springer Nature, The International Journal of Advanced Manufacturing Technology, Fault detection and classification in automated assembly machines using machine vision, Vedang Chauhan et al. 2016

missing O-rings in hopper occur which were detected by a machine vision system with maximum of 94% accuracy for necessary correction [73].

The process zones described above with their importance of application are summarized in Table 4.1 to introduce to readers the possibility of image processing in manufacturing applications.

So, real-time inspection and process control are two major areas in manufacturing, where machine vision plays an important role due to its flexibility, repeatability, and reliability. Applicability of machine vision for unmanned process inspection and control make it a tool for cyber-physical system [76]. The requirement of image

Table 4.1 Applicability of machine vision in various manufacturing methods for real-time process control and inspection

Process	Application	Process zone	References
Real-time process control			
***Manufacturing method**: Machining*			
Profile grinding	Control of tool position for profile error compensation	Tool and profile	Xu et al. [1]
μ-EDM	Tool position control or gap compensation from tool wear information	EDM tool	Yan and Lin [2], Malayath et al. [3]
μ-ECM	Control of vacuum gap from geometric information of produced cavity	Cavity	Guo et al. [4]
	Control of electrode's centre line to maintain the gap between machined surface and tool in laser ECM	EDM tool centreline	Jia et al. [5]
***Manufacturing method**: Additive manufacturing*			
DED process	Control of laser power from molten pool geometry	Molten pool	Rodriguez-Araujo et al. [6], Hofman et al. [7], Moralejo et al. [8]
	Control of scan speed from the thermal image of molten pool path to extract time–temperature history for microstructure control	Molten pool path	Farshidianfar et al. [9]
***Manufacturing method**: Welding*			
Fusion welding (mainly arc welding)	Control of process parameters from weld pool geometry	Weld pool	Wang et al. [10], Ma et al. [11], Na et al. [13]

(continued)

Table 4.1 (continued)

Process	Application	Process zone	References
	Tracking of the weld seam, adjusting offset between electrode centre and weld path centre, and updating process parameters	Weld seam and weld pool	Zhou et al. [12], Ma et al. [14], Shen et al. [15], Ye et al. [16], Chen et al. [17]
Real-time inspection			
***Manufacturing method**: Casting*			
Casting	Surface defect (bleed) detection during continuous casting	Cast surface	Pan et al. [18]
***Manufacturing method**: Forging*			
Hot forging	Dimension measurement of forged part during forging	Whole part	Liu et al. [19], Bračun et al. [20]
***Manufacturing method**: Forming*			
Hot rolling	Detection of various defects and their classifications during hot bar rolling	Surface of hot bar	Agarwal et al. [21]
	Detection of various defects and their classifications during hot rolling of flat products	Hot rolled surface	Song and Yan [22], Luo and He [23]
Cold rolling	Detection of cracks in presence of coolants' patches on cold-rolled surface	Cold rolled surface	Zhao et al. [24]
	Detection of surface roughness of rolls during reconditioning	Roll surface	Fischer and Stöbener [25]
***Manufacturing method**: Machining*			
Machining	Direct inspection or monitoring of cutting tools' flank wear width placed in tool holder	Tool wear region	Fernández-Robles et al. [29, 30], García-Ordás et al. [31], You et al. [32], Duan et al. [33], Lins et al. [34]
	Indirect prediction of tool wear from machined surface images	Machined surface	Kassim et al. [35, 36, 39, 40], Bhat et al. [37, 38], Gadelmawla et al. [42], Dutta et al. [41, 44–47]

(continued)

Table 4.1 (continued)

Process	Application	Process zone	References
	Grinding depth prediction in belt grinding	Spark field	Ren et al. [48]
Machining	Detection of surface roughness, machining chatter	Machined surface	Dhanasekar et al. [49], Al-Kindi et al. [50], Bamberger et al. [51], Samtaş et al. [52], Tootooni et al. [53], Sun et al. [54]
Robotic machining	Automatic identification of job in 3-D	To be machined job	Diao et al. [55]
	Automatic identification of hole's position	Aircraft's panel for assembly	Mei et al. [56], Yu et al. [57]
***Manufacturing method**: Welding*			
Fusion welding	Surface and internal defects from a sequence of radiographic images	Radiographic images	Shao et al. [58], Zou et al. [59]
	Defect and weld dimensions in 3-D	Weld surface	Chu and Wang [60], Li et al. [61]
	Just-in-time detection of defects	Weld bead	Alfaro et al. [62], Chu and Wang [63]
	Fuel-injector weld defect	Weld surface	Sassi et al. [64]
Friction Stir Welding	Surface defect detection	Weld surface	Bhat et al. [65], Sudhagar et al. [66], Ranjan et al. [67]
	3-D reconstruction of defects and their detection	Weld surface	Parikh et al. [68]
Resistance welding	Detection of weld defects of canister in production line	Weld surface	Sun et al. [69]
Laser welding	Prediction of porosity from the geometry of keyhole	Keyhole	Kim and Ahn [70], Luo and Shin [71]
Machine vision metrology			
Automatic assembly	Use of machine vision-based digital twin in the automatic assembly of components	Components and assembly on conveyor belt	Eckhart and Ekelhart[75]
Automatic assembly	Fault detection in the automatic assembly of O-rings in carriers	Detection of O-rings in various steps of assembly	Chauhan and Surgenor [73]

processing for process digital twin and inspection digital twin is required to improve the process seamlessly in the fourth industrial revolution [75–77].

4.3 Image Acquisition

Good image quality in terms of resolution, contrast, in-motion features for real-time detection with considerable field-of-view (FOV) is the primary requirement for applying image processing in manufacturing to develop a digital twin for process monitoring and online inspection. As a digital twin is envisaged to have accurate seamless monitoring, a proper machine vision system is required to acquire a perfect image having the above-mentioned quality, so that required information can be extracted for precise or repeatable inspection. A machine vision system, generally, consists of a camera, lens and illumination, and a proper combination of these three can allow the system for the acquisition of images having the required quality. A brief introduction of camera, lens, and illumination system and their major parameters to acquire good images in manufacturing, for precise quality control and process monitoring, are described in the following subsections.

4.3.1 Digital Camera

In a digital camera, an object or scene is acquired by image sensor and associated electronics in digital form, i.e., the whole image is represented in form of pixels. The pixels are the basic elements of the image sensor. One pixel is the basic unit of an image. Variation of brightness is represented by the pixel value. At each of the elements, lights from an object are incident on and the light energy or photons are converted into charges by the required electronic hardware for producing the digital value or pixel value. Image sensors have majorly two types of structures (1) Charged Couple Device (CCD), and (2) Complementary Metal Oxide Semiconductor (CMOS), based on the hardware for the photon to voltage and digital signal conversion process hardware. Both the sensors have some pixels arranged like a matrix, where incident photons are converted into charges with the help of semiconductor-based on the photo effect. Four sequential processes i.e., photon to charge conversion in each of the pixels, accumulation of charges, transfer of charges, and analogue to digital conversion occur in the sensor [78].

In CCD, lights from the scene or object are incident on each pixel (Fig. 4.14a), and the photons of the incident light are converted into charges or electrons by the semiconductor electronics inside the pixels (Fig. 4.14a). Then the charges are transferred to the adjacent CCD registers (vertical) through the transfer gates and accumulated in horizontal registers. Charges are transferred to voltage converter from the horizontal register and after amplifying voltage within the CCD sensor, it is converted to a digital signal in analogue-to-digital converter (ADC), situated outside

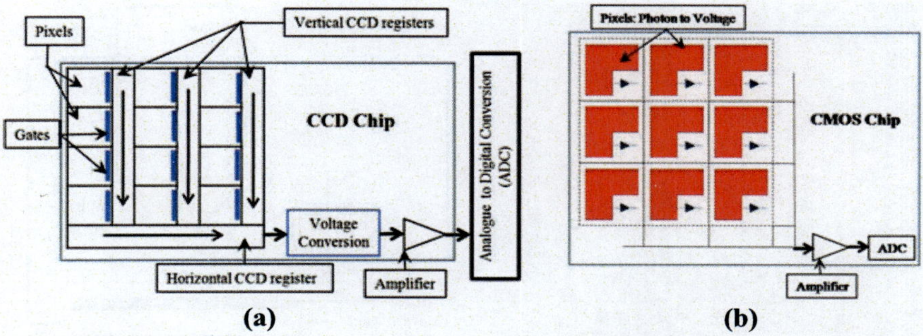

Fig. 4.14 Schematic representations of **a** CCD chip and **b** CMOS chip

CCD. The transfer of charges along with sequential readout is expressed as charge transfer efficiency (CTE), especially in high-resolution CCDs [78].

On the other hand, in the CMOS sensor, photon to charge and then voltage conversion occur in each pixel, due to their inherent structures. Transfer of charges along with sequential processing is not required, unlike the CCD camera. Digital signal is generated by the conversion of analogue signal through ADC in the CMOS sensor itself (Fig. 4.14b). Thus, the time of processing is reduced in CMOS.

There are some basic parameters involved at the time of a camera selection process for a particular monitoring or inspection purpose in manufacturing. They are mainly, sensor resolution, sensor size and format, type of scanning, exposure or shutter time and acquisition speed, spectral transmissivity range, and dynamic range.

4.3.1.1 Sensor Resolution

The resolution of an image sensor is expressed in terms of $M \times N$ pixels, where M and N represent the number of rows and columns, respectively. Pixel to pixel distance or pixel pitch (unit in μm/pixel) in a high-resolution sensor is much lower than a low-resolution sensor having the same sensor size (in length unit). So, high-resolution sensors are always preferred to detect objects or features having a very small size for better precision. However, a high-resolution sensor requires more time than that of a low-resolution sensor to convert photons into charges due to the presence of more pixels (i.e., semiconductor and associated electronics) in it. So, resolution enhancement reduces the speed of acquisition. At this point, the selection of resolution of the camera needs to be performed in an optimized way.

For example, a high-resolution sensor i.e., Genie GIGE HM1024, Dalsa, with 7.4 μm pixel to pixel distance was used to acquire the machined surface images (Fig. 4.15a), where the tool flank wear was monitored from the features extracted from the surface images. The surface became rougher with an increase in tool flank wear [27], progressively. However, a low-resolution camera can be used for classifying the unworn tool and worn tool from the machined surface images. On the

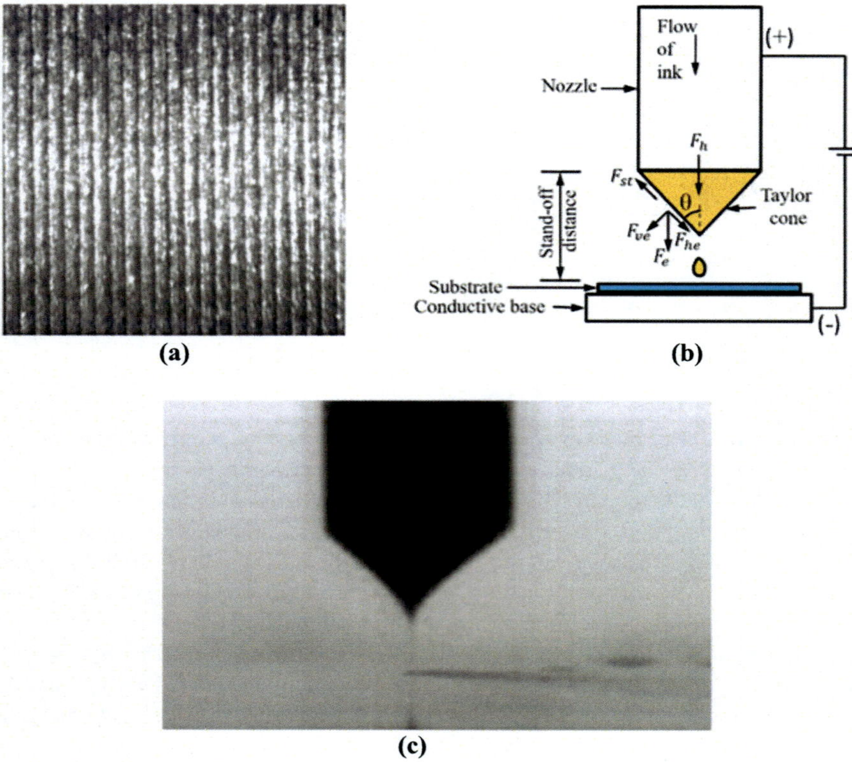

Fig. 4.15 **a** Image of a machined surface acquired using a high-resolution image sensor (7.4 μm pixel pitch) and **b** Taylor cone formation phenomenon in EHD jet printing and **c** corresponding image of Taylor cone acquired using a low-resolution image sensor (20 μm pixel pitch)

other hand, in electrohydrodynamic (EHD) jet printing, a type of micro/nano additive manufacturing, conductive ink having nanoparticles is supplied with a controlled flow rate (in the range of picolitre or nanolitre) via a nozzle to a flexible substrate to produce a pattern. It is a process to fabricate flexible sensors. The sensitivity of the sensor is dependent on the width and resolution of that pattern. The flexible substrate is placed on a conductive base over a controllable translational stage, by which the required pattern can be generated keeping the nozzle in a fixed position. Micro/nano resolution printing can only be achievable by reducing the size of drops falling on the substrate, as shown in (Fig. 4.15b). The size of the drop and its uniformity throughout the process is dependent on the formation of a Taylor cone due to the EHD action. Taylor cone is formed due to the application of electrostatic forces between nozzle and substrate [78]. In this case, hydrostatic, electrostatic, and surface tension forces are responsible for the geometry of the Taylor cone or Taylor cone angle (θ). All these forces are varied due to the variation of various online controllable process parameters, such as nozzle to substrate distance, applied voltage, and pulse frequency and offline parameters, such as surface tension, viscosity, and conductivity of the ink. The quality of EHD jet printing is partially dependent on the Taylor cone angle. So, monitoring Taylor cone angle using image processing can be useful for controlling the process in real-time. However, there is a need for high-speed acquisition with comparatively low-resolution images that can be used. As an example, a high-speed camera with a low-resolution sensor (Model: Photron Mini AX50), having 20 μm/pixel resolution, was used to acquire the formation of Taylor cone in 2000 frames per sec (fps) speed (Fig. 4.15c) to evaluate Taylor cone using image processing.

However, the resolution can also be enhanced by using a high magnification lens. But the cost of a high-resolution camera and high magnification lens need to be considered at the time of selection. Speed of acquisition and illumination also need to be considered in this case.

4.3.1.2 Sensor Size and Sensor Format

Sensor size is the physical size (i.e., in mm × mm) of the image sensor, where sensor format is the diagonal of the rectangular image sensor expressed in 1/3″, 1/2″, 2/3″, 1″, etc. However, the 1″ here is represented as 16 mm. The size of the camera is dependent on the sensor size. Pixel to pixel distance is dependent on the sensor size.

Various lens formats are shown in Fig. 4.16. Lens selection is dependent on the sensor format so that no pixel can be wasted, i.e., the selected lens format should be greater than or equal to the sensor format of the imaging sensor.

The incompatibility issue between lens format and sensor format is shown in Fig. 4.17, where a lens (Navitar 7000 zoom lens) having 1/3″ format is used with a camera having a square sensor format. Information of many pixels outside the lens format is lost due to format mismatching.

Fig. 4.16 Dimensions of sensors with various sensor formats (all dimensions are in mm)

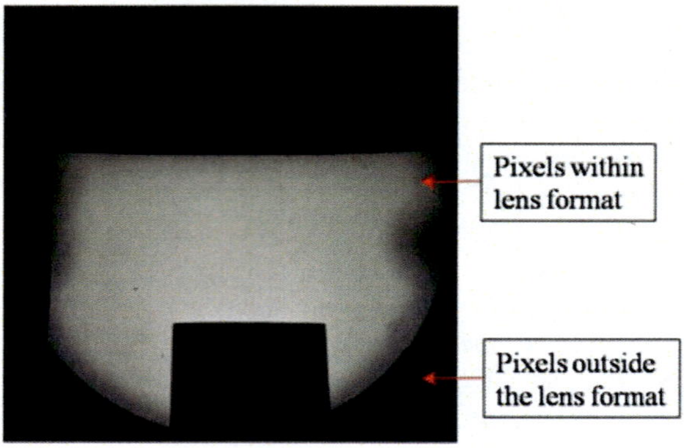

Fig. 4.17 Compatibility of lens and sensor formats

4.3.1.3 Type of Scanning

There are two types of camera, based on the type of scanning—line-scan and area-scan. In a line-scan camera, a single row of pixels is exposed, wherein a matrix of pixels is exposed in an area-scan camera at the time of image acquisition. Thus, a relative motion between the object and the line-scan camera is required at the time of acquisition. The acquisition speed of the line-scan camera is, generally, more as compared to the area-scan camera as less exposure time is needed due to the presence of a smaller number of pixels in the line-scan camera than that of an area scan camera. Thus, the line-scan camera can work in a large range of reset timing and synchronization conditions. Flexibility in on-site imaging for industrial inspection is more in line-scan camera, as it is free from perspective distortion due to the angle between the camera and object. It has also less depth of view requirement for focusing on a site. Due to these advantages, a line-scan camera is highly applicable in industrial inspection scenario, where, the object is at high speed as sheets in the hot rolling process [23].

4.3.1.4 Exposure or Shutter Time and Acquisition Speed

Exposure time is dependent on the intensity of light that falls on each pixel. A high-intensity light needs to be exposed for less duration to convert photons into charges. Also, a large number of pixels require more exposure time with a particular level of illumination. So, acquisition speed i.e., frames per second for area scan camera and Hz for line scan camera, is dependent on the exposure or shutter time. In the exposure time, a frame is scanned.

CCD camera has a high dynamic range, as the well capacity to store charge or electron in each pixel is high. However, less charge transfer efficiency (CTE) is a challenge in CCD camera for high-speed image acquisition. Motion blur, i.e., a faint image of a previous frame getting superimposed over the current frame due to the remaining residual charges due to image lag, is a major image distortion that occurs while acquiring images of a moving object. Exposure time (t_e) can be determined by Eq. 4.1 [78].

$$t_e < \frac{D_B}{v \cdot \cos\alpha \cdot M} \tag{4.1}$$

where D_B, v, α, and M are allowable blur circle diameter, the relative speed between camera and object, the angle between the image plane and direction of motion, and magnification, respectively. Therefore, to acquire a deblurred or properly focused image of an object moving with velocity,v, maximum exposure time, t_e can be allowed. In manufacturing, weld features in robotic welding, surface inspection in continuous casting and rolling etc., acquisition speed matters a lot, where real-time monitoring and control is predominant. In these cases, an optimized selection of acquisition speed, illumination, and resolution is needed for maintaining precision in the results.

To show the effect of exposure time or shutter speed, an example in Fig. 4.18 is given, where the same object i.e., surfaces of a cylinder after turning operation, is imaged with two shutter speeds i.e., 5000 and 10,000 fps, keeping all other parameters same. It can be seen that the contrast of Fig. 4.18a is higher than that of Fig. 4.18b due to an increase in shutter speed, however, more minute details are depicted in the image captured with higher shutter speed, as seen in Fig. 4.18. Higher light intensity is also required to acquire an image with higher shutter speed, as with higher shutter speed, the time of exposure of each pixel reduces.

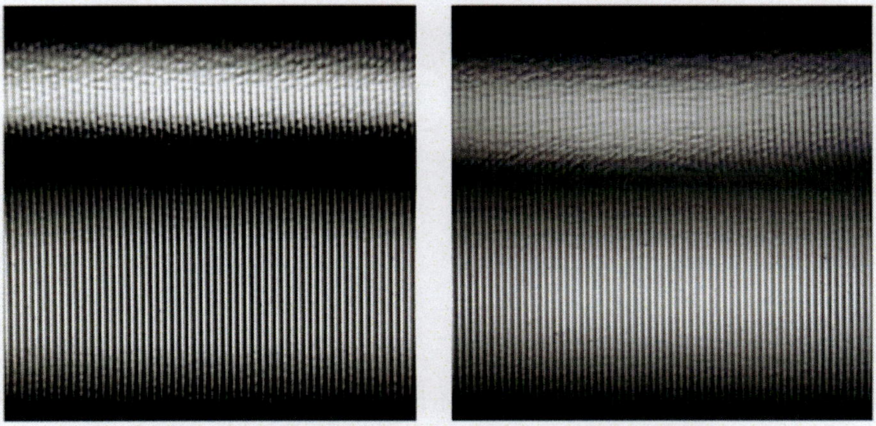

(a) (b)

Fig. 4.18 Surface image of a rotating cylinder (i.e., 30 rpm) of 100 mm diameter acquired with **a** 5000 fps and **b** 10,000 fps

4.3.1.5 Spectral Response

Image sensors have relative sensitivity in the illumination of various wavelengths i.e., spectral response. A CCD camera is sensitive to low light in the visible range. Some image sensors are sensitive to infrared range (wavelength above 700 nm), which are generally used as a thermal camera for sensing temperature fields. In some manufacturing processes, such as fusion welding, hot rolling and DED, properties of produced materials depend according to the rate of change in temperature at the time of manufacturing. Defects may also be produced due to improper thermal distribution at the process zone. As an example, the heat transfer rate can be monitored via a time–temperature history of molten pool in the DED process [9]. Instead of using a thermal camera, a pyrometer could also be used in this case as a low-cost system. But, the time–temperature information with molten pool geometry cannot be extracted with only involving a pyrometer. So, to monitor thermal information along with the geometry of the process zone, a thermal or infrared camera is always preferred.

If the spectral response of the image sensor is irregular, then the blooming effect may occur to acquire a highly bright scene due to the overflowing of charges to the neighbouring pixels. To reduce blooming, some photosensitive areas are assigned, where the area is quantified by fill factor i.e., percentage of photosensitive area.

4.3.1.6 Dynamic Range

The ratio of peak signal to the minimum signal detectable in the image sensor is termed as dynamic range. Thus, the contrast capability (i.e., the ratio of the maximum brightness to the least brightness detectable) in an image can be enhanced for a high dynamic range camera. Also, the blooming effect can be reduced for high dynamic range sensors, such as CMOS sensors. Dynamic range is expressed in decibel (dB). For example, the applications where high contrast details with low noise are necessary, a good dynamic range camera needs to be applied. In molten pool imaging during the DED process, a camera with a 100 dB dynamic range is necessary [6].

4.3.2 Lens

Acquisition of object image with proper sharpness and magnifying the distant object are major two requirements for which lens need to be selected optimally. The major parameters of a lens are focal length, format, aperture, depth of view to obtain high contrast and sharp image with the required information.

4.3.2.1 Focal Length

Proper focusing at the time of image acquisition is the prime criterion to obtain a sharp image, which needs to detect the edge of the target object, accurately, for the subsequent inspection purpose. Hence, the focal length, F, of the lens is the primary parameter in lens selection. Focal length can be determined by using the Gaussian lens formula, as shown in Eq. 4.2.

$$\frac{1}{d_O} = \frac{1}{F} - \frac{1}{d_I} \tag{4.2}$$

where d_O and d_I are the distance of object and image from the principal plane of the lens, respectively, and F is the focal length. The focal length of a lens is dependent on the refractive power. High focal length lenses are telephoto lens, where compound lenses are used in the lens system. To select a lens focal length, parameters such as the image sensor's size, the field of view (FOV), and d_O are need to be known and by using Eq. 4.3, focal length can be chosen.

$$F = \frac{\text{Image Sensor Size in } mm \ \times d_O}{\text{FOV}} \tag{4.3}$$

Generally, a small focal length (for example, 12 mm) is prone to aberrations or distortions, wherein high focal length lenses are preferable in terms of sharpness but with high illumination. There are fixed and controllable focal length lenses, where the controllable focus is preferable in terms of manoeuvrability.

Acquired images of a turned surface with 70 and 108 mm focal length are shown in Fig. 4.19a, b, respectively, keeping all other acquisition factors constant. An increase in focal length causes an increase in the optical zoom, so optical magnification is visible in Fig. 4.19.

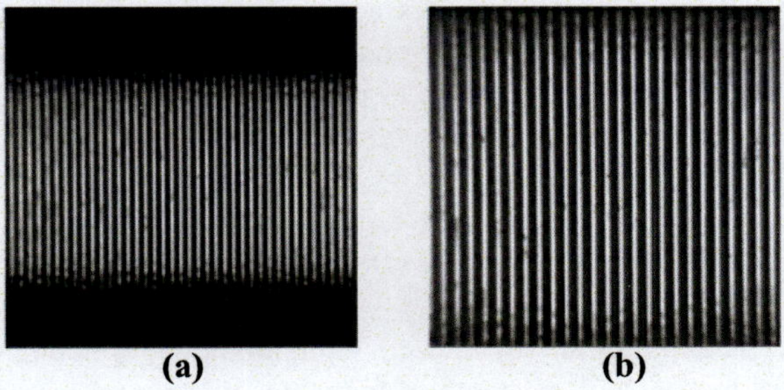

Fig. 4.19 Turned surface image acquired with **a** 70 mm and **b** 108 mm focal length using the same object to camera distance

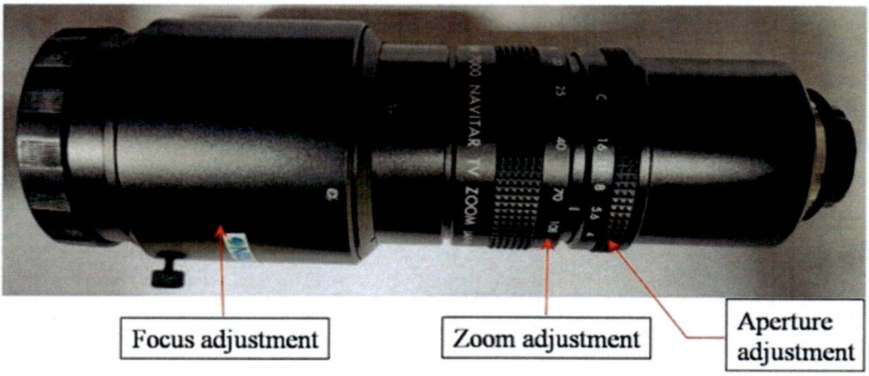

Fig. 4.20 Picture of a lens having focus, zoom, and aperture adjustment

4.3.2.2 Aperture

The aperture of the lens is the iris opening diameter (D_A) to allow the light into the lens. It is generally expressed through f-number or f-stop or relative aperture ($f_{\#}$) i.e., a ratio between the focal length to D_A or F/D_A. So, higher F needs higher D_A for proper illumination to image an object. On the other hand, where illumination is less, a large D_A i.e., a low f-number is needed. However, a very low f-number is not preferable to obtain a sharp image, and there is a limit on high f-number as well to resist diffraction (diffraction limit) at the edges of aperture or iris. So, an optimized aperture selection is necessary for low diffraction blur and a proper sharp or focused image. In terms of acquisition time, a smaller aperture or higher f-number requires higher acquisition or exposure time with the same illumination.

Figure 4.20 shows the picture of a telephoto lens having adjustment capability of focus, zoom, and aperture adjustment. However, this kind of lens is costlier and the size is also bigger.

Turned surface images acquired using different aperture settings are shown in Fig. 4.21, where it can be observed that some pixels are nearly saturated in Fig. 4.21a i.e., least f-number with highest D_A. Saturation level, as well as contrast, reduces with an increase in f-number as seen in Fig. 4.21.

4.3.2.3 Depth of View (DoV)

The depth of an object (i.e., depth along the optical axis) in focus is known as depth of view (DoV) or depth of focus. DoV is expressed as shown in Eq. 4.4 [78].

$$\text{DoV} = 2D_B \times f_{\#} \times M(M + 1) \tag{4.4}$$

where D_B represents allowable blur circle diameter (generally, 1–2 pixels).

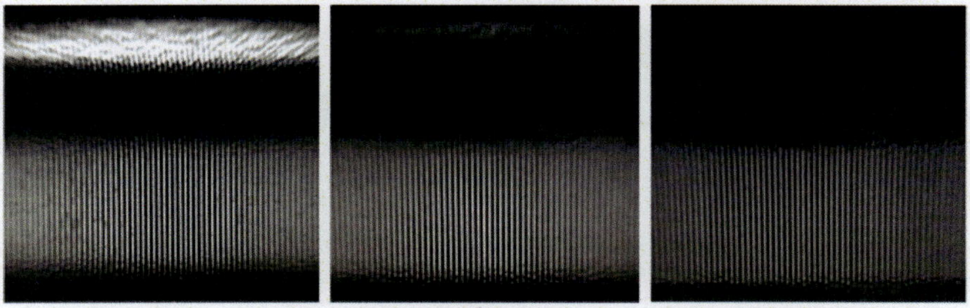

Fig. 4.21 Turned surface images acquired using f-number **a** 11, **b** 16 and **c** Maximum available (near close)

If required magnification M and $f_{\#}$ are high (for high illumination), then a large DoV is needed. The Telecentric lens has a larger DoV, as compared to other lenses, but a high illumination is needed for proper attenuation. If the amplitude of the displacement along the optical axis is greater than DoV, motion blur can occur. In telecentric lens, range of DoV is higher, thus objects or scene prone to vibrations can be acquired using the telecentric lens. If an acquisition system is required to attach onto a robotic manipulator, such as in robotic welding [10], e.g., for seam tracking or weld pool monitoring, then the telecentric lens is preferred.

4.3.2.4 Lens Resolution

The resolving power of lenses is generally expressed by the modulation transfer function (MTF) curve, where the measure is discrimination between the two adjacent lines and expressed in lines per mm (lpmm). Better quality lens or high-resolution lens can distinguish between finely spaced lines i.e., high lpmm. The resolution of the lens should be chosen according to camera sensor resolution. A high-resolution camera can work properly only with high-resolution lenses.

4.3.3 Illumination

In an active machine vision system, an illumination source is required, for acquiring still or moving objects, enhancing the contrast between the process zone and its background, overcoming the fluctuation due to ambient lighting present in a harsh industrial environment. However, no external illumination source is required if the illumination from the process zone is utilized to image the process zone such as in the monitoring of weld pool in arc or laser welding [10, 11, 15, 16, 74]. This type of machine vision system is known as a passive machine vision system. Mainly two types of lighting i.e., front and back-lighting are used to illuminate process zone in

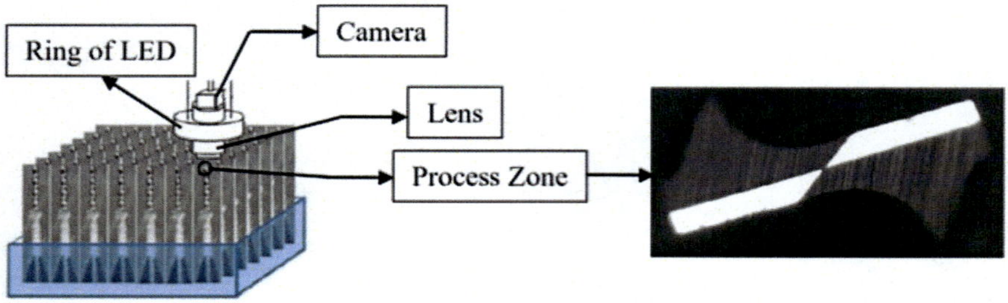

Fig. 4.22 Co-axial active front illumination to acquire an image of micro-drill bit for wear detection [33]. Reprinted by permission from Springer Nature Customer Service Centre GmbH: Springer Nature, Machine Vision and Applications, Automatic optical flank wear measurement of microdrills using level set for cutting plane segmentation, Guifang Duan et al. 2010

an active machine vision. In front lighting, the illumination source is placed in front of the process zone to illuminate it, where the light from the illumination source is falling on the zone and the reflected light from the zone is acquired via lens and camera system. In most of the cases in manufacturing, such as the acquisition of machine surface images for surface finish evaluation, cutting tool wear images for tool flank wear measurement; even in the inspection of defects in rolling or in casting or in welding, the front lighting system is used. In front lighting, there is an option for coaxial illumination, where the illumination axis and optical axis are aligned together. This type of illumination system is termed bright field illumination. In cases, such as, wear monitoring of drill bits or mills, a coaxial illumination system has been used, where the object is illuminated by a ring of LED [33], as shown in Fig. 4.22.

In the inspection of machined surfaces, the undulations due to abrasive wear or flank wear of the cutting tool, roughness component is mainly experienced on machined surfaces at an inclined direction parallel to the flank of the cutting tool. A proper image of these undulations can be obtained if the illumination source is kept normal to the inclined plane, as shown in Fig. 4.23a, and the camera needs to be kept

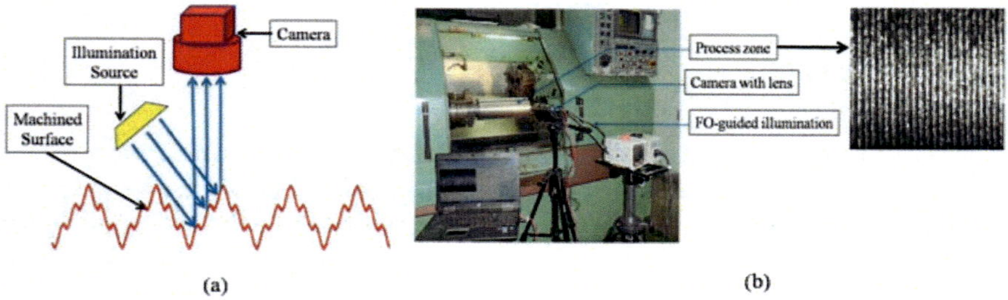

Fig. 4.23 Directional front illumination **a** schematic representation, **b** actual experimental set-up in turning and acquired image of process zone, i.e., turned surface with fibre optic (FO)-guided illumination

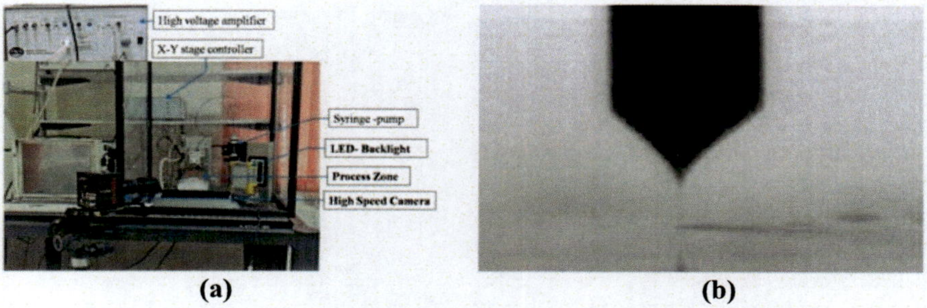

Fig. 4.24 Back illumination system **a** set-up of image acquisition in EHD jet printing, **b** image of process zone i.e., EHD jet

normal to the process zone. The image acquisition set-up to acquire turned surface images is also shown in Fig. 4.23b. The acquired image of the turned surface is also shown in Fig. 4.23b. This type of front illumination system is popularly known as dark field front illumination.

A back-lighting system is used in cases, where the boundary or the outer profile of the targeted object only needs to be processed for the required information. Examples include control of offset in profile grinding [1], tool wear compensation in micro-EDM [2, 3], imaging of Taylor cone formation in EHD jet printing for micro or nano additive manufacturing needs backlighting to get sharper and high contrast object, which can be easily delineated from the background [79]. In back illumination, the object is situated between the illumination source and camera system, where all three are lying on the optical axis.

The back illumination system is shown in Fig. 4.24, as an example. In the EHD jet printing process, as already mentioned in Sect. 4.3.1.1, detection of the geometry of jet i.e., Taylor cone geometry is important for updating the process parameters or predicting the quality of printing. So, in this case, backlighting is used, as only object boundary information is enough.

4.3.3.1 Sources of Illumination

Various sources of illumination are briefly mentioned here. One of the low-cost illumination sources is a fluorescent lamp that is employed to obtain uniform lighting, where frequencies in illumination source and shutter should be matched through a rectifier to avoid the problem of interference in acquired images. Array or ring of light-emitting diodes (LEDs) is another kind of low-cost illumination source, which are very popular in industrial imaging due to their small size and weight, monochromatic nature, cool or diffused lighting and longer lifecycle. LEDs are widely used as ring and array structures to make homogeneous illumination. However, reflective metal surfaces are prone to more specular, where diffuser or diffused illumination is recommended for acquiring A uniformly illuminated surface image to avoid shadows [32].

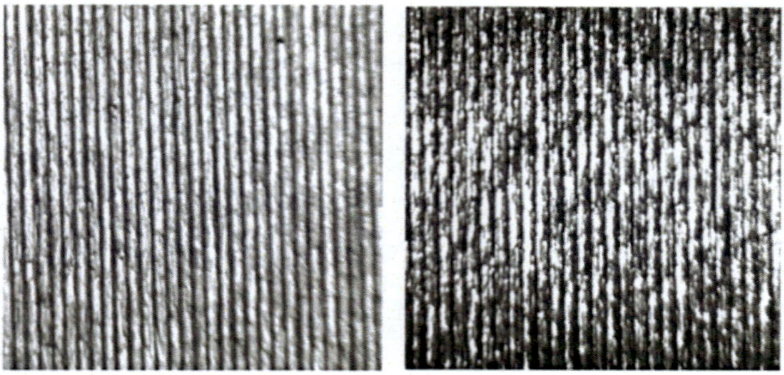

(a) (b)

Fig. 4.25 Turned surface image acquired with **a** a diffused illumination, and **b** FO-guided directional illumination

Turned surface images, acquired with a diffused illumination and with FO-guided directional illumination, are shown in Fig. 4.25a, b, respectively. The diffused light provides homogeneous illumination at the surface as compared to the directional light. However, directional light is more robust to ambient lighting.

In real-time and high-speed image acquisition, stroboscopic lighting with LEDs is recommended [78]. In stroboscopic lighting, a controller is attached, where the camera trigger is synchronized with stroboscopic lighting frequency. In this case, the exposure is triggered, only when stroboscopic light is on. Another class of small and highly directional light source with a long lifecycle and less power input is the quartz halogen lamp, which can be recommended as an illumination source in the industrial environment. Though the temperature generation of this lamp is very high, due to the presence of quartz glass, the service life is not affected because of the increase in temperature. This kind of lighting can be integrated with an FO-guided system, where directional lighting is necessary. This kind of directional lighting is needed in the harsh industrial environment, such as tool condition monitoring in machining.

Laser sources are also used for lighting, where monochromatic and coherent light is required. Structured lighting, in the form of an array, line or circles is used with a laser source for projecting lasers on some profile to obtain 3-D information, such as depth. Structured lighting can be used in the 3-D inspection of weldment [60]. Laser illumination is also utilized to generate a speckle pattern on the manufactured surface to monitor its surface finish [25].

4.3.4 Optical Filters and Special Optics

In some special cases like in welding or laser-induced manufacturing, high heat, as well as high illumination at the process zone, may cause glaring and blooming in the acquired image. It can also affect the image sensor, adversely, by the continuous overflowing of charges. This can also cause damage to the lens optics. In these cases, the use of optical filters is extremely necessary while capturing images in some particular cases with features, such as weld pool, weld bead, metal transfer, plasma plume, keyhole, etc.

Optical filters are mainly of two kinds, wavelength-dependent and wavelength-independent. Most of the filters such as band-pass, band-stop or notch, edge, colour, dichroic, etc. are wavelength dependent. Wavelength dependent filters allow for transmitting light having a particular range of wavelength. Light having a band of wavelength can only pass through the band-pass filter. The range of wavelength that can pass through this filter is the main criterion for filter selection. If one needs to monitor the weld pool in inert gas welding of aluminium alloy, then any filter within 600–700 nm range is preferable with some good attenuation percentage [10, 11]. If one needs to extract the exact process zone, especially for weld seam tracking or monitoring of penetration in arc welding or laser welding processes, then a narrowband filter having a very small bandwidth (20–30 nm) is preferred. Parameters to select a band-pass optical filter are central wavelength (wavelength at which the transmission is the maximum) and bandwidth (generally, full width half maximum or FWHM). Band-stop or notch filters are used to stop light having a range of wavelength, i.e., just vice-versa of the band-pass filter. Edge filters are of two types—low-pass and high-pass. Low-pass filters attenuate low wavelength light and high-pass filter is just the opposite of low-pass filter. A neutral density (ND) filter is a wavelength-independent filter through which a flat range of wavelength can attenuate uniformly. ND filter usually reduces the light intensity to avoid glares or blooming or oversaturation.

4.3.4.1 Image Acquisition System in Various Manufacturing Applications

Image acquisition systems used in various manufacturing applications are shown in Table 4.2. From this table an overall idea on the type of camera, lens, optical accessories, and illumination system can be gained.

An overall idea on the selection of camera sensor resolution and shutter speed or exposure time can be obtained from Table 4.2. For instance, a camera with high sensor resolution in the range of 1–3 μm/pixel can be selected for detection of micro-tool wear used in micro-EDM as the size of the tool is very small (in 100 μm range), where the speed of the process is lower (up to 12 mm/s) [2–4]. On the other hand, the shutter speed can be limited in the range of 15–30 fps, for acquiring process zone in a low-speed process. For instance, at the time of tool wear inspection, just after machining, the process is stopped and the process zone is very small (within 300 μm) [29–31]. In that case, a high-resolution camera with a low shutter speed is recommended. In profile grinding, position coordinates of the contour affect the shape of the ground profile in the micron range. But the speed of the process is very low i.e., 0.02 mm/s. In this case, 5–10 μm/pixel resolution camera with 10–15 fps shutter speed is enough for process control or inspection purpose. Moderate speed

Table 4.2 Image acquisition system used in real-time process control and process inspection in various manufacturing processes

Process and application	Camera sensor and resolution	Process speed and acquisition speed	Lens and other optical accessories	Illumination	References
Real-time process control					
Manufacturing method: *Machining*					
Control of tool position for profile error compensation in *profile grinding*	CCD sensor, 5 MP resolution 6.982 μm/pixel	0.02 mm/s process speed 15 fps acquisition speed	Telecentric lens	Back illumination LED with 5 W power	Xu et al. [1]
Tool position control or gap compensation from tool wear information in *micro-EDM*	CCD with 640 × 480 pixels resolution 1.3 μm/pixel	0.25–0.75 mm/min process speed	Zoom lens	Back illumination	Yan and Lin [2]
Tool position compensation in *μ-EDM drilling of a blind hole*	CCD with 800 × 600 pixels resolution	0.2 mm/min	Zoom lens	Back illumination	Malayath et al. [3]
Control of vacuum gap from geometric information of produced cavity in *ECM*	Dino-Lite Edge AM4115ZT microscope (1.3 MP) 1.43 μm/pixel	30 fps	220X magnification with polarizer or anti-reflection glass	LED front illumination	Guo et al. [4]
Control of electrode wire position in *ECM* with respect to centre of the laser drilled micro holes	CCD sensor, 1600 × 1200 pixels 5.7 μm/pixel		0.8X	Ring LED	Jia et al. [5]

(continued)

Table 4.2 (continued)

Process and application	Camera sensor and resolution	Process speed and acquisition speed	Lens and other optical accessories	Illumination	References
Manufacturing method: *Additive manufacturing*					
Control of laser power from molten pool geometry in *DED* for controlled dilution	CMOS with 120 dB dynamic range Resolution: 800 × 600 CameraLink interface	>100 fps	Dichroic mirror/beam splitter, Visual range bandpass and neural density (ND) filters	Coaxial front illumination	Rodriguez-Araujo et al. [6]
	CMOS with 42 dB dynamic range Resolution: 800 × 600 Spectral range: 400–1000 nm	200 Hz	Dichroic mirror/ eam splitter, band-pass filter	Passive	Hofman et al. [7]
	CMOS with 25 μm/pixel	400–800 mm/s process speed and 150 fps	Notch filter with 1064 nm having 73 nm FWHM 700 nm central wavelength band-pass filter	10 ms with 3.5 pixel error	Moralejo et al. [8]
Control of scan speed from thermal image of molten pool path to extract time temperature history for microstructure control	Infrared camera monochrome 384 × 288 pixels resolution Spectral range: 750 nm Pixel equivalent: 0.5 to 0.65 mm/pixel				Farshidianfar et al. [9]

(continued)

Table 4.2 (continued)

Process and application	Camera sensor and resolution	Process speed and acquisition speed	Lens and other optical accessories	Illumination	References
***Manufacturing method**: Fusion welding*					
Control of process parameters from weld pool geometry in TIG	CCD camera	125 fps	560–700 nm bandpass Dimmer glass with 30% attenuation Depth of view = 1/1000	Passive	Wang et al. [10]
Weld pool and seam monitoring in GTAW with Argon gas shielding	Two CCD cameras with 60 × 60 pixels area	2.6 mm/s is process speed and 20 fps according to the pulse frequency	590–680 nm (depending on the Ar gas shielding due to less intensity) ND filter with 10% transmission DoV = 1/1000	Passive	Ma et al. [11]
Weld pool monitoring and control of laser current in laser welding using active vision	CCD sensor			Active with laser strobe illumination having 5 ns pulse duration, 330 nm wavelength and 7 kW peak power	Na et al. [13]
Weld seam tracking during robotic welding in GTAW process	CCD sensor	20 mm/s speed		Passive vision	Zhou et al. [12]
Welding voltage rectification based on the information of offset between weld pool centre and seam centre line	CCD sensor	20 fps	590–680 nm band-pass filter with 4% transmission ND filter with 90% attenuation	Passive vision	Ma et al. [14]

(continued)

Table 4.2 (continued)

Process and application	Camera sensor and resolution	Process speed and acquisition speed	Lens and other optical accessories	Illumination	References
Weld seam tracking during welding of curved line from feedback information of offset between torch and seam centreline in GTAW	CCD 768 × 576 pixels resolution	2.67 mm/s welding speed	590–710 nm bandpass filter, two dimmer glasses with 90 and 70% attenuations, respectively	Passive vision	Shen et al. [15]
Weld seam tracking during welding from feedback information of offset between torch and seam centreline in pulsed MAG welding	CCD sensor	5 mm/s welding speed	620–680 nm bandpass filter, Dimmer glass with 94% attenuation	Passive vision	Ye et al. [16]
Weld seam tracking from offset information between seam and pool centres in GMAW with Ar shielding	CCD sensor	7 mm/s welding speed	660 nm central wavelength bandpass filter, Dimmer glass with 6% transmission	Passive vision	Xu et al. [74]
Tracking of 2 welding heads of robot for weld seam of 0.2 mm gap in presence of CO_2 gas for shielding	CCD sensor; 60 μm/pixel	1000 mm/s welding speed		White LED light	Chen et al. [17]
Real-time inspection					
Manufacturing method: *Casting*					
Surface defect (bleed) detection during continuous casting	512 × 2048 pixels	25.4 mm/s			Pan et al. [18]

(continued)

Table 4.2 (continued)

Process and application	Camera sensor and resolution	Process speed and acquisition speed	Lens and other optical accessories	Illumination	References
***Manufacturing method**: Forging*					
Dimension measurement of forged part during hot forging	CCD sensor			Projected light stripes	Liu et al. [19]
Dimension measurement of forged part during hot forging from projected laser stripes on hot forged parts	CCD sensor with 1290 × 960 pixels resolution; 0.35 mm/pixel	50 fps	530 nm band-pass filter; Lens with 35 mm focal length	530 nm laser	Bračun et al. [20]
***Manufacturing method**: Forming*					
Detection of various defects and their classification during hot rolling	200 × 200 pixels	3–18 m/s		LED	Song and Yan [22]
Detection of various defects and their classification during hot rolling (650–950 °C)	Line scan camera Resolution: 4096 pixels/line Optical fibre communication	0.5–20 m/s 40 kHz line rate		Array of halogen lamps Strobe triggering depending on speed of rolled surface	Luo and He [23]
Detection of cracks in presence of coolants' patches on cold rolled surface	CCD sensor with 640 × 480 pixels resolution, 33.5 μm/pixel	90 fps		LED light	Zhao et al. [24]

(continued)

Table 4.2 (continued)

Process and application	Camera sensor and resolution	Process speed and acquisition speed	Lens and other optical accessories	Illumination	References
Surface roughness detection of rolls' surface during reconditioning	CCD sensor	200 Hz and acquisition of speckle pattern	Collimator, mirror, lens for creating a spot on the surface	Pulsed Laser with 532 nm wavelength Gaussian beam profile creation Synchronous illumination	Fischer and Stöbener [25]
Manufacturing method: *Machining*					
Tool wear localization and detection on profile milling head inserts	CMOS camera with 1280 × 960 pixels resolution	Offline but around 30 inserts within 4 min	25 mm focal length lens	LED	Fernández-Robles et al. [29, 30], García-Ordás et al. [31]
Milling tool wear detection using wide FoV camera	CMOS, 2048 × 2048 pixels resolution, 4.125 μm/pixel		Telecentric lens	Ring LED	You et al. [32]
Monitoring of micro drill wear in PCB manufacturing	CCD, 512 × 680 pixels resolution			LED	Duan et al. [33]
Drill wear measurement integrated with cyber-physical system	CCD			LED	Lins et al. [34]
Tool wear prediction from machined surface texture image	CMOS with 300 × 300 pixels subject area	On-machine tool	Telecentric Macro zoom lens	FO guided directional light	Dutta et al. [41, 44–47], Gadelmawla et al. [42], Datta et al. s[43]
Grinding depth in belt grinding	Color CCD	100 fps			Ren et al. [48]
Detection of surface roughness in moving images with blur	CCD	25.4 mm/s speed	2X magnification	Diffused white light	Dhanasekar et al. [49]

(continued)

Table 4.2 (continued)

Process and application	Camera sensor and resolution	Process speed and acquisition speed	Lens and other optical accessories	Illumination	References
Detection of surface roughness with low cost set-up for process parameters updation with camera connected in the machine tool	Webcam with 2592 × 1944 pixels resolution				Al-Kindi et al. [50]
Machining chatter mark identification on automotive valve seat surface	Camera			Ring LED	Bamberger et al. [51]
Surface roughness detection on face milled surface	Microscope with 1280 × 1024 pixels resolution		240X magnification		Samtaş et al. [52]
Surface roughness measurement with camera connected in the machine tool	CCD sensor with 2448 × 2048 pixels resolution Image resolution (actual): 600 × 300 pixels 5 µm/pixel	15 fps upto 256 rpm speed	Telescopic lens with 12 mm focal length	diffuse white LED with 20,000 lx intensity	Tootooni et al. [53]
Position detection in robotic drilling, especially in assembly	Baumer industrial camera				Mei et al. [56]
Inspection of countersinks in aircraft's panel	CCD camera		Telecentric lens	Ring LED (coaxial)	Yu et al. [57]

(continued)

Table 4.2 (continued)

Process and application	Camera sensor and resolution	Process speed and acquisition speed	Lens and other optical accessories	Illumination	References
***Manufacturing method**: Welding*					
3-D mapping of weld for defect detection	CCD sensor with 768 × 576 pixels resolution			Structured laser light	Chu and Wang [60]
Weld pool geometry detection in TIG welding	Infrared camera Spectral range: 7.5–13 μm	60 fps, 4–16 mm/s	12 mm focal length lens with adjustable aperture		Alfaro et al. [62]
Weld quality inspection of shell tube heat exchanger	CCD camera		Band-pass filter having 590 ± 10 nm wavelength		Chu and Wang [63]
Weld defect detection of fuel injector	CCD camera with VGA resolution, 4 μm/pixel		0.75X–1.0X	White LED array controlled by micro-controller	Sassi et al. [64]
Defect detection on FSW surface	CCD camera with 640 × 480 resolution	Offline	Macro zoom lens	Diffused white light	Bhat et al. [65]
Defect detection in welding of canister on a production line	CCD camera with 1024 × 768 pixels resolution	35 cm/s, 15 fps	Telecentric lens with 55 mm focal length	Coaxial LED light	Sun et al. [69]
Monitoring of keyhole in Laser welding	CCD camera with 644 × 239 pixels resolution	30 fps	band-pass filters with 532, 660, 800, 810, and 1000 nm central wavelength ND filter (ND16) Coaxial monitoring with dichroic mirror	Active vision with 532 nm wavelength laser and 1.6 W power	Kim and Ahn [70]

(continued)

Table 4.2 (continued)

Process and application	Camera sensor and resolution	Process speed and acquisition speed	Lens and other optical accessories	Illumination	References
Manufacturing method: *Machine vision metrology*					
Inspection and fault detection in automatic assembly line	CCD camera with 1928 × 1448 pixels resolution	26 fps	Lens: 8.5 mm focal length with 2/3″ format and working distance >1100 mm; Polarizer	Dark-field diffused LED	Chauhan and Surgenor [73]
Identification and inspection of parts and assembly in vision based digital twin	CMOS camera with 4208 × 3120 pixels resolution; 1.1 μm pixel size	19 fps (Camera integrated with GPU platform)	Wide angle lens		Eckhart and Ekelhart [75]

processes, such as molten pool monitoring in DED, higher shutter speed is needed with the sacrifice of the resolution, as the accuracy in molten pool width is required in the range of sub-millimetre. In that case, 25–60 μm/pixel resolution with 100–200 fps shutter speed is good enough [6–8]. If the object under inspection is very large, such as dimension measurement of a hot-forged object during forging, the resolution can be considered as 350 μm/pixel [20]. Real-time inspection of surface defects such as in hot or cold rolling requires very high shutter speed as the speed of the process zone (i.e., speed of rolled surface) is in the range of 1–20 m/s [23, 24]. In this case, a line scan camera can also be employed, as in a line scan camera, pixel size or resolution need not be compromised with an increase in acquisition speed even with a high line rate such as 40 kHz. Also, the images are required to be processed as fast as possible for a process having quick variation. The time required for processing an image using an image processing hardware board is lesser as compared to a dedicated PC. Also, portability is another advantage of image processing hardware, integration of which makes an imaging system smarter. A GPU, utilizing several cores in smart cameras integrated with a GPU based developer board was used to develop the machine vision-based digital twin to inspect part and assembly [72]. FPGA is another solution for performing faster computation due to the presence of semiconductor logic circuits in it for real-time embedded image processing.

The role of telecentric lens or zoom lens is required to magnify the process zone, where a large distance between object and camera is needed to avoid manufacturing hazards, but a high-resolution image is essential to extract micron level features. For instance, in profile grinding, a sufficiently large distance between object and camera was required to avoid machining hazards, however higher resolution is needed to reduce the error between theoretical and actual contour to 3 μm [1]. Magnification is required to enhance the imaging of micro-tools and electrolyte droplets in micro-EDM and ECM, respectively, where a telecentric lens is indispensable [2–4]. Detection of the coordinate of the centre of laser machined hole was required to perform further ECM operation as discussed in Sect. 4.2.1.1, where hole size was in the range of 1–2 mm diameter to limit the maximum error within 12 μm [5]. In that case, the zoom lens should be incorporated during image acquisition. At the time of tool wear measurement and surface roughness evaluation, precision should be in the sub-micron range. However, the inaccessibility issue is there, so that the camera to object distance is maintained 100–250 mm. Therefore, a telecentric macro zoom lens with high magnification is used for this purpose [29–32, 49, 52, 53].

Special optical accessories, such as band-pass filter, neutral density filter, dimmer glass, beam-splitter, and dichroic mirrors are required in the manufacturing processes where process zone are highly intense and shielding gases are present.

Also, coaxial imaging of molten pool in DED is used as a passive vision system [6–8]. In this case, a laser beam is focused to melt the blown powder for deposition on a substrate to produce the required layer in a layered deposition. Just after the deposition, a molten pool is formed, whose image needs to be acquired co-axially to avoid any angular deformation for controlling laser power. A special optical arrangement is needed to avoid imaging of laser beam. Thus, a beam splitter or dichroic mirror is placed, which deviates the light direction to 90° and only allows the light from

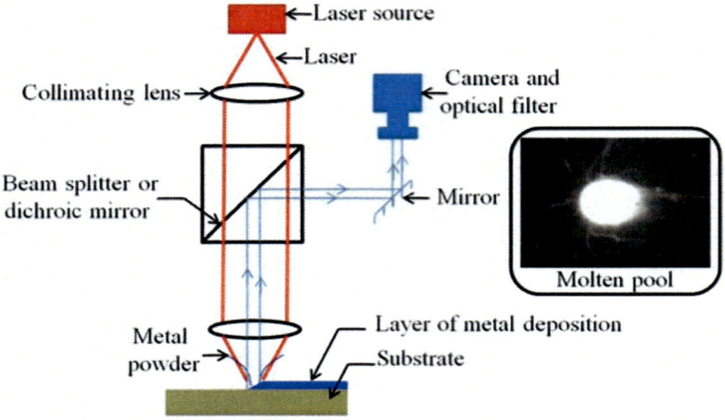

Fig. 4.26 Co-axial imaging by passive illumination to acquire the image of molten pool in metal additive manufacturing

the molten pool as shown in Fig. 4.26. A similar arrangement for the acquisition of keyhole image is also made in the laser welding process [70].

For process monitoring of weld pool and seam, a band-pass filter is required to transmit the light with a particular central wavelength [10, 11, 14–16, 74]. ND filter or dimmer glass is used to attenuate extra light to avoid pixel saturation. In DED, optical filters are used for the same reason. In forging application, as the laser light falls on the forged surface, it creates lines or stripes. The acquisition of the stripes using a band-pass filter is required where the central wavelength of the filter should be the same as that of the laser [20].

Back illumination is always preferred, where the external boundary is to be detected, subsequently reducing processing activity. Detection of tool offset in profile grinding [1], tool wear in micro-EDM [2, 3], and monitoring of Taylor cone in EHD jet printing are some of the applications where back illumination is required. In front lighting, LED, either in array form or in ring form is used for illuminating electrolyte droplets in ECM [4], to identify the centre of micro-hole in laser ECM [5], and also to detect surface defects in cold rolling [24], tool wear in machining [29–31], surface quality inspection [51], weld defects identification [64, 69] etc. Surface inspection with diffused white light gives uniform illumination over the reflective surface where non-homogeneity or defects on the surface is highlighted [49, 53, 65]. Structured light is employed to gather 3-D surface information in welding [60]. Laser triangulation is required for extracting dimensions of hot forged parts during forging [20]. Synchronous strobing with high-intensity halogen light was employed to acquire hot rolled surface moving with very high speed i.e., 20 m/s, with 40 kHz line acquisition rate without motion blurring [23]. In-situ detection of surface roughness with 0.5 nm resolution was realized by creating a synchronous speckle pattern on rolls' surface, where the speed of acquisition was 200 Hz without blurring [25].

4.4 Image Enhancement

Digital image and its representation are first discussed here, before discussing image enhancement. A grayscale image consists of pixels in the form of a matrix, where gray-level values of the pixels are the matrix elements. Each of the matrix elements or pixels has some value, in the case of an 8-bit or grayscale image, which is equivalent to the intensity of light, falls on the sensor. Higher grayscale values result from the brighter portion, whereas lower grayscale values result from the darker portion of the object to be imaged.

For example, an image of a surface turned by an unworn tool is shown in Fig. 4.27a, and an image fragment, i.e., the rectangular zone (shown by a red-coloured rectangle in Fig. 4.27a is cropped. Then the pixel values or gray-level values of the pixels are shown in Fig. 4.27b in the form of a matrix, where *r* and *c* are row and column numbers, respectively. The pixel values of the brighter region, i.e., first 3 columns are on a higher side as compared to the pixel values of the darker region, i.e., 6th and 7th columns. It can be observed that according to the variation in the intensity of incident light, the gray level values change. Therefore, spatial information of gray-level values is the main source to extract informative features from an image. In the spatial domain, an image is represented as, *i*, in which, the gray-value of coordinate or pixel (*x*, *y*) is represented as, *i*(*x*, *y*). As shown in Fig. 4.27b, the gray level value of pixel coordinate (5, 5) is 132, i.e., *i*(5, 5).

Neighbourhoods of the pixel (x, y) are pixels located at the co-ordinates, $(x - 1, y - 1)$, $(x - 1, y)$, $(x - 1, y + 1)$, $(x, y - 1)$, (x, y), $(x, y + 1)$, $(x + 1, y - 1)$, $(x + 1, y)$, $(x + 1, y + 1)$. As an example, neighbourhood pixels of (5, 5) are (4, 4), (4, 5), (4, 6), (5, 4), (5, 6), (6, 4), (6, 5), and (6, 6) and their gray-level values are 204, 130, 90, 196, 89, 196, 129, and 85, respectively. The gray-level values of neighbourhood pixels are required in most of the spatial image processing operations,

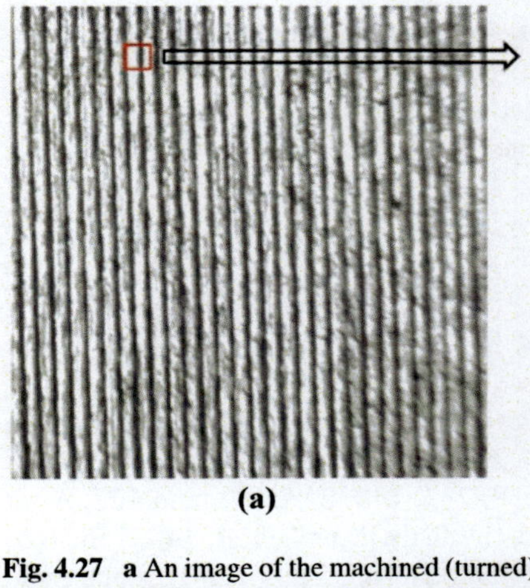

(a)

r \ c	1	2	3	4	5	6	7	8	9	10
1	250	255	252	223	129	89	95	147	222	251
2	255	255	252	224	133	89	92	145	225	252
3	255	255	247	195	124	86	93	148	208	249
4	255	255	249	204	130	90	96	152	208	249
5	255	255	247	196	132	89	92	145	211	250
6	255	255	248	196	129	85	94	151	228	252
7	255	255	247	192	130	93	101	153	214	225
8	255	255	246	189	129	91	91	144	224	241
9	255	255	247	190	121	85	90	137	225	252
10	255	255	247	192	123	86	89	136	226	252

(b)

Fig. 4.27 **a** An image of the machined (turned) surface with an unworn tool, and **b** the pixel values of the image fragment of the red-coloured rectangular zone of Fig. 4.27a starting from enhancement to features extraction. Processing in the spatial domain or the frequency domain is performed, mainly, to extract the informative features.

An image is enhanced by reducing the noise content and/or by enhancing the contrast or edges in spatial or the frequency domain before subsequent processing, such as image segmentation and feature extraction. In spatial domain processing, the image is filtered in the spatial domain through the neighbourhood or convolution operation using a filter for noise reduction in the image. Image can also be enhanced in terms of its contrast by modification of its histogram, in the spatial domain. Frequency domain transformation is performed by Fourier transformation on an image to enhance it in the frequency domain by using frequency domain filters, where, element to element multiplications or array operations are only performed to reduce computational time. After performing the operations in the frequency domain, inverse Fourier transform is done to obtain the processed or enhanced image. Various spatial and frequency domain operations are described as follows with examples of images of various manufacturing operations.

4.4.1 Image Enhancement in the Spatial Domain

Neighbouring operations with or without convolution is the main key to perform spatial filtering of an image. Image, affected by noise, results in distorted features, which may be detrimental to decide on the selection of process parameters in a manufacturing process.

4.4.1.1 Image Denoising in the Spatial Domain

Neighbourhood operation and convolution of an image by a spatial smoothing filter are two main operations for image denoising in the spatial domain. Filtering operations are explained on a turned surface image having salt and pepper noise. The noise is created on an original image of turned surface (Fig. 4.28a) by applying salt and pepper noise having a mean of 0 and standard deviation of 0.02 (Fig. 4.28b).

Neighbourhood Filter

A median filter is a kind of neighbourhood filter, where the median value of the neighbourhood (including the target pixel value) is replaced at the target pixel.

As an example, the 3rd row and 5th column of Fig. 4.29a is a pepper noise, which is the target pixel value to be filtered by median filtering. The neighbourhood values of pixel (3, 5) of Fig. 4.29a are 165, 191, 158, 237, 140, 248, 182, and 127. After sorting them including the (3, 5) pixel value, the list will be 0, 127, 140, 158, 165, 182, 191, 237, and 248, where the median value is 165. Thus 165 is replaced in target pixel (3, 5) as shown in Fig. 4.29b, d is the median filtered image of Fig. 4.29c.

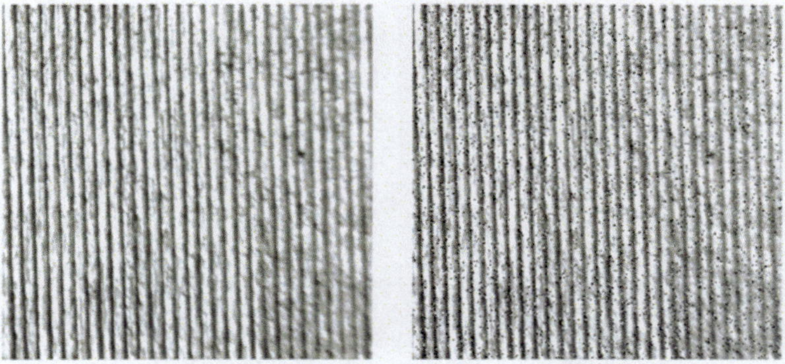

(a) (b)

Fig. 4.28 Turned surface image **a** original, **b** salt and pepper noise (mean = 0, variance = 0.02) applied on it

233	179	197	222	203	145	88	118	216	243
218	168	189	165	191	158	89	119	214	5
231	217	237	237	0	140	93	98	193	214
255	231	255	248	182	127	107	94	156	229
253	225	255	238	201	110	107	82	113	205
236	255	255	0	149	132	77	82	122	0
241	225	254	255	180	119	251	80	103	213
4	224	251	225	0	139	72	78	115	183
231	238	255	247	177	141	82	79	116	203
255	241	240	203	154	138	95	60	101	219

(a)

0	179	168	189	158	89	89	89	118	0
179	217	197	197	165	140	118	118	193	193
217	231	231	191	165	127	107	107	156	156
225	237	237	237	182	110	107	107	156	156
231	255	248	238	149	127	107	107	113	113
225	253	254	238	149	132	107	103	103	103
224	241	251	225	139	132	82	82	103	103
224	238	247	247	177	139	82	82	115	115
224	240	240	225	154	138	82	82	115	115
0	238	238	177	141	95	79	79	79	0

(b)

(c) (d)

Fig. 4.29 **a** Image fragment of a noisy image, Fig. 4.28b and **b** its median filtered part, **c** Noisy image and its **d** median filtered image

0.0625	0.125	0.0625
0.125	0.25	0.125
0.0625	0.125	0.0625

(a)

118	145	144	150	141	110	83	101	134	102
158	198	197	186	160	134	111	132	162	111
169	220	222	192	137	118	109	122	161	130
181	238	242	219	163	124	106	110	155	148
183	243	231	200	167	131	101	99	124	111
181	244	218	168	146	136	115	102	104	81
148	225	231	187	141	136	133	108	113	104
117	207	243	204	138	119	113	97	125	126
149	225	242	210	156	122	95	86	125	129
138	182	178	158	128	99	72	62	92	100

(b)

Fig. 4.30 Example of convolution. **a** Filter, f, and **b** output of convolution of Fig. 4.29a with filter, f

Median filtering is a non-linear filter, which is mainly used to overcome salt and pepper noise.

Filtering by Convolution

Convolution is the key in spatial filtering operation for image smoothing and image sharpening for noise elimination and boundary or edge enhancement, respectively. The convolution operation is defined mathematically in Eq. 4.5.

$$o(x, y) = \sum_{m=-a}^{a} \sum_{n=-b}^{b} f(m, n) i(x + m, y + n) \tag{4.5}$$

where an image i is convolved with a filter, f to obtain an output image, o. This operation is explained with an example.

The values of filter coefficients of a filter f (Fig. 4.30a) are $f(-1, -1), f(-1, 0), f(-1, 1), f(0, -1), f(0, 0), f(0, 1), f(1, -1), f(1, 0)$ *and* $f(1, 1)$ are 0.0625, 0.125, 0.0625, 0.125, 0.25, 0.125, 0.0625, 0.125, and 0.0625, respectively. For instance, if the target pixel is (3, 5) for convolution, then according to Eq. (4.5), the output pixel value is obtained by masking the target pixel and its 8-neighbourhoods with the filter, f.

$o(3, 5) = f(-1, -1) \times i(2, 4) + f(-1, 0) \times i(2, 5) + f(-1, 1) \times i(2, 6) + f(0, -1) \times i(3, 4) + f(0, 0) \times i(3, 5) + f(0, 1) \times i(3, 6) + f(1, -1) \times i(4, 4) + f(1, 0) \times i(4, 5) + f(1, 1) \times i(4, 6) = 0.0625 \times 165 + 0.125 \times 191 + 0.0625 \times 158 + 0.125$

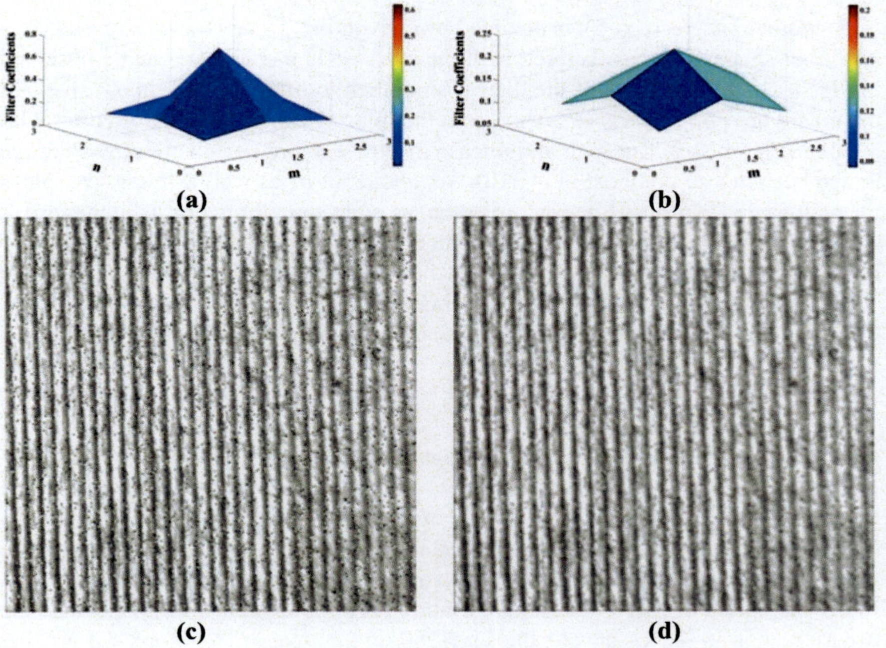

Fig. 4.31 **a**, **b** Surface plot of 3 × 3 Gaussian spatial filter with **a** σ = 0.5 and **b** σ = 1, **c**, **d** smoothen image of Fig. 4.26c with filter shown in Fig. 4.31a and with filter in Fig. 4.31b, respectively

× 237 + 0.25 × 0 + 0.125 × 140 + 0.0625 × 248 + 0.125 × 182 + 0.0625 × 127 = 137.

It is shown by a square in the filtered or convolved matrix in Fig. 4.30b. The other values of the image are convolved with the filter by striding the filter from left to right.

Gaussian Filter

It is a smoothing filter, which is convolved with the noisy image, where its parameters are its size and standard deviation values. In a smoothing filter, the low frequency features in the image are only passed; so smoothing filters are also called low-pass filters (LPF). Characteristics Gaussian filter (f_g) is dependent on Eq. 4.6 [80].

$$f_g(m,n) = \frac{1}{2\pi\sigma} e^{\frac{(m^2+n^2)}{2\sigma^2}} \tag{4.6}$$

where σ is the standard deviation. The variation of Gaussian filter with σ, and its effect on images are shown in Fig. 4.31.

Smoothen image, $o(x, y)$, is obtained by convolving $f_g(x, y)$ with $i(x, y)$. Gaussian filter shape becomes flattered with the increase in σ-value, as can be observed in Fig. 4.31a, b. The effect of the filter is smoother with the increase in σ-value. All the noises are not properly removed from the image by using a Gaussian filter with σ =0.5 (Fig. 4.31c). But with an increase in σ to 1, more noise is removed but the image becomes smoother (Fig. 4.31d). So, choice of σ, as well as filter size, plays the major role for proper filtering operation, as more smoothing results in the loss of many important high-frequency details. For example, the variations in pixel value due to the presence of feed mark help to detect the edges or feed lines in Fig. 4.31d. Due to smoothing, the contrast between the feed mark and the background reduces. It affects the minute details present due to surface roughness. This is discussed elaborately in Sect. 4.4.1.3.

4.4.1.2 Image Enhancement by Histogram Processing

Image histogram represents the contrast of an image, where the number of occurrences of a gray-level value is plotted, as shown in Fig. 4.32. In the image histogram plot, X-axis and Y-axis are gray-level intensity and number of pixels, respectively. Image histogram of Fig. 4.32a is represented in Fig. 4.32b, as an example, where the occurrence of 50–55 Gy-level values is 4150, in the image. Therefore, the number of occurrences of low gray-level values is more in the darker image, which can be seen in Fig. 4.32b.

Therefore, Fig. 4.32a is a low-contrast image. Histogram equalization is adopted for contrast enhancement. In histogram equalization, the image is mapped to a desired contrast-enhanced histogram. Ideally, the histogram of a perfect contrast image should be a horizontal straight line, i.e., occurrences of all the gray-level values are the same. Histogram equalized image of Fig. 4.32a is shown in Fig. 4.32c, and the corresponding histogram is shown in Fig. 4.32d. However, the contrast cannot be enhanced properly in histogram equalization as it is a global operation. So, local non-uniformity due to highly uneven illumination cannot be fully overcome through this technique. To mitigate this issue, there is a requirement to employ some technique where local histogram processing can be possible and contrast can also be controlled. Contrast limited adaptive histogram equalization (CLAHE) technique can serve both purposes [81]. CLAHE image of Fig. 4.32a is shown in Fig. 4.32e. It can be observed that the inhomogeneity in the image due to uneven illumination is corrected, which is also reflected in the corresponding histogram (Fig. 4.32f).

4.4.1.3 Image Sharpening in the Spatial Domain

Image features, such as edges, point, and line extraction are highly required for further segmentation operation. In surface finish inspection, small undulations due to tool wear can be observed at the feed marks. So, a clear distinction between feed marks, undulations, and other features is required to extract correct information.

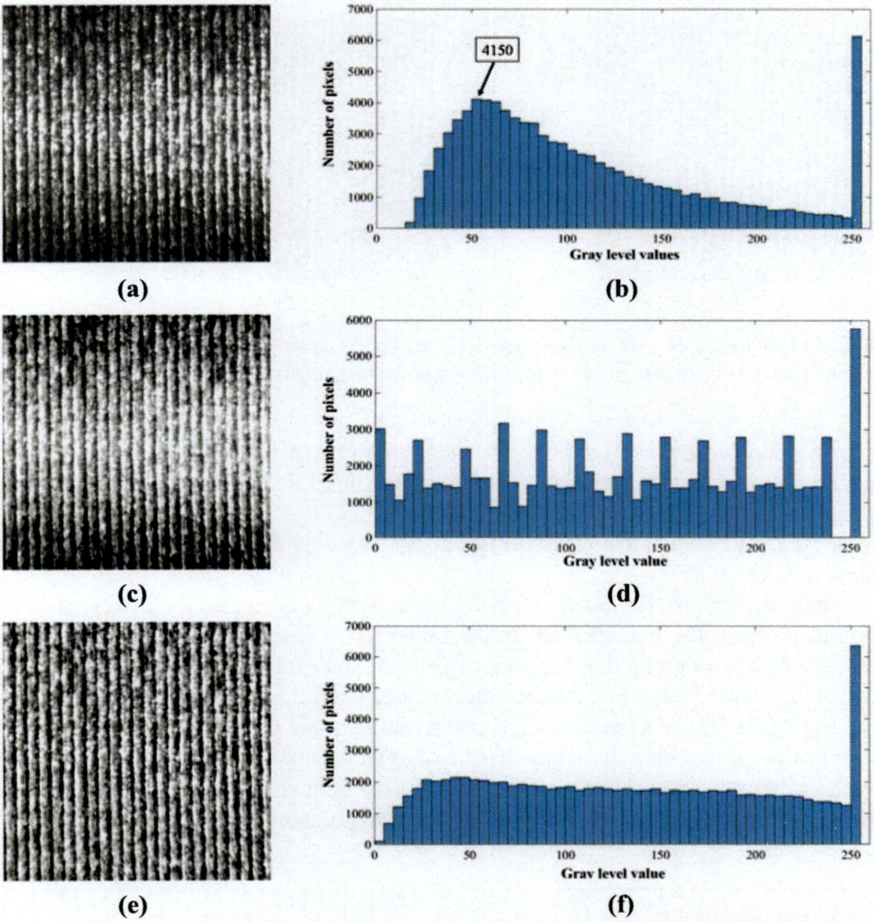

Fig. 4.32 **a** Turned surface image acquired using FO-guided front illumination system, and **b** its histogram, **c** histogram equalized image of Fig. 4.30a, and **d** the corresponding histogram, **e** contrast limited adaptive histogram equalized image, and **f** the corresponding histogram

Undulations on the feed marks due to tool wear are visible in the magnifying fragment part of Fig. 4.33a. However, the difference between the gray-level values of edges of undulations and the adjacent background is required to be more prominent (i.e., sharpen the image) for making the subsequent segmentation accurate. The sharpened image of Fig. 4.33a is represented in Fig. 4.33b. Unsharp masking and image sharpening filters are used in image sharpening, which are described as follows.

In unsharp masking, the original image, $i(x, y)$ is first smoothened with some low-pass filter and a smoothened image, $s(x, y)$ is obtained. Then high-frequency features (i.e., edge, lines, corners, etc.) are extracted by subtracting $s(x, y)$ from $i(x, y)$, to obtain $e(x, y)$. Then this image $e(x, y)$, with the high-frequency feature is

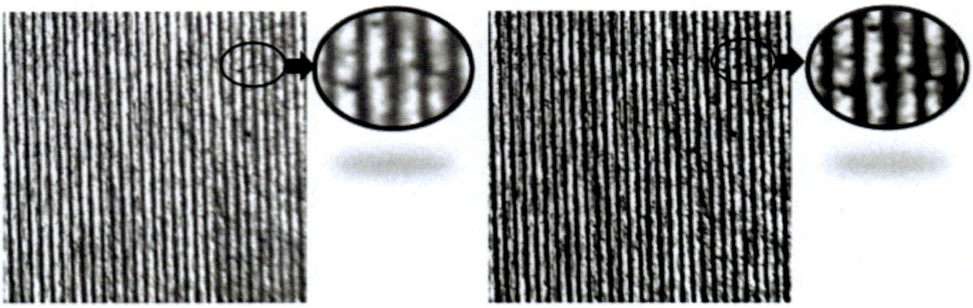

(a) (b)

Fig. 4.33 **a** Image of a turned surface and a fragment shown within elliptical boundary by magnifying it, **b** sharpened image of Fig. 4.33a and the magnified portion

simply added with $i(x, y)$ to obtain the unsharp mask image, $o(x, y)$. The algorithm of the unsharp masking operation is shown in Fig. 4.34a and the unsharp masked image of Fig. 4.33a is shown in Fig. 4.34b.

Unsharp masking of images obtained from various manufacturing processes is shown in Fig. 4.35.

Enhancement of Taylor cone boundary is extremely necessary to segment the boundary from the background, if the object is to detect the Taylor cone angle, automatically, for predicting the print quality. It can be observed from Fig. 4.35b that the Taylor cone image is enhanced, substantially, after applying the unsharp mask on Fig. 4.35a. There dynamic range of the radiographic image (Fig. 4.35c) of FSW is very low from which the segmentation of defect is challenging. So, it is enhanced with unsharp masking (Fig. 4.35d). Similarly, the surface defect of FSW can also be enhanced using unsharp masking for further segmentation as shown in Fig. 4.35f.

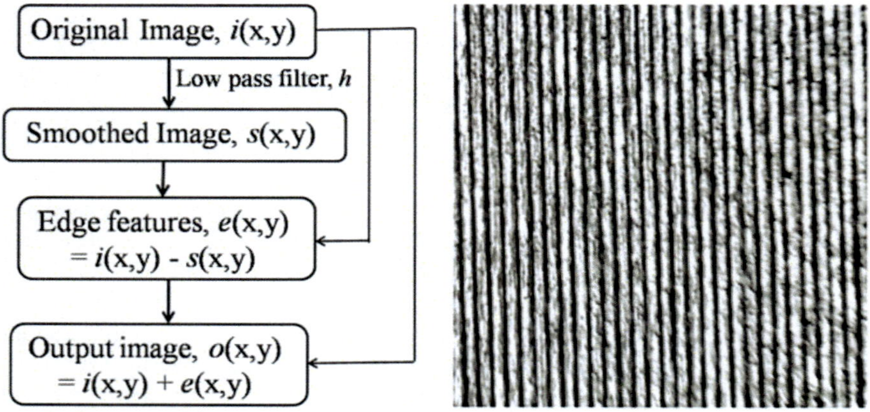

(a) **(b)**

Fig. 4.34 **a** Algorithm of unsharp masking and **b** unsharp mask image of Fig. 4.33a

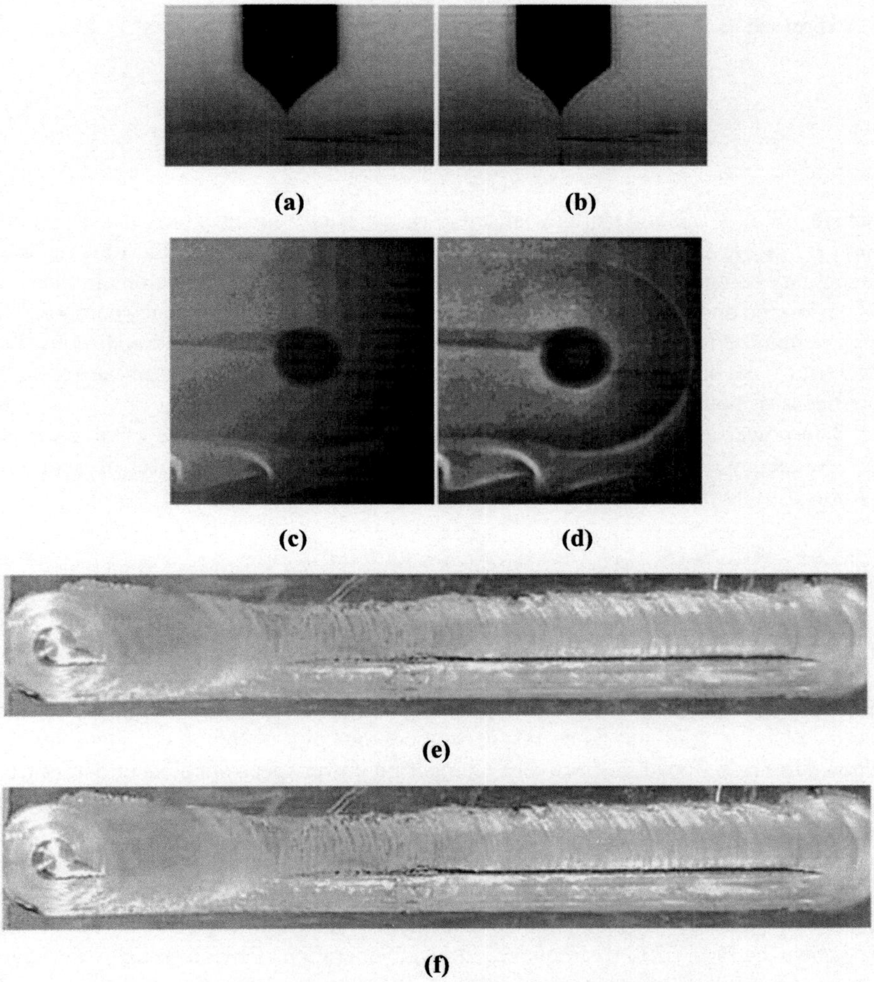

Fig. 4.35 Original images of **a** Taylor cone in EHD jet printing, **c** X-ray image of FSW defect, **d** surface defect of FSW and **b**, **d**, **f** are the corresponding unsharp mask images, respectively

Unsharp mask is only used to enhance the high-frequency features i.e., sharpening the overall image.

4.4.2 Image Enhancement in the Frequency Domain

The frequency information of an image is mapped, when an image is converted into the frequency domain by discrete Fourier transform (DFT). DFT of an image, $i(x, y)$

is expressed as $I(u, v)$, as shown in Eq. 4.7.

$$I(u, v) = \sum_{x=0}^{R-1} \sum_{y=0}^{C-1} i(x, y) e^{-j2\pi(\frac{ux}{R} + \frac{vy}{C})} \tag{4.7}$$

where $j = \sqrt{-1}$, u and v are two discrete variables or frequency axes in 2-D, with $u = 0, 1, 2, \ldots, R-1$ and $v = 0, 1, 2, \ldots, C-1$ for image size $R \times C$. So, there are real and imaginary components after DFT. After performing DFT, frequency domain filtering is performed on frequency transformed image, $I(u, v)$, where frequency components can be modified directly. Also, the frequency domain filter is not required to convolve to $I(u, v)$; the only element to element multiplication or array operations are enough with less computational cost.

The power spectrum is represented as the power of the image obtained from the frequency domain. This is expressed in Eq. 4.8, where the power of both real ($Re(u, v)$) and imaginary components ($Im(u, v)$) are involved.

$$I^2(u, v) = Re^2(u, v) + Im^2(u, v) \tag{4.8}$$

From the information of power spectrum, energy or power of the image is represented.

The smooth or defect-free surface of FSW and its corresponding contour plot of the power spectrum are shown in Fig. 4.36a, c, respectively. Also, a defective weld surface having void and its corresponding contour plot of the power spectrum are shown in Fig. 4.36b, d, respectively. Comparing power spectrum plots of defect-free and defective surface, it can be observed that the power at the central frequency of defect-free, smooth surface is more than that of the defective surface, due to more periodic distribution of spatial components. So, the power spectrum plot is useful to know about the nature of the surface.

A block diagram regarding the enhancement of image through frequency domain is shown in Fig. 4.37. In frequency domain operation, first an image, $i(x, y)$ is transformed into the frequency domain, $I(u, v)$, through DFT. Then the elements of an enhancement filter (frequency domain filter), $F(u, v)$, which is real and symmetric about the central frequency, is multiplied with $I(u, v)$, through array operations for obtaining, $O(u, v)$, i.e., $O(u, v) = F(u, v)I(u, v)$. Then inverse discrete Fourier transform (IDFT) is performed to reconstruct the enhanced image, $o(x, y)$, in the spatial domain. Here, two types of frequency domain filters, i.e., Butterworth and Gaussian, for both smoothing and sharpening are discussed in the following sub-sections.

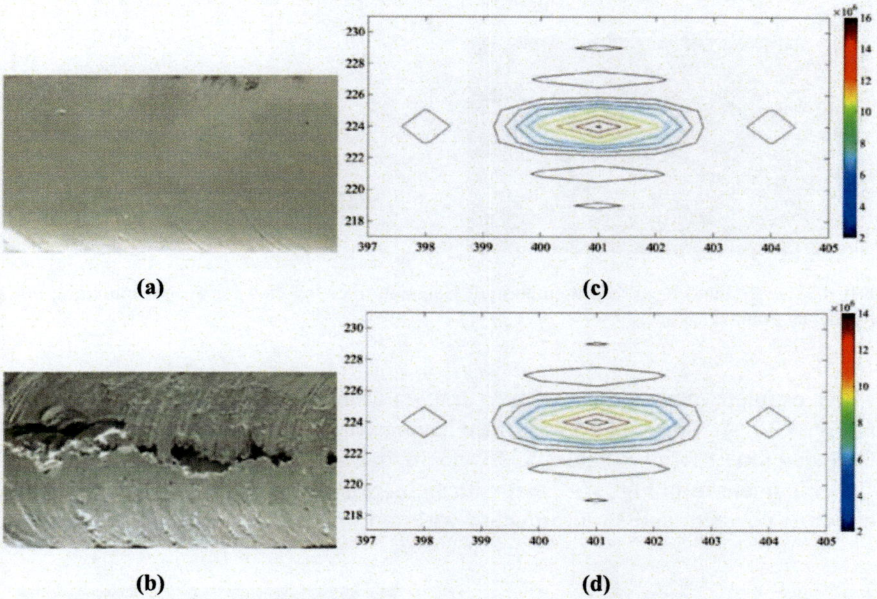

Fig. 4.36 FSW image **a** without surface defect and **b** with surface defect i.e., voids, and **c**, **d** the corresponding contour plots of the power spectrum

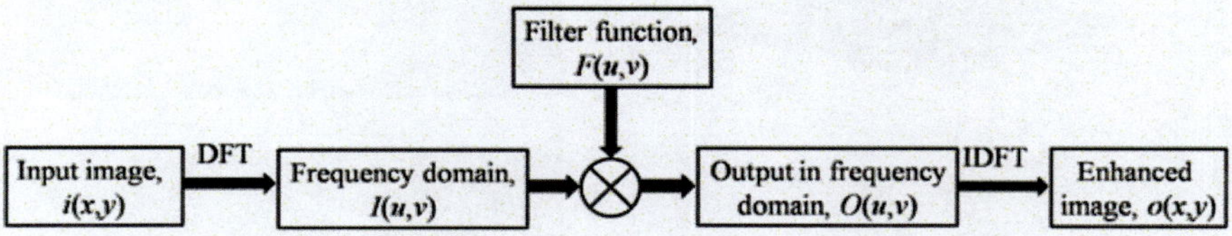

Fig. 4.37 Block diagram of image enhancement through frequency domain

4.4.2.1 Image Smoothing in the Frequency Domain

The first technique is Gaussian filtering, whose function, $F(u, v)$ is expressed in Eq. 4.9.

$$F(u, v) = e^{-D^2(u,v)/2\sigma^2} \tag{4.9}$$

where $D(u, v)$ is the distance of the point (u, v) from central point i.e., (0,0) in frequency axis, and σ is the spread at bottom of the filter, as shown in Fig. 4.38a, whose 3-D representation is shown in Fig. 4.38b.

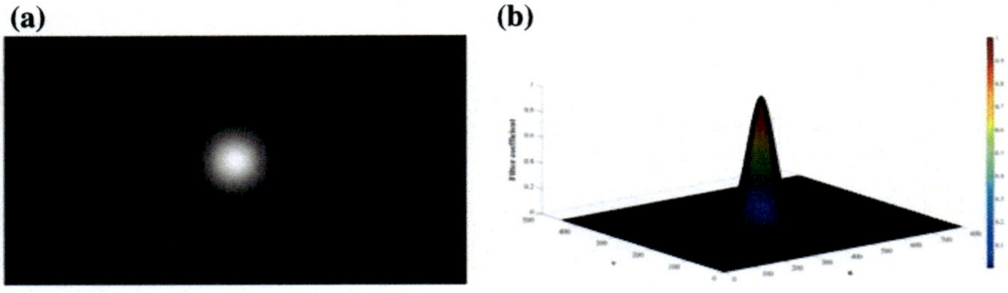

Fig. 4.38 **a** 2-D and **b** 3-D representation of Gaussian filter function for image smoothing in the frequency domain

An example of smoothing with a variation of the controlling factor, σ, is shown in Fig. 4.39. An image of FSW surface having voids (Fig. 4.36b) is smoothened by Gaussian filter having σ-values 5, 30 and 50, in Fig. 4.39b, d, f, respectively.

It is evident from Fig. 4.39, that with an increase in spread or σ, more frequency region is covered, and so information with more frequency can be passed. As an

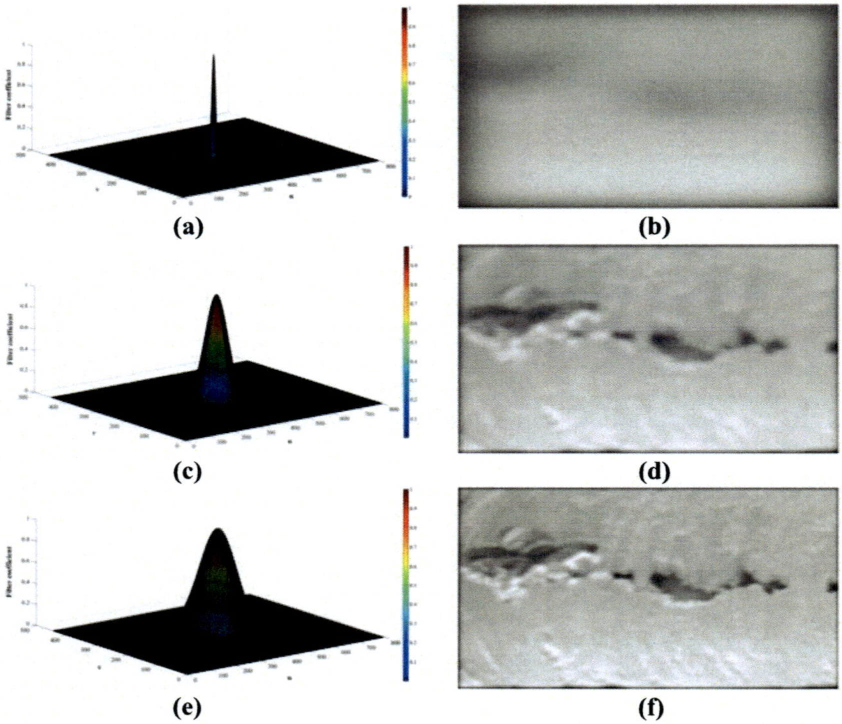

Fig. 4.39 Gaussian filters with **a** $\sigma = 5$, **c** $\sigma = 30$, and **e** $\sigma = 50$, and the corresponding smoothed images of Fig. 4.36b represented in Fig. 4.39b, d, f, respectively

example, in Fig. 4.39a, less frequency can pass, (almost at the central frequency or DC component and some small neighbourhood of it), so the smoothed image (Fig. 4.39b) becomes almost averaged. Therefore, no edge of void regions can be visible, whereas the result of smoothing is better in Fig. 4.39b with spread = 30. With more spread i.e., at 50, some details on the surface other than the void boundary are also visible. If one wants to segment the void region from the background, then those details on the surface may fall into the segmented defect portion. Within these three cases, spread = 30 will be the most proper choice.

The second technique is the Butterworth filter, which is expressed as follows (Eq. 4.10):

$$F(u, v) = \frac{1}{1 + [D(u, v)/\sigma]^{2n}} \tag{4.10}$$

where the new term, n is known as the order of the filter.

Low-order Butterworth filter (Fig. 4.40a) have less sharp transition at the bottom as compared to higher-order Butterworth filter (Fig. 4.40c). Therefore, ringing, an unwanted artefact, can be observed, where Butterworth filter having higher-order i.e., order 50 is applied as shown in Fig. 4.40d. Otherwise, comparatively, smoothing with lower order (i.e., 1) Butterworth filter can result in a better smoothing as shown in Fig. 4.40b. Thus, order and spread are the two factors of the Butterworth filter. In Gaussian filtering, ringing is absent as there is no sharp transition at the bottom.

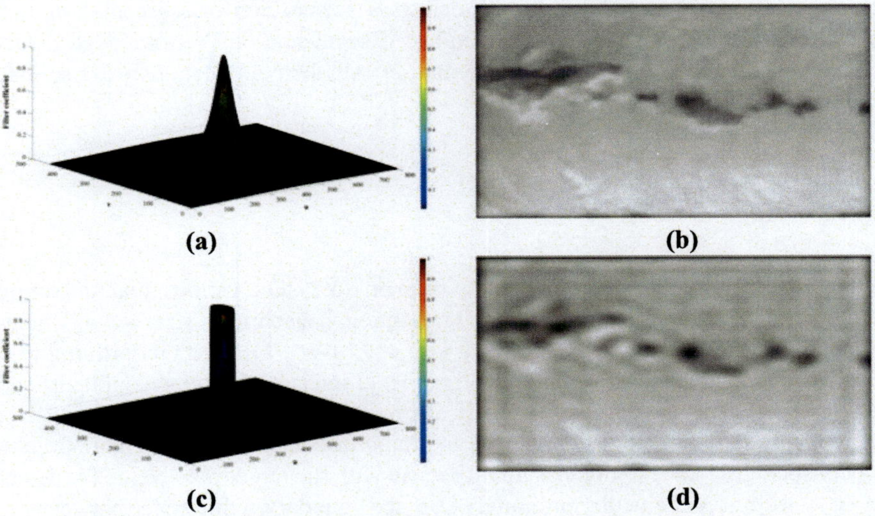

(a) (b) (c) (d)

Fig. 4.40 Butterworth filters of spread 30 with **a** order 1, and **c** order 50, and **b, d** the corresponding filtered images, respectively, of Fig. 4.36b

4.4.2.2 Image Sharpening in the Frequency Domain

A high-pass filtering operation, where high-frequency features are passed, is mainly used for image sharpening operation. The high-pass filter function in the frequency domain is expressed in Eq. 4.11.

$$F_{\mathrm{HP}}(u,v) = 1 - F_{\mathrm{LP}}(u,v) \quad (4.11)$$

where $F_{\mathrm{HP}}(u,v)$ and $F_{\mathrm{LP}}(u,v)$ are high-pass and low-pass filter functions, respectively, in the frequency domain.

Image sharpening is majorly used in enhancing the boundary, edges for further segmentation or texture analysis in defect detection or surface finish evaluation. Image of turned surface (Fig. 4.33a) is enhanced by using a Gaussian filter of three different spread values (i.e., 5, 30, and 50), as shown in Fig. 4.41. It is evident from Fig. 4.41 that with the increase in spread most of the low-frequency components in the image are obstructed, whereas much more high-frequency components such as feed marks, edges are enhanced.

4.5 Image Segmentation

In an image, regions are segmented for further processing by image segmentation technique. Edge, line, and point detection is one kind of segmentation, and another type of segmentation is region-based segmentation. Various edge detection techniques and region-based segmentation with their applications are described below.

4.5.1 Edge Detection

Edge detection techniques are widely applied to detect various manufacturing applications, such as, in rolling [24], conventional machining [29, 30, 43], non-conventional machining [1, 4], welding [10, 11, 14–16, 74], EHD jet printing [73] etc. Sudden change or slow change in gray-level intensity from its neighbourhood along a line or curve is termed as edge. In an image, there are three areas i.e., constant gray level, sharp or step discontinuity, and ramp discontinuity. The neighbourhood of constant gray level region has no difference with the target pixel value for detection. At the onset of any discontinuity, i.e., at the boundary, a difference between the target pixels with two neighbourhood pixels i.e., at two directions is present. If the discontinuity is one or two pixels, wide or sharp discontinuity with connected pixels in other direction, then the edge is thin. In another case, the ramp type discontinuity, where intensity values changing linearly along a certain width of pixels, are mainly

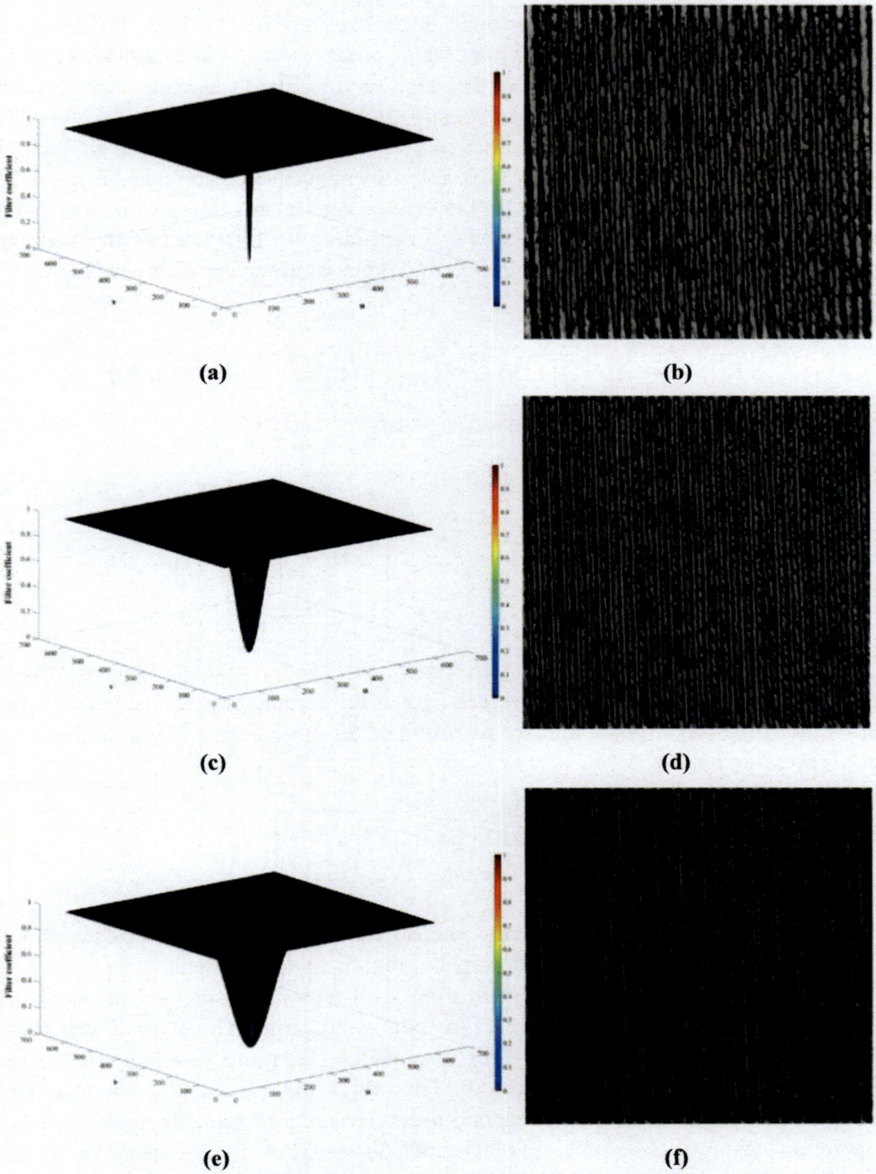

Fig. 4.41 Gaussian high pass filters of spread **a** 5, **c** 30, and **e** 50 and **b**, **d**, **f** the corresponding filtered images, respectively, of Fig. 4.33a

thick edges. There are three steps in edge detection. They are (1) image smoothening to reduce spurious noise, (2) detection of edge points or valid candidate pixels (here thresholding operation is performed for assigning some pixels, having values greater than a threshold gray level value, as value 1, and for others as value zero), (3) edge localization based on the true candidate pixels. The basis to find a valid edge point is a derivative operation. This operation helps to enhance the difference between the edge features and the background. In the case of vertical edge detection, horizontal neighbourhood points are considered for determining the difference with the target pixel, $i(x, y)$. The difference equation to find gradient along the X-direction, ∇i_x, is shown in Eq. (4.12).

$$\begin{aligned} \nabla i_x = \frac{\partial i}{\partial x} &= \{i(x+1, y) - i(x, y)\} + \{i(x-1, y) - i(x, y)\} \\ \Rightarrow \nabla i_x &= i(x+1, y) + i(x-1, y) - 2i(x, y) \end{aligned} \quad (4.12)$$

Similarly, the gradient along Y-direction, ∇i_y, is expressed as shown in Eq. 4.13.

$$\begin{aligned} \nabla i_y = \frac{\partial i}{\partial x} &= \{i(x, y+1) - i(x, y)\} + \{i(x, y-1) - i(x, y)\} \\ \Rightarrow \nabla i_y &= i(x, y+1) + i(x, y-1) - 2i(x, y) \end{aligned} \quad (4.13)$$

The gradient of an image, ∇i has two components i.e., ∇i_x and ∇i_y. The magnitude of the gradient, $m_\nabla(x, y)$ for target pixel, (x, y) is the parameter to determine the edge strength of the target pixel, and the direction of the edge at that pixel, i.e., $\alpha(x, y)$ are represented in Eq. 4.14.

$$\left.\begin{aligned} m_\nabla(x, y) &= \sqrt{(\nabla {i_x}^2 + \nabla {i_y}^2)} \\ \alpha(x, y) &= \tan^{-1}(\nabla i_y / \nabla i_x) \end{aligned}\right\} \quad (4.14)$$

The magnitude value is required to choose valid edge pixels and the direction is needed to search for the connected neighbourhoods to detect edges.

The first derivative of the constant gray-level region is zero, and the onset of discontinuity is non-zero, also non-zero along the ramp. The second derivative becomes zero at constant gray-level region and along the ramp, however non-zero at the onsets of step and ramp discontinuity. Thus, thick and fine edges or boundary can be detected by first-order and by second-order derivatives, respectively. Some edge operation techniques like Sobel, Laplacian of Gaussian (LoG), and Canny edge detection techniques are discussed here as these are mainly used to detect manufacturing process zones.

4.5.1.1 Sobel Operation

In Sobel operation, the first derivative operation is mainly used, where the kernel or filter is used to find the edge pixel as shown in Fig. 4.42a, for detecting horizontal edges. The total of the filter coefficients is zero so that zero response can be received for the constant intensity region.

Sobel operation on two parallel lines (Fig. 4.42d) is shown in Fig. 4.42e, where some spurious points are observed around the edges, as smoothing operation is missing here.

Sobel operation is mainly a first derivative operation where thick edges are also enhanced, as the first derivative along a ramp is non-zero. The kernel of the Sobel

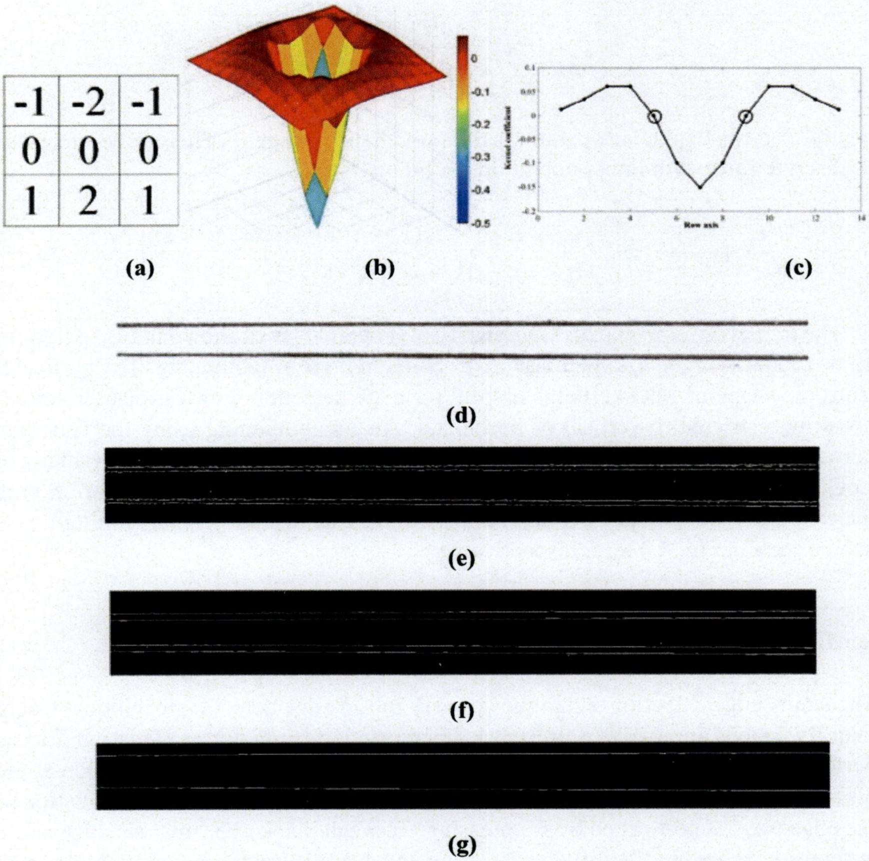

Fig. 4.42 **a** Kernel of Sobel operator; **b** plot for kernel (of size 13 × 13 with $\sigma = 2$) of Laplacian of Gaussian (LoG) operation, and **c** zero crossing of LoG shown by circles, **d** image of hand-drawn two parallel lines, edge detection by **e** Sobel operator **f** Laplacian of Gaussian operator, **g** Canny edge detector of Fig. 4.42d

operator shown in Fig. 4.42a is used to enhance horizontal edges and the transpose of Fig. 4.42a is used to enhance vertical edges. In edge detection techniques, after enhancing the edges, thresholding operation is performed, and so a thresholding parameter needs to be selected properly to obtain the accurate output.

4.5.1.2 Laplacian of Gaussian

In this operation, a Gaussian smoothening operation is used first on images to reduce noise and then the Laplacian operation is used on the smoothened image for edge enhancement. Laplacian operation is a second derivative operation, as shown in Eq. 4.15, especially useful for enhancing thin edges and find zero-crossing.

$$\nabla^2 i_g = \frac{\partial^2 i_g}{\partial x^2} + \frac{\partial^2 i_g}{\partial y^2} \tag{4.15}$$

where ∇^2 is the Laplacian operator in the smoothened image,i_g. This can be expressed as discrete form with linear operations in Eq. 4.16:

$$\frac{\partial^2 i_g}{\partial x^2} + \frac{\partial^2 i_g}{\partial x^2} = \begin{bmatrix} i_g(x+1, y) + i_g(x-1, y) + i_g(x, y+1) \\ +i_g(x, y-1) - 4i_g(x, y) \end{bmatrix} \tag{4.16}$$

The kernel of Laplacian of Gaussian (LoG) operator is as shown in Fig. 4.42b for 13×13 size with $\sigma = 2$. Increase in σ results in more smoothening. The particular isotropic shape of LoG kernel is helpful for edge detection with various directions, not only restricted to vertical or horizontal. Also smoothening using the Gaussian kernel can be effective to avoid ringing. Zero-crossing is also another advantage of LoG, which is shown in Fig. 4.42c, for accurate edge detection. However, in both the cases, i.e., Sobel and LoG operations, double line detection is one limitation, as can be seen in Fig. 4.42e, f, respectively.

4.5.1.3 Canny Edge Detection

The canny edge detection technique is a very robust edge detection technique, where majorly 5 steps are involved. In step 1, image smoothening with a Gaussian filter is performed to reduce noise. Then, in step 2, gradient magnitude and direction at the discontinuity are detected for sorting strong and weak edges. In step 3 localization of the edge pixel is performed by suppressing the weak pixels. For this, the edge angle is determined first. In Canny edge detection, the overall circular region is divided into eight regions having equal angle. The regions are from $-22.5°$ to $+22.5°$, $+22.5°$ to $+67.5°$, $+67.5°$ to $+112.5°$, and so on. For example, the horizontal edge direction will fall into the region $-22.5°$ to $+22.5°$. If the actual edge direction is $15°$, then the edge direction is approximated as horizontal as it falls in the region $-22.5°$ to $+22.5°$. Then the edge strength of the points lying in the direction is compared, locally, to suppress the points whose edge strength is less than its neighbours. Therefore, the third step is termed non-maximal suppression. In the fourth step, false edge points are reduced by hysteresis thresholding operation, with two thresholds, one is high to detect a few points (e_h) having high edge strength, and the another is low to allow more points (e_l) for candidate pixels for strong and weak edge points, respectively. If only (e_h) are selected for an edge then edge gaps can result. This gap is eliminated by intelligent edge linking operation in the last or at the fifth step. Here, the neighbourhood of (e_h) having gaps are searched from e_l or weak pixels. If connected pixels from e_l are found then those pixels are selected and others are suppressed. Therefore, a less spurious response by noise suppression, proper localization, and thin edge detection by the process of non-maximal suppression, hysteresis thresholding, and edge linking are the major advantages of the Canny edge detection method. It can be evident from Fig. 4.42g that the double-edge effect can be avoided through the Canny edge detection technique. However, the quality of edge detection is dependent on two parameters like LoG i.e., threshold value and σ.

It can be evident from Fig. 4.43, that more accurate edge detection can be performed with the Canny algorithm as compared to LoG. Detection of non-uniform

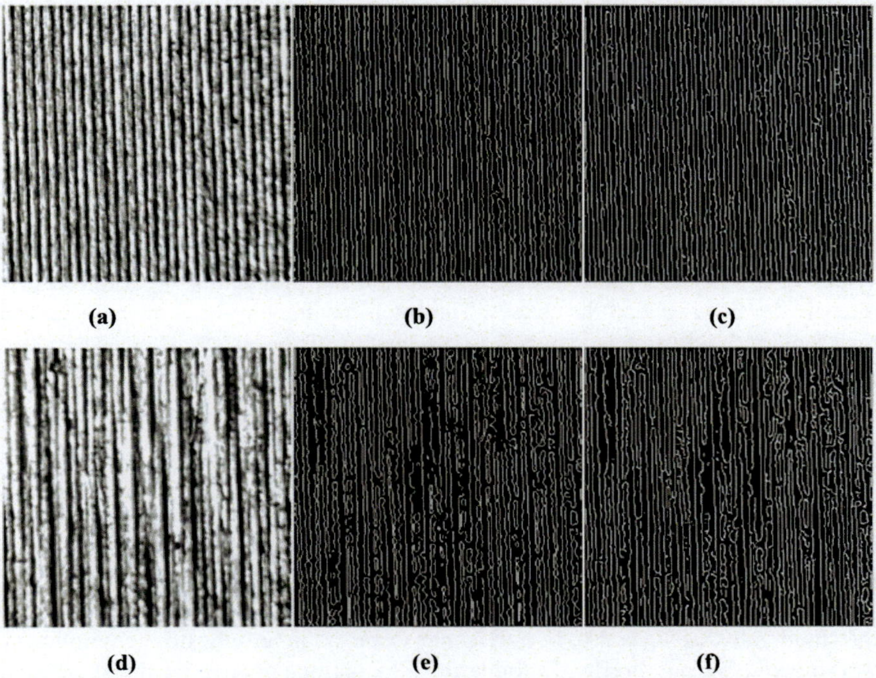

Fig. 4.43 **a**, **d** Images of surface machined with an unworn tool and a worn tool, respectively and its corresponding edge detected images with **b**, **e** LoG operator, and **c**, **f** Canny operator

edges on the surface can be detected better with Canny edge detection as compared to LoG.

4.5.2 *Thresholding*

Image thresholding is a segmentation operation, where a region or object is segmented from its background based on the difference in their intensity levels. In manufacturing, thresholding operations are mainly used to segment the zone or regions, whose geometric or texture features are needed to be extracted for inferring some semantic decisions related to the process or part. In laser cladding, thresholding is used to segment the molten pool to calculate its geometry for controlling laser power [6–8]. Tool wear regions are needed to be thresholded to determine the tool wear width for tool condition monitoring [24, 25]. Indirect monitoring of belt wear in the abrasive belt grinding process was performed by evaluating spark field geometry, where thresholding took a lead role for accurate segmentation [48]. In surface defect detection, thresholding plays a major role in segmenting the defect region from its background in continuous casting [18], welding [58] etc. Identification of keyhole and its evaluation plays a major role in the laser welding process in determining the penetration in real-time [71]. Here, thresholding plays a major role in keyhole segmentation. In this section, two basic thresholding methods are discussed for fundamental understanding.

4.5.2.1 Global Thresholding

In thresholding operations, pixels are classified into two classes based on a threshold value. This threshold value is determined from the probability of occurrence of intensity levels of each of the classes. The pixels having gray-level intensity greater than the threshold value is assigned as 1-valued pixels or object, whereas the other pixels are assigned as 0-valued or background as represented in Eq. 4.17 [80].

$$o(x, y) = \begin{cases} 1 & if i(x, y) > t \\ 0 & if i(x, y) \leq t \end{cases} \tag{4.17}$$

where t is the threshold level, $o(x, y)$ and $i(x, y)$ are the output and input images, respectively.

However, with the change in lighting or illumination of the process zone, the threshold value changes. So, in a dynamic scene, as in an industrial environment, working with a fixed threshold is infeasible. So, optimal or adaptive thresholding is required here.

4.5.2.2 Optimal Thresholding

Optimal thresholding is also termed as Otsu's thresholding, where the threshold value is selected optimally based on the maximization of between-class-variance of object and background. The optimization problem is represented as to find optimal t i.e., t^*, for which [80]:

$$\left.\begin{array}{c}\max\limits_{0\leq t\leq L-1} {s_R}^2\\ s.t., 0<\mathrm{P}_1(t)<1\end{array}\right\} \tag{4.18}$$

where $[0, L-1]$, s_R, and $P_1(t)$ are the gray level range, between-class-variance, and the cumulative summation of the probability of occurrence of pixels up to the gray level,t, respectively.

Between class variance is expressed in Eq. 4.19:

$${s_R}^2 = \frac{\{m_i P(t) - m(t)\}}{P(t)\{1 - P(t)\}} \tag{4.19}$$

Where cumulative average up to 't' gray level, $m(t) = \sum\limits_{g=0}^{t} g p_g$ and $p_g = \frac{n_g}{MN}$ (p_g is the probability of occurrence of the pixels with gray level, g, in an $M \times N$ image.

The average intensity of the image, $m_i = \sum_{g=0}^{L-1} g p_g$.

Cumulative sum or probabilities of occurrence of gray level values up to t, $P(t) = \sum_{g=0}^{t} p_g$.

In automatic detection of surface defects, the thresholding is a common primary step of image segmentation. Here an example of surface defect segmentation through thresholding operation is shown in Fig. 4.44, where the comparison between manual and optimal thresholding is also shown. In manual thresholding, the threshold is generally chosen from the image histogram's peak position. The peak position occurs for Fig. 4.44a is for the gray-level value, 174. Thus it is selected as the manual threshold, where, the gray level value below and above 174 is considered as 1 and 0, respectively. The manually thresholded image of Fig. 4.44a is shown in Fig. 4.44b, where the particular defect portion is not exactly segmented. On the other hand, Fig. 4.44c represents the optimally thresholded image with an optimal threshold value as 114, where a more accurate segmentation of defects is depicted. However, in some regions pixels' continuity is missing, for which morphological operations are described next.

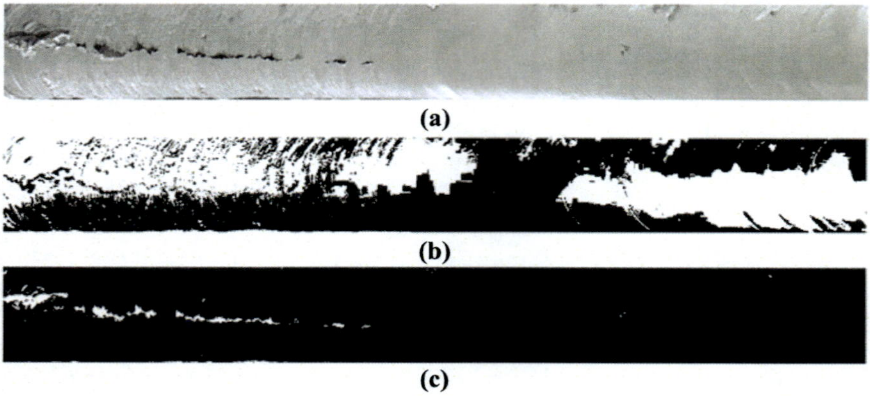

Fig. 4.44 **a** Original image of FSWed surface and its global thresholded image with **b** manually chosen threshold from image histogram's peak, and **c** optimally chosen threshold by Otsu's algorithm

4.5.3 Morphological Operations

Morphological operations are generally used to modify the segmented image by closing broken portions or by removing already thresholded spurious noise for obtaining more accuracy. The boundary of the object is thinned or thickened by structuring elements in morphological processing. Two major operations in morphological processing are erosion and dilation.

4.5.3.1 Erosion

The objective of erosion is to thin the object so that the spurious boundary becomes smooth. In erosion, a structuring element, E is translated along the boundary of the object, to erode the object's boundary, as described in Eq. 4.20.

$$i \Theta E = \{ p | (E)_p \cap i^c = \varnothing \} \tag{4.20}$$

where the boundary of i is eroded by the structuring element, E to thin the image. E_p is the translation of E by p amount, i.e., the element is shifted from pixel (x, y) to the pixel,$(x + p1, y + p2)$.

A schematic representation of the morphological erosion mechanism is shown in Fig. 4.45. As depicted in Fig. 4.45a, the origin of the structuring element, E, is first placed on the boundary. If there is an intersection between any of the elements of E and the background (i.e., outside the object) as shown in Fig. 4.45a, then intersecting or common pixel of the object and the element E is removed or eroded, as shown in Fig. 4.45b. It can also be reflected in Eq. 4.20, that when any element of E is

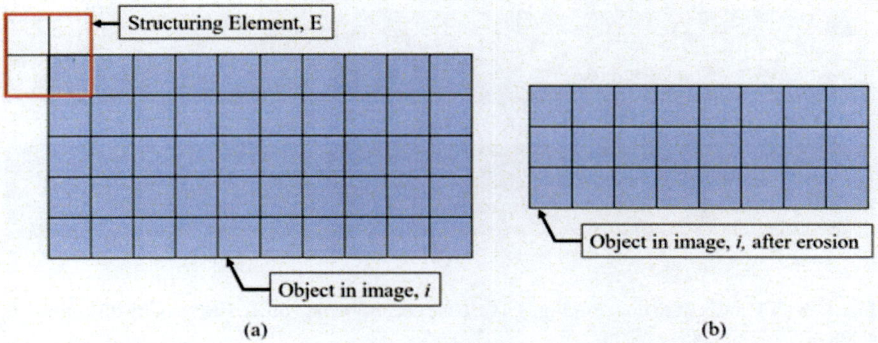

Fig. 4.45 **a** Mechanism of Erosion and **b** Object after erosion

intersected with the background of the object then the set becomes null. Then after completing the erosion at one position, the structuring element is slide to the next boundary pixel in a horizontal or vertical direction. This way, the erosion operation is completed on the object for removal of spurious pixels or to thin the object.

4.5.3.2 Dilation

The objective of dilation is to thick the object boundary so that some noisy small holes can be filled or the object can be thickened. As opposed to the erosion, the reflection of a structuring element, E is translated along the boundary of the object, so that the intersection of elements of E and the object's background should not be a null set, as described in Eq. 4.21.

$$i \oplus E = \{p|(\widehat{E})_p \cap i \neq \varnothing\} \tag{4.21}$$

Other morphological processing like closing, opening, skeletonization, thinning, linking, etc. exist in image processing, where dilation and erosion are the basic operations.

An example of dilation operation is shown in Fig. 4.46, where the thresholded image of a welded surface is dilated to close some of the boundaries where the required pixels are missing due to thresholding operation.

It can be seen from Fig. 4.46b that the discontinued boundaries are filled in along with the filled holes.

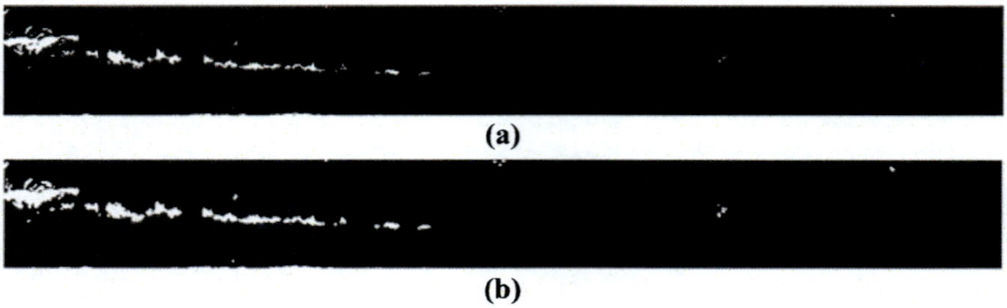

Fig. 4.46 a Thresholded image of Fig. 4.44a, b the corresponding dilated image with disk structural element

4.6 Feature Extraction and Object Recognition

The most important and the final part of the image processing is the feature extraction and object recognition. For example, geometry extraction of weld pool [11] or molten pool in DED process [6–8] is important for controlling welding current/voltage or laser power for obtaining high-quality weld or precise component. Another example is tracking the seam and weld pool so that the offset between the weld torch centre and weld line can be minimized for high accuracy [11]. So, extractions of weld line and weld pool centre are the primary requirement here. Detection of defect location, type, and sometimes geometry is important to monitor in real-time for cast surface, rolled surface, welding in production line [18, 24, 69]. Extraction of tool flank wear width and extraction of machined surface texture features in direct and indirect monitoring of tool wear, respectively, requires utilizing various object recognition and texture analyses methods [31, 35–40]. Therefore, in the feature recognition or extraction step, various geometrical or shape features are extracted from the process zones in case of extracting shape features. Texture analysis is mainly utilized to extract features for further classification or discrimination between various surface textures or defects. Texture analysis and shape recognition are discussed in the following sections.

4.6.1 Texture Analyses

A pattern of repetitive pixels or primitives is the main feature of a texture. A machined surface is an example of a texture image. Such as in Fig. 4.43a, a turned surface image is shown where feed marks occurred in a periodic manner. So, it has some particular pattern. Texture analyses are used to detect texture pattern in the image. The texture of a surface machined by an unworn tool or texture of a smooth surface is different from the rough surface. Texture features extracted from the image through texture analyses can discriminate between two types of surfaces. Three broad types of texture analyses, such as statistical, structural, and spectral are widely used in manufacturing applications.

7	7	7	7	6	6	5	4
8	7	7	7	6	5	5	4
8	8	8	8	6	5	4	4
8	8	8	8	6	5	4	4
8	7	8	8	6	5	4	4
8	7	7	7	6	4	4	4
7	8	7	7	6	5	4	4
7	8	8	8	6	5	4	4

(a)

n \ m	1	2	3	4	5	6	7	8
1	0	0	0	0	0	0	0	0
2	0	0	0	0	0	0	0	0
3	0	0	0	0	0	0	0	0
4	0	0	0	7	0	0	0	0
5	0	0	0	8	2	0	0	0
6	0	0	0	0	7	0	0	0
7	0	0	0	0	1	6	14	4
8	0	0	0	0	0	1	5	1

(b)

Fig. 4.47 **a** Image fragment, and **b** corresponding GLCM

4.6.1.1 Statistical Texture Analysis

In statistical texture analysis, features having statistical importance are extracted. A co-occurrence probability map or matrix of similar patterns is created from the original image, and then features are extracted from this map in statistical texture analysis. Gray level co-occurrence matrix (GLCM) based and run-length statistical (RLS) analysis techniques are discussed here, as these are very popular techniques.

Gray Level Co-occurrence Matrix (GLCM) Based Texture Analysis

The probability of occurrence between two gray level intensity valued pixels having a particular spacing or pixel pair spacing (pps) along a direction (or orientation) is evaluated as a matrix, termed as gray level co-occurrence matrix (GLCM).

The formation of GLCM from an image matrix is explained lucidly in Fig. 4.47, where m and n are the gray levels present in the image. Co-occurrence of 4 with 1-pixel pair spacing in the horizontal direction can be found as 7. The occurrence of gray level pixels having 6 and 5 intensities can be found as 7, also in Fig. 4.47b. Due to this, it becomes an element (encircled) of the GLCM, as shown in Fig. 4.47b. GLCM can be constituted by considering the co-occurrence of pixels having specified gray level values for vertical and diagonal directions also.

The surface turned by an unworn tool is smoother than that of the worn tool. Therefore, periodicity or repetitiveness of the feed marks or texture primitives are more in Fig. 4.43a than that of Fig. 4.43d. So, the GLCM of Fig. 4.43a is more diagonal (Fig. 4.48a) as compared to the GLCM of Fig. 4.43d (i.e., Fig. 4.48b), due to the presence of better texture. Co-occurrence between the same or nearly same gray level pixels are more with a particular spacing in the horizontal direction, as

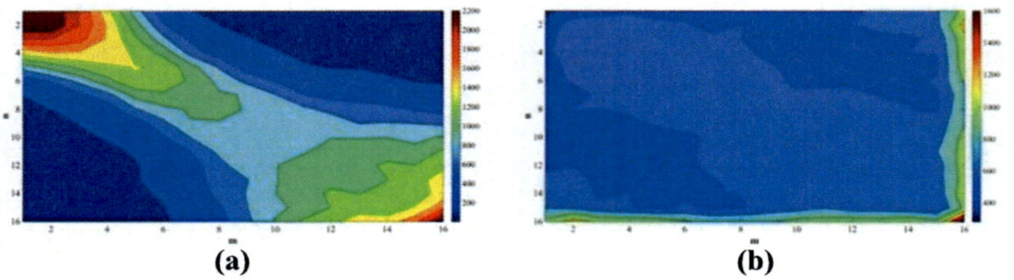

Fig. 4.48 GLCMs of images of surfaces turned by **a** unworn tool, i.e., Fig. 4.43a, and by **b** worn tool, i.e. Fig. 4.43d

evident in Fig. 4.48a, due to the presence of a more repetitive pattern in Fig. 4.43a. Hence, the texture features, such as contrast, homogeneity, energy, and correlation give clear discrimination between Fig. 4.43a, d, from GLCM maps (Fig. 4.48a, b).

The features are represented in Eqs. 4.22–4.25 [28, 41].

$$\text{Contrast} = \sum_m \sum_n (m-n)^2 p(m,n) \tag{4.22}$$

$$\text{Homogeneity} = \sum_m \sum_n \frac{p(m,n)}{(1+|m-n|)} \tag{4.23}$$

$$\text{Correlation} = \sum_m \sum_n \frac{p(m,n)(m-a_x)^2(n-a_y)^2}{s_x s_y} \tag{4.24}$$

$$\text{Energy} = \sum_m \sum_n p^2(m,n) \tag{4.25}$$

where $p(m, n)$,s_x, s_y, a_x and a_y are the GLCM elements i.e., probability of occurrence for pair of pixel values *m* and *n*, the standard deviation of pixel values present in row *x*, the standard deviation of pixel values present in column *y*, an average of pixel values present in column *x*, and the average of pixel values present in column *y*, respectively.

According to Eq. 4.22 and contour plot of the image of surface machined by an unworn tool i.e., Fig. 4.48a, *contrast* is lesser due to lesser scatter, i.e., $(m - n)^2$ than that of surface machined by a worn tool i.e., Fig. 4.48b. *Contrast* for Fig. 4.48a, b are 11.92 and 44.70, respectively. Similarly, according to Eq. 4.23, *homogeneity* is higher in Fig. 4.48a as lesser scatter i.e., $|m - n|$ and more $p(m, n)$ than that of Fig. 4.48b, where *homogeneity* for Fig. 4.48a, b are 0.40 and 0.27, respectively. According to Eq. 4.24, if $p(m, n)$ is higher, and s_x and s_y are lesser, then *correlation* should be higher. In Fig. 4.43a, pixels are more uniform along the column *y* than that in Fig. 4.42b, due to the presence of a more uniform feed mark in Fig. 4.43a for machining with an unworn tool. Therefore, s_y is lesser in Fig. 4.43a than that in Fig. 4.43d. Also, a more repetitive pattern can be observed in Fig. 4.43a than that in

Fig. 4.43d. Thus, the pixel values along the rows are more periodic in Fig. 4.43a as compared to Fig. 4.43d. So, the s_x value should be lesser in Fig. 4.43a than that in Fig. 4.43d. So, correlation features for Fig. 4.48a, b are 0.72 and 0.0035, respectively, i.e., it is higher for the surface machined by an unworn tool. As $p(m, n)$ value is more in Fig. 4.48a than that in Fig. 4.48b, the *energy* value for the surface machined by an unworn tool should be more than that of the surface machined by a worn tool, according to Eq. 4.25. So, the *energy* values for Fig. 4.48a, b are evaluated as 0.0066 and 0.0042, respectively. Therefore, *contrast*, *homogeneity*, *correlation,* and *energy* can be selected as GLCM features to discriminate between the tool condition based on the machined surface images. In another example, i.e., evaluation of weld quality from weld surface images can be performed by using GLCM based texture analysis. Here, three weld surfaces and their corresponding GLCMs are shown in Fig. 4.49.

From Fig. 4.49, the four features are shown in Table 4.3.

For classification of these three types of weld surface, the GLCM features, *contrast, homogeneity,* and *energy* are working well, but *correlation* gives ambiguous result according to Table 4.3. However, if the objective is to classify between defective and defect-free FSW welds from their surface images, then all four features are working good. So, the selection of proper features is important based on targeted application in GLCM-based texture analysis. The accuracy or performance of GLCM is dependent on two parameters—pixel pair spacing (PPS), and orientation. Proper or optimal selection of these two parameters is required to obtain an accurate result.

Run-Length Statistics (RLS) Based Texture Analysis

Gray level runs or placing of same gray level pixels are analyzed through RLS analysis. Consecutive pixels of the same gray value or level, in a given direction (i.e., horizontal, vertical or diagonal), constitute a run. The number of occurrence of the same gray level values in a particular direction is the length of run of that particular gray level.

The formation of RLM is shown in Fig. 4.50, as an example. Gray level value, 4, occurs consecutively in three pixels along the vertical direction, which is marked in Fig. 4.50a. Therefore, in this case, the length of run for 4 (as encircled) is 3. Also, gray level value, 4, with run length 3 occurs one time in Fig. 4.50a. So, in the run-length matrix (RLM), shown in Fig. 4.50b, the element (as encircled) for gray level value (***g***), 4, with 3 length of run (***r***) is 1. In the same manner, the occurrence of 4 with run length 1 is three times, so in RLM the corresponding element is written as 3. RLS analysis is mainly applicable to the cases, where continuity of texture pattern needs to be evaluated.

Run-length matrices of Fig. 4.43a, d, i.e., image of machined surfaces turned by unworn and worn tools, respectively, are represented as contour plots in Fig. 4.51a, b, respectively. Surface machined by worn tool becomes rougher as compared to the surface machined by the unworn tool. As surface roughness increases, the illumination becomes more diffused, so feed marks with high gray levels are more evident in Fig. 4.43d as compared to Fig. 4.43a. Due to this reason, the spread of

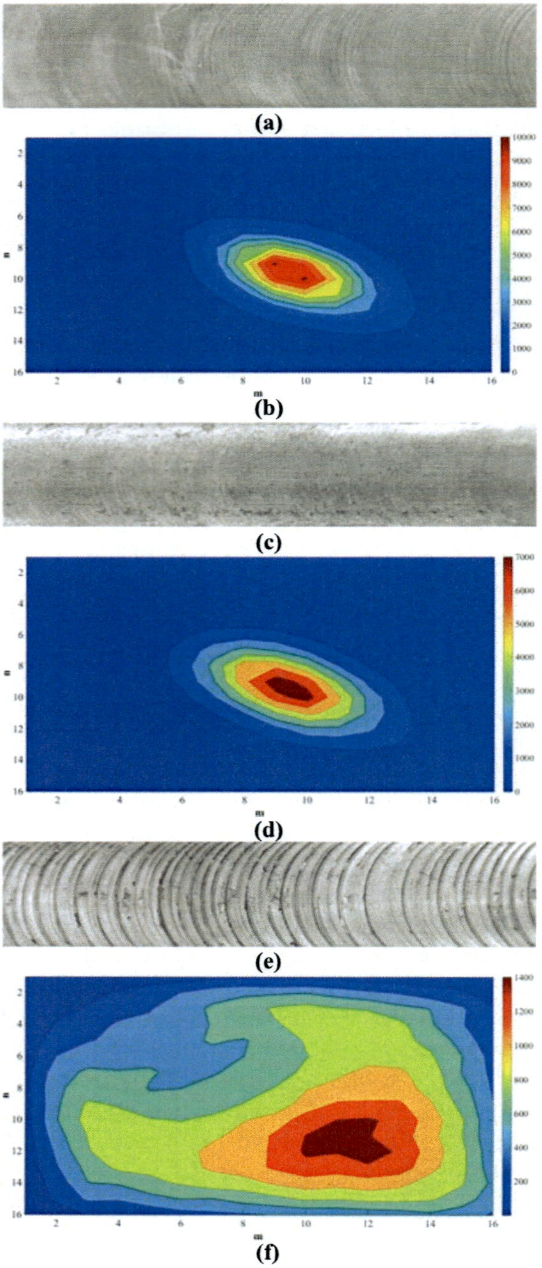

Fig. 4.49 Surface of FSW **a** smooth and defect-free, **c** smooth having minute cracks and **e** rough having cracks, and **b**, **d**, **f** their corresponding GLCMs, respectively

Table 4.3 Features extracted from GLCMs of Fig. 4.49

Features	Smooth and defect-free FSW weld surface	FSW weld surface having minute cracks	Rough weld surface having defects
Contrast	3.54	5.24	29.25
Homogeneity	0.54	0.50	0.29
Correlation	0.42	0.50	0.02
Energy	0.0301	0.02	0.005

3	1	4	2
3	1	4	1
3	1	4	1
4	2	3	2
3	1	3	2
4	1	4	1

(a)

g \ r	*1*	*2*	*3*
1	1	2	1
2	2	1	0
3	1	1	1
4	3	0	1

(b)

Fig. 4.50 **a** An image fragment and **b** the corresponding run-length matrix (RLM)

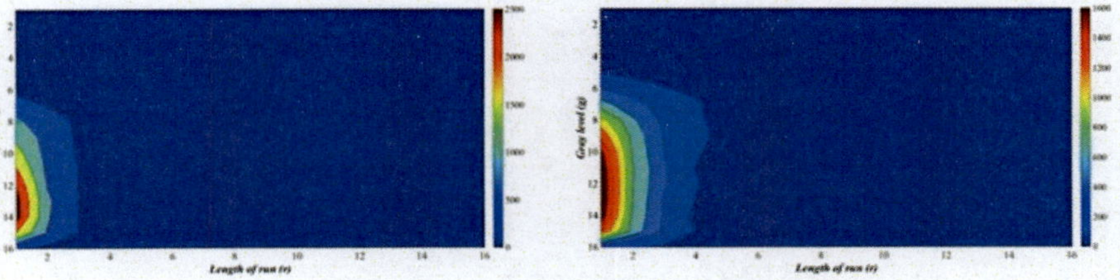

(a) (b)

Fig. 4.51 Contour plots of RLMs for images of surfaces machined by **a** unworn tool (Fig. 4.43a) and **b** worn tool (Fig. 4.43d)

the runs of various lengths in various gray levels are more in Fig. 4.51b than that in Fig. 4.51a. As the tool becomes worn, more abrasive and ploughing actions occurs instead of shearing. Therefore, the feed marks become smaller and discontinuous on the surface machined by a worn tool, whereas long, continuous feed marks are observed on the surface machined by the unworn tool. Therefore, the number of runs having smaller lengths should be more in the image of surface machined by the worn tool (Fig. 4.43d) than that in the image of surface machined by the unworn tool (Fig. 4.43a). This phenomenon is evident in contour maps presented in Fig. 4.51.

On this basis, the features from RLM can be extracted to monitor the tool condition. Various features of RLM are represented in Eqs. 4.26–4.36.

$$\mathrm{RP} = \frac{n_R}{n_p} \tag{4.26}$$

$$\mathrm{SRE} = \frac{1}{n_R}\sum_{g=1}^{M}\sum_{r=1}^{N}\frac{R(g,r)}{r^2} \tag{4.27}$$

$$\mathrm{LRE} = \frac{1}{n_R}\sum_{g=1}^{M}\sum_{r=1}^{N}R(g,r)\cdot r^2 \tag{4.28}$$

$$\mathrm{GLN} = \frac{1}{n_R}\sum_{g=1}^{M}(\sum_{r=1}^{N}R(g,r))^2 \tag{4.29}$$

$$\mathrm{RLN} = \frac{1}{n_R}\sum_{r=1}^{N}(\sum_{g=1}^{M}R(g,r))^2 \tag{4.30}$$

$$\mathrm{LGRE} = \frac{1}{n_R}\sum_{g=1}^{M}\sum_{r=1}^{N}\frac{R(g,r)}{g^2} \tag{4.31}$$

$$\mathrm{HGRE} = \frac{1}{n_R}\sum_{g=1}^{M}\sum_{r=1}^{N}R(g,r)\cdot g^2 \tag{4.32}$$

$$\mathrm{SRLGE} = \frac{1}{n_R}\sum_{g=1}^{M}\sum_{r=1}^{N}\frac{R(g,r)}{g^2\cdot r^2} \tag{4.33}$$

$$\mathrm{SRHGE} = \frac{1}{n_R}\sum_{g=1}^{M}\sum_{r=1}^{N}\frac{R(g,r)\cdot g^2}{r^2} \tag{4.34}$$

$$\mathrm{LRLGE} = \frac{1}{n_R}\sum_{g=1}^{M}\sum_{r=1}^{N}\frac{R(g,r)\cdot r^2}{g^2} \tag{4.35}$$

$$\mathrm{LRHGE} = \frac{1}{n_R}\sum_{g=1}^{M}\sum_{r=1}^{N}R(g,r)\cdot g^2.r^2 \tag{4.36}$$

where n_R, n_p= total number of runs and pixels and RP is run percentage, SRE is short run emphasis, $R(g,r)$ = element of RLM, LRE is long run emphasis, GLN is gray-level non-uniformity, RLN is run-length non-uniformity, LGRE is low gray-level run emphasis, HGRE is high gray-level run emphasis, SRLGE is short run low

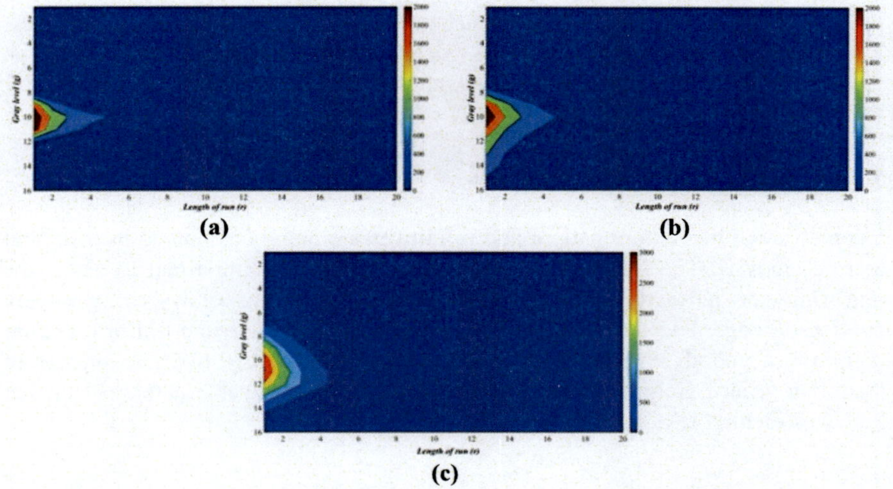

Fig. 4.52 Contour plots of RLMs for weld surface with **a** no defect, i.e., Fig. 4.49a, b minute cracks, i.e. Figure 4.49c, and **c** rough texture with defects, i.e., Fig. 4.49e

gray-level emphasis, SRHGE is Short run high gray-level emphasis, LRLGE is long run low gray-level emphasis, and LRHGE is long run high gray-level emphasis.

In SRE, the $R(g, r)$ value is more in Fig. 4.51b than that in Fig. 4.51a, due to the presence of more short runs (r of small values), which is evident from the SRE values i.e., 0.0096 and 0.0111 for Fig. 4.43a, d, respectively. In LRE, the $R(g, r)$ value is more in Fig. 4.51a than that in Fig. 4.51b, due to presence of more long runs (r of large values), which is evident from the LRE values i.e., 138.62 and 128.83 for Fig. 4.43a, d, respectively. Due to more brightness for diffused reflection on the surface machined by worn tool, the number of runs with high gray-level is more than that of the surface machined by the unworn tool. So, LGRE of Fig. 4.43a is more (i.e., 0.55) than that in Fig. 4.43d, (i.e., 0.52). Due to the same reason, the HGRE of Fig. 4.43a is lesser than that of Fig. 4.43d. Combined effects of length of runs with gray-level can be observed in the features SRLGE, SRHGE, LRLGE, and LRHGE. Short runs in both low and high gray-levels are more in Fig. 4.43d, as compared to Fig. 4.43a. Therefore, SRLGE and SRHGE are more for the image of surface machined by the worn tool. For a similar phenomenon, the values of LRLGE and LRHGE are more for Fig. 4.43a than that in Fig. 4.43d.

Weld surface image quality can be assessed by using RLS-based texture analysis technique. Contour plots of RLMs for Fig. 4.49a, c, e are represented in Fig. 4.52a–c, respectively.

Smooth and defect-free FSW surface image has run in limited gray levels, whereas, in rougher and defective weld surface image has run with various gray levels, which are evident in Fig. 4.49a, e. The variation of gray levels and short runs are evident in FSW surface with small cracks (Fig. 4.49c). This can also be depicted in the contour plots of RLMs of Fig. 4.49a, c, e, which are shown in Fig. 4.52a–c, respectively. Therefore, in this case, GLN, HGRE, SRHGE, and LRHGE can be selected as features for discriminating between the weld surfaces to determine weld quality.

4.6.1.2 Structural Texture Analysis

Geometries of the texture pattern and primitives are mainly evaluated in structural texture analysis. The structure of the texture primitives is important to detect the uniformity in the texture. Structural or geometrical texture analysis is generally applied on edge detected images. For example, the repetitive and uniform texture pattern in a smooth surface results in smoother geometrical results as compared to the rough surface. Hough transform and Voronoi tessellation techniques are discussed here as structural texture analysis techniques.

Hough Transform

Detection of lines through boundary tracing or contour tracing is computationally expensive. Therefore, the concept of Hough transform came to perform faster detection of lines or continuous boundary tracing [80]. Straight lines passing through a point (x_i, y_i) is dependent on the parameters, slope, a, and intercept, b, as per straight line equation, $y_i = ax_i + b$. Representation of all the straight lines passing through a particular point in xy-plane can be mapped in ab-plane or in parameter space through the equation, $b = -x_i a + y_i$, i.e., as a straight line. A line passing through points (x_i, y_i) and (x_j, y_j) in xy-plane can be represented as an intersection point between two mapped straight lines in parameter space, where the intersection coordinate, (a', b') is representing the slope and intersection of the straight line in xy-plane. Therefore, the number of lines or edges along a particular direction can be known from the parameter space. From the manufacturing perspective, this information is helpful to take a decision regarding surface roughness analysis or to detect tool wear. For example, the surface machined by the unworn tool (Fig. 4.43a) is smoother as more vertical long edges are observed in its edge detected image (Fig. 4.43d). However, the surface machined by the worn tool (Fig. 4.43d) has non-uniform feed marks having non-uniform, broken, short edges as evident from Fig. 4.43f. Therefore, the number of vertical or nearly vertical edges should be more in Fig. 4.43c than that in Fig. 4.43f, and quantification of this number is required to discriminate between these to surfaces for assessing tool condition, indirectly. This quantification can be done in parameter space. This parameter space is modified to avoid the problem for mapping infinite lines, by taking edge normal in $\rho - \theta$ space by converting the xy map through the equation, $\rho = x\cos\theta + y\sin\theta$, where, ρ and θ are the length of normal and angle of normal with the X-axis, respectively, as shown in Fig. 4.53.

Therefore, information of ρ and θ is represented in $\rho - \theta$ space by performing Hough transform. This works efficiently in terms of computation by dividing $\rho - \theta$ space into some accumulator cells, where closest values of calculated ρ and θ

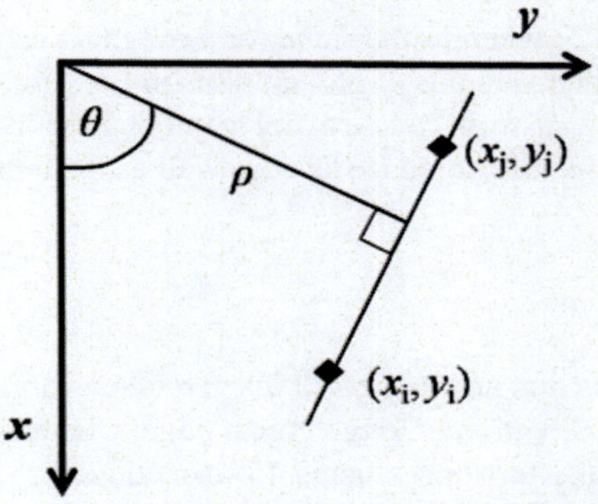

Fig. 4.53 Representation of an edge normal of an edge in xy -plane

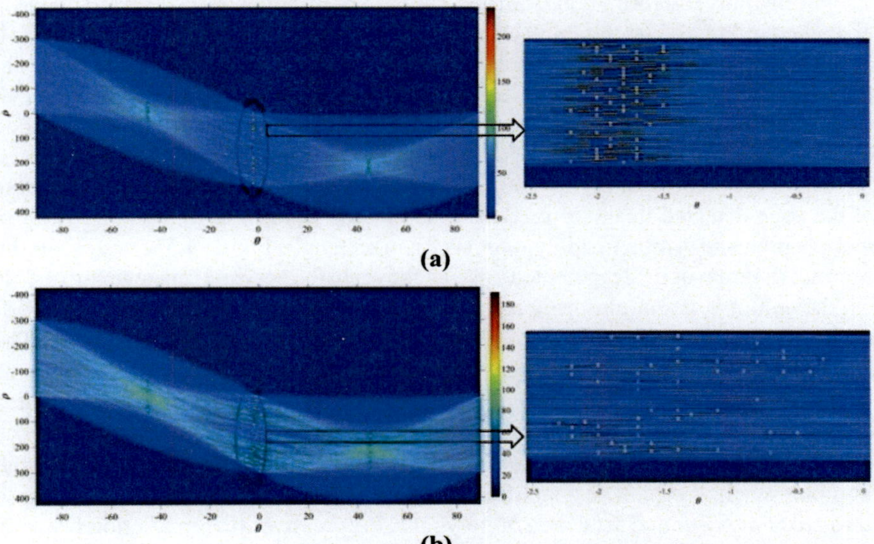

Fig. 4.54 Hough transform map of the edge detected image of surface turned by **a** unworn tool (Fig. 4.43c) and **b** worn tool (Fig. 4.43f)

are mapped in the space. Hough plots of Fig. 4.43c, f are shown in Fig. 4.54a, b, respectively.

A clear discrimination between the Hough transform plots or maps of the edge detected images of surface turned by unworn and worn tools can be evident in Fig. 4.54a, b, respectively. The peaks, i.e., the number of near-vertical edges are more in Fig. 4.54a than that in Fig. 4.54b, as expected. Also, it can be quantified that the angular ranges of near-vertical lines in smoother surface lie in the range between −2.2 and −1.5, whereas the same for a rougher surface is −2.3 to −0.2 as evident from the zoomed versions of Hough plots in Fig. Fig. 4.54a, b, respectively. Therefore, the Hough transform can be a tool to get quantified information regarding the conditions of the manufactured surface, as well as the cutting tool.

Voronoi Tessellation

Edges consist of points that are connected to each other. More number of points are well connected in 90° direction in long vertical edge, whereas less number of points are connected diagonally for short edge in 45° direction. The geometric structure of a texture is reflected in the neighbourhood arrangements of points. In Voronoi tessellation, these points are decomposed into regions by generating polygons following some rules for neighbourhoods of a point.

Voronoi polygon formation from decomposed points of an edge detected image is explained in Fig. 4.55a. P2, P3, P4, P5, and P6 are neighbourhood points of P1 [43]. There are two conditions of Voronoi polygon formation. Firstly, all the points should lie in one plane and secondly at least three of the points should be non-collinear. Perpendicular bisectors of connecting lines from P1 to other points (i.e., 5 dashed lines as shown in Fig. 4.55a) are formed and the polygon created by the bisectors, i.e., *abcdea*, is the Voronoi polygon. Points placed in a proper pattern, as in the case of the edge detected image of surface turned by the unworn tool (Fig. 4.55b), form more regular and symmetric polygons, such as a rectangle. Figure 4.55c represents the Voronoi diagram of the fragmented area, as shown in the zoomed image in Fig. 4.55b. On the contrary, a non-uniform pattern of points (i.e., zoomed image in Fig. 4.55d) create non-uniform and asymmetric Voronoi polygons, as shown in Fig. 4.55e. The *cross-moment* of the symmetric polygons should be zero and thus, more polygons with zero *cross-moment* are present in Fig. 4.55c than that in Fig. 4.55e. The area of polygons in Fig. 4.55e is greater than that in Fig. 4.55c, due to the presence of more number of bigger polygons. Hence, the discrimination of texture, through Voronoi tessellation, can be quantified via the features i.e., number of polygons with zero *cross-moment* and total area of polygons. Progressive wear prediction from the surface texture images in machining was performed by using Voronoi tessellation from the information of change in feed marks [43]. However, the computational complexity is more in this technique.

4.6.1.3 Spectral-Based Texture Analysis

In spectral analysis based texture analysis, the frequency map or space-frequency map of a textured pattern is analysed, and the corresponding frequency features are extracted. In the case of tool condition monitoring from machined surface images, the image of a surface machined by an unworn tool has a more repeated pattern i.e., uniformity as compared to the image of surface machined by a worn tool. A more uniform pattern should consist of a more concentrated central frequency as reported in

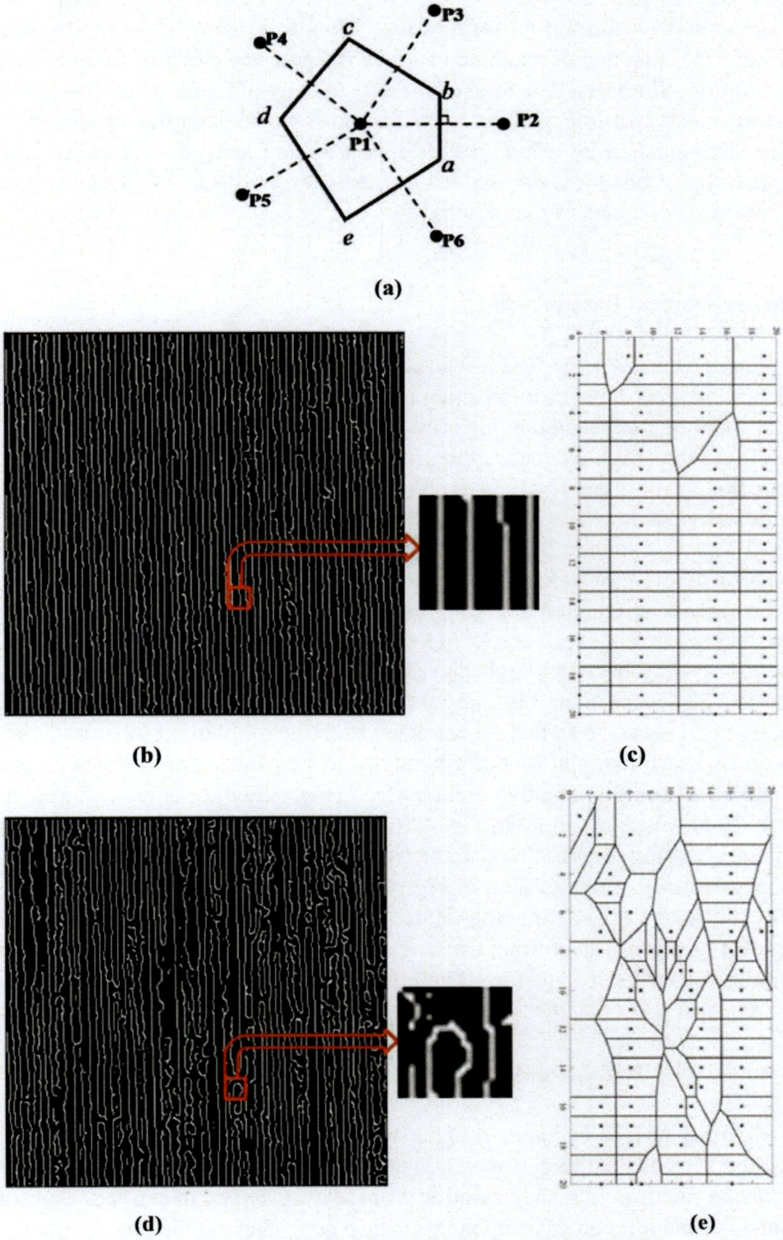

Fig. 4.55 **a** Voronoi polygon formation from points, **b** edge detected image of surface turned by unworn tool and **c** its corresponding Voronoi diagram, **d** edge detected image of surface turned by worn tool, and **e** its corresponding Voronoi diagram

[82], as compared to the non-uniform surface. Therefore, from the area and magnitude of the central frequency of machined surface images, one can discriminate between tool condition. Similarly, detection of finish or quality of manufactured surfaces can be distinguished from this point of view. Defect-free and defective weld surface can also be distinguished by using spectral-based texture analysis. Frequency-domain and space-frequency domain-based texture analyses are discussed in this sub-section with practical examples of manufacturing.

Frequency-Domain Based Analysis

Fast Fourier transform or FFT of an image gives the information regarding the frequency features present in an image as described in Sect. 4.4.2. The power of an image can be represented along with the frequency from its FFT plot. In defect-free weld surface, high-frequency features are less as compared to the surface having defects. So, power or energy (i.e., expressed in Eq. 4.8) at central frequency is more in the defective surface as compared to the defect-free surface. It can also be depicted in the FFT plots shown in Fig. 4.56.

Smooth, defect-free weld surface contains only feed lines with low gray levels in a periodic manner as depicted in Fig. 4.56a. This periodic structure becomes disturbed due to the presence of small cracks in the weld surface as seen in Fig. 4.56b. Hence, frequency spacing becomes higher in the FFT plot (Fig. 4.56f) of Fig. 4.56b than that of the frequency plot (Fig. 4.56e) of Fig. 4.56a. Hence, area of frequency or frequency spacing can be a feature extracted from the frequency plots for delineating between two surfaces. On the other hand, more frequencies are present in the plot (Fig. 4.56h) of rough, defective weld surface (Fig. 4.56d). Also due to roughness, the gray level values are more in Fig. 4.56d, and thus, the power of the FFT is more as evident from Fig. 4.56h. The defective root surface has less gray level value with the non-uniform texture, as seen in Fig. 4.56c. So, the power of frequency is very less for the image of weld having defective root as depicted in the corresponding FFT plot (Fig. 4.56g). Therefore, the area of frequency circle, frequency spacing and power can be extracted as features from FFT plots for surface quality monitoring.

Discrete Wavelet Transform (DWT) Based Texture Analysis

The machined surface becomes rougher with the increase in tool wear, and hence, the regular surface texture gets deviated with the increase in roughness. Hence, tool wear can be monitored, if the deviation from texture can be quantified. The surface roughness of manufactured components is also dependent on the process parameters and the change in surface texture can be quantified for detection of surface roughness from its images. Surface defects can also be detected via texture quantification in manufactured components, such as in cast, formed or weld surfaces. Through this information, process parameters can be updated to achieve batch production of components having acceptable product quality. However, detection of the roughness

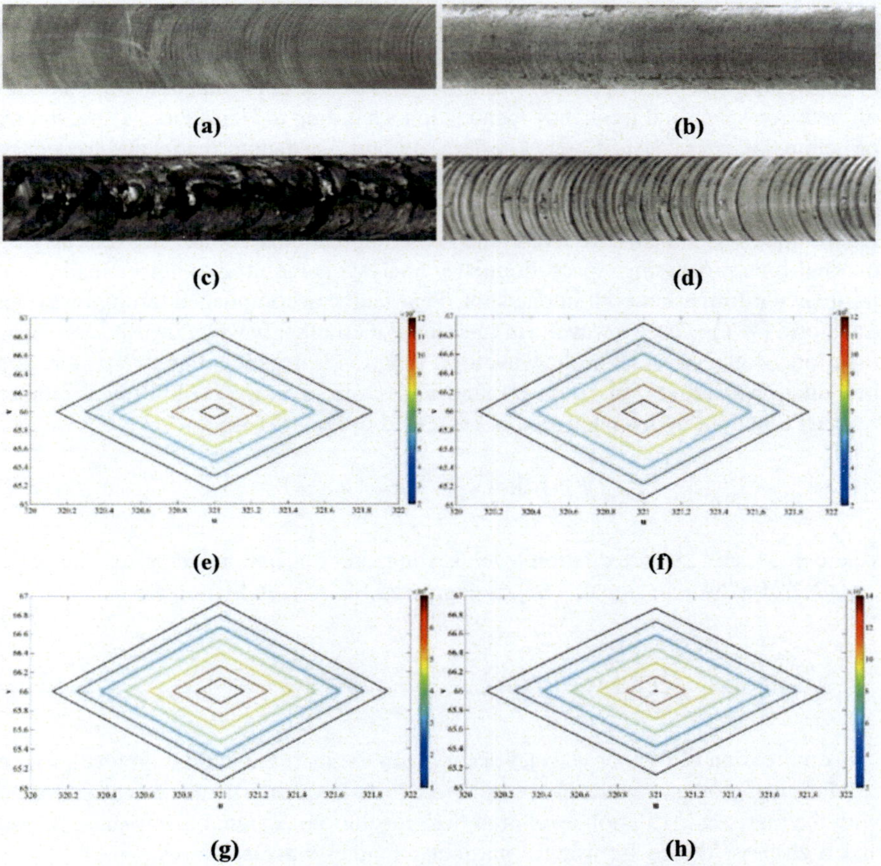

Fig. 4.56 Surface images of FSW weld having **a** smooth, defect-free, **b** small cracks, **c** root defect, **d** rough, defect prone, and **e–h** the corresponding contour plots of the FFT of Fig. 56a–d, respectively

and the defects are required to perform at a micro-scale, which can be fulfilled by using a multi-resolution technique. The non-uniformity in texture for any manufactured surface images can be detected through the power spectrum density, and corresponding frequency information through fast FFT, as shown in the previous section. Detection of information of only global frequency distribution in the image can only be possible via FFT as the basis function is only a sinusoidal function, as explained in Chapter 3. On the other hand, in natural images i.e., images of manufactured surfaces, frequency is varied locally due to the presence of local, non-stationary signals. Therefore, it is very challenging to detect local frequency information in non-stationary images by using FFT. One way of this kind of solution is short-time Fourier transform (STFT), where, the image is windowed into some segments to detect stationary information content in that windowed portion, by using sinusoidal bases of FFT [83].

But, the selection of proper window size concerning the unknown image information in real-time is the main limitation of applying STFT in real-time process monitoring or inspection. Selection of proper window size is highly important as there is a trade-off between space and frequency localization according to Heisenberg's uncertainty principle, i.e., $(\Delta s)^2(\Delta\omega)^2 \geq 1/4$, where, Δs and $\Delta\omega$ denote spatial and frequency resolutions, respectively [84]. In STFT, every time during analysis, the spatial resolution i.e., window size needs to be changed for extracting the features based on local frequency contents. To overcome this computational complexity for carrying out real-time processing, space-frequency analysis is required to detect spatial and frequency information both in one shot. This can be accomplished through wavelet transform (WT), where a waveform function (i.e., mother wavelet) is used as a basis function. Mother wavelet or the waveform function is dependent on two parameters i.e., dilation (*d*) and translation (*t*), and it has zero average [85]. Typical mother wavelet function, in discrete form, is expressed in Eq. 4.37 [85].

$$\psi_{q,k}[a] = 2^{-q/2}\psi(2^{-q}a - k) \tag{4.37}$$

where q, 2^q, and $2^q k$ are dilation level, scaling, and translation factor, respectively.

DWT of a discrete signal, $f[a]$, is expressed for q-level dilation as:

$$\mathrm{DWT}\, f_{q,k}[a] = \sum_{a=1}^{A} f[a] 2^{-q/2}\psi(2^{-q}a - k) \tag{4.38}$$

Compression of mother wavelet occurs with the increase in dilation level, so the resolution of mother wavelet increases. More high-frequency details can be extracted with the increase in dilation level or at high resolution. A signal is low-pass filtered and high-pass filtered by using a particular mother wavelet function followed by down-sampling (by 2) to extract approximate and detailed coefficients, respectively. Therefore, this is also termed wavelet decomposition as a signal is subdivided into approximate and detailed parts, in a single level decomposition. However, the total energy of all the coefficients remains nearly the same as that of the total energy of the original signal. Hence, it can be used for signal or image compression for reducing the computational complexity in subsequent processing steps. The approximation can also be used for denoising using wavelet transform. The detailed coefficients are mainly used to detect fluctuations or deviations from the signals acquired in the usual process, as explained in Chap. 3. The efficiency of detection of fluctuations is also dependent on the choice of a proper mother wavelet function. An application of DWT on an image is represented in Fig. 4.57.

Image at decomposition level, $n-1$, i.e. i^{n-1}, is low-pass filtered by the scaling function, l_r, and then downsampled by 2, row-wise ($2 \downarrow 1$), followed by another column-wise low-pass filtering (l_c) and column-wise downsampling ($1 \downarrow 2$) to obtain the approximate coefficients, ll^n, in the next level (i.e., n). Similarly, lh^n is constituted, where row-wise approximate coefficients and column-wise detailed coefficients are

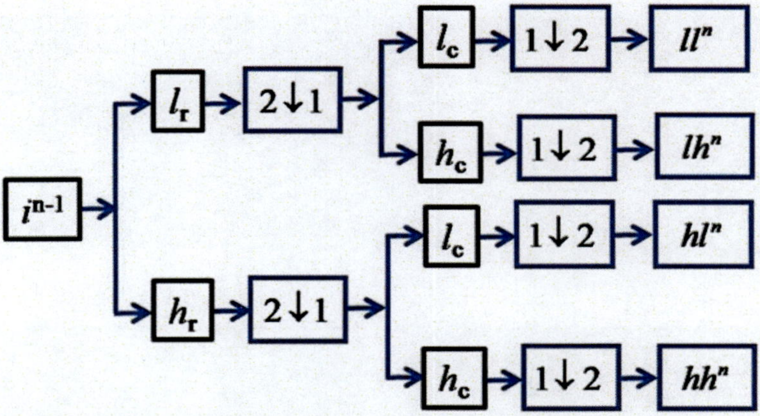

Fig. 4.57 Block diagram of DWT of an image from $(n-1)$th level to nth

present. Here, row-wise low-pass filtering and column-wise band-pass filtering are performed by scaling and wavelet functions, respectively, followed by corresponding downsampling (i.e., $2 \downarrow 1$ and $1 \downarrow 2$) in each case. Row-wise band-pass filtering by wavelet function, h_r, and corresponding downsampling ($2 \downarrow 1$) followed by column-wise low-pass filtering by scaling function, and respective down-sampling ($1 \downarrow 2$) process constitutes hl^n coefficients at level, n. Two applications of lh^n and hl^n coefficients are discussed here. In end-milled surfaces, feed marks are distributed horizontally. So, the fluctuations are present in the vertical direction and for the same reason, surface roughness is measured with the stylus type surface profiler along the vertical direction on the end-milled surface. Analogous to this, column-wise detailed coefficients need to be extracted for enhancing the fluctuation map present in the surface. Therefore, in this case, lh^n coefficients are useful for discrimination between the surface features.

An image of an end-milled surface and its corresponding contour plot is shown in Fig. 4.58a, b, respectively, where horizontal feed marks are evident. But some vertical marks are also seen. Detailed coefficients, lh^1 and hl^1 are represented in contour plots, Fig. 4.58c, d, respectively. The horizontal details are enhanced in the contour plot of lh^1 coefficients (Fig. 4.58c), which is not evident in Fig. 4.58d. So, the fluctuations in a change of roughness along vertical direction can only be evaluated from lh^1 coefficients for end-milled surfaces. It is also seen that the size of detailed coefficients become half in both row and column due to downsampling, i.e., image size, $2R \times 2C$ at level 0 becomes $R \times C$ after level 1 decomposition.

Application of DWT based texture analysis is shown in Fig. 4.59, where, images of surfaces machined by unworn and worn tools are depicted in Fig. 4.59a, b, respectively. Their corresponding contour plots are also shown in Fig. 4.59c, d, respectively. Detailing and scaling along the row and column i.e., hl coefficients should be useful here as the feed marks are lying vertically. Hence, decomposition up to level 5 is shown for example to show the increase in the difference in micro-scale with the

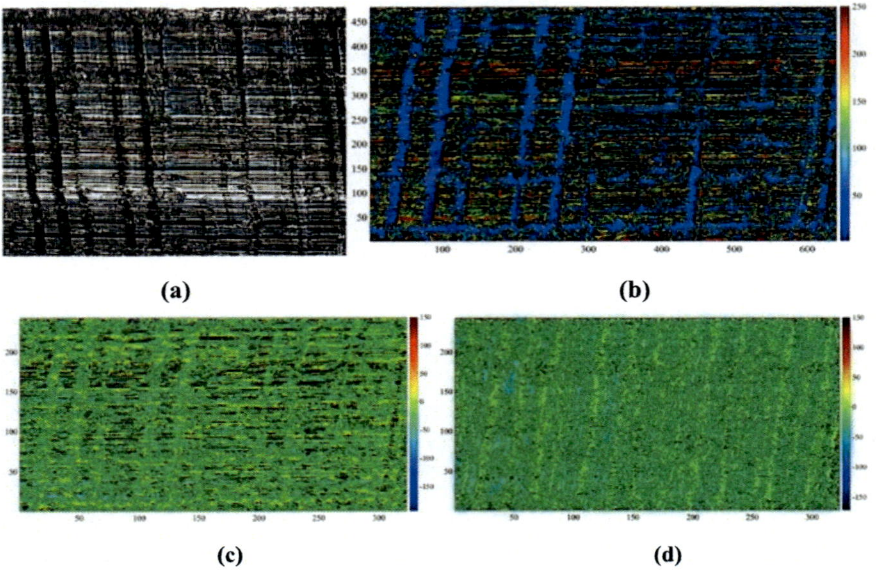

Fig. 4.58 **a** Image of end-milled surface, **b** its contour plot, **c** contour plot of detailed coefficients, lh^1 and **d** contour plot of detailed coefficients, hl^1 of end-milled surface image

increase in resolution. It can be evident from the colourmap of contour plots (shown in Fig. 4.59e, f) that the difference between the lowest and the highest coefficients is much more in the case of Fig. 4.59f than that in Fig. 4.59e. This difference is not achievable in contour plots (Fig. 4.59c, d) of original images. Extraction of features, such as root mean square (RMS) and energy (ENE) can be extracted from the hl^5 coefficients to quantify the surface finish, directly and tool condition, indirectly. The equations of RMS and ENE are given in Eqs. 4.39 and 4.40.

$$\text{RMS} = \sum_{x=1}^{R_n}\sum_{y=1}^{C_n}\sqrt{\frac{(i_{hl}^n(x, y) - AVE)^2}{R_n \times C_n}} \tag{4.39}$$

$$\text{ENE} = \frac{1}{R_n \times C_n}\sum_{x=1}^{R_n}\sum_{y=1}^{C_n}(i_{hl}^n(x, y))^2 \tag{4.40}$$

where AVE is the average value of the coefficients, i.e. $i_{hl}^n(x, y)$ obtained in nth decomposition level.

The deviation of coefficients from the mean is determined by the feature root mean square or RMS, analogous to R_q value in surface roughness measurement. As the surface becomes rougher, non-uniformity in the surface increases. Therefore, the deviation of pixels from the mean should increase. Due to this reason, RMS can be

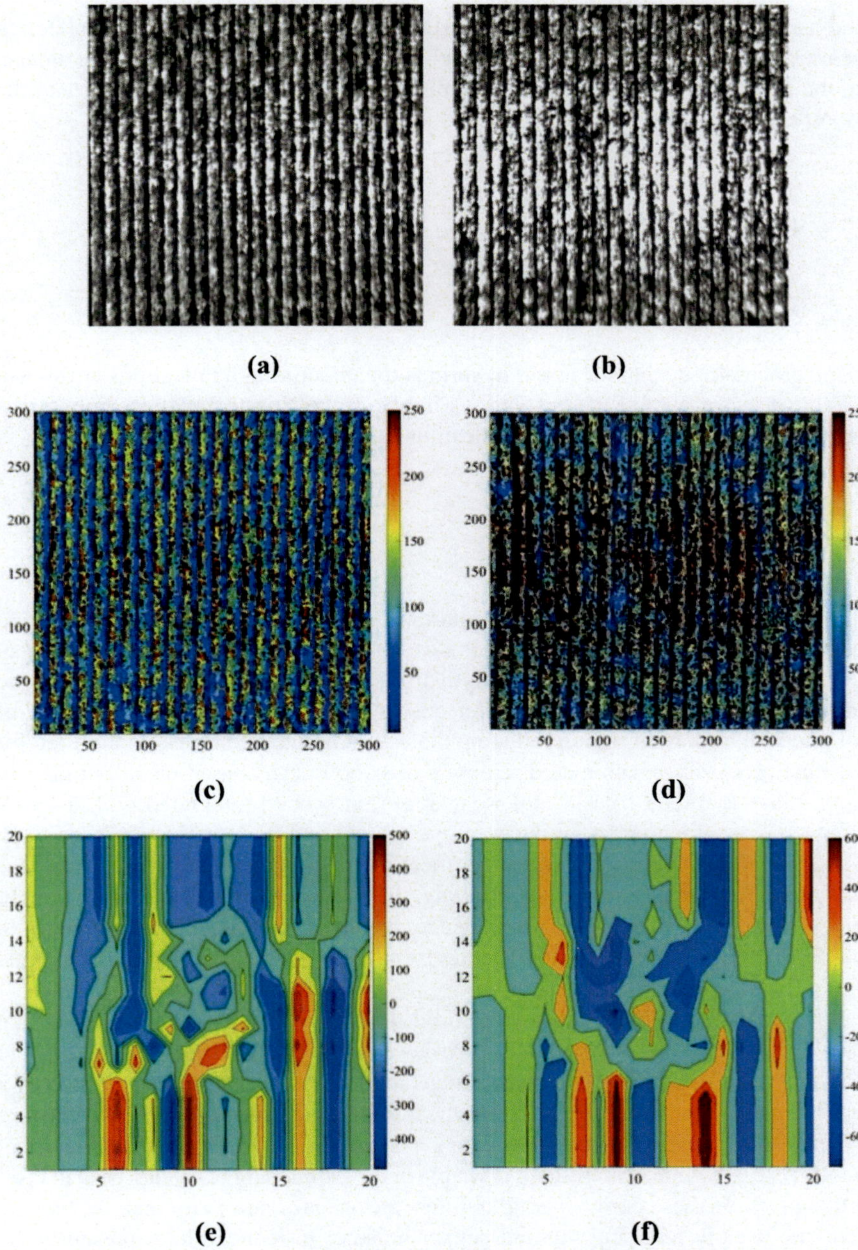

Fig. 4.59 Image of surface machined by **a** unworn and **b** worn tool, and (**c**–**d**) their corresponding contour plots, respectively; Contour plots of hl^5 coefficients of **e** Fig. 4.57a, b, f, respectively

used as a feature for determining the surface finish. Also, more diffused reflection occurs on a rougher surface causing high brightness or pixel value in the image. So, the feature, energy or ENE should more for a rougher surface than that of the smoother surface.

4.7 Summary of Various Image Processing Techniques Used for Real-Time Process Control and Inspection in Manufacturing

Image processing techniques used in various manufacturing applications are shown in Table 4.4. An overall idea of various image processing techniques applied in a particular type of manufacturing application has been summarized for readers.

4.7.1 Image Enhancement

Automatic selection of RoI is very important in an unmanned monitoring system, where humans are relying on the cyber system. In real-time control of vacuum gap by monitoring the diameter of electrolyte droplet [4], weld seam tracking in robotic arc welding [11, 12, 25], tool wear detection [29–31], etc. automated selection of RoI plays an important role in reducing the computational complexity and time for real-time processing. Automated selection of good images based on self-similarity index, plays a crucial role in tool wear detection system integrated with a CPPS to enhance accuracy and to reduce computational and communication time [34]. In some cases, image resolution cannot meet the resolution requirement of process resolution. For example, in profile grinding, the position of grinding wheels need to be controlled in sub-pixel precision. Hence, resolution enhancement by 4 times to obtain 1.746 μm/pixel resolution was performed by image interpolation technique [1].

Low-pass filtering techniques were used to reduce noise in the acquired image when the process zone is prone to disturbances due to process illumination. In weld pool and seam tracking, these disturbances are observed due to the presence of a high-intensity electric arc, even after optical filtering, as shown in Fig. 4.4a b. Due to the high-intensity weld pool, the seam gap is not visible in the case of weld offset control [14, 16]. A similar problem is visible in the monitoring of molten pool in laser welding [70]. In these cases, averaging filters are used for noise elimination. Median filters are used in reducing salt and pepper noise or to reduce glares present in an image. Salt and pepper noise may be present in radiographs where the median filter was used in preprocessing for the detection of weld defects [58]. Defect detection in cold rolled surface [24], countersink inspection [57], and post-weld inspection of weld surface [60, 63] also requires precision inspection where the presence of noise

Table 4.4 Image processing used in real-time process control and process inspection in various manufacturing processes

Process and application	Image enhancement	Image segmentation	Feature extraction	Computing hardware and time	References
Real-time process control					
***Manufacturing method**: Machining*					
Control of tool position for profile error compensation in profile grinding	Resolution enhancement by 4 times	Edge detection	Contour error detection between edge image and the theoretical profile	OpenCV and Qt performed fastest, Multithread seamless control, 135.2 ms including control	Xu et al. [1]
Tool position control or gap compensation from tool wear information in micro-EDM					
Control of vacuum gap from geometric information of produced cavity in ECM	Automatic region of interest (RoI) selection	Thresholding Morphological hole filling or Shi-Tomashi corner detection method	Diameter of cavity from the detected corners of electrolyte droplets	OpenCV and Qt library of C++ Time < 1.1 s	Guo et al. [4]
Control of electrode wire position in ECM with respect to centre of the laser drilled micro holes		Eight neighbourhood based boundary tracking, boundary tracing, Circle fitting	Centre of circle detection Co-ordinate transformation from camera reference frame to the workpiece reference frame	13.35 s	Jia et al. [5]

(continued)

Table 4.4 (continued)

Process and application	Image enhancement	Image segmentation	Feature extraction	Computing hardware and time	References
Manufacturing method: *Additive manufacturing*					
Control of laser power from molten pool geometry in ***DED*** for controlled dilution		Thresholding, morphological erosion	Molten pool width	FPGA for embedded computation with 100 Hz control sampling	Rodriguez-Araujo et al. [6]
	Gaussian PSF to remove bright part around the molten pool	Thresholding	Molten pool width from principal moment and molten pool area	0.4 ms in C++	Hofman et al. [7]
		Thresholding	Molten pool width	10 ms	Moralejo et al. [8]
Control of scan speed from thermal image of molten pool path to extract time temperature history for microstructure control				OpenCV and Qt performed fastest, Multithread seamless control, 135.2 ms including control	Farshidianfar et al. [9]
Manufacturing method: *Fusion welding*					
Control of process parameters from weld pool geometry in TIG	Median filtering Edge enhancement	Thresholding Thinning Edge detection based on neural network	Length and width of the weld pool	industrial computer	Wang et al. [10]
Weld pool and seam monitoring in GTAW with Argon gas shielding	Median filtering	Edge detection	Line fitting for weld seam Contour fitting for weld pool		Ma et al. [11]

(continued)

Table 4.4 (continued)

Process and application	Image enhancement	Image segmentation	Feature extraction	Computing hardware and time	References
Weld pool monitoring and control of laser current in laser welding using active vision			Weld pool width		Na et al. [13]
Weld seam tracking during robotic welding in GTAW process	Setting of camera angle optimally Median filtering	Thresholding Thinning	Seam centre line		Zhou et al. [12]
Welding voltage rectification based on the information of offset between weld pool centre and seam centre line	Mid-value filter	Edge detection Morphological operation for noisy points removal	Line fitting Seam centre co-ordinate extraction Offset between weld torch and seam centres	1000 ms time including tracking control	Ma et al. [14]
Weld seam tracking during welding of curved line from feedback information of offset between torch and seam centreline in GTAW	Region division for weld pool and seam Median filtering	Thresholding Morphological removing Robert's edge det Thinning	Least square fitting	Less than 400 ms	Shen et al. [15]
Weld seam tracking during welding from feedback information of offset between torch and seam centreline in pulsed MAG welding	Mean filter for seam detection	Thresholding for weld pool Edge detection for seam Morphological operations to reduce spurious edges—dilation and erosion	Weld pool centre Hough transform for line detection Information from image calibration	Real-time control at 5 mm/s weld speed with less than 0.4 mm error	Ye et al. [16]

(continued)

Table 4.4 (continued)

Process and application	Image enhancement	Image segmentation	Feature extraction	Computing hardware and time	References
Weld seam tracking from offset information between seam and pool centres in GMAW with Ar shielding	Median filter	Improved Canny edge detection with anisotropy diffusion for edge enhancement; Otsu's optimal thresholding	Line and contour fitting Centre line for seam Weld pool Centre	Error: within ±0.45 mm	Xu et al. [74]
Tracking of 2 welding heads of robot for weld seam of 0.2 mm gap	Contrast enhancement for uneven illumination Noise elimination by min and median filtering ROI selection	Line extraction Hough transform	Least-square fitting	100 ms including control 3–4 pixels maximum error	Chen et al. [17]
Real-time inspection					
***Manufacturing method**: Casting*					
Surface defect (bleed) detection during continuous casting		Thresholding Morphological operation	Contour detection		Pan et al. [18]
***Manufacturing method**: Forging*					
Dimension measurement of forged part during hot forging from projected light stripes on hot forged parts	Top-hat transform of light stripes	Thresholding Edge detection Inflation and erosion	Centre points of stripe detection	0.7% error	Liu et al. [19]

(continued)

Table 4.4 (continued)

Process and application	Image enhancement	Image segmentation	Feature extraction	Computing hardware and time	References
Dimension measurement of forged part during hot forging from projected laser stripes on hot forged parts	Noise filtering Difference imaging	Point extraction	Circle fitting from extracted points	1 mm uncertainty	Bračun et al. [20]
Manufacturing method: *Forming*					
Detection of various defects and their classification during hot rolling			Local binary pattern (LBP)	98% accuracy	Song and Yan [22]
Detection of various defects and their classification during hot rolling (650–950 °C)	Contrast enhancement by homomorphic filtering		Variance, entropy, average gradient	92% accuracy FPGA based computing 12.5 μs computing	Luo and He [23]
Detection of cracks in presence of coolants' patches on cold rolled surface during cold rolling	Clustered of similar pixels by color information and required regions for processing Median filtering and contrast enhancement	Canny edge detection	Edge difference Clustered into industrial liquid regions and defect regions Area of liquid region and defect lengths are features	91% accuracy 10 ms time	Zhao et al. [24]
Surface roughness detection of rolls' surface during reconditioning	Speckle pattern image and its enhancement		Auto-correlation (AC) Derived optical surface roughness from AC	Resolution: 0.5 nm FPGA based computation Speed > 0.5 m^2/min	Fischer and Stöbener [25]

(continued)

Table 4.4 (continued)

Process and application	Image enhancement	Image segmentation	Feature extraction	Computing hardware and time	References
Manufacturing method:*Machining*					
Tool wear localization and broken insert detection on profile milling head inserts	CLAHE, insert localisation using circular HT and cropping	Canny edge detection HT for inserts' edges localization Dilation	Detection of broken insert	24 images within 3 min time with 0.92 precision	Fernández-Robles et al. [29, 30]
Tool wear detection on profile milling head inserts	CLAHE, insert localisation using circular HT and cropping		Local binary pattern and SVM	Raspberry Pi SBC 4 s for 30 inserts 90% accuracy	García-Ordás et al. [31]
Milling tool wear detection using wide FoV camera	Homomorphic filtering	Thresholding	Line fitting and tool flank wear width	91% accuracy	You et al. [32]
Monitoring of micro drill wear in PCB manufacturing	Histogram stretching	Contour segmentation using level set method and histogram	Tool wear	1.96 s	Duan et al. [33]
Drill wear measurement integrated with cyber-physical system	Best image selection by self similarity index	Thresholding, Sobel edge detection, Blob detection	Tool wear width (max)	3.1% error	Lins et al. [34]
Tool wear classification from machined surface texture image	Contrast enhancement or compensation of uneven illumination	Thresholding, Edge detection Hough transform Texture analysis	Hough features Texture features		Kassim et al. [36, 37, 40, 41], Bhat et al. [38, 39]
Tool wear prediction from machined surface texture image	CLAHE for compensation of uneven illumination	Edge detection Texture analyses	Texture features	50–100 ms for a single image	Dutta et al. [41, 44–47], Gadelmawla et al. [42], Datta et al. s[43]

(continued)

Table 4.4 (continued)

Process and application	Image enhancement	Image segmentation	Feature extraction	Computing hardware and time	References
Grinding depth prediction in belt grinding	Color image to gray level by extracting the required color	Spark field region segmentation Fourier transform Thresholding	Area, boundary, density, energy spectrum	95% accuracy	Ren et al. [48]
Detection of surface roughness in moving images with blur	Richardson-Lucy deconvolution algorithm for deblurring	Fourier based and statistical texture analyses	Texture features	0.85 to 0.92 correlation with measured surface roughness	Dhanasekar et al. [49]
Detection of surface roughness with low cost set-up for process parameters updationwith camera connected in the machine tool			Gray level mean	Maximum 20% accuracy	Al-Kindi et al. [50]
Machining chatter mark identification on automotive valve seat surface		Edge detection Hough transform for circle detection GLCM texture analysis	Circularity Texture features		Bamberger et al. [51]
Surface roughness detection on face milled surface		Thresholding	Binary images to ANN model for surface roughness prediction	0.99 test correlation	Samtaş et al. [52]
Surface roughness measurement with camera connected in the machine tool		Algebraic graph theoretic	graph theoretic invariant, Fiedler number	0.5–10 μm surface finish measurement Accuracy: ±2 μm Computation time 50 ms	Tootooni et al. [53]

(continued)

Table 4.4 (continued)

Process and application	Image enhancement	Image segmentation	Feature extraction	Computing hardware and time	References
Position detection in robotic drilling especially in assembly			Co-ordinate transformation from workpiece to the robot and the control of position		Mei et al. [56]
Inspection of countersinks in aircraft's panel	Median filter	Edge detection Morphological operation Contour fitting			Yu et al. [57]
Manufacturing method: *Welding*					
Weld defect detection in sequential radiographic images	Median and averaging filter for noise elimination	Binarization Modified Hough transform	Defects	15.9 ms time and 5 m/min speed of welded pipe	Shao et al. [58]
Weld defect detection in spiral pipes			Kalman filtering		Zou et al. [59]
3-D mapping of weld for defect detection	Median filter	Thresholding Morphological operation	Centreline, feature points Dimensions such as bead width, undercut, reinforcement, plate displacement etc		Chu and Wang [60]
Weld pool geometry detection in TIG welding	Isotherm selection	Area segmentation	Weld pool geometry from segmented area		Alfaro et al. [62]
Weld quality inspection of shell tube heat exchanger	Median filter	Thresholding Morphological operation	Area extraction and defects detection	100 ms computation time	Chu and Wang [63]

(continued)

Table 4.4 (continued)

Process and application	Image enhancement	Image segmentation	Feature extraction	Computing hardware and time	References
Weld defect detection of fuel injector	Histogram equalization	Internal and external circle segmentation	Points extraction	97% accuracy with deep learning	Sassi et al. [64]
Defect detection on FSW surface	CLAHE	Windowed wavelet analysis	Energy, variance and entropy SVM classification	99% accuracy	Bhat et al. [65]
Defect detection in welding of canister on a production line		Modified optimal thresholding to segment weld area	Brightness Curve and edges	30 to 35 ms 96% accuracy	Sun et al. [69]
Monitoring of keyhole in Laser welding	Low pass filtering	Thresholding Edge detection	Area		Kim and Ahn [70]
Quality check:*Metrology*					
Inspection and fault detection in automatic assembly line	Adaptive Gaussian filter, Histogram equalization,	Gaussian mixture model based segmentation, Running average based segmentation, thresholding, morphological	Blob analysis and blob area Running average area	Time is 2 s including communication	Chauhan and Surgenor [73]

may produce ambiguous results. White Gaussian noise was reduced by Gaussian low-pass filter in DED [7]. Contrast enhancement or histogram equalization is required for reducing uneven illumination, especially in the detection of surface texture in machining [35–47] and in friction stir welding [65], where the CLAHE technique was used. In tool flank wear measurement [29–31, 33], weld seam tracking [17] and weld defect detection in fuel injector [64], the images of process zones are generally low-contrast or low dynamic range images, where the delineation between the object and background is challenging. In this situation, the histogram equalization technique is used for contrast enhancement. In some cases, homomorphic filtering was used to homogenize the image affected by uneven illumination artefact. Hot rolled surface images are affected by uneven illumination due to improper surface flatness at the time of rolling [23]. Tool wear acquired by a low-cost, wide field of view lens is also affected by uneven illumination and low contrast [32]. In these two cases, homomorphic filtering gave a good image enhancement result as compared to the histogram equalization technique.

4.7.2 Image Segmentation

Image segmentation tasks are mainly thresholding, contour, and edge segmentation. These steps are performed before feature extraction. Optimal thresholding technique to segment the object from the background are used widely. In control of vacuum gap in ECM, electrolyte droplet portion was segmented from its background to monitor the droplet diameter [4]. Molten pool in DED was thresholded accurately to monitor molten pool width or aspect ratio for controlling the laser power [6–8]. Welding current was controlled from the information of weld pool dimension where the weld pool was segmented from the background [10]. Then the weld pool width was extracted for further feedback information to current control. Tracking of weld seam in robotic welding was accomplished through the information of offset between weld pool and seam centres, where these two were segmented separately from two windows [12, 15, 16, 74]. Inspection of bleed from its noisy background was segmented by using an intelligent thresholding operation in continuous casting [18]. Thresholding of light stripes projected on the hot forged parts was performed in determining the dimensions of forged parts during forging [19]. Extraction of tool flank wear portion from its background in direct tool wear monitoring requires an accurate thresholding operation [32, 34]. Segmentation of spark field from its background was required for real-time monitoring of grinding depth [48]. Though thresholding did not play a much role in surface texture analysis in surface finish monitoring, binarization of textures for feeding it into the artificial neural network was performed to predict the machined surface roughness [52]. In laser welding, proper penetration was monitored from the size of the keyhole, where the thresholding technique took an important role [70].

Edge detection techniques are used to detect the semantic edges from the object to find the required dimensions. Detection of edges from tool and profile contours are required to extract tool offset for controlling the tool position in profile grinding [1]. Shi-Tomashi corner detection technique was used to detect two opposite corners by avoiding shadows from the thresholded images of electrolyte droplet to extract its diameter [4]. Detection of contours of weld pool was required to determine its width or the centre point [10, 11, 14]. Detection of edges in weld seam was employed for further detection of seam centreline for control of the position of end-effector in robotic welding, where the accuracy of edge detection matters a lot [11, 14–16, 74]. In the measurement of part dimension during hot forging, the stripes' edges need to be detected for further processing [19, 20]. Cracks or defect detection during cold rolling requires segmentation of defects from the region of industrial liquid patches, thus, an important role is played via defect segmentation [24]. Segmentation of tool wear boundary edges and removal of spurious edges requires indirect monitoring of tool wear in machining [29, 30, 34]. Detection of automotive valve seal geometry required the detection of edges from its image [51]. Position of the countersink is required in robotic drilling and rivetting operation, where the detected edges of the countersink helped to determine the centre point coordinate for further path planning [57].

Morphological operations were used for the enhancement of thresholded image for further processing accurately n various manufacturing processes starting from casting to welding. Spurious noises were removed by selecting disconnected pixels from the thresholded image of the continuous cast surface for separating the bleed portion [18]. The diameter of the job during the hot forging process was known from the information of light stripes, where some bright spots or areas were coming in the image as noise. These areas were removed by erosion operation [19]. RoI of tool insert is automatically selected after performing dilation in recognition of broken insert in profile milling operation [29, 30]. In post-weld inspection, 3-D mapping of weld bead was performed, where the closing operation was used for smoothening the structured line pattern from the optimally thresholded image [60]. In defect detection on the weld surface, morphological opening and closing operations were used to smooth the contour of the object by removing noise on boundaries and by linking the narrow gaps, respectively, present in the thresholded image [63].

4.7.3 Feature Extraction and Decision Making

The final part of image processing is the extraction of features from segmented or pre-processed images of the manufacturing process zone. Proper or optimal feature extraction is required here for accurate decision making i.e., process detection or inspection. Hough transform is used to detect line or contour in this phase. Circular Hough transform was used to locate inserts from profile milling tool head, however, standard Hough transform was used for detecting the inserts' edges, accurately, in automatic detection of broken inserts [29, 30]. In literature, extraction of Hough length and angle of the edge detected machined surface images were used in a multi-layer perceptron neural network model to predict tool condition [39]. In automatic recognition of circularity on ring pattern of valve seals, Hough transform was used to detect circular profiles [51]. Major weld defects in the radiographic image sequence were tracked through the Hough transform as well [58].

In profile grinding operation, the surface having a contour or profile is ground for finishing operation. Here, the actual contour was compared with the theoretical contour or CAD data, where the coordinates of points on the actual profile were extracted from the acquired images. Subsequently, the work table was controlled via image-based feedback to reduce the error between the coordinates of the theoretical profile and the actual profile to compensate. Consequently, the average profile error obtained was 3 μm [1]. In micro-EDM operation, spark gap compensation was performed by real-time processing of the images of the electrode having 100 μm diameter as shown in Fig. 4.2. Here, a maximum of 10 μm deviation from the CAD profile was obtained in the machined surface after compensation of tool position using image processing [2]. Tool wear in micro-EDM drilling was also detected to obtain a maximum error of less than 1% for machining a blind hole having 200 μm depth [3]. The diameter of the electrolyte droplet was detected to control process parameters of an ECM process (as shown in Fig. 4.3) for producing micro-gravity of 900 μm diameter within 4.8 μm error [4]. In laser-electro chemical processing, ECM was performed to finish the laser-drilled holes, where the centre position of the laser-drilled hole was detected using image processing for accurate positioning of ECM electrode. With this position compensation, a maximum 12 μm error was obtained for drilling a hole of 200 μm diameter [5]. Weld pools were detected by fitting the contours and after that weld pool centre coordinates were detected for tracking of weld path [11, 16, 74]. Seam centrelines were detected at the time of weld seam tracking by line fitting in [11, 12, 14–17, 74]. Detection of the extent of the defect (bleed width) on the cast surface during continuous casting was performed as a real-time inspection [18]. The width of the molten pool in the DED process was extracted to control laser power [6–8]. Tool flank wear width was evaluated, automatically, in tool condition monitoring [32–34]. Weld defects like reinforcement and undercut were evaluated as a post-weld real-time inspection technique [60]. Weld pool width was extracted to control welding current [10]. The areas of industrial liquid patches on rolled surfaces were determined automatically during cold rolling operation [24]. The area of spark field was extracted from the thresholded image in belt grinding application to predict the grindng depth in belt grinding operation [48]. The keyhole area was determined in laser welding to monitor penetration [70].

Statistical information such as variance, entropy, and gradients was used for defect recognition from images of the hot-rolled surface by using FPGA-based parallel computation with an average recognition rate of 92.11% in real-time [23]. GLCM features were extracted from turned surface images to predict classify tool condition into three classes using support vector machine [37] and hidden Markov model [38] with at least 94% classification accuracy. Tool wear was detected from turned surface images using GLCM features, where parameters of GLCM were selected optimally [41]. In this case, GLCM features i.e., contrast and homogeneity had a good correlation (adjusted R^2-value of 0.93) with tool flank wear. Chatter mark identification using extracted GLCM features from the surface images of automotive valve seals were performed [51]. RLS-based texture analysis was performed to detect tool condition from the texture of turned [35, 40] and milled surfaces [44]. Geometrical features extracted from the Voronoi tessellation map of turned surface images were utilized to predict tool flank wear in [43]. In this case, nine experiments were performed on turning operation to evaluate the effect of extracted features, i.e., number of polygons with zero cross-moment and void area as mentioned in Sect. 4.6.1.2, where the first feature had a better linear correlation with tool flank wear. Use of GLCM based, Voronoi tesselation based and DWT-based features were extracted from machined surface images and fed to SVM regression model for monitoring tool flank wear from machined surface images in turning [45, 46], where average prediction efficiency was improved to 99%. Features from FFT of images of milled and ground surfaces were extracted and fed to a neural network to predict surface roughness with 92.3 and 84% accuracies, respectively [49]. FFT features were also extracted from the thresholded spark field to monitor grinding depth in belt grinding operation [48]. DWT-based features were extracted from images of friction stir weldments for classification of defective and sound welds [65] with 99% accuracy by using SVM. It can be observed that most of the surface quality assessment or tool wear assessment from manufactured surfaces, texture analyses techniques are preferred due to their robust statistical texture features for predicting the phenomena, reliably.

4.8 Summary

This chapter gives an overview of common image processing methods applicable to various manufacturing processes for real-time process control and inspection. Type of manufacturing process zones to be imaged for extraction of proper information in process control and inspection are first discussed. Readers will have an idea about the importance of image processing in various manufacturing applications from this section. Components of the image acquisition system are mentioned with manufacturing examples. Then image processing methods including image enhancement, segmentation, and feature extraction for pattern recognition are described from very basics to the manufacturing applications. However, details on machine learning techniques with the extracted features are not in the scope of this chapter as this will be dealt with in Chap. 6. Image acquisition systems and image processing methods are varying for different manufacturing processes. Thus, they are summarized in tabular form for easy understanding purpose, so that readers can get an idea for their targeted manufacturing applications.

4.9 Questions for Practice

1. *What are the major components required in a machine vision system?*

2. *If you want to apply machine vision in a high precision measurement application, then what are the two most important features for image acquisition? Name two types of illumination systems with one application each. What is F-number? How does the CCD sensor work in a CCD camera at the time of image acquisition?*
3. *Name the process zones to be imaged for real-time control of robotic arc welding process.*
4. *In a cold rolling technique, the rolled surface needs to be inspected by using machine vision system in real-time, when the speed of the surface is 5* m/s. *The minimum width of the defect need to be imaged without motion blur is 1* mm. *The camera needs to be placed in such a way that the angle between the planes of image and the rolled surface is zero. Considering the pixel pitch of image sensor as 7.5* μm, *find out the maximum exposure times are required to acquire good images for defect detection by varying magnification as 0.2, 0.4 and 0.8.*
5. *What is the output image g(x, y) obtained from convolution of input image (a fragment of machined surface), f(x, y) and the kernel h(x, y), where: f(x, y) =*

8	7	4	5	7
7	8	3	4	8
8	7	3	4	6
8	7	4	3	8
8	7	4	4	7

and, $h(x, y) = \begin{bmatrix} -2 & -2 & -2 \\ -2 & 16 & -2 \\ -2 & -2 & -2 \end{bmatrix}$

Name the type of kernel or filter, h(x, y).

8	8	3	1	7
7	8	3	1	8
8	7	3	3	6
8	7	4	3	8
8	7	4	4	7

8	8	3	1	7
7	8	3	8	2
5	4	5	3	6
8	5	8	3	5
8	7	4	6	7

Mention the features selected for the classification.

6. *Which segmentation technique should be preferred in terms of computational time to identify defect in weld?*
7. *Classify between a smooth and a rough surface using statistical texture analysis. The image fragment matrices are given below:*

8. *How do you apply Hough transform and Voronoi tessellation to classify the images shown in the problem 7?*

References

1. Xu LM, Fan F, Zhang Z, Chen Y, Hu DJ, Shi L (2019) Methodology and implementation of a vision-oriented open CNC system for profile grinding. Int J Adv Manuf Technol 100(5):2123–2131
2. Yan MT, Lin SS (2011) Process planning and electrode wear compensation for 3D micro-EDM. Int J Adv Manuf Technol 100(5):2123–2131
3. Malayath G, Katta S, Sidpara AM, Deb S (2019) Length-wise tool wear compensation for micro electric discharge drilling of blind holes. Measurement 134:888–896
4. Guo C, Qian J, Reynaerts D (2018) Deterministic removal strategy for machine vision assisted scanning micro electrochemical flow cell. J Manuf Process 34:167–178
5. Jia G, Dong X, Huo Q, Wang K, Mei X (2018) Positioning and navigation system based on machine vision intended for laser-electrochemical micro-hole processing. Int J Adv Manuf Technol 94(1):1397–1410
6. Rodriguez-Araujo J, Rodriguez-Andina JJ, Farina J, Vidal F, Mato JL, Montealegre MA (2012) Industrial laser cladding systems: FPGA-based adaptive control. IEEE Ind Electron Mag 6(4):35–46
7. Hofman JT, Pathiraj B, Van Dijk J, De Lange DF, Meijer J (2012) A camera based feedback control strategy for the laser cladding process. J Mater Process Technol 212(11):2455–2462
8. Moralejo S, Penaranda X, Nieto S, Barrios A, Arrizubieta I, Tabernero I, Figueras J (2017) A feedforward controller for tuning laser cladding melt pool geometry in real time. Int J Adv Manuf Technol 89(1–4):821–831
9. Farshidianfar MH, Khajepour A, Gerlich A (2016) Real-time control of microstructure in laser additive manufacturing. Int J Adv Manuf Technol 82(5–8):1173–1186
10. Wang JJ, Lin T, Chen SB (2005) Obtaining weld pool vision information during aluminium alloy TIG welding. Int J Adv Manuf Technol 26(3):219–227
11. Ma H, Wei S, Lin T, Chen S, Li L (2010) Binocular vision system for both weld pool and root gap in robot welding process. Sens Rev
12. Zhou L, Lin T, Chen SB (2006) Autonomous acquisition of seam coordinates for arc welding robot based on visual servoing. J Intell Rob Syst 47(3):239–255
13. Na X, Zhang Y, Liu Y, Walcott B (2009) Nonlinear identification of laser welding process. IEEE Trans Control Syst Technol 18(4):927–934
14. Ma H, Wei S, Sheng Z, Lin T, Chen S (2010) Robot welding seam tracking method based on passive vision for thin plate closed-gap butt welding. Int J Adv Manuf Technol 48(9–12):945–953
15. Shen H, Lin T, Chen S, Li L (2010) Real-time seam tracking technology of welding robot with visual sensing. J Intell Rob Syst 59(3–4):283–298
16. Ye Z, Fang G, Chen S, Zou JJ (2013) Passive vision based seam tracking system for pulse-MAG welding. Int J Adv Manuf Technol 67(9–12):1987–1996
17. Chen H, Liu K, Xing G, Dong Y, Sun H, Lin W (2014) A robust visual servo control system for narrow seam double head welding robot. Int J Adv Manuf Technol 71(9–12):1849–1860
18. Pan E, Ye L, Shi J, Chang TS (2009) On-line bleeds detection in continuous casting processes using engineering-driven rule-based algorithm. J Manuf Sci Eng 131(6)
19. Liu W, Jia Z, Wang F, Ma X, Wang W, Jia X, Song D (2012) An improved online dimensional measurement method of large hot cylindrical forging. Measurement 45(8):2041–2051
20. Bračun D, Škulj G, Kadiš M (2017) Spectral selective and difference imaging laser triangulation measurement system for on line measurement of large hot workpieces in precision open die forging. Int J Adv Manuf Technol 90(1):917–926

21. Agarwal K, Shivpuri R, Zhu Y, Chang TS, Huang H (2011) Process knowledge based multi-class support vector classification (PK-MSVM) approach for surface defects in hot rolling. Expert Syst Appl 38(6):7251–7262
22. Song K, Yan Y (2013) A noise robust method based on completed local binary patterns for hot-rolled steel strip surface defects. Appl Surf Sci 285:858–864
23. Luo Q, He Y (2016) A cost-effective and automatic surface defect inspection system for hot-rolled flat steel. Robot Comput-Integr Manuf 38:16–30
24. Zhao YJ, Yan YH, Song KC (2017) Vision-based automatic detection of steel surface defects in the cold rolling process: considering the influence of industrial liquids and surface textures. Int J Adv Manuf Technol 90(5–8):1665–1678
25. Fischer A, Stöbener D (2019) In-process roughness quality inspection for metal sheet rolling. CIRP Ann 68(1):523–526
26. ISO 3685: 1993(E), Tool life testing with single point turning tools
27. Dutta S, Pal SK, Mukhopadhyay S, Sen R (2013) Application of digital image processing in tool condition monitoring: a review. CIRP J Manuf Sci Technol 6:212–232
28. Dutta S, Pal SK, Sen R (2014) Digital image processing in machining. In: Davim JP (ed) modern mechanical engineering—Research, development and education. Springer, Berlin, pp 369–412
29. Fernández-Robles L, Azzopardi G, Alegre E, Petkov N (2015) Cutting edge localisation in an edge profile milling head. In: International conference on computer analysis of images and patterns. Springer, Cham, pp 336–347
30. Fernández-Robles L, Azzopardi G, Alegre E, Petkov N (2017) Machine-vision-based identification of broken inserts in edge profile milling heads. Robot Comput-Integr Manuf 44:276–283
31. García-Ordás MT, Alegre-Gutiérrez E, Alaiz-Rodríguez R, González-Castro V (2018) Tool wear monitoring using an online, automatic and low cost system based on local texture. Mech Syst Signal Process 112:98–112
32. You Z, Gao H, Guo L, Liu Y, Li J (2020) On-line milling cutter wear monitoring in a wide field-of-view camera. Wear, 460:203479
33. Duan G, Chen YW, Sukegawa T (2010) Automatic optical flank wear measurement of microdrills using level set for cutting plane segmentation. Mach Vis Appl 21:667–676
34. Lins RG, de Araujo PRM, Corazzim M (2020) In-process machine vision monitoring of tool wear for cyber-physical production systems. Robot Comput-Integr Manuf 61:101859
35. Kassim AA, Mannan MA, Jing M (2000) Machine tool condition monitoring using workpiece surface texture analysis. Mach Vis Appl 11:257–263
36. Kassim AA, Mian Z, Mannan MA (2006) Tool condition classification using Hidden Markov model based on fractal analysis of machined surface textures. Mach Vis Appl 17:327–336
37. Bhat NN, Dutta S, Vashisth T, Pal S, Pal SK, Sen R (2016) Tool condition monitoring by SVM classification of machined surface images in turning. Int J Adv Manuf Technol 83:1487–1502
38. Bhat NN, Dutta S, Pal SK, Pal S (2016) Tool condition classification in turning process using hidden Markov model based on texture analysis of machined surface images. Measurement 90:500–509
39. Kassim AA, Mian Z, Mannan MA (2004) Connectivity oriented fast Hough transform for tool wear monitoring. Pattern Recogn 37:1925–1933
40. Kassim AA, Mannan MA, Mian Z (2007) Texture analysis methods for tool condition monitoring. Image Vis Comput 25:1080–1090
41. Dutta S, Datta A, Chakladar ND, Pal SK, Mukhopadhyay S, Sen R (2012) Detection of tool condition from the turned surface images using an accurate grey level co-occurrence technique. Precis Eng 36:458–466
42. Gadelmawla ES, Al-Mufadi FA, Al-Aboodi AS (2013) Calculation of the machining time of cutting tools from captured images of machined parts using image texture features. Proc Inst Mech Eng Part B: J Eng Manuf 228:203–214
43. Datta A, Dutta S, Pal SK, Sen R (2013) Progressive cutting tool wear detection from machined surface images using Voronoi tessellation method. J Mater Process Technol 213:2339–2349

44. Dutta S, Kanwat A, Pal SK, Sen R (2013) Correlation study of tool flank wear with machined surface texture in end milling. Measurement 46:4249–4260
45. Dutta S, Pal SK, Sen R (2016) On-machine tool prediction of flank wear from machined surface images using texture analyses and support vector regression. Precis Eng 43:34–42
46. Dutta S, Pal SK, Sen R (2016) Tool condition monitoring in turning by applying machine vision. J Manuf Sci Eng 138
47. Dutta S, Pal SK, Sen R (2018) Progressive tool condition monitoring of end milling from machined surface images. Proc Inst Mech Eng Part B J EngManuf 232:251–326
48. Ren L, Zhang G, Wang Y, Zhang Q, Wang F, Huang Y (2019) A new in-process material removal rate monitoring approach in abrasive belt grinding. Int J Adv Manuf Technol 104(5–8):2715–2726
49. Dhanasekar B, Ramamoorthy B (2010) Restoration of blurred images for surface roughness evaluation using machine vision. Tribol Int 43:268–276
50. Al-Kindi G, Zughaer H (2012) An approach to improved CNC machining using vision-based system. Mater Manuf Process 27:765–774
51. Bamberger H, Ramachandran S, Hong E, Katz R (2011) Identification of machining chatter marks on surfaces of automotive valve seats. Trans ASME-J Manuf Sci Eng 133:0410031–0410037
52. Samtaş G (2014) Measurement and evaluation of surface roughness based on optic system using image processing and artificial neural network. Int J Adv Manuf Technol 73:353–364
53. Tootooni MS, Liu C, Roberson D, Donovan R, Rao PK, Kong ZJ, Bukkapatnam ST (2016) Online non-contact surface finish measurement in machining using graph theory-based image analysis. J Manuf Syst 41:266–276
54. Sun H, Gao D, Zhao Z, Tang X (2017) An approach to in-process surface texture condition monitoring. Robot Comput-Integr Manuf 48:254–262
55. Diao S, Chen X, Luo J (2018) Development and experimental evaluation of a 3D vision system for grinding robot. Sensors 18(9):3078
56. Mei B, Zhu W, Ke Y (2018) Positioning variation analysis and control for automated drilling in aircraft manufacturing. Assem Autom
57. Yu L, Bi Q, Ji Y, Fan Y, Huang N, Wang Y (2019) Vision based in-process inspection for countersink in automated drilling and riveting. Precis Eng 58:35–46
58. Shao J, Du D, Chang B, Shi H (2012) Automatic weld defect detection based on potential defect tracking in real-time radiographic image sequence. NDT and E Int 46:14–21
59. Zou Y, Du D, Chang B, Ji L, Pan J (2015) Automatic weld defect detection method based on Kalman filtering for real-time radiographic inspection of spiral pipe. NDT and E Int 72:1–9
60. Chu HH, Wang ZY (2016) A vision-based system for post-welding quality measurement and defect detection. Int J Adv Manuf Technol 86(9–12):3007–3014
61. Li Y, Li YF, Wang QL, Xu D, Tan M (2009) Measurement and defect detection of the weld bead based on online vision inspection. IEEE Trans Instrum Meas 59(7):1841–1849
62. Alfaro SCA, Vargas JAR, de Carvalho GC, de Souza GG (2015) Characterization of "humping" in the GTA welding process using infrared images. J Mater Process Technol 223:216–224
63. Chu HH, Wang ZY (2017) A study on welding quality inspection system for shell-tube heat exchanger based on machine vision. Int J Precis Eng Manuf 18(6):825–834
64. Sassi P, Tripicchio P, Avizzano CA (2019) A smart monitoring system for automatic welding defect detection. IEEE Trans Industr Electron 66(12):9641–9650
65. Bhat NN, Kumari K, Dutta S, Pal SK, Pal S (2015) Friction stir weld classification by applying wavelet analysis and support vector machine on weld surface images. J Manuf Process 20:274–281
66. Sudhagar S, Sakthivel M, Ganeshkumar P (2019) Monitoring of friction stir welding based on vision system coupled with Machine learning algorithm. Measurement 144:135–143
67. Ranjan R, Khan AR, Parikh C, Jain R, Mahto RP, Pal S, Pal SK, Chakravarty D (2016) Classification and identification of surface defects in friction stir welding: an image processing approach. J Manuf Process 22:237–253

68. Parikh C, Ranjan R, Khan AR, Jain R, Mahto RP, Chakravarty D, Pal S, Pal SK (2017) Volumetric defect analysis in friction stir welding based on three dimensional reconstructed images. J Manuf Process 29:96–112
69. Sun J, Li C, Wu XJ, Palade V, Fang W (2019) An effective method of weld defect detection and classification based on machine vision. IEEE Trans Industr Inf 15(12):6322–6333
70. Kim CH, Ahn DC (2012) Coaxial monitoring of keyhole during Yb: YAG laser welding. Opt Laser Technol 44(6):1874–1880
71. Luo M, Shin YC (2015) Estimation of keyhole geometry and prediction of welding defects during laser welding based on a vision system and a radial basis function neural network. Int J Adv Manuf Technol 81(1):263–276
72. Židek K, Piteľ J, Adámek M, Lazorík P, Hošovský A (2020) Digital twin of experimental smart manufacturing assembly system for Industry 4.0 concept. Sustainability 12(9):3658
73. Chauhan V, Surgenor B (2017) Fault detection and classification in automated assembly machines using machine vision. Int J Adv Manuf Technol 90(9–12):2491–2512
74. Xu Y, Lv N, Fang G, Du S, Zhao W, Ye Z, Chen S (2017) Welding seam tracking in robotic gas metal arc welding. J Mater Process Technol 248:18–30
75. Eckhart M, Ekelhart A (2019) Digital twins for cyber-physical systems security: State of the art and outlook. In: Security and quality in cyber-physical systems engineering. Springer, Cham, pp 383–412
76. The process digital twin: a step towards operational excellence. https://info.microsoft.com/rs/157-GQE-382/images/Digital%20Twin%20Vision.pdf. Accessed 10 Jan 2021
77. Digital twin for Industry 4.0 inspection systems. https://www.pleora.com/markets/machine-learning/digital-twin-for-industry-4-0-inspection-systems/. Accessed 10 Jan 2021
78. Burke MW (1996) Image acquisition: handbook of machine vision engineering. Chapman and Hall, London
79. Saba MH, Mukherjee S, Dutta S, Mallisetty PK, Murmu NC (2020) Electrohydrodynamic jet printing for desired print diameter. Mater Today: Proc
80. Gonzalez RC, Woods RE (2002) Digital image processing. Prentice-Hall, NJ, USA
81. Pizer SM, Amburn EP, Austin JD, Cromartie R, Geselowitz A, Greer T, Romeny BH, Zimmerman JB, Zuiderveld K (1987) Adaptive histogram equalization and its variations. Comput Vis Grap Image Process 39(3):355–368
82. Lee BY, Yu SF, Juan H (2004) The model of surface roughness inspection by vision system in turning. Mechatronics 14(1):129–141
83. Petrosian AA, Meyer FG (2001) Wavelets in signal and image analysis: from theory to practice. Kluwer Academic, Boston
84. Petrou M, Sevilla PG (2006) Image processing dealing with texture. Wiley, UK
85. Mallat S (1999) A wavelet tour of signal processing. Academic Press, London, UK

Chapter 5
Data Communication-Edge, Fog, and Cloud Computing

5.1 Overview

Industry 4.0 is aimed at connecting the machine, materials, and other infrastructure through data collected from the sensors. The data acquired from the sensors will be utilised to monitor, diagnose, predict, control, optimize plant operations, etc. Because the number of connected devices is going to be huge, the amount of data will also be large. Thus, it is imperative to understand and utilise data communication technologies such as edge, fog, and cloud. Further, it is also essential to understand the issues of transporting data to the cloud and its implications in final business outcomes. The research question in this direction includes: how to make the network and computing infrastructure scalable enough to handle large data and still churn out meaningful analytics for the business of a manufacturing plant. The infrastructure scalability issue will be discussed in-depth in this chapter.

5.2 IoT and Network

Let us begin the discussion with the famous quote by *Richard Hamming*, "*The purpose of (scientific) computing is insights, not numbers*". People often think of computing to be the numbers generated from a machine. However, it signifies the insights. It all depends upon the quality of the data, processing technique, analysis environment, and so on. The success lies in proper and efficient utilisation of these things, which will lead to meaningful insights fulfilling the purpose of computing.

To start with the discussions on data communication, which is critical, difficult, and includes too much stuff, let us recap and re-discuss Internet of Things (IoT). Chapters 2, 3, and 4 of this textbook provided the technical background and practical applications. Hence, as a reader, one is entirely consumed by sensor and signal (1D and 2D) processing, which, are the building blocks of a digital twin application. In

S. K. Pal et al., *Digital Twin – Fundamental Concepts to Applications in Advanced Manufacturing*, Springer Series in Advanced Manufacturing,
https://doi.org/10.1007/978-3-030-81815-9_5

this chapter, discussions are going to take a different direction. Different direction, as in, it is going to be entirely different from the discussions that have been so far. So, let us get started recapping IoT, and justifying the need for this chapter at this point.

IoT refers to the network of all connected devices. Consider a manufacturing plant, it will be having machines on the shop floor, instruments fitted in the machine, instruments in the pipelines all around the plant, human, and so on. Imagine, every component of the plant is connected, which is enabling seamless exchange of data amongst them. There are a lot of things, which are about to get connected to the internet, and as discussed earlier, 'network' is the main component in Industry 4.0. The idea is to connect the unconnected. Technically, this means connecting the devices over a network to control the physical objects by sensing and information processing. This is the overall idea of IoT. Already, manufacturing plants may be having controllers programmed for triggering/regulating some activity, but it does not provide a total picture of the process or the health of the components. The notion of getting things connected is to obtain, process, and decipher useful information in real-time about the assets. The successful implementation of this idea will make them smarter. In this regard, it is essential to understand the technologies for data communication across the smart devices, which is discussed in this chapter.

5.3 IoT Framework

Figure 5.1 depicts the general architecture of an IoT framework, which includes three layers namely, *perception, network*, and *application*. The *perception layer* is the physical layer which consists of the physical devices such as machines, sensors, actuators, controllers, etc. These devices are responsible for gathering the data for a particular process. The *network layer* comes next, which connects other smart

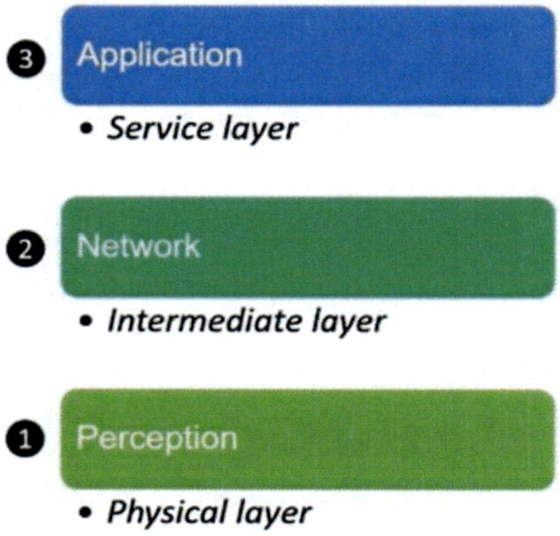

Fig. 5.1 IoT framework

things such as edge device, cloud server, etc. with the physical devices via suitable networking protocols or data communication technologies. The network devices will be responsible for receiving the data from the physical layer. Finally, the stack ends with the *application layer*, which is responsible for delivering the application-specific services to the physical layer.

Let us understand the framework in more details considering an example of manufacturing. Figure 5.2 depicts another schematic diagram for this purpose. The physical layer is the first that will comprise of the machines kept on the shop floor. In the schematic picture, three machines are depicted resembling the case of an industry, where such machines will be kept on the shop floor. These machines will be instrumented with sensors to acquire data about the process and the health of the machine. It will also consist of actuators and controllers so that any external data can be fed to the machine for controlling it. For example, consider a machining process such as turning, the turning machine fitted with a force sensor will form the physical layer. The machine will perform the machining operation and the force sensor will acquire the prevailing force generated during the process. So, for the case of industries, several such turning machines will be there. The data from the sensor is received

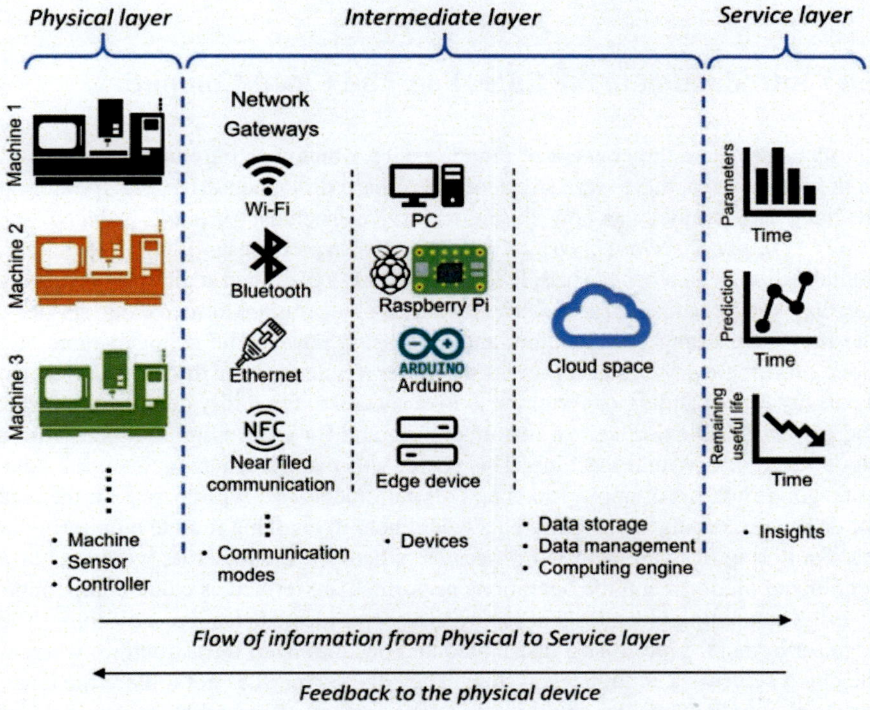

Fig. 5.2 Schematic for IoT framework

and stored in a device. As shown in the picture, these devices could be a computer, Arduino board, Raspberry Pi, or some other edge device.

For communicating with these devices, a suitable networking technology will be required to be followed. Finally, the data is sent to a cloud server which stores and manages the data, and also acts as a computing engine by providing the required resources for computational work. Thus, the intermediate layer, which is also referred to as the network layer, begins from the source of data generation to making it available at the cloud server. The server consists of application models, which upon receiving the data, acts on it, and derives the insights. In the example of the machining process, the objective could be to monitor the health of the cutting tool in real-time. Thus, the application/service layer will monitor the condition of the cutting tool via the data received from the force sensor engaged with the turning machine. These insights are sent to the physical device via the same intermediate layer for controlling the device in real-time. For the example of the machining process considered here, the details of the condition of the cutting tool will be sent periodically to the shop floor, which will help the operators to take decisions in real-time.

The next section gives a brief overview of the industrial communication system and networks followed by a discussion on edge, fog, and cloud computing.

5.4 Introduction to the Edge, Fog, and Cloud Computing

Let us recapitulate the concept of cloud service which was introduced in Chap. 1. In the earlier days, there were frequent use of hard drive, pen drive, etc. for storing the data. The practice has now shifted to services such as "*pCloud*", "*OneDrive*", "*Box*", "*Dropbox*", "*Google Drive*", etc. This shift in practices is possible because of digitalization, which has led people to wish to access their data at all times, from any location, and be secured. These aforementioned platforms are termed cloud services, and they have storage, application, and processing power. The major features of a cloud service are providing: (a) access whenever it is needed and through the web, (b) flexibility for expanding or shrinking a drive space, (c) flexibility of sharing a space, and (d) usage statistics, etc. In manufacturing and for developing a digital twin, a cloud server is a virtual machine. This server will manage the manufacturing data. Managing, in terms of applications and digital models that will operate on the data for deriving meaningful insights. This would help in realizing remote monitoring or telemonitoring of manufacturing processes, where the cloud server would act as a centralized platform and the operations performed are termed as cloud computing.

Edge computing indicates computing in a server located closer to a machine. The edge server is referred to as a distributed and decentralized infrastructure, whereas the cloud server is a centralized system. The edge server does not replace the cloud server, it complements the cloud instead. The fog server resembles multiple nodes that are located closer to the physical machine. They are used for short-term edge analysis due to their instant responsiveness. This enables data to be received at the monitoring platform faster, and also reduces traffic in the cloud. Thus, the fog server acts as a mediator between a physical device and a cloud server. Referring to Fig. 5.2, the edge server/edge device will be useful for pre-processing of the data. Pre-processing may include small computations. For instance, it can be filtering of the data (signal/image acquired during a process) or making a quick prediction, etc.

The routers or switches of the network, the sensor nodes, and gateways can also act as compute nodes for analytics. Utilising the edge device for computing can be an elegant way to address both the network and compute scalability problem of IoT systems. The sensor nodes and gateway devices are used for edge computing, which is usually known as device edge computing—they run application-specific analytics on the sensor data. On the other hand, the network switches and routers can run analytics in a sensor and application-agnostic way on top of the network packets running through them—this paradigm of computing is known as fog computing [1]. Fun fact here is the term 'fog', which in general represents a layer of tiny water droplets near the earth's surface. 'Cloud' on the other hand, represents the condensed water vapour floating in the atmosphere, high above the ground level. The computing paradigms are entitled on basis of their general meaning, where fog computing refers to running analytics as close to the source of the data generation. It embodies edge computing. This also ensures the data remain in the local network only.

To implement fog computing systems, it is imperative to create a virtualized distributed computing infrastructure across the edge devices and cloud [2]. Each device in the network can run a software agent capable of dynamically downloading and executing analytics code. However, there are quite a few challenges in creating a practically deployable fog computing framework. They stem mainly from the huge heterogeneity of the edge devices in terms of computational capability, memory, and battery power, and also from the unpredictable availability of such devices.

Figure 5.3 depicts the IoT computing stack as per the discussions made in this section. This includes edge computing at the bottom-most and cloud at the top of the stack. The stack also represents the sensitivity of the data to be analysed. Like, the most sensitive data, is thought to be analysed at the edge of the network. Similarly, the analysis which can be delayed by a bit can be taken to the higher levels of the

Fig. 5.3 IoT computing stack

stack. Also, it is not the case that one of the levels of the stack could be sufficient. Two layers or all three layers may be utilised together. For manufacturing and the development of digital twin application, a chunk of the data will be required to be analysed at the edge itself, and some portion could be sent to the cloud. Discussion on the utility of these computing paradigms will be throughout this chapter.

5.5 The Necessity of Cloud and Edge Computing in Industry 4.0 Perspective

With the rising attempts of industries at making their production-system Industry 4.0 compliant, there will be an increase in the usage of sensing and monitoring devices. The amount of data generation is going to be huge. In an industry, there may be numerous manufacturing machines configured with a large number of sensors [3–5]. In this case, a central server will help in getting the data in one platform and the same would make it easier for computations. However, from the networking perspective, "*scalability*" is a point of concern, which means having sufficiently high bandwidth/capacity/coverage and sufficiently low latency for transporting a sensor's data to the cloud for given application requirements. This calls for the right choice of (a) communication technology, (b) network topology, and (c) network protocol [3]. However, it is the meaningful analytics of the collected sensor's data that drives the value of such systems, often under the constraints of real-time requirements. With a large amount of sensor data, this leads to a compute scalability problem. To address these two issues, analysis of the data needs to be performed at the edge of the network itself—this paradigm of computing is known as edge computing.

5.6 Edge Versus Cloud Computing

Industry 4.0 in smart manufacturing introduces several interesting use cases that add value to business [6, 7]. These use cases demand and pose a lot of requirements on the infrastructure of the network, storage, and computing. To understand these infrastructural requirements, it is important to understand the use cases. In smart manufacturing, the applications are mainly in: defect detection, quality control of products/raw materials, and predictive maintenance/diagnosis/prognosis for machines [8]. The sensors deployed range from simple vibration, pressure, or load sensors to more complex 2D/3D visions systems, polarimetric systems, and radar systems. There are four main reasons why Industry 4.0 or smart manufacturing applications are demanding support of edge computing [9]. This is illustrated in Fig. 5.4 and discussed in the following sections.

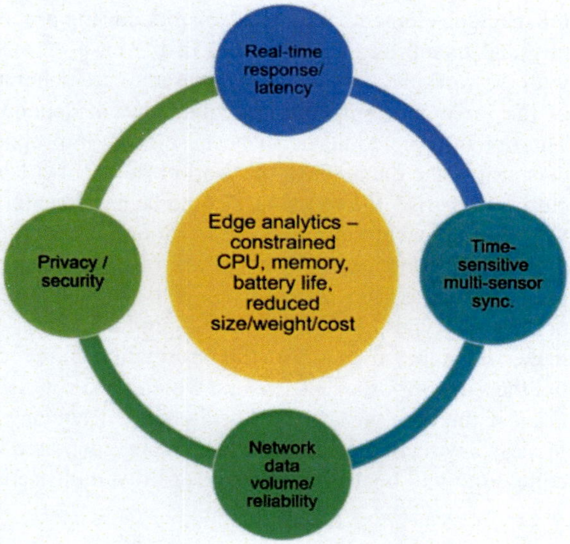

Fig. 5.4 Edge computing—drivers and constraints

5.6.1 Real-Time Response/Latency

All applications that need real-time monitoring and control/low latency response cannot depend on internet delays. Manufacturing operations are certainly the ones that cannot bear delayed responses. It is of utmost importance to derive meaningful insights from the sensors' data in real-time and subsequently control the process. This is doable at a cloud server, but it may come with a network lag as the amount of data grows. The most advantageous point of edge computing is the "*speed*", which it provides. It increases the performance of a network by reducing latency. As the cloud server is a central platform, data will be needed to travel some distance to reach its final destination. Besides, if an edge device is utilised for this purpose, i.e., if the data is processed near the source itself, the distance that data needs to travel is reduced. Considering that even a single moment of latency or downtime can cost companies a huge capital loss, the speed of edge computing cannot be overlooked.

5.6.2 Multi-sensor Synchronisation

When the analytics demands a fusion of multi-sensor data i.e., time-sensitive or there is a cost associated with a large amount of sensor data transfer to the cloud, again the fluctuating latency of the internet may cause a problem. In manufacturing and also for developing a digital twin, multiple sensors will be engaged with the physical device. This is so because a machine is made up of a large number of components and different types of sensors will be required to capture the dynamics/condition of the machine elements. Similarly, manufacturing processes also depend upon several physical quantities. For instance, in the case of machining, as the tool interacts with the workpiece, force and temperature are generated. The amplitude and nature of the force and temperature signatures will depend upon the cutting conditions. The rise in force could occur because of an improper combination of tool rotation concerning the diameter of the workpiece or maybe because of the higher depth of cutting. The rise of temperature could be an indicator of insufficient flow of cutting fluid during machining. Thus, information fusion is often required in manufacturing which yields better results than the information acquired by a single sensor. However, a challenge in the sensor fusion task is to deal with heterogeneous data. The detection of different types of phenomena and producing measurements for several attributes makes it cumbersome to fuse the information gathered. Besides, the sampling rate is not the same for different sensors, which makes it difficult for real-time application. Even if the data are sampled at the same frequency, a delay is introduced in the process, owing to the volume of data being generated from these sensors. Thus, edge computing will be useful for time-sensitive multi-sensor synchronisation.

5.6.3 Network Data Volume/Competency

Edge computing also provides reliability. The manufacturing system can be coupled with several edge devices owing to their low cost. At times, the central cloud system may fail, thus, in such a scenario, the edge device will be useful. Further, multiple edge devices can help channelize the data to flow through alternate devices, if one device fails. Thus, they can ensure the reliability of the data. In the development of a digital twin, it is crucial to obtain real-time data, and thus, multiple edge devices will be required. The edge devices may be engaged for pre-processing activities such as data filtering, prediction of product quality etc.

5.6.4 Privacy/Security

Finally, organisations/individuals may be reluctant to transfer data to the cloud due to privacy regulations/security reasons. This brings the usage of edge devices. However, edge devices are always constrained in computing power, memory, and battery life, and usually have reduced size/weight/cost, which makes deployment of IoT analytics on edge devices a challenge.

5.7 Application Classification

Based on the aforementioned drivers, various application requirements of smart manufacturing are cited and those are mapped to cloud/edge computing, as desired. The deployment and analytics requirements from an edge computing perspective vary widely among the applications.

As it is obvious from Tables 5.1 and 5.2, each application use case, based on its classification, will pose different edge requirements. Understanding these requirements is important for the creation of the right design for the system architecture—there is no "one-size-fits-all" solution.

From the discussions made above, edge computing can be said to be essential for realising the implementation of real-time manufacturing process monitoring and control. Figure 5.5 depicts a schematic representation for the use of edge and cloud computing for the digital twin application. In the diagram, it can be observed that a machine is communicating with an edge device. Considering the machine is instrumented with a large number of sensors, multiple edge devices can be deployed for this application. As the edge devices will be located closer to the machine, there will be no latency and thus, monitoring of the process or health of the machine would be much faster. This has been represented as '1' in the diagram. Multiple edge devices would also be beneficial if one of the devices fails.

However, one may advocate that from the health monitoring perspective, one may need to collect and store a large amount of data. For instance, the health of the cutting tool during a machining process may degrade at a faster rate than that of the machine itself. The machine here indicates the mechanical components. Thus, evaluation of the health of the cutting tool may be accomplished via edge computing. However, for evaluating the degradation of the machine elements, a large amount of data will be collected as the history of the machine is an important aspect. In this case, cloud computing will be required to store and manage the historical data and also perform computations on them. This is why, the vertical dotted line which is shown in between the edge and cloud server is different from that of the one shown in between machine and edge device, and it means that only a portion of the data is sent to the cloud. Besides, the edge devices, to provide service to the machine, the cloud servers will contain various applications. This will include mathematical models for working on the data. These models may be self-learning or may be trained offline and deployed in the edge device. In many cases, the second option could be in use because they are easy to build. The computational power of the cloud server could also prove advantageous in this case, as a user may train the models offline here, and will deploy them in the edge devices. This could be a periodic activity. In some cases, it may also happen that a majority of the monitoring and evaluation of a process will occur on the edge itself, and only a small part is being sent to the cloud server. This is represented as '2' in the diagram. The essence is that both edge and cloud are going to be beneficial for manufacturing and digital twin applications with the fact that edge computing will be the primary requirement.

Table 5.1 Requirements mapping for monitoring applications

Application use case	Sensors used	Analytics goal	Real-time response/latency requirement	Time-sensitive multi-sensor sync. requirement	Data volume/network reliability requirement	Privacy/security concerns to send data to public cloud
Raw material assembly line monitoring	RGB camera, polarimetric camera	Raw material impurity detection	Near real-time/low	No	High/high	No
Manufacturing process monitoring	Vibration, temp, plant-specific sensors	Process anomaly	Near real-time/low	Yes	Low/high	No
Product quality monitoring	RGB camera, polarimetric camera	Output product defect detection	Near real-time/low	No	High/high	Yes

sync.—Synchronisation

Table 5.2 Requirements mapping for control/optimisation applications

Application use case	Sensors used	Analytics goal	Real-time response/latency requirement	Time-sensitive multi-sensor sync. requirement	Data volume/network reliability requirement	Privacy/security concerns to send data to public cloud
Hazardous scenario detection and response	RGB camera, RF radar, Ultrawide band, wearable	Geo fencing	Near real-time/low	No	High/high	Yes
Defect detection and quality control	RGB camera, Polarimetric camera	Real-time control of production line to prevent defect	Real-time/very low	No	High/high	Yes
Predictive maintenance and diagnosis for machines	Vibration, temperature, sound, machine-specific sensors	Machine prognosis	Non real-time /high	No	High/medium	No
Manufacturing process optimization	Vibration, temperature, sound, plant-specific sensors	Process improvement	Non real-time /high	Yes	High/medium	No

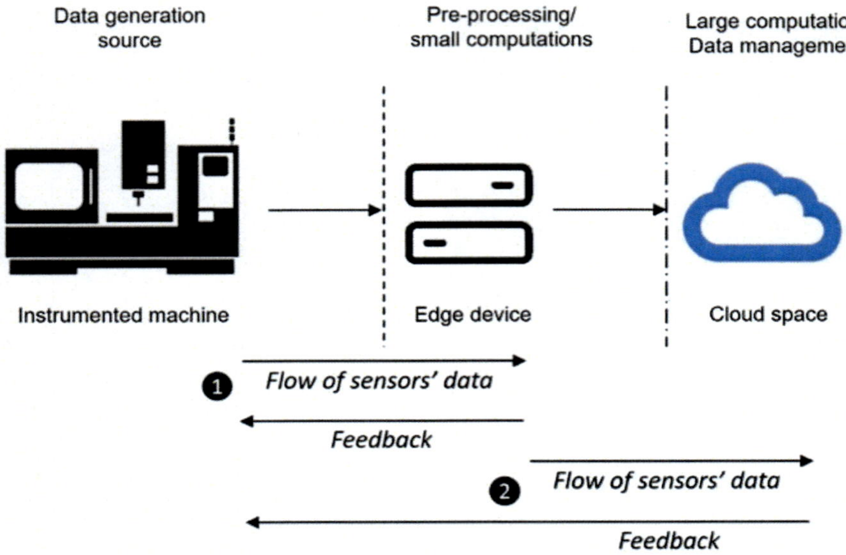

Fig. 5.5 A possible deployment of edge and cloud computing in manufacturing

The discussions so far in this chapter have covered the essentials of edge and cloud computing and their needs in the Industry 4.0 perspective for developing a digital twin. The next important thing is to understand and know about the communication technologies to establish the links between the physical device, edge device, and cloud server. The next section lists the data communication technologies, network topologies, and network protocols needed for edge and cloud computing.

5.8 Data Communication Technologies

Till this point of the chapter, discussions have been made on the computing paradigms such as cloud and edge, differences between them, and the importance as well. They can be referred to as services, which can be leveraged as a platform for analysis of the data. Data here refers to the one acquired from the sensors prevalent for monitoring the process and the health of the machine. The question here is how to send data from the sensory device to the edge/cloud platform. For this purpose, an understanding of the data communication technologies is required. Let us begin with the basic structure, which is the open system interconnection (OSI) model, as shown in Fig. 5.6. The OSI model shows seven layers that are used for communicating data over a network.

The bottom-most three layers of the OSI model are collectively referred to as the hardware layers. It begins with the *Physical layer,* which connects the nodes

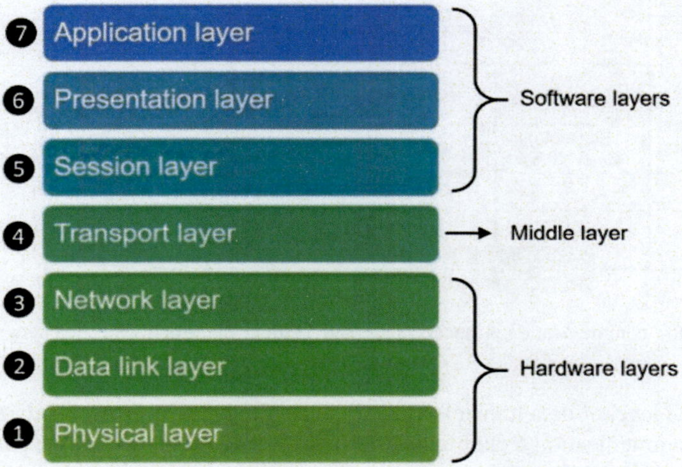

Fig. 5.6 Schematic representation of the OSI model

either via cable i.e., wired-mode or wirelessly. It transports the raw data, which is in the digital format, i.e., 0 s and 1 s. The *Data link layer* is responsible for connecting and terminating nodes over a network. Upon receiving the raw data, it forms frames and transmits the same. It is composed of two layers, one is the *Logical Link Control (LLC)*, and the second is the *Media Access Control (MAC)*. The first defines the network protocols, checks for errors, and synchronises the frames. The second utilises the MAC address of devices to connect and define permissions to transmit and receive data. The *Network layer* routes the packets by discovering the best possible path across a physical network. The *Transport layer* is responsible for the end-to-end delivery of the data. It ensures the order of the data as it is sent by a sender and also ensures that there is no duplication of the data. The two protocols utilised in this layer are transmission control protoco*l (TCP)* and *User Datagram Protocol (UDP)*. Details about these two protocols will be discussed in a later section. The *Transport layer* is the middle one and is the heart of the OSI model as it connects the hardware layer with the software ones. The *Session layer* establishes and controls the communication between devices. It opens sessions for data transfer, ensures that the transmission channel remains active and open during the transfer of data, and closes the connection once the data is received. It also includes check-points while transmission of the data. In case of occurrence of any error, the transmission will occur again from the check-point. This process of placing the check-points is referred to as synchronisation and recovery. Next is the *Presentation layer*, which translates the data as desired by an application. At last, the *Application layer* is present, with which a user interacts. This is how a user interacts with the computer-based services over a network. Thus, this knowledge is going to be the fundamental knowledge for the transfer of data over a network.

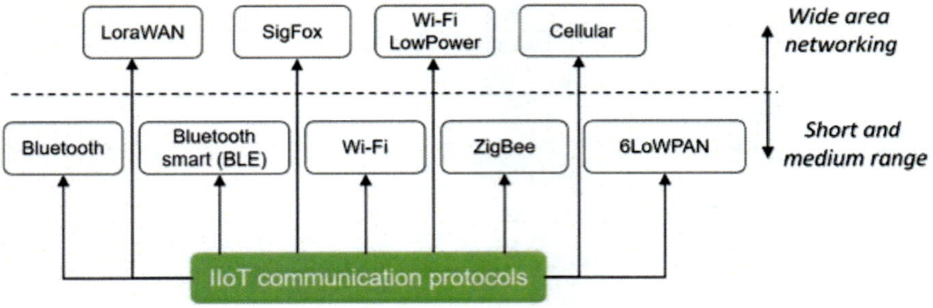

Fig. 5.7 IIoT communication protocols: short and medium-range and wide area networking

At each level of data transmission, i.e., from the physical to the final application, several communication technologies are involved. From this point, the communication technologies will be discussed for the transmission of data from the physical layer to the application layer. There are several communication technologies and access level technologies used in industrial IoT (IIoT) systems [3]. They may be categorized majorly into two namely wireless personal/local area networks (WPAN/WLAN) and wireless wide area networks (WWAN). The following section discusses these two in details.

Figure 5.7 depicts the IIoT communication technologies, which are segregated into short and medium-range and wide-area networking protocols. Range refers to the area up to which the signal is to be communicated and accordingly, it can be sub-divided into three namely, short, medium, and long ranges. Long-range is also termed wide-area networking. Short-range signifies communication to a few metres; this includes communication technologies such as Bluetooth in the wireless category and short-range cable communication in the wired category. The medium-range may go up to 1000 m and covers technologies such as Wi-Fi in the wireless category and Ethernet in the wired one. The long-range signifies communication required over more than 1000 m and includes cellular technology such as 2G, 3G, and 4G, and outdoor Wi-Fi in the wireless category.

Also, there are a few more features that are crucial while connecting the smart objects, and those are frequency band, power consumption, constrained devices, etc. Let us discuss these features briefly. The frequency band represents a band allotted for communication. The IoT technologies mostly cover the licensed and unlicensed band, where the licensed one is for the long-range access technologies and the unlicensed band is for the short and medium-range technologies. Next is the power consumption, which brings a concern for the IoT devices, whether they are provided with a continuous power supply or are battery operated. The battery-operated devices will certainly provide more flexibility for the deployment of IoT. In this regard, it becomes crucial for communication technology to address the low-power requirements of battery-operated devices. A low-power wide-area network is one of the important selections for fulfilling this requirement. Yet another important consideration is the constrained devices, which refers to the limited resources associated with smart devices. These resources include computing power, battery, storage, etc., and accordingly, the communication technology needs to be chosen. Besides these aforementioned features, network topology also plays an important role and it includes star, peer-to-peer, and mesh types. The short and long-range mostly employ star topology, which consists of a central station allowing communication with others. The medium-range may employ any of the topologies. The following section discusses the communication technologies and explores these features in more details.

5.8.1 Personal/Local Area Networks

WPAN is a kind of network technology that is used for interconnecting devices around a person's workspace. The network technology works within a range of 10 m. A few examples of devices that are networked using this network technology are desktops, printers, mobile phones, microphone, speaker, barcode readers, sensors, etc. The major features of WPAN include (a) short-range communication, (b) low-power consumption, (c) communication within personal space, and (d) low-cost. One of the key challenges of the IIoT system is the connectivity inside the factory. There are several wireless technologies available based on two core standards namely (a) IEEE 802.15 for WPAN and (b) IEEE 802.11 for WLAN. Within WPAN, 802.15.1 is for Bluetooth and 802.15.4 is for low data rate WPANs. Within WLAN, WiFi is the mostly used technology. The following section discusses these technologies in details.

5.8.1.1 IEEE 802.15.1 WPAN/Bluetooth

Bluetooth is not a new technology. Almost everyone has heard and probably used Bluetooth technology, which was available in the normal media player phones and now in smartphones. It is a known communication technology for years, even much before the advent of the IoT concept, but it is still employed owing to its evolution and what is known today as the smart Bluetooth technology. Bluetooth has grown with the popularity of mobile phones. The conventional form is termed the Bluetooth classic, which is known as a short-range wireless technology enabling point-to-point communication. However, conventional Bluetooth does not apply to the IIoT world owing to huge battery consumption and slow transfer of data. For the IIoT application, a recent version has arisen named Bluetooth Low Energy (BLE), also referred to as Bluetooth 4.0. The major difference between the classic Bluetooth and BLE is the low-power consumption. The classical Bluetooth can handle a large amount of data and that is why it consumes large power and is expensive. On the other hand, BLE can handle small amounts of data and thus consumes low power and is inexpensive. It is crucial for applications requiring an exchange of small amounts of data periodically. For IIoT applications, this technology can be utilized in a periodic update of some

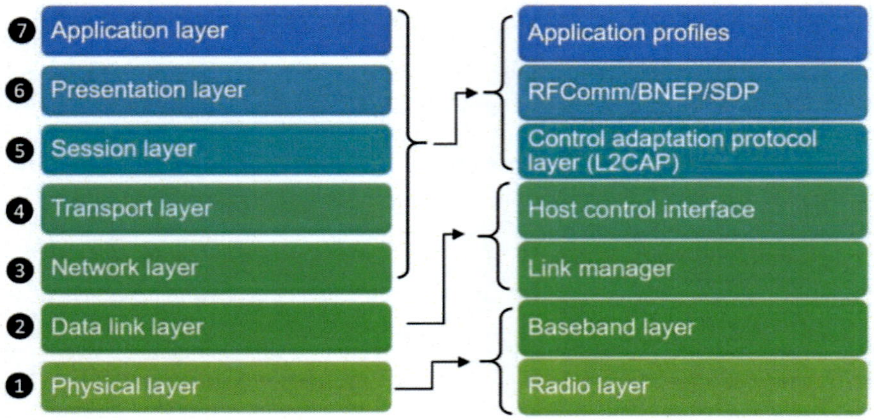

Fig. 5.8 Block diagram for Bluetooth protocol stack

parameters. The architecture of the Bluetooth protocol in comparison to the OSI model is depicted in Fig. 5.8.

Let us understand the modules of the stack as shown in Fig. 5.8. At the bottommost level of the stack, radio and baseband layers represent the physical layer of the standard model. For Bluetooth, the physical layer is responsible for carrying out the modulation and demodulation of the data into radio frequency signals. The information sending process via Bluetooth employs the use of radio waves propagating through the air with the help of modulation and demodulation. Data can be audio, video, or categorical data. The process of modulation refers to the change in the amplitude, frequency, or phase so that the information can be prepared to be transmitted. Similarly, demodulation refers to the process engaged at the receiving end. The baseband layer establishes communication within a piconet. If you are wondering what a piconet is, it refers to the devices connected via Bluetooth. Usually, the connection is between two devices. The link-layer manages the established connection. The management includes checking the erroneous message, forwarding, etc. The host control interface enables the communication between a host and a controller. Next is the control adaptation protocol layer (L2CAP), which is also known as the heart of the Bluetooth protocol stack because it acts as the mid-point and connects the upper and lower layers of the architecture. Essentially, it is responsible for multiplexing, segmenting, and reassembling data. These tasks can be understood as combining the smaller data packets, decomposing larger ones into smaller ones, and re-assembling the smaller ones into a full data packet across the layers so that the signal is transmitted. RFComm refers to radio frequency communication protocol, SDP refers to the service discovery protocol, and BNEP refers to Bluetooth network encapsulation protocol; all three protocols are built on top of the L2CAP. The first protocol provides the serial interface for performing tasks such as multiplexing and reassembling the data. The second protocol allows to announce and discover services provided by a Bluetooth device. The third one is for the transmission of data packets over other networking protocols such as TCP/IP. Finally, at the highest level, the application profiles are present that enable the visibility of one device to the other for communication and transmission of the data.

802.15.1 i.e., Bluetooth, defines *Physical layer* and *MAC* specification for wireless connectivity with fixed, portable, and moving devices within or entering personal operating space based on the Bluetooth technology. It is mainly used in a point-to-point or a star network topology for short-range communications. It supports data throughput up to 2 MBps. Bluetooth operates at frequencies between 2.400 and 2.4835 GHz, which is in the globally unlicensed industrial, scientific and medical (ISM) band. The technology is fairly low-power but high compared to other WPAN communications. Bluetooth is well suited for consumer products where a larger amount of data is exchanged every moment, i.e., for continuous streaming data applications.

5.8.1.2 802.15.4 (LR-WPAN)

LR-WPAN stands for low-rate wireless personal area networks and it is a two-layer standard protocol, as depicted in Fig. 5.9. The said protocol defines only the *Physical layer* and *MAC* of the standard OSI model, while the upper layers make use of secondary communication protocols, such as ZigBee or 6LoWPAN or WirelessHART. Discussions will be around these secondary communication protocols, but let us first get a glimpse of the IEEE 802.15.4. This protocol is known for its features such as low-cost, low-power, and short-range communications. It is suited for applications, where protocols like Bluetooth is costly and low-range communication is desired. Highlighting the term low-rate, which means that this protocol is not designed for applications requiring high bandwidths, and this is why it is limited within a personal area only. One major application of this protocol is in the wireless sensor networks, which brings down the cost of sensor installation and data

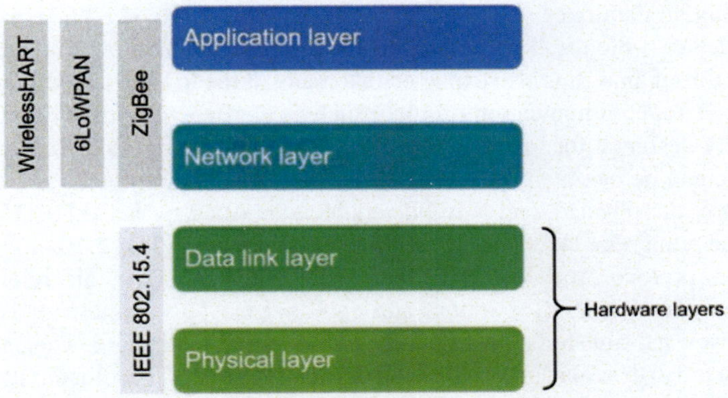

Fig. 5.9 Schematic representation of 802.15.4

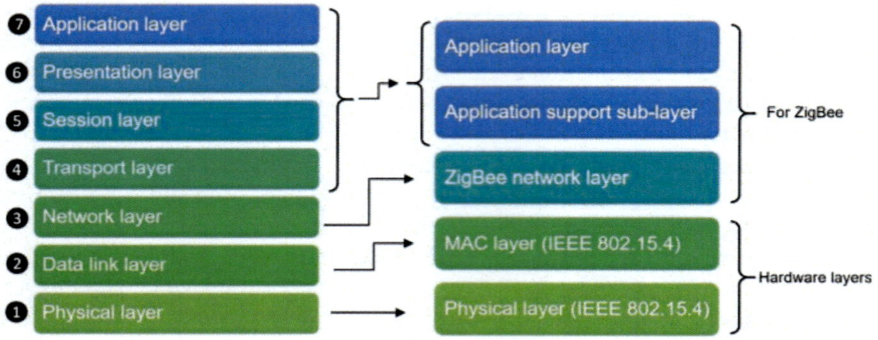

Fig. 5.10 Schematic representation of ZigBee architecture

acquisition. Things such as video analytics could not be supported via this communication protocol. It communicates with a data transfer rate of 250 kbit/s. Important features include real-time support via guaranteed time slots, collision avoidance via carrier sense multiple access (CSMA/CA) and integrated security support. It has power management for low power. It operates in three possible frequency bands (868/915/2450 MHz). A brief idea of the secondary protocols is discussed in the following points.

ZigBee

The ZigBee standard from the ZigBee Alliance was developed for short-range communication in the order of 10 to 100 m based on IEEE 802.15.4. Its main feature is its low energy. It can maintain very long sleep intervals or operate with low duty cycles, thereby enabling battery life of years. So, what is the problem with classic Bluetooth? As pointed out in the previous section, Bluetooth is mainly designed to transmit higher data rates, which requires a high-power rate of 1 W. It may happen that the device which will be employed for the IIoT application, may not require high data throughput. It is also expected that many of the IoT devices will be battery-operated. Here, communication protocol such as ZigBee will prove beneficial as it is specially designed for low data throughput, requiring only 10 to 100 mW. Thus, the device can be operated for more hours. This is where ZigBee is more reliable. For ZigBee, the physical and MAC layers are supported by IEEE 802.15.4, and it defines the remaining layers of the OSI model, as shown in Fig. 5.10.

For this protocol, the device includes three types and those are referred to as ZigBee coordinator, ZigBee router, and end-devices. The coordinator is the one, who receives and transmits data with other devices in the network. The end devices are connected to the coordinator. To relay the message to the coordinator, the router is present as an intermediate device. Thus, the coordinator in the network identifies the end devices and assigns a unique identity to each one of them for transmission/receipt

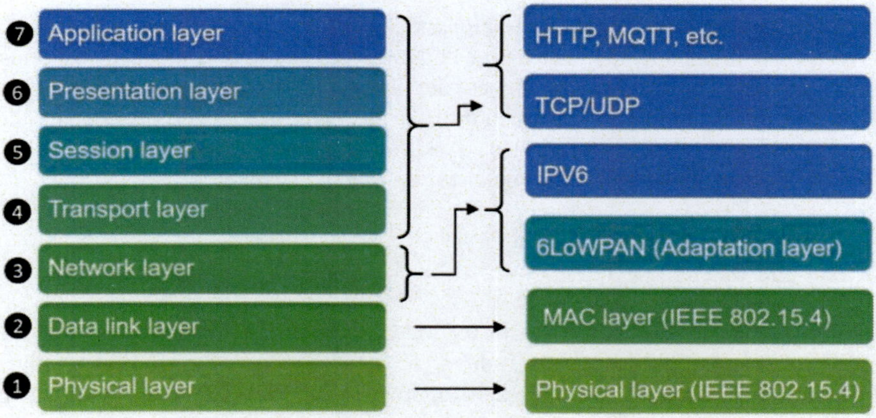

Fig. 5.11 Schematic representation of 6LoWPAN architecture

of data. The bottom-most layer of the stack is responsible for modulation and demodulation of the message, i.e., during sending and reception of data, respectively. The second one from the bottom is responsible for the reliable transmission of the data. In the network layer, tasks included are starting a network, managing end-devices, joining/leaving a network, routing, configuring new devices, addressing, and security. The application support sub-layer runs over the network layer and it enables the services required for the ZigBee objects and application objects to interface with the network layer. It looks into the things to make two devices communicate, which includes generating end-to-end acknowledgement, maintaining local binding table, local groups table, and local address group. This application support sub-layer is contained within the application framework.

6LoWPAN

6LoWPAN or IPv6 over low power wireless personal area networks is a standardisation driven by Internet Engineering Task Force (IETF). As the name implies, it is low-power, IP driven and has a large mesh network support for establishing the connection between devices. It has a goal to define a suitable adaptation layer between the 802.15.4 link layer and TCP/IP stack. The major problem with the previously discussed Bluetooth or ZigBee protocol enabled devices is that they require complex application gateways to connect to the internet. In contrast to this, the use of an IP-based router in the 6LoWPAN protocol remains agnostic to the application level protocol. This also lowers the burden on the router and that is why they have simpler architecture, low power consumption, and easy configuration. Thus, the major advantage of 6LoWPAN is that it enables IP support on devices running on different heterogeneous physical networks, thereby ensuring interoperability. Figure 5.11 represents the architecture of this protocol.

As discussed in the previous sections, 6LoWPAN has the physical and MAC layer as per IEEE 802.15.4 and defines other layers of the OSI model. Observing Fig. 5.11, IP is the only way to connect the physical and MAC layers to the higher layers of the OSI model. However, in the case of 6LoWPAN, this is replaced by the 6LoWPAN stack, which is referred to adaptation layer in combination with the IPV6. The protocol employees an edge router for the same.

5.8.1.3 802.11

IEEE 802.11 is the WLAN protocol for MAC and physical layer that extends the IEEE 802.3 wired LAN protocols. It should be noted that wired LAN protocols cover from normal Ethernet LAN to Gigabit Ethernet LAN. 802.11 WLAN, also known as Wi-Fi, uses various frequency bands such as 2.4, 5 and 60 GHz. Based on the IEEE 802.11 standard, Wi-Fi is quite popular for IIoT applications. The system can leverage the existing wireless network infrastructure and can work with common off-the-shelf Wi-Fi enabled devices such as smartphones, thereby making it cost-effective and deployment friendly. Power consumption can be an issue for low power IoT devices using Wi-Fi, which needs carefully designed power management. But Wi-Fi allows extremely high bandwidth (100 Mbps+), which can be useful in high data rate transmissions like video.

Table 5.3 summarizes the features of the notable WPAN and WLAN networks.

5.8.2 Technologies for Wide Area Networks (WAN)

The uses for wide-area IoT technology are predominant in civic infrastructure systems—they are also finding applications in IIoT systems including applications like asset monitoring and tracking. In cases where there are low power constraints, low-power wide-area networks (LPWAN) are being used. Out of these, LoraWAN

Table 5.3 Short and medium-range communication technologies

Technologies	Standards body	Frequency band	Maximum range (m)	Maximum data rate	Maximum power
Bluetooth	Bluetooth SIG	2.4 GHz ISM	100	1–3 Mbps	1 W
Bluetooth smart (BLE)	IoT interconnect	2.4 GHz ISM	35	1 Mbps	10 mW
ZigBee	IEEE 802.15.4, Zigbee alliance	2.4 GHz ISM	160	250 kbps	100 mW
Wi-Fi	IEEE 802.11 g/n/ac/ad	2.4 GHz 5 GHz 60 GHz	100	6–780 Mbps 6 Gbps @60 GHz	1 W

and SigFox have found good applications in the IIoT segment while WiFi LowPower is not that popular. LoraWAN operates on varying frequencies, a range up to 2.5 km in urban and 15 km for rural areas, and a data rate of 3–50 kbps. Current cellular 2G, 3G, and 4G technologies are also popular for IIoT applications. They are suitable for applications that require wide-area connectivity, mobility, and where mains power supply is available or frequent re-charging is possible. The 3GPP LTE-M profile for machine-type communications (MTC) is specifically designed for IIoT that has battery life, cost, and coverage optimizations but does not need a high data rate. It reduces the cost and complexity of devices and can improve coverage for machines deployed in radio-challenged locations like inside-buildings or underground. Currently, 5G technologies are under development, which provides a very high rate and low latency, making them somewhat suitable for industrial control applications. 5G has a special standardisation for IIoT under the name narrow-band IoT (NB-IoT), which serves for a high number of IoT devices with limited data rates and low latency. Major technologies applicable for WAN are summarized in Table 5.4.

5.8.3 Applications Level Protocols

This section will discuss the application-level protocols. The interest of this section is to explain the things which enable the transport of messages amongst computing devices. In this regard, TCP and User Datagram Protocol (UDP) are the two widely employed communication standards. TCP, as mentioned, it is used for sending data from one device to the other. It enables the transfer of data between a computing device and application programs over the internet. It does so by sending packets of data over the network and it also ensures the successful delivery of the data and the message. The advantage of employing TCP is that it organises the data that is to be sent to a client from a server and it also guarantees the integrity of the data that is being communicated. To transmit the data between a server and a client system, it establishes the connection between the two, and then breaks the message into multiple segments, and transfers them to the desired destination. The IP is the method that is employed for data transmission. Every device on the network can be identified by a unique IP, which enables data communication across the devices over the network. In a nutshell, the IP provides identification to a device which is required to identify where the data is to be sent, and then the TCP protocol sends the data to that device. In this way, the TCP/IP suite exchanges information between devices.

UDP, on the other hand, is employed for establishing low-latency connections between applications. This helps in speeding the transmission. It is an alternative suite to the TCP. Comparing the two, TCP is more expensive as it includes error messages, ensures the delivery of the data by sending acknowledgements, connection start-up, and flow control. UDP does not provide an error message for connection detachment. It also does not sequence the data that needs to be delivered and it does not signal the

Table 5.4 Low power wide area networking (LPWAN) technologies

Technology	Standards/governing body	Frequency	Range	Data rate	Devices per access point
LoraWAN	LoRa alliance	433/780/868/915 MHz ISM	2.5–15 km	0.3–50 kbps	1 M or more
SigFox	SigFox	Ultra narrow band	30–50 km (rural) 3–10 km (urban)	100 bps	1 M
WiFi LowPower	IEEE P802.11ah expected in 2016	Sub GHz license exempt bands	1 km (outdoor)	150 Kbps–340 Mbps	8191
UMTS(3G) HSDPA/HSUPA	3GPP	Cellular	27/10 km	0.73–56 Mbps	Hundreds/cell
LTE-3G/LTE-M	3GPP R-13	Cellular	2.5–5 km	2–12 Mbps	20,000+
LTE-advanced or 4G	3GPP	Cellular	2.5–5 km	50–100 Mbps	20,000+
5G	3GPP	Cellular	2.5–5 km	50–1000 Mbps	20,000+
5G NB-IoT	3GPP	Cellular	2.5–5 km	Upto 200 Kbps	20,000+

delivery point as well. These features make it less expensive as compared to TCP, but at the same time, makes it less reliable.

For digital twin application, a successful connection between a physical device and the virtual server will be the prime requirement. This connection will enable the flow of sensors' data from the machine to the application points, where the data will be harnessed to decipher the information. Thus, a proper data communication protocol should be chosen for this purpose. While UDP is less expensive with low latency, it can lead to loss of data if the communication breaks. Ensuring data delivery is to be taken care of.

Let us understand the TCP/IP communication in more details. Figure 5.12 represents the links included in the TCP/IP suite, which are, from the bottom, *Datalink, Internet, Transport,* and *Application* layers. The *Datalink* layer defines how the data is to be sent/receive between the communicating devices on a network. The *Datalink* layer shown here is a combination of the *Physical* and *Datalink* layers of the OSI architecture. The OSI model has these two as its lowest layers. The *Physical* layer is responsible for the actual physical connection between the devices. This layer contains information in the form of bits and it transfers individual bits from one node to the other. As the data is received, this layer converts it into 0 and 1 s and sends it to the next layer, which is the *Datalink* layer. This layer of the OSI model is responsible for the node-to-node delivery of the message. It is responsible for framing the data, which means providing patterns for proper understanding of the data that is to be sent. It also adds physical addresses to the data and defines mechanisms for error-free delivery of the data. This helps in ensuring the lost data packets, if any, during a communication. It also specifies the data flow rate for avoiding corruption of the data.

The *Internet* layer in the TCP/IP communication is responsible for the transmission of data from one device to the other device on a network. It takes care of

Fig. 5.12 Links in TCP/IP connection

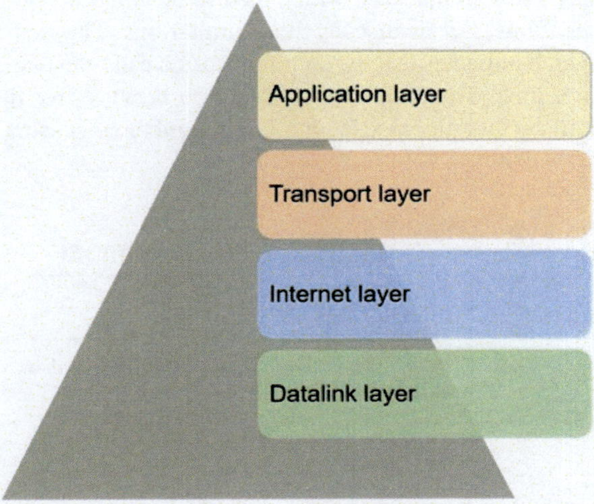

the routing of the data packets, where it selects the shortest path for transmitting a packet of data. Thus, it may take more than one route available on the network. The main function of this layer is routing and logical addressing. Routing refers to the route to be opted for sending the data to the destination. Logical addressing refers to the unique identities of the sender and receiver, which are placed in the header by this layer. The *Transport* layer comes next in the sequence and it is responsible for providing a reliable data connection between the source device and the destination. It provides service to the *Application* layer and takes service from the *Network* layer. Once the data is transmitted, the *Transport* layer acknowledges the same and re-transmits the data, if any error is found. Finally, the *Application* layer refers to the programs that need TCP/IP to help them communicate with each other. It is a combination of the *Session, Presentation,* and *Application* layers of the OSI model. It provides services to the user such as handling the data, editing, logging, etc. The users interact with this layer.

Though IP connectivity is not mandatory for industrial machine-to-machine communication, the use of IP makes it easy to connect things to the Internet. But there will still be issues towards interoperability at the application level. Since the communication nodes in IoT are constrained in memory, CPU, and battery, there is a need for efficient data transmission mechanisms above the IP layer. In this regard, the following section will discuss a few relevant application-level protocols.

5.8.3.1 Hypertext Transfer Protocol (HTTP)

The first application-level protocol to be discussed is HTTP. It follows a request-response model and works on the TCP/IP layer, as depicted schematically in Fig. 5.13.

It means that the client requests the server each time and the server responds. This protocol was mainly devised for the World Wide Web applications. HTTP has a few limitations, which lead to unsuitability for machine-to-machine communications. Let us discuss those limitations. The first is one-to-one communication, which indicates that the communication can be established between two devices only at a time. This will pose a limitation because, for the IIoT applications, machines will be instrumented with a large number of sensors. As discussed in the previous

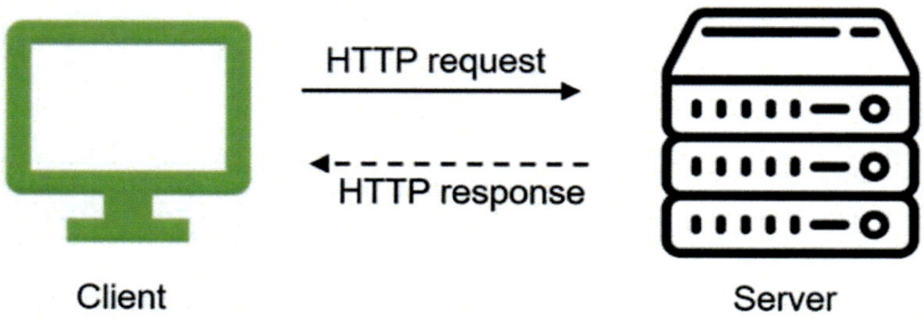

Fig. 5.13 Schematic representation of HTTP protocol

sections, numerous edge computing devices will be required for pre-processing of these sensory data. For the digital twin application, sensors will include both engaged for process and health monitoring, and HTTP cannot establish such communication. The second limitation is the timeliness of the data transmission. Since the protocol follows a request-response model, a client has to wait after raising a request, to which the server has to respond. This does not happen synchronously. The third being the single-directional flow of data, which means either the client or the server can only communicate at one time, owing to the request-response model. For monitoring application, multiple sensors will be communicating to the server, and the applications running on the server needs to respond based on the decisions taken on the data receipt. The fourth is event-based communication, which is not fulfilled by HTTP. A large number of decisions in the digital twin application may be event-based. For instance, in the PLC-based applications in process industries which was discussed in the previous chapters, a decision is triggered based on the sensory data. This is event-based, however, since, HTTP follows a request-response model, it could not fulfil this IIoT application. As an alternative, i.e., to support the two-way messaging system, *WebSockets* can be employed. The *WebSockets* opens a TCP connection between a client and a server and lets the data flow between them. This is not fulfilled by the HTTP because of the uni-directional data flow. The *WebSockets* establishes point-to-point communication and can send/receive messages to/from one client or a group of clients without making requests for open and close unlike it happens in case of the HTTP. Thus, they prove suitable for push and pull communications. One problem is that they may become overhead when a lot of IoT devices will be present at the same time.

5.8.3.2 Message Queue Telemetry Transport (MQTT)

The second application-level protocol to be discussed is the MQTT. It follows a publish-subscribe messaging system that relies on a broker model, as depicted schematically in Fig. 5.14.

This protocol also works over the TCP/IP communication mode. It also has the client and the server, like in the case of the HTTP protocol. In general, the client is the one who utilises the services provided by a server. In MQTT protocol, the server is the broker, which is depicted in the schematic figure (Fig. 5.14). This broker handles the requests of the clients. Two operations are involved in this protocol namely, publishing and subscribing. When a client sends data to the broker, which is the

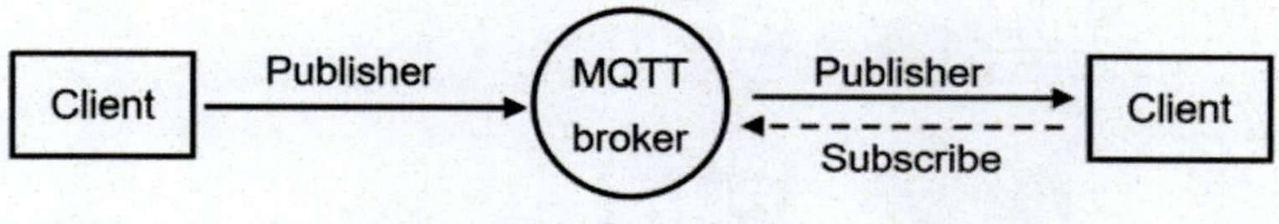

Fig. 5.14 Schematic representation of MQTT protocol

server, the operation is referred to as 'publish', and in the case a device wants to receive data from the server, it is called 'subscribe'. The protocol has a very small overhead that can work under lossy and intermittently connected networks. MQTT is useful in large networks of small devices that need monitoring and control from a back-end server on the Internet—however, it is not designed for device-to-device communication. When compared to the HTTP model, MQTT works faster because it takes small packets of data for communicating with the server. Unlike HTTP, multiple devices can communicate amongst each other. The timeliness of the data transmission is also high because MQTT keeps the transmission channel open by staying online. On the other hand, HTTP closes the connection once the data is sent, thus, it is slower. MQTT further gives options to set the priorities such as the transmission of data without confirmation, with confirmation, or with a handshake. These features enable more security which is not available with HTTP or *WebSockets*.

5.8.3.3 Constrained Application Protocol (CoAP)

CoAP is the third application layer protocol. It is designed to provide a RESTful application protocol modelled on HTTP semantics, but with a much smaller footprint. Requests and responses are exchanged asynchronously using CoAP messages over UDP. Usage of UDP provides the necessary low footprint. XMPP is its persistent need for a TCP connection for which it is not suitable for low-power and lossy networks. There are IoT-centric variants of XMPP such as XEP-0322, XEP-323, and XEP-324 which try to address the above issues.

The whole connectivity/networking/protocols described in this section is summarized in Fig. 5.15.

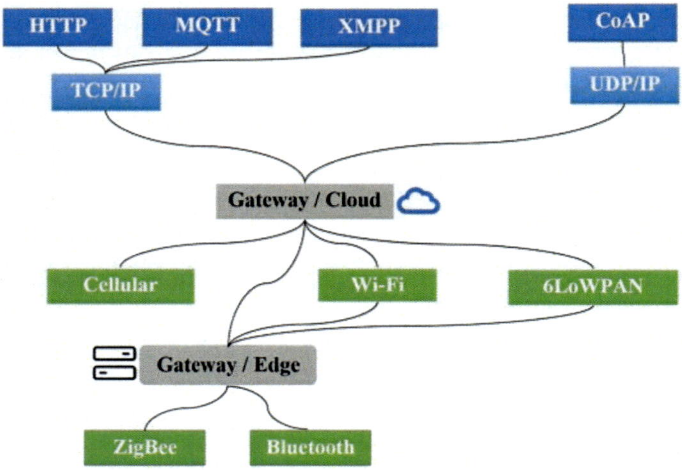

Fig. 5.15 Connectivity/networking protocols

5.9 Network Architectures for Edge/cloud Computing

Traditional information and technology systems successfully used the client–server model and cloud services for deploying applications. However, for IIoT applications, a client–server approach may not be the best model. Recalling the discussion held at the beginning of this chapter, any IIoT deployment would need a set of sensors connected to the internet cloud via some gateway device. It has been discussed that when there is a requirement to make some decisions locally and trigger an instant action, computing at a remote server may not meet the required response time. Rather, computing at the gateway device will make more sense, which is known as the edge computing architecture. A practical solution may need to use a judicious mix of edge and cloud architectures. For example, consider a large sensor network deployed on machines in a factory providing some computing and storage resources at the IoT WAN gateway level. This can enable such a system to do some edge computing to reduce the communication bandwidth requirements.

One of the biggest problems of cloud-connected networks and analytics thereof, in the context of IIoT, is the issue of network delay and network reliability. This is the biggest driver for edge computing. Many communication protocols require continuous connectivity to be kept alive and communicating over intermittently connected networks poses a challenge for many applications involving sensor nodes. Re-transmission, store, and forward technologies are used to mitigate this issue, but they do not work for applications that need a real-time response from analytics and hence, a low consistent latency. Delay-tolerant-networking (DTN) is one way to address this requirement by using a bundled overlay network on top of multiple available networks trying to provide a best-effort delay performance. If such options are not available, one needs to look for analytics at the edge itself. The subsequent section outlines edge analytics.

A typical IIoT deployment architecture consists of a set of sensors connected via a gateway device to the cloud [10–12]. Application services are deployed both in edge and cloud on top of this sensor data transport layer [13]. This architecture is depicted in Fig. 5.16. Several implementation practices can affect an IIoT System. Their requirements stem from the need for real-time and power considerations, utilizing

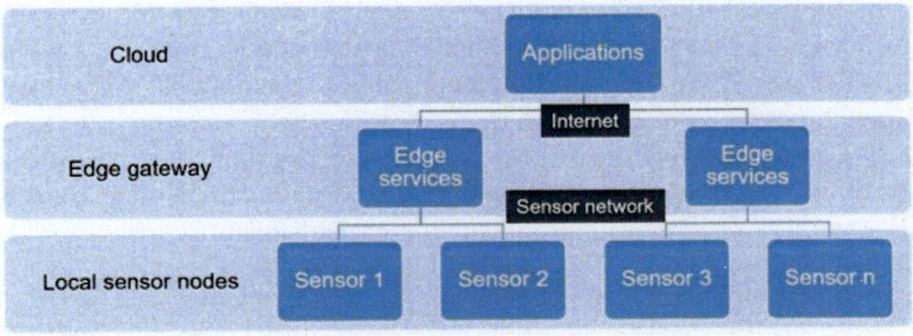

Fig. 5.16 IIoT deployment architecture

the edge devices for analytics, and using a service-oriented platform for application deployment. These are explored in detail in the sections below.

5.9.1 Cloud Sub-System

In a classical IIoT based system, the cloud system receives the sensor data over the internet and runs analytics on the sensor data. Inherently, the sensor data flow is fluctuating as they are linked to sensing real-world events. The elastic nature of the cloud handles the non-uniform demand on processing power and storage capacity. The data analytics can be done "in-stream" or "in-storage". In-storage systems require to be big data-enabled—there may be a limited number of large files (say from a video camera) or a large number of relatively small files (generated from a multitude of sensors). There is a whole lot of cloud-enabled software platforms to support these activities [5].

5.9.2 Edge Considerations for IIoT Systems

As outlined previously, many of the IIoT use cases require real-time computation and presentation of analytics insights. A typical data flow of an IoT system involves sensing the event, extracting information, analysing the extracted information to derive actionable insights and drive suitable response. The insight and response can range from simple descriptive visualisation to diagnostic/prognostic/prescriptive systems. This is elaborated in more details in Chap. 6. However, this chapter itself tries to briefly bring out the edge considerations stemming from such analytics for such systems.

IIoT systems are typically constrained by low configuration sensor data acquisition and gateway devices at the edge of the network with potentially infinite compute capability available in the cloud. There are some conflicting requirements that typically characterise such systems [3, 14].

- For the ideal real-time system, all processing of sensor data can be done at the edge device so that suitable decision/responsive action can be executed there itself. However, due to limited computational power and memory, this may result in lowering of the performance in terms of accuracy.
- The less computation is done at the edge device, the more data needs to be sent to the cloud which, in turn, results in higher congestion, higher data transfer cost, higher battery power consumption, and increased latency.

To take care of all the above conflicting requirements, referring to Tables 5.1 and 5.2, the following directions/suggestions could be taken up for an efficient edge-based IIoT system design.

- For low latency and real-time response from analytics, as much computation as possible should be done on the edge device, leaving only the computations requiring aggregate sensor information from multiple systems for the cloud. The partitioning of the computation between edge and cloud can be done to balance data flow between edge and cloud while draining less battery on the edge device. This can be formulated as an optimisation problem.
- For an electrically powered edge device that has a high bandwidth and low-cost network at its disposal, a lot of computation can be off-loaded to the cloud, provided the real-time requirements are not stringent enough compared to the available network latency.
- For small systems with a battery-powered gateway, even if there is no real-time requirement, one should do as much computation as possible in the edge and send only the insights and results to the cloud.
- For systems not having low latency requirements but having highly computationally complex analytics, it is better to do as much computation as possible in the cloud.
- If there are privacy concerns for the data to go to the cloud, edge analytics is a better option.

Further, multiple edge devices can also be utilised for a practical IIoT system. These devices may be connected to a cloud server. For instance, if one considers the case of monitoring a CNC machine, instruments such as force, speed, position, temperature, etc. may be engaged with the machine. Typically, for this scenario, multiple edge devices may fulfil different requirements. Say, the force data can be processed in one edge device for monitoring the condition of the cutting tool and may be the quality of the machined surface. Since the data sampling rate for temperature signal is much lesser as compared to the force data, it can be utilised to infer other features of machined surface on a different edge device with low power requirements. The data acquired from position and speed sensors may be utilised from a health-monitoring perspective and may directly be sent to the cloud server after pre-processing at an edge device. In this manner, the load can be distributed across multiple edge devices and real-time requirements can be met. Also, the security of the data is achieved.

The above scenario puts multiple edge devices for realising the IIoT applications. In addition to this, interaction among the cloud servers could also be realised. This is referred to as cloud-to-cloud interaction. It will be particularly useful for interaction amongst the machines on a shop floor. Consider the following example, where on a shop floor, several similar machines are performing one type of operation. Industries do mass production, thus, to have higher productivity, multiple machines are engaged. In this case, as numerous machines are performing a similar operation, the parameters that will be in use for operating, will also be the same. However, the rate of degradation of the health of those machines may not be the same. This means that one machine out of the series on that shop floor may degrade faster than the other. Reasons for this could be the quality of power coming to the machines, level of vibrations in parts of the floor, friction amongst the mechanical components of the machine, rate of use of the machine, quality of the coolant, hydraulic oil, etc. to the machine, and so on. Realising the IIoT deployment, these instrumented machines may be having an individual edge and cloud infrastructure for process monitoring and control application. So, in the case of a faulty situation, with the help of the sensors' data, the machine can correct itself. Till this point, the objective is to monitor, take the decision, and control the process with the help of sensors' data. The next idea here is to ask whether one machine can decide by looking at the decision of another machine. In that case, machines on the shop floor will understand and learn from each other's actions, just like the way humans in a society learn from each other. In that case, interaction may be required between the servers of those machines to inherit one's behaviour in a particular type of situation. Of course, the machine, to inherit the action/behaviour of another machine would require to understand its health condition and judge the suitability of the action.

5.9.3 Utilizing the Edge Devices for Artificial Intelligence

In addition to the above, edge computing systems needs to support running for artificial intelligence learning and inferencing. Inferencing at the edge is characterised by model compression/pruning and artificial intelligence acceleration and partitioning. Though this is covered in details in the next chapter, for sake of completeness, the main requirements of utilising edge devices for artificial intelligence are introduced in this chapter itself. There are two kinds of application in this space—inferencing at the edge and learning at the edge.

For inferencing at edge, models are trained on the cloud with large compute capacity. But they are deployed on the edge—for such systems, the main considerations are two-fold. Model compression focuses on the reduction of the artificial intelligence inferencing model size to fit into the reduced memory size of constrained edge devices. Artificial intelligence acceleration and partitioning take advantage of artificial intelligence specific hardware accelerators on-chip and distributed computing cores to execute inferencing models [15, 16].

Learning at the edge additionally involves learning the model in the edge itself—many use cases warrant such requirement. Such systems require some additional features of artificial intelligence learning like starting from a pre-trained base model and improving the base model with incremental streaming data on edge itself (referred to as *Incremental Learning*) [17], or learning in a semi-supervised way with active user feedback (referred to as *Semi-supervised Active Learning*) [18], or learning in a distributed manner using multiple connected edge devices (referred to as *Federated Learning*) [19]. All the signal processing and artificial intelligence techniques described above in the edge context will be covered in detail in the next chapter.

5.10 Real-Life Example in Manufacturing

In this section, an example is discussed to demonstrate the use of edge and cloud computing in the case of a manufacturing process. An example is of friction stir welding (FSW) process, where the core idea was to develop a real-time monitoring and control system for the welding process. In this chapter, a part of the developed system is discussed i.e., the use of edge and cloud computing in the developed monitoring and control system. The details of the execution of monitoring of the process along with real-time control are discussed later. Figure 5.17 shows a block diagram of the developed system. A few sensors such as force, torque, and power are engaged with the FSW machine. These sensors collect the data during welding. The acquired data is transmitted to a computing unit referred to as 'edge'. As discussed in the preceding sections, the edge devices generally have low configuration hardware systems attached directly to machine or IoT devices through protocols such as TCP/IP, MQTT, HTTP, etc. The main purpose of using an edge device is to filter, analyse, and process data locally i.e., close to the point of its generation. It acts as an intermediate layer between conventional centralised cloud architecture and IoT devices.

The sensors used in this work generates a large volume of data. To take an intelligent decision for automation of the FSW process, the data is being analysed, and useful information is extracted by applying signal processing and machine learning techniques. The edge device initiates the developed architecture by setting up a connection amongst the different units shown in the picture i.e., FSW machine, cloud device, and itself. The connection protocol utilised here is TCP/IP and *WebSockets*. Figure 5.18 depicts the connection between them schematically. For the *WebSockets*, the IP of the machine and a specific port number are assigned in the architecture. It should be noted that most of the times, the assigned IP within a network are dynamic. It has to be static. After specifying the IP and port number of the machine, the remote

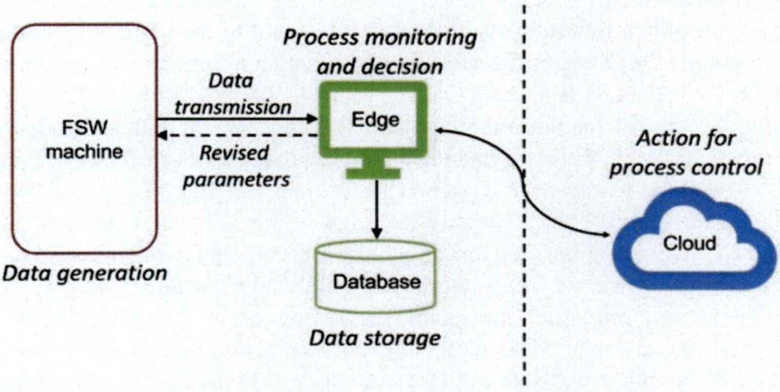

Fig. 5.17 Use of edge and cloud for process monitoring and control

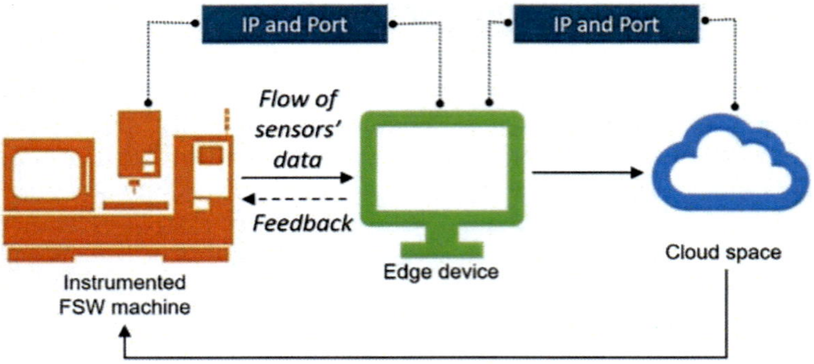

Fig. 5.18 Use of *WebSockets* for communication

server i.e., the edge device pings the machine and the connection is established. Similarly, the IP and port number of the cloud server is also specified for establishing the communication. The IP of the cloud server is already a static one. In this way, the *WebSockets* is employed in this case for communication.

Once the connection is established, the edge device receives data from the sensors engaged with the machine. The received data is first pre-processed i.e., cleaned and then it is fed to a predictive model. This model is a neural network that predicts the quality of the weld in real-time when fed with the sensors' data.

The edge device is continuously comparing the predicted value of the weld quality with the desired value. The desired value is configured by the user based on the application for which welding is being performed. If the predicted value is found to be less than the desired value, then the data is sent to the cloud through TCP/IP. The cloud server consists of another neural network model that predicts the revised parameters to achieve the desired weld quality in real-time. In this manner, the process is controlled. Thus, the edge device serves as a local system connected to the FSW machine through a LAN for monitoring the welding process in real-time. It can be observed that only a few samples of data are only sent to the cloud server, that also if it is required. This ensures the security of the data within the local network itself, thus, it prevents data breach.

Figure 5.19 shows the throughput of the data transmission to the edge device and cloud server. Through the experiments, it is found that the transmission of data from the FSW machine to the edge device is faster than the transmission of data from the edge to the cloud. This is because the edge device is located close to the FSW machine and they are connected through a local server. On the other hand, the cloud server and edge device are connected through a global network.

The total operational time (throughput) is the sum of the time taken for the transmission and processing of data among different blocks of the developed architecture. The key idea behind using edge in this architecture is to process the data at the very point of its generation. This reduces the latency caused by slow connectivity and bandwidth issues.

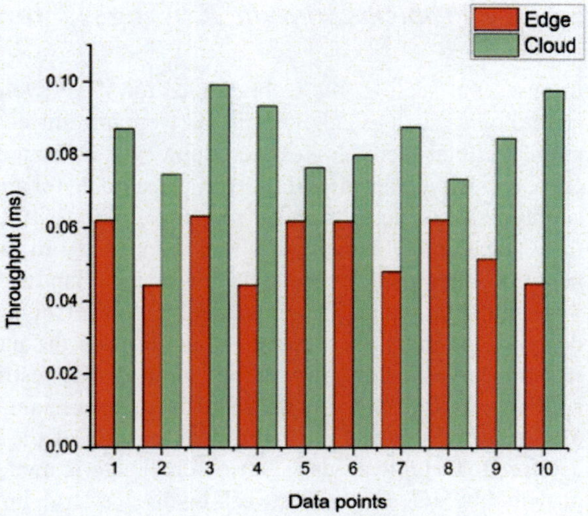

Fig. 5.19 Throughput of data transmission to edge device and cloud server

It should be noted one cannot avoid usage of the cloud. Instead, the objective is to derive the best from both edge and cloud servers. Considering the example of an industry, numerous machines will be operating simultaneously. These machines may or may not be performing similar type of operations. Thus, the sensors to be engaged for monitoring the process and the health of the machine will be different. In such a scenario, the high computational power of the cloud server will be essential for controlling these machines in real-time. At the same time, edge devices can be attached to an individual machine for real-time data streaming, pre-processing, and performing small computations on the data. This will help in eliminating time lag. The edge system would also prove beneficial if the cloud server faces downtime. In this case, the already connected edge device will still work and process the data locally.

5.11 Futuristic Concept—5G in Manufacturing

This section will discuss the potential of a futuristic technology, namely, 5G in manufacturing.

5.11.1 Problems with the Existing Network Technology

In an earlier section, the technologies for WAN were discussed, wherein, cellular technologies such as 2G, 3G, and 4G were introduced. Here, 'G' refers to the generation. Cellular technologies are popular for IIoT applications owing to their wide-area connectivity and mobility. They are suitable for applications, where mains power supply is available or frequent recharging is possible. The most advanced amongst these are the 4G networks, which is currently in use in India. Conventional 4G networks operate in the 2.4 GHz band with a bandwidth of 20 MHz.

Manufacturing applications involve control of machines, which demand zero delays. This indicates the need of controlling the machines in real-time, not near real-time. Network configuration, such as 4G, restricts the potential of the IIoT applications as they do not facilitate the increasing number of heterogeneous sensors on the factory floor. Let us discuss the cons of the 4G network, which limits its use for IIoT applications. The first challenge is owing to the bandwidth and speed provided by 4G. The 4G network involves a large infrastructure that allows a wide coverage area. For communication purpose, a 4G device has to communicate with the infrastructure first, then to a switch, and finally with a second device. This incurs a delay in the network and this is one of the reasons which is limiting the use of 4G in applications such as real-time control of the machines and machine-to-machine communication. The second challenge is that it is not flexible for facilitating service to a large number of clients because of the limited bandwidth. The 4G infrastructure i.e., the cell towers have a fixed number of connections to support, and thus, with the increasing number of devices, the traffic in the network increases. This also incurs delay in communication. A few manufacturing applications may include the use of both single sensor and multimedia data, which needs to be processed in real-time to infer the desired characteristics of the process. If the network poses a delay in the transmission of the data, then processing and feedback to the machine will not be in real-time, thus, will not be useful. Typically, the latency of a 4G network is ~50 ms. The third and crucial factor is the reliability of the 4G network. As highlighted, the 4G network purely relies on the infrastructure, which means that at the time of failure of the same, the device utilising that infrastructure will no more be able to communicate. In such a scenario, the IIoT deployment will be hampered. The fourth challenge is scalability. The industrial scenario will include a plethora of sensors, actuators, and other devices that need to communicate constantly for real-time monitoring and control action, and machine-to-machine communication. However, in the case of a 4G network, expansion means installing new towers for communication, which is expensive. Besides, the towers have a static configuration, which means they will support a limited number of devices only.

The challenges can be addressed by introducing 5G in manufacturing applications, which is more reliable and offers several advantages over the 4G network. This is elaborated in the following section.

5.11.2 Introduction to 5G Network

In contrast to the 4G network, 5G is a software-driven technology, which is virtualised and also exploits cloud technologies. It utilises 5G new radio, which is a new global standard for the wireless air interface. 5G is based on orthogonal frequency-division multiplexing (OFDM), a wireless baseband technology that is used since WiFi. OFDM is a type of digital transmission and a method of encoding digital data over multiple carrier frequencies. The 5G technology provides a low and medium frequency range that spans from 1 to 6 GHz, and a high-frequency range that spans from 6 GHz to ~above 24 GHz. The high-frequency range is because of the use of millimetric wave (mmWave), which is a specific part of the radio frequency spectrum. As the high-frequency band is unutilised, it indicates that the traffic in this band is almost negligible, thus, 5G ensures ultra-high data rates. With an ultra-high data rate, the speed of 5G may go more than 20 Gbps, which in the case of 4G is only 1 Gbps. Thus, it is 20 times faster than the speed of a 4G network.

A challenge in the implementation of 5G network is the involvement of mmWave, which is having a high-frequency range. As the frequency increases, the wavelength or the transmission distance decreases. At higher frequencies, the mmWaves can be blocked by tree, human, building, even they may get absorbed in the air. In contrast to this, at lower frequency ranges, a signal can travel more distances. To address the issue, 5G network will make use of small cells, which can be thought as base stations. The small cell technology is not new, but the ones which are designed for 5G are capable of handling ultra-high data rates, connecting millions of devices, and ultra-low latency. This is being made possible by the use of technologies such as mmWave, multiple-input-multiple-output (MIMO) antennas, and beamforming. The MIMO technology targets to facilitate several devices for communication. It is the use of antennas with multiple ports that will be engaged with the radio transmitters of the 4G network. An extension to the MIMO is massive MIMO (mMIMO) that will employ multiple transmitters and receivers enabling the transfer of a larger amount of data simultaneously. Beamforming is a technique that focuses a signal towards a specific receiver, rather than spreading the signal in all directions. Thus, the connection is faster and more reliable.

There are three main technology variants for 5G namely: (a) enhanced mobile broadband (eMBB)—this is the natural extension of current 4G cellular technology with high throughput, (b) ultra-reliable low latency communications (URLLC)—this is for ultra-low latency networks with medium throughput, and (c) massive machine type communications (mMTC)—this is for connecting a massively large number of IoT devices with low data rate. The next section will discuss a few applications of manufacturing, wherein, these features/variants of 5G are going to be apt.

Thus, 5G is designed to not only deliver faster, better mobile broadband services compared to 4G LTE, but can also expand into new service areas such as mission-critical communications and connecting the massive IoT. This is enabled by many new 5G New Radio air interface design techniques. It also includes network slicing enabling dedicated virtual networks with specific functionality, distributed edge-cloud that allows execution of workloads closer to the edge—such workloads include real-time machine learning and artificial intelligence to drive business value-add. As 5G is software-driven, sub-networks can be created in software architecture termed network slicing, which will enable a network administrator to provide network functionality based on user and devices. In the next section, a few potential applications of 5G in manufacturing will be discussed.

5.11.3 Applications of 5G in Manufacturing

5G technologies, by their inherent design, provide the required network performance that is essential for manufacturing. For example, low latency and high reliability are needed to support critical applications while high throughput and a large number of devices are needed for high sensor density plants [5]. 5G will allow manufacturers to move from fixed line to wireless thereby enabling faster yet higher flexibility at a lower cost for reconfigurable factory setups. This is applicable for industrial control and automation systems like using smart robotics, and IIoT and planning and design systems using factory digital twins [6]. There are several high impact application use cases of 5G that demand high throughput, low latency, high reliability and a large number of device support followed by the need for edge-cloud computing for sophisticated artificial intelligence/machine learning inferencing and forecasting. In the next paragraph, let us discuss a few possible use cases for the application of 5G in manufacturing.

5.11.3.1 Remote and Real-Time Monitoring, and Predictive Analytics for Real-Time Decision Making and Control—Use of URLLC

With digitalisation, manufacturing industries are now becoming intelligent and smart. The attribute 'intelligent' is aimed at identifying the condition of a process in real-time. It also aims at identifying the health of the critical machine components in real-time. Both of these will be achieved by engaging a variety of sensors. These sensors will acquire the data, which needs to be processed in real-time using signal and image processing and machine learning techniques. The derived analytics will help in making real-time decisions that will be profitable for the process as well as for keeping the machine healthy and up all the times. In this way, the machine and the process become intelligent. This means that they will be able to gaze at their conditions in real-time and make optimised decisions. As they grow, with the sensors' data, they will become smart. However, the real-time need is of utmost importance. Let us see the importance through an example.

Consider a series of robots working in an industry, as shown in Fig. 5.20. These robots perform different operations, say they execute different welding techniques. Thus, during the manufacturing of a product, the material passes through these robots,

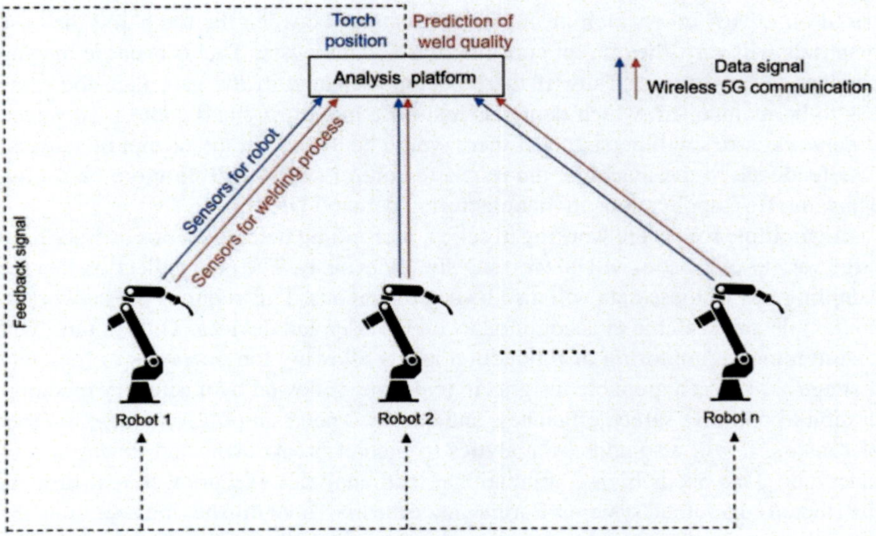

Fig. 5.20 Schematic of robots working in series

where, each robot executes the intended fabrication. Depending on the welding technique, several sensors will be required to monitor the process. For instance, if it is a metal inert gas welding technique, sensors such as current, voltage, sound, temperature, and so on will be essential for monitoring the process. Different sensors will be utilised to capture different features of the process. Say, for example, the sound sensor's information is useful for identifying the occurrence of blowhole defects, and it is also useful for investigating the arc characteristics. Similarly, current and voltage information will be useful for studying the arc characteristics during root, hot-pass, and final pass of the welding. Information from the temperature sensor will help in investigating the nature of the arc and information deciphered from the current and voltage sensors. However, it is very crucial to decipher the information in real-time, so that, the defects, if any, could be remedied in real-time. Again, to control the welding parameters, the feedback needs to reach the welding controller in real-time. Thus, the receipt of sensors' information from the machine to the analysis platform and feedback from the platform to the machine has to be fast, without any delay. Moreover, the robot will also be engaged with a variety of sensors such as position, velocity, current, etc. The position sensor will help to track the position of a robot in a given workspace, similarly, the velocity of the robot will be known from the velocity sensor. The current sensor will be utilized to track the current consumption by the motors engaged in each axis of the robot. A key point, in the case of conventional welding technique, such as metal inert gas welding is to maintain a constant gap between the welding torch and the base materials. This enables the proper and adequate flow of the molten material from the electrode to the joint line. From the sensors' data, such as from the current and voltage, this gap can be tracked. This is based on *Ohms' law*, which means as the distance between the torch and the base materials will vary, the current consumption will also vary. This is because the gap changes the resistance. Thus, if there occurs a change in the gap, then the same has to be maintained, which demands real-time monitoring and control. However, if network issues will persist, or if there would be a delay in the receipt of sensors' data/feedback to the machine, the requisite control action will no more be useful. Thus, the IIoT applications in manufacturing demand URLLC.

As multiple robots are working in series, performing heterogeneous operations, a large variety of sensors will be working simultaneously. The data collection i.e., the sampling rate of these data will also be heterogeneous. This requires the network to be flexible and scalable to accommodate all the user and devices. This, in turn, will ensure remote monitoring of production assets allowing the operators to remotely manage assets via actionable insights in real-time, followed by machinery reconfiguration to optimise safety, efficiency, and quality—need support for a large number of sensors. It will also address analytics to predict breakdowns and downtimes of machines. This needs high computational load analytics execution in real-time on distributed edge-cloud systems. Further, once the first robot finishes the operation, the job will move to the second robot, thus, there needs to be coordination amongst the machines. This is also needed to be performed in real-time to ensure synchronization.

5.11.3.2 Use of Single and Multimedia Data for Real-Time Monitoring and Control—Use of eMBB

In this section, the potential of 5G in manufacturing will be discussed for image and video analytics. Refer to Fig. 5.21, wherein, a robot is shown, which is intended to perform a welding operation. The conventional method teaches the robot to follow a path for welding. This needs a skilled operator, who can write programs, teach the robot, and finally, the welding will be done. However, an interesting problem is detecting the edge of the base metals to be welded, determining the mid of the detected edges, and directing the welding torch mounted on the robot to follow the mid of the detected edges for welding. For this purpose, an industrial camera can be mounted and integrated with the robot, as shown in Fig. 5.21. The camera will acquire the image/video, and the same would be processed in real-time for edge detection. The successful tracking of the edge will help in eliminating the need of teaching the robot. Besides, the base materials might be having some undulations on their edges or irregularity. With the manual configuration, which does not involve any intelligence, the robot will perform the welding, even if, the undulations or irregularities are present. The objective is to develop an intelligent system through the help of real-time data analysis and control via real-time decision making and feedback to the robot. Inferences regarding the presence of undulations/irregularities can be achieved by processing the image/video, and accordingly, the welding parameters must be changed in real-time to achieve a sound weld.

The challenge here is the processing of the streaming image/video. Industrial cameras are rugged and are built for industrial environments. They can capture a

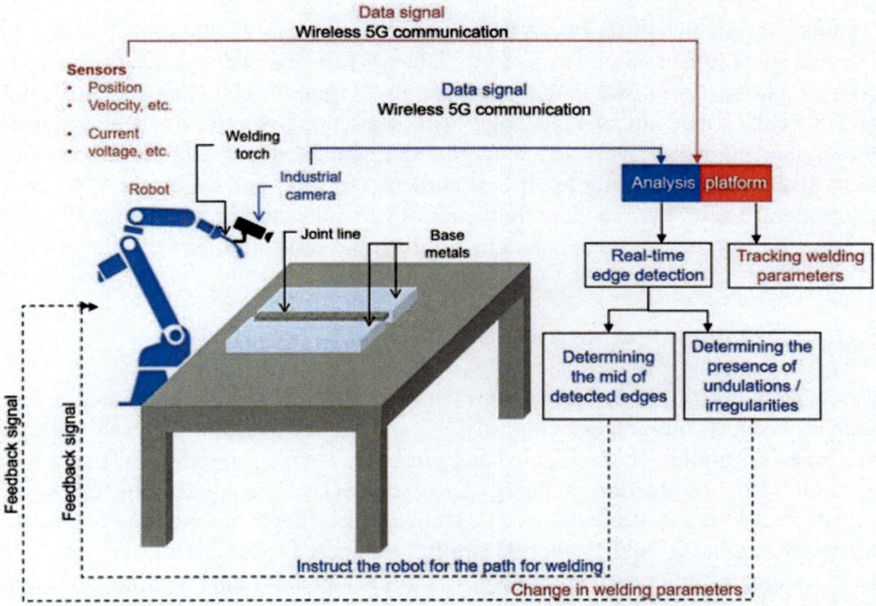

Fig. 5.21 Schematic of robot for welding

large number of frames per second, so that, minute details can be tracked from the image/video. Thus, a large amount of data is involved when it is a multimedia. Also, other sensors such as position and velocity will be engaged with the robot to track its position and velocity during axes movement. It should also be noted that the camera, in this example, is mobile, which means, as it is fixed on to the robot, it can be treated as if it is also moving with movement of the robot manipulator. 5G will provide a massive uplink bandwidth and guaranteed low latency, which is required for real-time image/video analytics. This will be fulfilled by eMBB service. As the data size is large, a conventional monitoring system may include only sensors' data and not a camera. This will generate relatively fewer data. However, for problems such as edge detection and determining the presence of undulations/irregularities on a surface can be fulfilled via image and video processing only. With 5G, a large density of cameras can be connected because of the high uplink bandwidth and the quality of service, which supports 4 and 8 K video analyses. This is aimed at increasing the throughput with reliability.

For addressing this problem, one may need to fix the camera some distance ahead of the welding torch. Since, the problem involves image/video analyses, fixing the camera in-line with the welding torch will make it cumbersome for taking the feedback. It may happen that by the time feedback action is executed, the welding torch had already crossed the location. Thus, the camera will be required to be fixed some distance ahead of the welding torch using a suitable fixture. However, this distance will depend upon the latency incurred in the analysis and the network. With 5G communication, this distance can be minimized and requisite control action will be faster. Thus, the system will be more reliable. While this can be a relatively easier task for a linear type of weld, the problem will be complicated when it comes to 2D or 3D welds. Especially, the challenge will be at corner positions or while moving through an arc, respectively. In such a situation, the focal length of camera must be such that when the welding torch is at corners, the weld line is not out of focus of the camera. Therefore, reduction in the latency by the network will definitely prove beneficial for fixing the camera and monitoring and control of the welding process.

5.11.3.3 Multiple Machines on Shop Floor—Use of mMTC

Referring to Fig. 5.20, wherein, three robots are shown working in a production line, such a system is common to an industrial shop floor. A shop floor will have multiple machines of similar or dissimilar configurations working together. In the shown example of the production line (Fig. 5.20), the sensors are spelled out for 'Robot 1', which has been considered to be executing metal inert gas welding technique. Similarly, 'Robot 2' and 'Robot 3' can be considered to be performing spot and laser welding techniques, respectively. The thing, which is being highlighted here is the increasing number of devices that are required to be connected. This problem is termed scalability. As the number of machines grows, sensors and actuators that will sense the physical quantity and act on the machine, respectively, will also increase. The scalability issue can be rightly applicable for industrial scenarios as they consist of numerous machines. The conventional 4G network can facilitate a fixed number of devices, while in the case of 5G, a large number of devices can get connected simultaneously via mMTC technology. It includes four success factors namely, coverage, cost-efficiency, low-power consumption, and long-time availability.

Other examples for the implementation of 5G in manufacturing applications include collaborative robotics, augmented reality, additive manufacturing, etc. In the case of collaborative robots, industrial robots will collaborate among themselves, and work alongside human safely. This needs high throughput, high reliability alongside low latency. Augmented reality is envisaged for assistive repair and maintenance, which needs very high, reliable throughput with low latency. Finally, additive manufacturing will require real-time supply chain optimisation via large scale IoT sensing and predictive analytics, where a large number of sensors will be engaged, and they will require low latency and high reliability. Thus, 5G is going to be highly essential for IIoT applications of manufacturing industries.

5.11.4 Challenges to Be Addressed for Implementation of 5G in Manufacturing

Manufacturing industries consist of several machines that work simultaneously, and they include a large number of sensors, actuators, data acquisition devices, PLCs, etc. These devices shall operate seamlessly for enabling real-time monitoring and control of the process. However, as the industrial shop floor will have several machines, the 5G signal may get blocked because of the presence of metallic devices, taller structures, office building, etc. To address this issue, 5G will employ coordinated multiple access points (CoMP) technique, in which, multiple base stations coordinate dynamically to support the transmission of a signal. Imagine a scenario, where the edge system for computations is located in a certain region, which is away from the source machine. The objective is to ensure that all the data packets generated from the source must reach to the destination (edge system). However, the signals may get blocked due to the hinderances that are discussed above, if there is only one base station i.e., a single access point. The 5G software architecture may be configured to account for the loss of data packets, that were not able to reach the destination. To deal with the same, multiple access points can be fixed in the industrial shop floor, and the software architecture can help understand how the signal is transmitted from the source to the destination. At time, it may happen that the signal is taking different routes i.e., through different access points, it is reaching the destination.

5G also involves dynamic selection of channels for transmission of the signals from a source machine to a destination. In the case of manufacturing systems involving scalar sensors, vector sensors, and multi-media, the volume of data is going to be large. Sensors such as temperature, current, power, etc. can be referred to as scalar sensors, in which, the output of the sensor is a quantity proportional to the magnitude of the quantity to be measured. Vector sensors consider both magnitude and direction. Example can be accelerometer, sound, force, etc. Finally, multi-media refers to image and video. 5G communication system can dynamically select and allot channels depending upon the volume of data to be sent in one packet. This will reduce the traffic and enhance the synchronisation.

5.12 Summary

The power of IIoT is its capability to create intelligent factories and plants for monitoring, control and optimisation applications. Many applications will require near-real-time responses with a high computing load for analytics. To achieve the required robustness and real-time response for control action, some computing might have to be done at the edge gateway level. This chapter established the need for edge computing in IIoT systems, outlined the different communication and networking technologies at play, and finally introduced artificial intelligence analytics at the edge for such systems. A case study is also presented at the end of the chapter, which demonstrated the usefulness of edge and cloud server in manufacturing. The chapter also introduced the futuristic communication technology, 5G, and also presented a few possible use cases which will require implementation of the 5G. The next chapter will discuss the machine learning and artificial intelligence techniques required for decision-making in manufacturing.

5.13 Questions for Practice

1. *What is the difference between cloud, edge, and fog computing?*
2. *What is indicated by "learning on the go"?*
3. *Which is more important for manufacturing—edge or cloud? List the features of edge and cloud.*
4. *List the communication technologies required for IIoT applications.*
5. *What is meant by application level protocol? List the available application level protocols.*

References

1. Bonomi F, Milito R, Zhu J, Addepalli S (2012) Fog computing and its role in the internet of things. In: Proceedings of the first edition of the MCC workshop on mobile cloud computing—MCC '12, p 13. https://doi.org/10.1145/2342509.2342513
2. Chowdhury NMMK, Boutaba R (2010) A survey of network virtualization. Comput Netw 54:862–876. https://doi.org/10.1016/j.comnet.2009.10.017
3. Pal A, Purushothaman B (2016) IOT Technical challenges and solutions. Artech House
4. Pal A (2015) Internet of things: making the hype a reality. IT Prof 17:2–4. https://doi.org/10.1109/MITP.2015.36
5. Purusottam B, Misra P, Pal A (2013) Software platforms for internet of things and M2M. J Indian Inst Sci Multidiscip Rev J 93:487–497
6. Zheng P, Wang H, Sang Z et al (2018) Smart manufacturing systems for Industry 4.0: conceptual framework, scenarios, and future perspectives. Front Mech Eng 13:137–150. https://doi.org/10.1007/s11465-018-0499-5
7. Xiaoli X, Yunbo Z, Guoxin W (2011) Design of intelligent internet of things for equipment maintenance. In: 2011 Fourth international conference on intelligent computation technology and automation. IEEE, pp 509–511
8. Nagorny K, Lima-Monteiro P, Barata J, Colombo AW (2017) Big data analysis in smart manufacturing: a review. Int J Commun Netw Syst Sci 10:31–58. https://doi.org/10.4236/ijcns.2017.103003
9. Bittman T (2019) Exploring the edge: 12 frontiers of edge computing
10. Gubbi J, Buyya R, Marusic S, Palaniswami M (2013) Internet of things (IoT): a vision, architectural elements, and future directions. Futur Gener Comput Syst 29:1645–1660. https://doi.org/10.1016/j.future.2013.01.010
11. Bonomi F, Milito R, Zhu J, Addepalli S (2012) Fog computing and its role in the internet of things. In: Proceedings of the first edition of the MCC workshop on Mobile cloud computing—MCC '12. ACM Press, New York, New York, USA, p 13

12. Georgakopoulos D, Jayaraman PP, Fazia M et al (2016) Internet of things and edge cloud computing roadmap for manufacturing. IEEE Cloud Comput 3:66–73. https://doi.org/10.1109/MCC.2016.91
13. Latvakoski J, Alaya M, Ganem H et al (2014) Towards horizontal architecture for autonomic M2M service networks. Futur Internet 6:261–301. https://doi.org/10.3390/fi6020261
14. Dey S, Mukherjee A, Paul HS, Pal A (2013) Challenges of using edge devices in IoT computation grids. In: 2013 international conference on parallel and distributed systems. IEEE, pp 564–569
15. Cheng Y, Wang D, Zhou P, Zhang T (2018) Model compression and acceleration for deep neural networks: the principles, progress, and challenges. IEEE Signal Process Mag 35:126–136. https://doi.org/10.1109/MSP.2017.2765695
16. Cheng Y, Wang D, Zhou P, Zhang T (2017) A survey of model compression and acceleration for deep neural networks
17. Giraud-Carrier CG (2000) A note on the utility of incremental learning. AI Commun 13:1–9
18. Goldberg AB, Zhu X, Furger A, Xu J-M (2011) OASIS: online active semi-supervised learning. In: Proceedings of the twenty-fifth AAAI conference on artificial intelligence. pp 362–367
19. Li T, Sahu AK, Talwalkar A, Smith V (2020) Federated learning: challenges, methods, and future directions. IEEE Signal Process Mag 37:50–60. https://doi.org/10.1109/MSP.2020.2975749

Chapter 6
Artificial Intelligence and Machine Learning in Manufacturing

6.1 Overview of This Chapter

The diagnosis of manufacturing processes and systems, prediction of machine health for corrective measures are mainly achieved through various machine learning techniques. In the previous chapters, discussions were held around the signal and image processing techniques, using which meaningful information was gathered from the raw data. The results are validated by correlating with the experiments. With this, a successful system is developed that can analyse the incoming signals. However, to build an automated system, machine learning comes into the picture, which takes the output drawn from the signal processing technique to learn a mathematical transform between input and output. This chapter will discuss the concept of machine learning, give a fundamental understanding of various static and dynamic machine learning techniques, along with their importance, and limitations. A few case studies are also discussed at the end of this chapter. Through the discussions in this chapter, readers will get a holistic way of how artificial intelligence can help in automating the processing pipeline, especially from signal morphology and sensor fusion context. The chapter will also cover artificial intelligence-driven automation that is necessary for edge computing. Finally, a few emerging concerns and challenges of artificial intelligence-driven automated systems will be addressed—in terms of trust, liability, and explainability. A few case studies are also included throughout the discussions for better understanding.

6.2 Introduction to Artificial Intelligence

Ever imagined how channels such as *Facebook* or *Netflix* or similar ones recommends or suggests videos and advertisements that one might be interested in. Ever checked the junk folder of your email inbox to find unread emails, which are characterised as

S. K. Pal et al., *Digital Twin – Fundamental Concepts to Applications in Advanced Manufacturing*, Springer Series in Advanced Manufacturing,
https://doi.org/10.1007/978-3-030-81815-9_6

spam, and have been automatically removed from the inbox? Ever thought about how a *Chatbot*, which is available on some of the websites or Government e-facilities, reply to your consents? What is happening here and what is that common attribute one could see in these examples? 'Intelligence', is what they have, and that is why they are dynamic.

The thing is that these examples are a system, which is possessing some intelligence. Of course, they are not born intelligent, they are not humans. But humans do program them and feed them with this attribute. To feed this attribute, various techniques or algorithms are utilized. As these are initiated, invoked, and controlled by humans, and the system is inheriting the techniques, the intelligence is artificial. Hence, the term artificial intelligence surrounds the world in some or another way, and these systems can be termed as intelligent agents. The examples cited here such as *Facebook*, *Netflix*, or the email service provider are not used by a single person. They have billions of customers and thus, each of the customers might be having different tastes. The spammers, who send fake emails do attempt new ideas. The thing is that these systems need to evolve and learn each time they encounter things.

For example, one person might be interested in watching thriller movies, so *Netflix* will suggest movies with 'thriller' tag. Someone may be interested to know about new places around the globe, so *Facebook* will show advertisements, which will include tags such as 'travels to new places' or 'adventure trips'. *Chatbots* are also referred to as conversational agents, which mimic written or human speech for simulating a conversation or interaction with a real human. They may be web-based applications or mobile Apps. They process the written message or speech of a visitor to deduce the meaning and reply to them with the most probable answer to the query. Technically, they do natural language processing. So, all these things boil down to what is being referred to as artificial intelligence.

6.3 Foundation of Artificial Intelligence

This section will highlight the foundation of artificial intelligence and associated topics. To begin, it is known that artificial intelligence is the field of developing computer programs for carrying out tasks, simulating the intelligent way of problem-solving by humans. The term "*problem solving*" refers to the method of reaching the desired goal or finding solutions to a given problem. As a human, whenever a problem is to be solved, a structured approach is followed. In the field of computer science, the problem-solving ability is the use of artificial intelligent techniques. For problem-solving, the first step is to make the problem definition, which will list the starting and ending situations. The second step is to select an appropriate solution technique for the defined problem. The third step is acquiring the domain knowledge and finally, the fourth step is implementing the control strategies to solve the problem for different situations. The following points are known to be the activities associated with intelligence: to respond to situations flexibly, to make sense out of ambiguous messages, to recognise the relative importance of different elements of a situation,

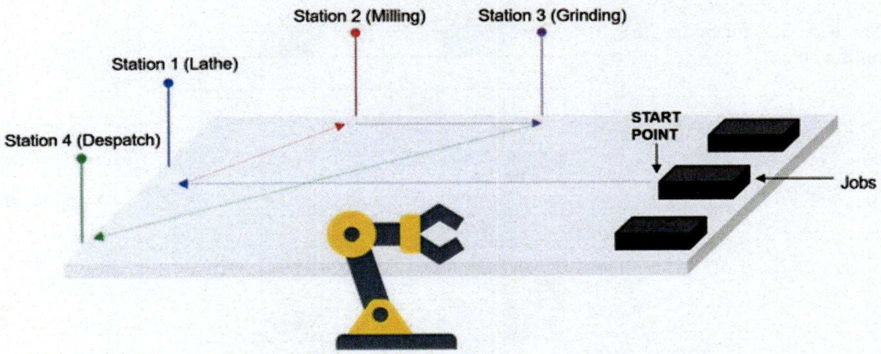

Fig. 6.1 Illustration of the importance of path planning in robot

finding similarities between situations despite differences that may separate them, and finding differences between situations despite similarities which may link them. While these activities of a human for solving a problem can be visualized, how do computers work or program to simulate the intelligence? To deal with this, one way is to make patterns followed by deducing inferences out of the patterns. The knowledge is presented as patterns of entities. Thus, the program *searches* for patterns in the deduced database and in the domain knowledge, which forms the foundation of artificial intelligence. Therefore, the core of an artificial intelligence program is the search through the knowledge.

The search algorithms in manufacturing are employed in path planning. The algorithms determine the shortest path between a start point and a destination point. It is also referred to as motion planning as it determines the motion of the object under study. For example, a robot, working in an environment can be thought of serving multiple machining stations, as shown in Fig. 6.1. In this picture, the robot is having a fixture on its end-effector for performing pick and place operations. Here, the robot has a start point, as shown in the picture, which is picking a job. The destination points are the machining stations.

Thus, the robot has to pick a job and place it at one of the machining stations, as desired. However, there may be obstacles in the path of the robot. Hence, to determine the optimal path, search algorithms are employed. The problem which is posed here can be considered in two approaches, one in offline and the other in an online manner. In the offline situation, the working space and the stationary obstacles are known a priori, then the search algorithm determines the optimal path for the movement of the robot. In an online situation, the robot will be integrated with sensors such as a camera for vision, proximity sensor for determining the presence of other objects. With the information acquired from these sensors and the known workspace, search algorithms will be deployed to deal with the dynamic obstacles. Various search algorithms such as A*, time-efficient A*, genetic algorithm, etc. can be employed. The problem-solving approach can be thought of as a grid, as depicted in Fig. 6.2.

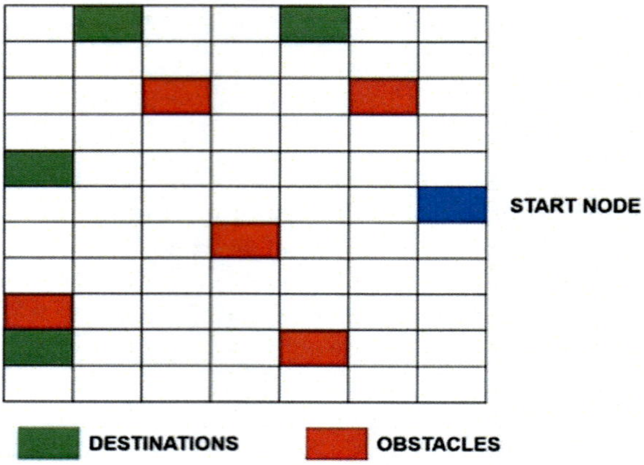

Fig. 6.2 Grid for finding the optimal path

In this grid, the start and destination points can be traced. The grid is equivalent to the workspace of the robot. The obstacles are then mapped and the search algorithms are deployed for finding the optimal path. It should be noted that there could be more than one solution to the problem. However, finding the most optimal one is treated as the best solution. This is where the importance of artificial intelligence and its search algorithms come in.

6.4 Requirement of Artificial Intelligence in a Digital Twin

With that introduction to the concept of artificial intelligence, let us immediately try to answer whether this would be at all required in a digital twin application.

A digital twin application is supposed to mimic the life of its physical asset, the one for which, it is built. Consider the manufacturing processes and machines, a lot of sensors will be engaged in collecting the data. Let us think in this manner, an experienced and age-old machine operator, who has been working on a process and a machine for quite some time, gets to know about the nitty–gritty of that process. He is also pretty much aware with the sounds that machine produces at different conditions. So, this operator can sometimes act as a twin for that machine or the process and can take matured decisions to come up with good quality products and also to keep the machine healthy. But this questions the reliability of the operator. Sensors are a good choice for this and they can track things they are intended for. Of course, sensor's data do not speak, but one needs to deduce inferences out of the data to track the events of the corresponding process or the machine on or for which they are engaged. This is also believed to be a task of intelligently processing the data to extract useful information.

Further, the digital twin will also need techniques of machine learning to automate the process of learning, providing analytics, all of which comes under the umbrella of artificial intelligence. So, here is a strong statement, *digital twins' success depends on how well the algorithms and analytics are involved in it.* Artificial intelligence will indeed be required for a digital twin in making intelligent choices, predictions, analyses, etc.

6.5 Deep into Artificial intelligence—The Knowledge Pyramid

Context is defined as "*The situation within which something exists or happens, and that can help explain it*" [1]. In IIoT deployments, transducers within sensors catch the signal of an event that is happening (for example, faults, failure, strain, quality, etc.). The role of signal/image processing and machine learning/deep learning is to analyse the collected signal to understand the corresponding context, and from there deduce the event that caused it. This contextual mapping from signal to physical phenomenon or events can help industrial manufacturing businesses in various ways.

- *Locate early signs and symptoms of failures, thereby enabling preventive maintenance*
- *Predict remaining useful life (RUL) leading towards proactive maintenance*
- *Monitor quality of the final product in real-time, thereby providing online quality control*
- *Real-time control and optimization of the whole manufacturing process*
- *Modelling the business processes and predicting future events—de-risk business and business planning.*

This whole process of context-driven analytics from sensor data to business insights depends on the physical deployment scenarios, type of sensors, and nature of insights and can be depicted as a knowledge pyramid (as shown in Fig. 6.3) [2]. This consists of four main components, which are:

- ***Signal informatics from sensors***—converts raw sensor data to information: This can be thought of as descriptive analytics that provides contextual information, trying to answer questions like "what", "where", and "when" did the event happen.
- ***Interpretation of processed information***—converts from information to knowledge: This can be thought of as diagnostic analytics that creates knowledge models, trying to answer questions like "how" did the event happen.
- ***Business insights from interpreted knowledge***—converts from knowledge to wisdom: This can be thought of as predictive analytics that creates system understanding, trying to answer questions like "why" did the event happen.
- ***Wisdom from business insights***—This is the holy grail of sensor analytics, which can be thought of as prescriptive analytics that tries to answer questions like "what to do" to make the event happen or prevent the event.

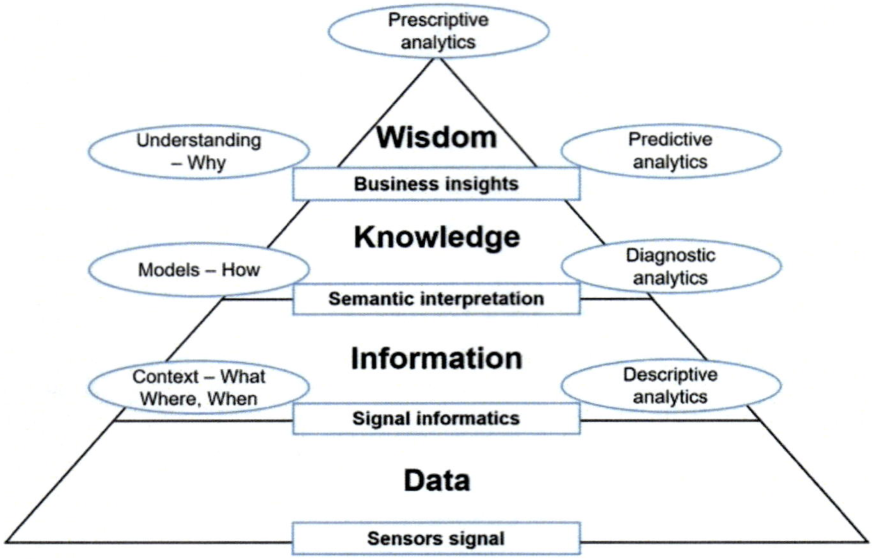

Fig. 6.3 The knowledge pyramid

The concept of artificial intelligence is divided into two namely *weak* and *strong*. *Weak artificial intelligence* refers to things that are already performed at a human level. The role of artificial intelligence, in this case, is to imitate human behaviour along with the ability to solve complex problems, execute things in a better way than human cognition, etc. The examples in this category include playing chess and other computer games, providing information to customers, data recording, etc. Contrastingly, *strong artificial intelligence* is about the ability to learn, optimize, reproduce, communicate, and even perform more advanced things. It includes machine learning and deep learning techniques for the same. For a digital twin application, this will be useful for information interpretation and business insights. The following sections will discuss the use of artificial intelligence in various domains. This will be followed by a discussion on various machine learning and deep learning techniques.

6.6 Sensor Signal Processing

Figure 6.4 depicts the major blocks in the sensor signal processing pipeline that maps to the analytics stack in form of the knowledge pyramid presented in Fig. 6.3. The image processing pipeline is very much similar except for the difference in the first three blocks. These pipelines are similar to the discussions held in Chaps. 3 and 4, respectively—but with a different abstraction. This abstraction is necessary for the

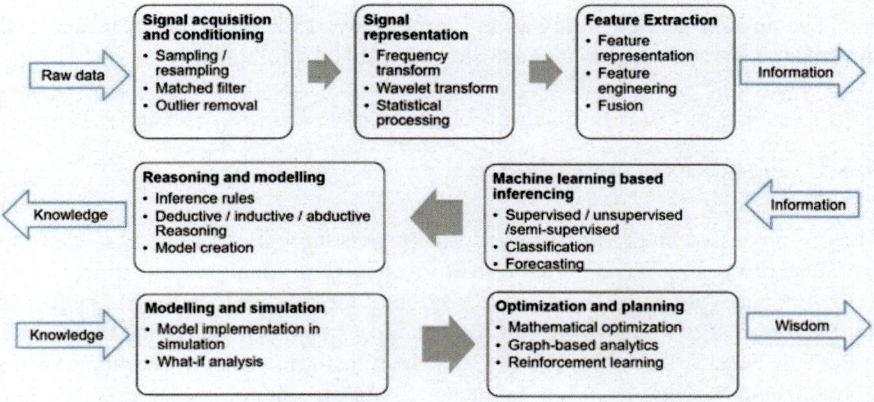

Fig. 6.4 Sensor signal processing pipeline

artificial intelligence-based automation of the pipeline. Let us understand the blocks present in Fig. 6.4 before discussing the pipeline automation methodology.

6.6.1 Signal Acquisition and Conditioning

Sampling is the first step for signal digitization. As per *Nyquist Sampling Theorem* [3], sampling frequency needs to more than twice the maximum frequency of a sensor's signal to take care of aliasing [4]. In real-world systems, the sampling rate is usually unequal—this unequal sampling rate may need to be made uniform using techniques of resampling and interpolation to present a uniform sampling rate to the subsequent blocks [5]. The sampling is followed by a matched filter, which is nothing but an optimal filter for maximizing the signal to noise ratio (SNR) [6]. The filter is tuned or matched to the sensor signal and any out-of-band noise is removed. Matched filters can be designed as low/high pass or band pass filter depending upon sensor signal characteristics. Digital filtering techniques like finite impulse response (FIR) and infinite impulse response (IIR) are useful in designing suitably matched filters. Frequency domain filter design techniques can also be useful here. Outliers are signal points that are statistically distant from other signal data points—usually, errors in the sensor measurements come out as outliers that need to be excluded [7–9]. There are many different methods of outlier removal, most of which try to model the statistical distribution of the signal or tries to find a set of features in the signal that can separate between regular values and outliers.

In Chap. 3, while discussing signal processing techniques, examples have been demonstrated for sampling, digitization, and filtering. The pipeline of signal acquisition and conditioning has also been discussed in detail. These techniques in signal acquisition and preparing them for the analysis for extraction of meaningful information are attributed to the inclusion of artificial intelligence.

6.6.2 Signal Representation

The pre-processed and cleaned signal usually goes through a set of transforms and statistical processing. The Fourier Transform converts a signal from the time domain to the frequency domain. The discrete version of the Fourier transform is known as the Discrete Fourier Transform (DFT). The most popular implementation of DFT is the Fast Fourier Transform (FFT) [10]. The transforms result in an output vector of frequencies ranging from low-frequency to high-frequency coefficients. Wavelet Transforms generalizes the signal representation by using generic basis functions, instead of sine and cosine as in Fourier Transform [11]. The discrete version is called Discrete Wavelet Transform (DWT). DWT is useful when the frequency of the signal varies with time. Using DWT, high-frequency components can be represented better in time than low-frequency components and low-frequency components can be represented better in frequency compared to high-frequency components. Statistical processing techniques can be used to effectively represent the information-bearing part of the signal [12]. The simplest form of statistical processing is to have order statistics like mean, median, mode, variance, correlation, skewness, kurtosis, etc. for a signal. There also can be windowed statistical processing techniques like Piecewise Aggregation Approximation (PAA) and Symbolic Aggregate Approximation (SAX), which can provide good representation for different signals [13].

In Chap. 3, these processing techniques have been discussed in details, where the utility, advantage, and also limitations of each technique are pointed out. Recap the case studies illustrated in Chap. 3, wherein, statistical features such as mean and kurtosis were extracted from force and sound signals, respectively. The mean of the force signal helped in characterizing signals corresponding to different process parameters employed for the friction stir welding. Kurtosis of the sound signal, collected in case of the gas metal arc welding, helped in identifying the portions of the signal corresponding to the occurrence of the welding defects. Moving further, because of the limitation of the time domain analysis, which is the absence of the frequency information, the Fourier Transform technique is employed to determine the frequencies present in a signal. As an example, a case of gas metal arc welding is presented, where the importance of frequency content is discussed. Towards the end of Chap. 3, the wavelet transform technique is discussed following the limitations of the Fourier transformation and windowed Fourier transformation techniques, which includes the absence of time information and fixed width of the window, respectively. Force signals corresponding to defective and defect-free welds have been analysed via DWT technique and useful information was extracted for differentiating the welds via the signals.

6.6.3 Feature Extraction

Features are interesting signal representations that correlate with the physical event of interest as captured by the sensor. Feature engineering describes the process of extracting representative features from the sensor data. Real-time feature representation and engineering are required to process the signal "in-stream"- this is called stream processing. Multi-senor fusion is required to effectively use the information acquired from multiple sensors [14, 15]. Sensor redundancy and sensor correlation in such fusion can help in making better inferences including better accuracy and noise tolerance. The fusion can happen at multiple levels—at the raw data level, at the feature level, or the decision/inference level. Multi-sensor fusion can become challenging in the face of different noise and non-stationary characteristics of different sensors.

As an example, let us take the case of a drilling process. Drilling in manufacturing is a cutting process. It employs a tool, which is termed a drill bit. The drill bit is mounted on the machine spindle, so it rotates and cuts a hole of circular cross-section in solid materials. The cutting tool is multi-point. Drilling is important in manufacturing, as it is mostly encountered during assembly operations. But often because of the wear of the drill bit, the quality of the drilled section gets poor. The success in drilling majorly depends on achieving increased drilling speeds or a faster rate of penetration. For this, the wear of the drill bit is inevitable. Manually tracking this phenomenon is quite a difficult task and also involves risk. So, sensor-driven analysis of the same is a popular task. Researchers have engaged current, force and torque, and vibration for monitoring the drill wear.

To present the importance of feature engineering, the following discusses a study involving a current sensor for monitoring the drill wear phenomenon [16]. The raw signals acquired from the process are shown in Fig. 6.5a, b, which corresponds to average drill wear of 0.145 mm and 0.295 mm, respectively. Though there is a little increase in the magnitude of the signal with an increase in the drill wear, it is hard to monitor the phenomenon. In such a situation, feature engineering is the best solution, which could represent the signal characteristics. In this case, the feature extracted from the signal is the RMS value. Details of this feature are discussed in Chap. 3.

Figure 6.6a depicts the RMS map with the average flank wear, which does not show a trend. Sometimes, customized solutions may track better. For this example, it was thought that there may be fluctuations in the supply voltage. Thus, the magnitude of the current was determined when no drilling operation was conducted. The RMS feature was again extracted, but this time, with the base current subtracted from the current consumed during the drilling operation. This was termed as the differential RMS, which depicted a clearer picture of the average flank wear. The same is depicted in Fig. 6.6b. From the plot, it can be observed that the feature value is increasing with an increase in the average flank wear.

The above example marks the importance of feature engineering. Also, feature engineering would be useful for reducing the amount of data. Considering the digital twin application, for tracking a process, the use of a single sensor would be rare. Often,

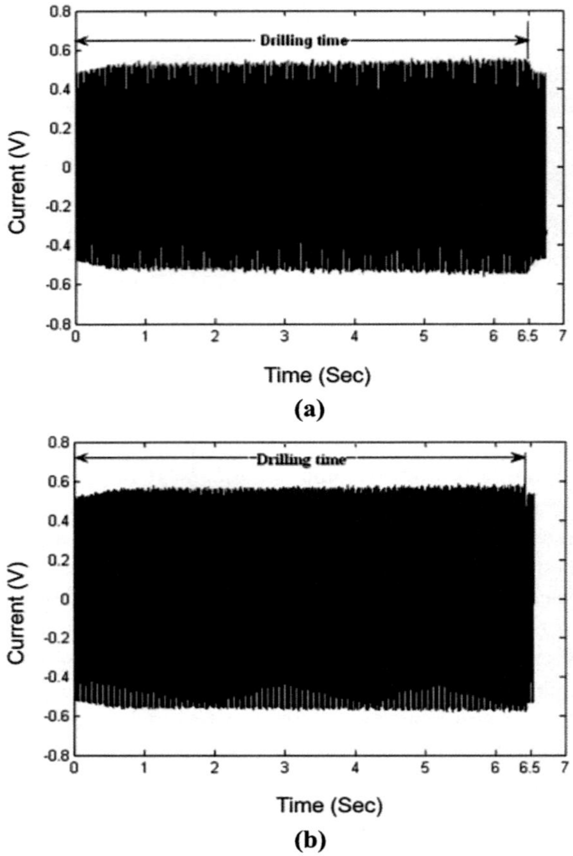

Fig. 6.5 Current signal: **a** 0.145 mm and **b** 0.295 mm average flank wear

multiple sensors will be required to decipher information for different phenomena of the process. Thus, the amount of data is also going to be huge. So, feature engineering will be useful to reduce the traffic of the data.

Let us dive further more into the usefulness of feature engineering. As observed, features are important, however, there might be a possibility of getting different features, each showing correlation with the process or depicting the true value of the signal. Here, the question is whether to include all those features or is there a need to look into which of the features are important. Of course, the second option is the real purpose of feature engineering, and this process is also referred to as feature ranking. The idea is to rank the features based on their importance with the considered phenomenon. In this example of machining, the idea is to classify the condition of the cutting tool, which is crucial because the quality of the job being machined depends upon the quality of the tool. Thus, early detection of tool condition will help in reducing or eliminating the rejection of the product.

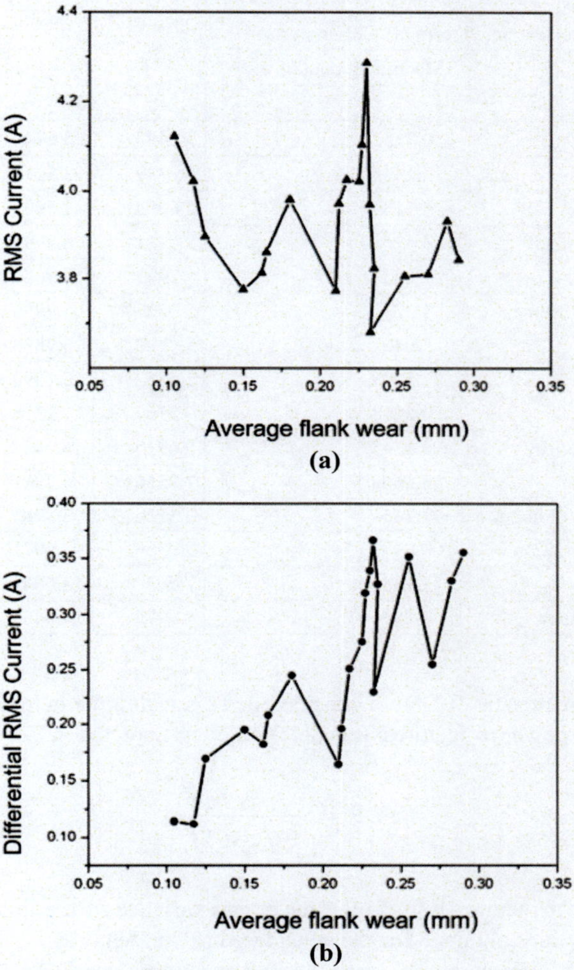

Fig. 6.6 Feature extracted from current signal: **a** RMS and **b** Differential RMS

For the above problem, monitoring is performed by analysing the machined surface images, which is an indirect monitoring technique. Gray-level co-occurrence matrix (GLCM) technique, which has been discussed in Chap. 4, is applied for analyzing the images. Features extracted out of the analysis are contrast, energy, correlation, homogeneity, entropy, inverse diagonal moment, dissimilarity, mean, variance, maximum probability, diagonal moment, second diagonal moment, coefficient of variation, cluster shade, and cluster performance [17]. So, this list consists of fifteen features. Four different machine conditions are considered. For each of the machining conditions, the final image is divided into two, where the first set was assigned to the sharp tool and the second to the dull tool. For ranking of the features,

Table 6.1 Features extracted from GLCM analysis

Features	Machining conditions			
	1	2	3	4
Contrast	0.7652	3.8457	5.644	0.7091
Energy	0.1485	4.9739	3.5219	3.4507
Correlation	5.5134	2.6093	3.199	1.1925
Homogeneity	4.4268	5.6981	6.2995	7.839
Entropy	1.243	5.2993	0.8659	4.9514
Inverse diagonal moment	1.3211	5.6227	6.2598	7.6882
Dissimilarity	5.0407	5.4793	6.983	4.9517
Mean	1.5098	2.9552	6.1892	4.2552
Variance	0.3963	0.5567	1.2216	7.5669
Maximum probability	4.8345	5.6915	6.242	6.4861
Diagonal moment	5.5669	**7.9556**	6.5941	7.7944
Second diagonal moment	7.3651	7.4126	6.5985	4.9517
Coefficient of variation	0.097	0.9366	3.6971	3.7844
Cluster shade	**11.1386**	7.503	**13.6841**	**14.67**
Cluster performance	8.241	4.9979	14.466	12.343

fisher discriminant ratio (FDR) is considered. This technique helps by obtaining a discriminatory power of features. It is determined as depicted in Eq. 6.1 [17].

$$\mathrm{FDR} = \frac{(m_1 - m_2)^2}{\left(\sigma_1^2 + \sigma_2^2\right)} \tag{6.1}$$

where m_1, m_2, σ_1, and σ_2 refer to the mean and variance of the images. Table 6.1 lists the FDR values obtained for the considered fifteen features.

From the table, it can be observed that for machining condition 1, cluster shade is the top-ranked feature with a ratio of 11.1386. Similarly, for condition 2, it is the diagonal moment, cluster shade is again for conditions 3 and 4. The second top-ranked features for the 4 conditions are cluster performance, cluster shade, cluster performance, and again cluster performance, respectively. Further, one could also see that there a few features whose ratio vary drastically with different machining conditions. The only features such as homogeneity, dissimilarity, maximum probability, and diagonal moment show fewer fluctuations amongst the conditions. Other features with a ratio less than 1 can be discarded straight away. Thus, amongst these 15 features, only six features are finally selected. This marks the importance of feature engineering and ranking of the feature.

6.6.4 Machine Learning and Inferencing

There are classical inferencing techniques like signal detection, signal identification, and signal estimation. Statistical signal processing techniques are used to detect the presence or absence of a particular pattern in the sensor signal. Identification can be considered as a detection problem that works on multiple patterns instead of one [18, 19]. Signal estimation can be thought of as inferring the value of unknown states of the sensor signal under noisy conditions. There are many estimation techniques out of which, the popular ones are Least Square Estimation (LSE) [20], Maximum Likelihood Estimation (MLE) [21], Bayesian Minimum Mean Squared Error (BMSE) [22], and Kalman Filtering (KF) [23].

With the advent of artificial intelligence, machine learning has gained traction as a tool for inferencing as compared to the classical inferencing techniques described above [24]. Machine learning assumes a construct for the relationship between input and output data and then tries to estimate the parameters of the construct (or models, as it is more popularly known). Machine learning can be supervised (meaning it has annotated examples to train on), unsupervised (meaning it has no annotated example to train on) or semi-supervised (meaning it has some annotated example but not all) [25]. A complete section dedicated to machine learning is given later (Sect. 6.7.1).

6.6.5 Reasoning and Modelling

Rule engines allow for putting logic combining detected events from sensor data through processing blocks described in previous sections to include or exclude a higher-level event [26, 27]. A reasoning system can generate conclusions from available knowledge using logic. They are an integral part of artificial intelligence and knowledge-based systems. There can be three kinds of reasoning systems—deductive (top-down logic), inductive (evidence-based), and abductive (infer a theory from observation) [28, 29]. Knowledge modelling means combining a set of inference models (classical or machine learning-based) with a set of rules and reasons to represent the behaviour of a complete end-to-end system with sensor data as input and knowledge as output.

6.6.6 Modelling and Simulation

The information and knowledge models (physical, mathematical, or logical) extracted from sensor data can form a basis for system simulations. It helps in understanding how the system will behave without actually implementing it in real life. Modelling and simulation can be used for what-if analysis during initial system design, followed by evaluation and optimization of designs [30].

6.6.7 Optimization and Planning

Sensor data analytics can help in designing better business processes in terms of cost, efficiency or features, thereby improving the return-on-investment (RoI). Optimization and planning, which is a mature field in operations research (OR) can be applied to information and knowledge models extracted from the sensor data to provide optimal or near-optimal solutions to complex decision-making problems and generate plans to achieve the same, thereby moving towards prescriptive analytics [31]. Combined with the simulation, optimization and planning can be a very potent tool for creating a digital twin of industrial IoT business systems [32, 33]. Off late, reinforcement learning-based systems are becoming good alternatives to classical linear programming-based optimization for dynamic environments, where the system continuously judges its closeness to the goal. Reinforcement learning uses a trial and error approach to discover, which actions yield the greatest rewards.

6.6.8 Image and Video Processing

The image/video processing pipeline is very similar to the signal processing pipeline except for the first three blocks [34]. The image/video acquisition block deals with pre-processing in form of format conversion, noise removal, and region of interest selection. The conditioning block takes care of filtering, masking, convolution, histogram equalization in both spatial (spatio-temporal for video) and frequency domains. The representation block comprises of the features like boundary/edge/shape/texture, etc. that represent the image/video. The output of this block can be considered as information that can be handled in an exactly similar way in the subsequent blocks as a signal processing pipeline.

6.7 Analytics Pipeline Optimization Strategies

It needs to be noted that both signal and image/video processing pipelines described in the previous two sections need a lot of manual intervention from a design and implementation perspective. It is also clear from the description of these systems that the manual intervention is of two forms—to choose between a host of algorithmic options and to find the optimal operating parameters of different algorithms [35, 36]. Let us start exploring how such manual processing tasks can be automated using artificial intelligence, which is the primary objective of this chapter.

6.7.1 *Machine Learning*

The term "machine learning" has been quite popular in the present scenario. This section introduces the basic concept of machine learning through a few examples. Later, a deeper context on the same will be discussed.

So, what you visualize when you hear the term machine learning? The formal definition of machine learning is the "*art of programming computers to learn from data*". The following sentences attempt to provide a general description of machine learning and what it is all about. A "machine" is something that takes input, processes it, and produces output. Here, in the interest of the subject, the machine in the term "machine learning" takes a soft form, i.e., it is available digitally on a computer. Data is the input to this machine and the derived output as well. This machine is intended to learn a transform from the given data as input. Thus, it is referred to as machine learning. The term "learning" means "the acquisition of knowledge or skills through study, experience, or being taught". A very childish example could be, asking a kid to differentiate between an apple and a mango. The kid observes the differences (colour, shape, size, etc.), and learns the same. So, humans learn from their experiences. The concept of machine learning is similar to this ideology, where a machine attempts to mimic the action of a human, i.e., it tries to learn from its experiences. While it learns, it improves. To make them learn, they are fed with data. The data is gathered from real-world objects. In manufacturing, the data refers to the information about the machine, process and health conditions, etc. Let us explore a few popular definitions of machine learning that are given by researchers and organizations around the globe:

- ***Definition by Arthur Samuel*** [25]—"*Field of study that gives computers the ability to learn without being explicitly programmed*"
- ***Definition by Tom Mitchell*** [25]—"*A computer program is said to learn from experience '**E**' with respect to some task '**T**' and some performance measure '**P**', if its performance on '**T**', as measured by '**P**', improves with experience '**E**'*".
- ***Definition by Kevin Murphy*** [37]—"*Algorithms that automatically detect patterns in data use the uncovered patterns to predict future data or other outcomes of interest*".
- ***Definition by NVIDIA*** [38]—"*Machine learning at its most basic is the practice of using algorithms to parse data, learn from it, and then make a determination or prediction about something in the world*".
- ***Definition by McKinsey & Company*** [39]—"*Machine learning is based on algorithms that can learn from data without relying on rules-based programming*".
- ***Definition by Stanford University*** [40]—"*Machine learning is the science of getting computers to act without being explicitly programmed*".
- ***Definition by Carnegie Mellon University*** [41]—"*The field of Machine learning seeks to answer the question "How can we build computer systems that automatically improve with experience, and what are the fundamental laws that govern all learning processes?*"
- ***Definition by the University of Washington*** [42]—"*Machine learning algorithms can figure out how to perform important tasks by generalizing from examples*".

So, the art of machine learning excludes explicit programming, i.e., it does not require line-by-line programming instruction. Figure 6.7 depicts the idea of a traditional programming system and it also shows the idea of machine learning. In the former case, both data and program are fed to the computer, which executes the same and generates the output. However, in machine learning i.e., in the latter case, data and output are fed to the computer. The computer generates the program using the data. Thus, the machine learning models are designed in such a manner that they learn by themselves upon working on the data. They build mathematical models using the input data to make the decisions. The most common example which everyone encounters in day-to-day life is the "filtering of spam-mail". The brain of mailing systems is devised in such a way that it always looks for certain keywords and attributes which help to recognize spam mails. Common keywords may be "cash bonus", "100% satisfied", "earn", "credit card", "4U", etc. The email system is having a machine learning model, which is trained for this activity.

Referring to Fig. 6.8, it is said that machine learning comes under the domain of artificial intelligence. Keeping aside other attributes of the signal and image/video processing pipeline, the domain of artificial intelligence covers machine learning and deep learning. It means to build artificial systems with intelligence. Machine learning encompasses several algorithms and artificial neural network (ANN) models for assisting computers in decision-making. Further, deep learning comes under the umbrella of machine learning. Deep learning is a subset of machine learning. It is about creating deep machine learning models. Details of deep learning will be discussed in a later section.

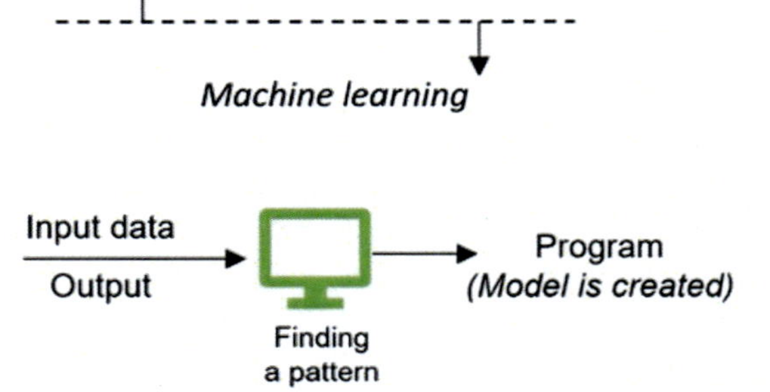

Fig. 6.7 Traditional programming versus machine learning

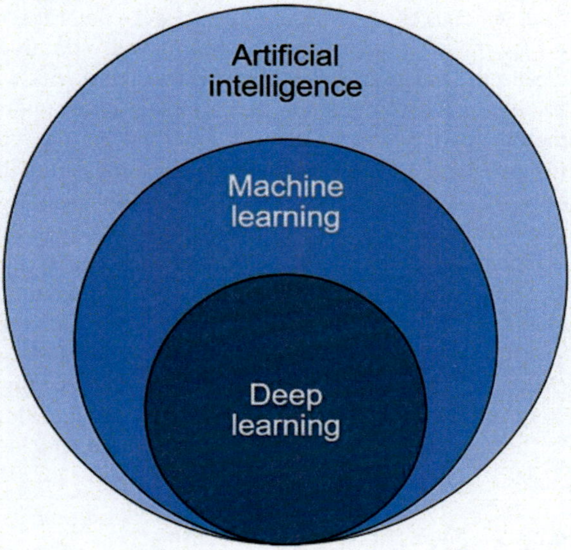

Fig. 6.8 Domains of artificial intelligence

6.7.1.1 Use of Machine Learning in Digital Twin

Let us answer the following questions. Is machine learning at all necessary? Can one not rely on traditional programming techniques? What difference could machine learning bring? Referring to the spam-mail example cited above, one could write a rule-based program. For this, a user needs to gather different words that could be found in a spam mail and accordingly, rules will be defined in the program to detect them. But wait, there is no limitation of such words. Spammers may use different words. Hence, it would be cumbersome to program new rules every time. This is where machine learning fits the best because, it could perform the task of detection and at the same time, it can learn as well. Thus, when a machine learning model will be fed with different keywords for detection of spam-mail, it learns from the data, and as it is deployed, it also learns new patterns, words, phrases, etc. that are used by spammers. So, it is learning from its experiences.

In manufacturing, processes will be surrounded with heterogeneous data that will be gathered by using different sensors at varying sampling rates. Machine learning can help in classifying faults, predicting product quality, determining the optimum set of parameters, predicting health conditions, etc. by learning from the sensors' data. Let us walk through a small example. Consider the case of the machining process in an industry. It is always crucial to ensure the quality of the product after machining, which requires the product to be tested via various quality tests. Often, the quality tests are time-consuming, expensive, and some of them may be destructive as well. Instead, machine learning along with signal processing techniques could help in the real-time prediction of the product quality. This would save a fortune for the industry. This discussion mostly surrounded the parametric conditions. But, what if, the cutting tool, which is used for machining has worn out. Most of the industries rely on offline inspection techniques. The operator may also observe a different sound coming out during the machining process while working with a worn tool. However, gauging something like this will only be possible if a larger portion of the tool has worn out. If the same will be in a smaller portion of the cutting tool, it will be next to impossible for the operator to detect during the machining. But, the degradation will hamper the quality of the product. Instead, if sensors such as force, vibration, temperature, etc. could be engaged in the process, then the collected information can be processed to decipher the cutting tool condition, product quality, and other features. Similarly, for welding processes, machine learning could help in the automatic identification and classification of welding defects, prediction of joint strength, etc.

Thus, machine learning best fits problems that require complex programming. It can learn a transform when fed with data, thus, saving time and effort that otherwise would have been put in writing programs. Machine learning could also be leveraged for finding hidden patterns in a dataset that otherwise could be hard to decipher.

6.7.1.2 Types of Machine Learning

Figure 6.9 show the general classification of machine learning, which includes supervised learning, unsupervised learning, and semi-supervised learning. These three learning techniques of machine learning are based on human supervision. The

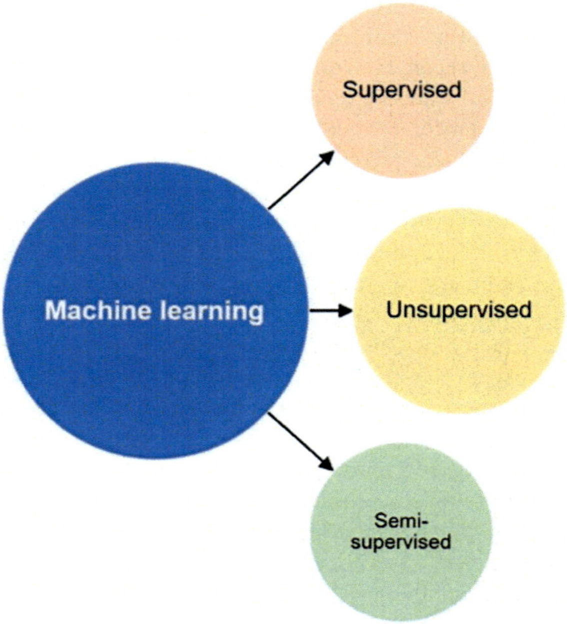

Fig. 6.9 Types of machine learning systems

following paragraph discusses these categories in brief highlighting the necessity of each technique. At the end, a few case studies are discussed.

As per the definition of machine learning, it is defined as the art of programming computers to learn from data. The data utilized for building a machine learning model is referred to as training data and the other on which it is tested is referred to as testing data. The said data resembles information gathered/collected/captured from real-world scenarios.

Supervised Machine Learning

Before going to the details of each type of machine learning model, it is important to know the basic ideology behind a machine learning model, which is playing with the data. So, essentially, there is nothing, and a model is formed only to process the data. Thus, to begin modelling, data is required. For understanding supervised machine learning, Fig. 6.10 depicts a schematic diagram. It shows two datasets, the first is referred to as training and the second is a test dataset. The training dataset consists of input data and labels. From the input data, feature engineering is carried out to extract meaningful features. These features certainly indicate the behaviour of the data. So, the features along with the corresponding labels are fed to a model. The model learns from the training dataset and is deployed to provide the solution for unseen data. The unseen data is the test data, which consists only of the input data. The learnt model makes a prediction based on its learning. Since the model is fed with both input and the labels for training, it is being referred to as a supervised training model. This indicates as if there is a supervisor (someone with data), who is trying to educate someone (a machine learning model) by giving various examples (data with inputs and labels).

So, the purpose of building a machine learning model could either be regression or classification. Say, for example, a dataset is containing continuous labels, such as the tensile strength values corresponding to different parametric conditions during

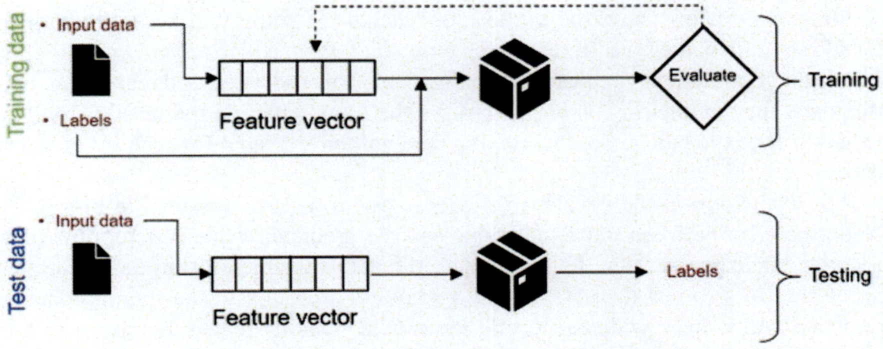

Fig. 6.10 Schematic of supervised learning

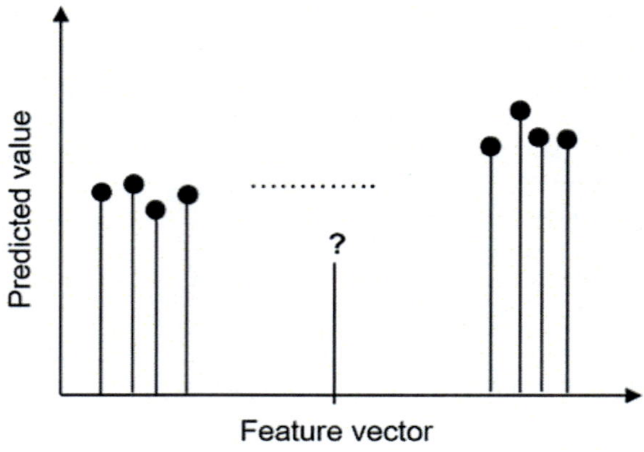

Fig. 6.11 Regression task

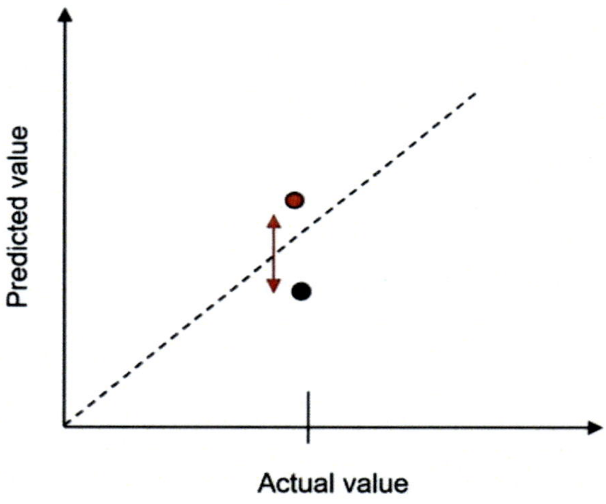

Fig. 6.12 Evaluation of the model

welding. So, for this case, the purpose of machine learning will be to predict the tensile strength value for different parametric conditions, which are not known. This idea is diagrammatically represented in Fig. 6.11. This task is termed regression. For this data, the parametric conditions will be the input data and the tensile strength values will be the labels. So, the training data will have a label for each set of input data.

The evaluation of the developed model is performed as presented in Fig. 6.12. Here, the error between the actual value and the predicted value is determined for a particular feature vector. As mentioned, the data on which testing is performed is referred to as a test dataset. During modelling, a portion of the training data is removed and is kept aside for testing the model. This is treated as unseen to the model, for which, the model will predict the labels (predicted value). These values

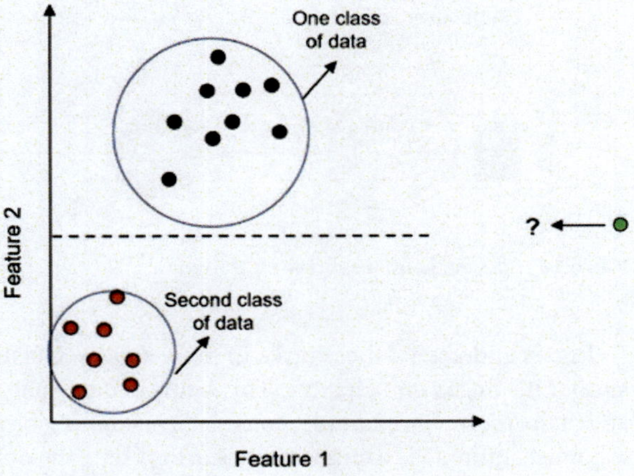

Fig. 6.13 Classification task

are compared with the actual values, which are experimentally verified using some instrument. Finally, the aggregated error indicates the ability of the model.

Often a validation dataset is also utilized in addition to training and testing datasets, which is optional. This is required after training to validate the model before testing it with the unseen data. The other task is classification, which implies that the dataset consists of discrete labels, also referred to as classes. This is depicted in Fig. 6.13. Consider, for example, let us say, a dataset contains input data corresponding to two classes, either defective (shown as red dots) or defect-free (black dots). In the picture, 'Feature 1' and 'Feature 2' are the explanatory variables that represent the object to be classified. The objective is to build a machine learning model to classify the new data (green dot) to either be defective or defect-free.

A few popular supervised machine learning algorithms utilised are linear regression, logistic regression, support vector machines, K-nearest neighbours, random forest, decision trees, neural networks, etc. Amongst these algorithms, neural networks have grabbed special attention because of their ability to solve complex problems. To discuss more on neural networks, a dedicated section is devoted towards the end (Sect. 6.7.2). Further to the above explanation, a case study is discussed in the following section, where a supervised machine learning algorithm is employed for decision-making.

Unsupervised Machine Learning

In an unsupervised machine learning algorithm, the training data is unlabelled. This learning system is not for regression or classification as the data is unlabelled. Rather, the algorithms under unsupervised learning are used for clustering, anomaly and novelty detection, visualization and dimensionality reduction, etc. Figure 6.14 depicts a schematic diagram of unsupervised machine learning.

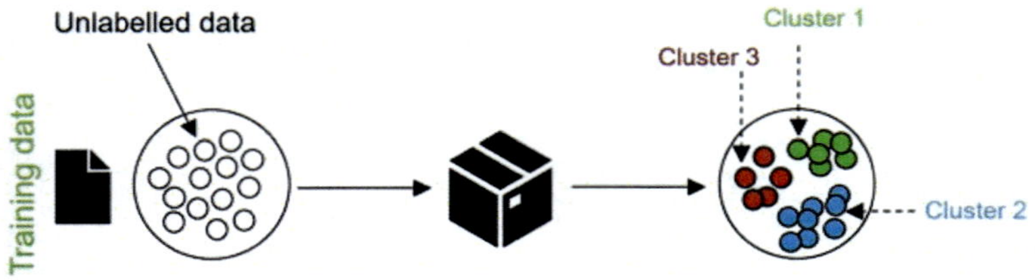

Fig. 6.14 Schematic of unsupervised learning

Let us understand these tasks in more details. Clustering refers to the grouping of similar instances into clusters. This helps in data analysis, dimensionality reduction, and many more. For example, consider a machining process in which, the temperature is being acquired via a temperature sensor. Here, the objective is to group the obtained temperature values into the ones corresponding to defect-free and defective products. So, a clustering algorithm can be applied to perform this task. The obtained outcome can help in developing a small algorithm, which will help in detecting the quality of the product. This can be visualized from the schematic diagram shown in Fig. 6.14. The training data, in this case, is unlabelled. This data upon feeding to the model attempts to build clusters around them.

In anomaly detection, the idea is to differentiate between normal and anomaly. Once the model is learnt on these instances, it can differentiate any anomaly, if occurred. The anomaly detection algorithms are beneficial for a production line that transports the products from one station to the other. Here, if in case, the desired product is not found, then an alarm can be raised. Visualization and dimensionality reduction are beneficial for easy visualizing a dataset. An enormous amount of data is going to be generated each day out of the industries, the processing of which will be very difficult. In this case, the visualization algorithms can be fed with unlabelled and complex data to detect unsuspected patterns and generate a simpler 2D or 3D visualization of the data. Dimensionality reduction helps in simplifying the data without losing much information. Some of the algorithms under this learning system are K-means clustering, principal component analysis, independent component analysis, etc.

Semi-Supervised Machine Learning

The semi-supervised machine learning system lies in between supervised and unsupervised types. In the case of supervised learning, data is labelled and learning begins by knowing what objects are in a dataset. This could particularly be time-consuming. The unsupervised learning algorithms have unlabelled data and learning starts by clustering them into groups. This is also referred to as self-learning. A semi-supervised learning system tries to take the benefits of both approaches. In this case, a part of the data is labelled and the rest is unlabelled. Thus, algorithms utilized for this purpose are combinations of supervised and unsupervised algorithms.

A Real-Life Example in Manufacturing—Support Vector Machine

In this case-study section, the example to be discussed is deciding upon whether a weld is defective or defect-free through a machine learning model [43]. The welds are fabricated using the friction stir welding technique. To introduce the problem, the objective is to develop an automated system that can classify welds into two categories, defective and defect-free. The base technique involved in studying the welds is image processing. As discussed, data is required to be generated for building a machine learning model. In this case, experiments were conducted at various processing conditions to come up with welded samples corresponding to defective and defect-free. Note the importance, the data consists of both defective and defect-free so that the model can learn from these instances to finally develop the program, i.e., a model. Here, the input data is the images of the welded samples. All the welded samples were annotated, either as defective or defect-free depending upon the presence/absence of surface defects, respectively.

Next, feature extraction was carried out. DWT has been applied to the images and the features extracted are energy, variance, and entropy of the approximation coefficients. DWT breaks down a signal into multiple scales and frequencies. It also preserves both time and frequency information, thus being beneficial over the time domain and frequency domain techniques. In this example, the approximation coefficients were extracted up to two levels and features such as energy, variance, and entropy of the coefficients were determined. Figure 6.15 shows the feature maps corresponding to the samples, where (a) corresponds to the map of energy, (b) corresponds to variance, and (c) corresponds to the entropy of the approximation coefficients computed for defective and defect-free welds. In the three maps, a clear separation can be observed between the features. For a good weld, the energy is supposed to be high owing to the lesser variation in the data and thus, the corresponding approximation coefficients have a higher amount of energy as compared to the defective weld samples. Similarly, variance is higher for the defective weld samples as compared to the defect-free welded samples. Lastly, entropy indicates the degree of randomness/information contained in a signal, and the same can be observed to be high for the defect-free welded samples. Thus, these features were considered for building a machine learning model.

The algorithm considered in this study is a support vector machine (SVM) model. It is a supervised machine learning algorithm. This algorithm can be utilized for both regression and classification tasks, and in this example, SVM is going to be utilized for the classification of the welding samples either into defective or defect-free. Let us discuss SVM in brief, followed by its implementation in the present example.

An SVM model tries to segregate data by creating a hyperplane in an n-dimensional space. This is the basic statement that could be found in many textbooks. Let us try to understand the same in a simple manner. Figure 6.16 depicts the idea of the SVM algorithm. Let us suppose the training data consists of two input features, namely, 'Feature 1' and 'Feature 2', termed 2-dimensional space, as shown in the picture. The hyperplane can be thought of as a separator, as shown in the figure, which attempts at splitting the input variables. Being a supervised algorithm,

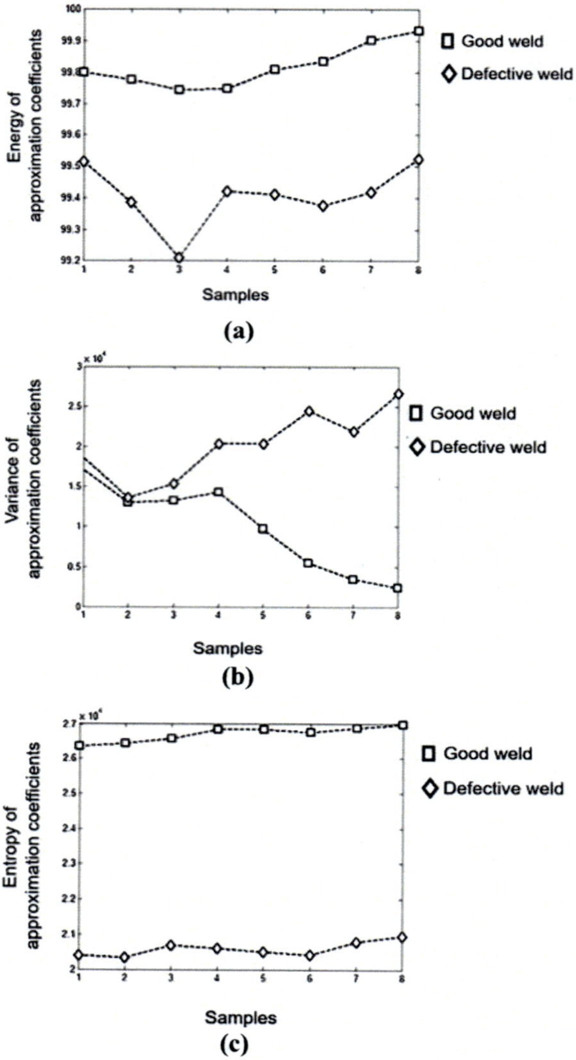

Fig. 6.15 Feature map corresponding to defective and defect-free welds: **a** energy, **b** variance, and **c** entropy of the approximation coefficients [43]. Reprinted from Journal of Manufacturing Processes, Volume 210, Nagraj N. Bhat, Kanchan Kumari, Samik Dutta, Surjya K. Pal, Srikanta Pal, Friction stir weld classification by applying wavelet analysis and support vector machine on weld surface images, 274–281, 2015, with permission from Elsevier

the data will have input features along with the class it belongs to. For instance, in the present example, there are two classes, defective or defect-free. Thus, one needs to specify the input features and the corresponding classes. There could be many possible hyperplanes for splitting the data, however, the algorithm tries to determine the hyperplane with maximum margin. Here, maximum margin refers to the maximum distance between the data point of the two classes, as shown in the figure. Finally, the support vectors are the data points that are closer to the hyperplane and

Fig. 6.16 SVM algorithm

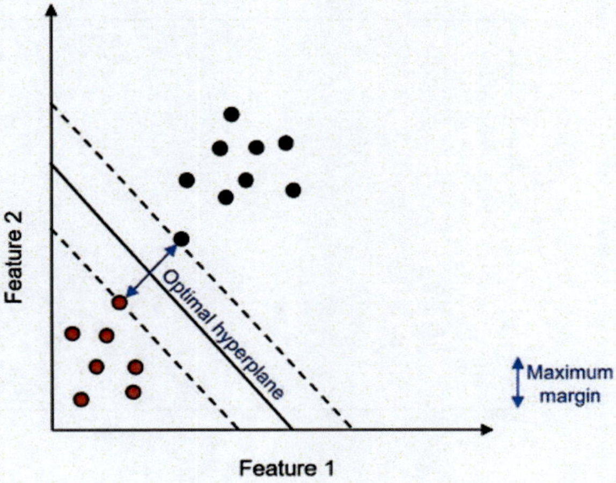

using these data points, the maximum margin is obtained. These data points also influence the position and orientation of the hyperplane.

But wait, the data (i.e., the input variables) shown in the figure is better to explain things more simply. However, the real-life data does not look so impressive. It is always messy and that is why machine learning models come in handy. For non-linear data, where a linear decision boundary cannot be determined, the data is mapped into a higher-dimensional space, where the decision boundary is found. This higher dimensional space, also referred to as feature space, is termed as kernel trick in SVM.

So, for the present case, two SVM models are built considering two features for each, energy and variance for one model, and the second with variance and entropy. The kernel trick utilised is a Gaussian kernel, which has been represented, as shown in Eq. 6.2 [43].

$$K(X_i, X_j) = \exp\left(-\frac{\|X_i - X_j\|^2}{2\sigma^2}\right) \tag{6.2}$$

where X represents the feature, i and j represent the two classes, and σ is the standard deviation. Figure 6.17 depicts the results obtained from the SVM model. In the picture, different markers can be observed, which represent training and testing data and the support vectors. In this case, the model was able to classify all the data points of the test set accurately, thus, having an accuracy of 100%.

A Real-Life Example in Manufacturing—Decision Tree

This section is dedicated to investigating the implementation of another supervised machine learning algorithm, a decision tree, in a manufacturing process. This model

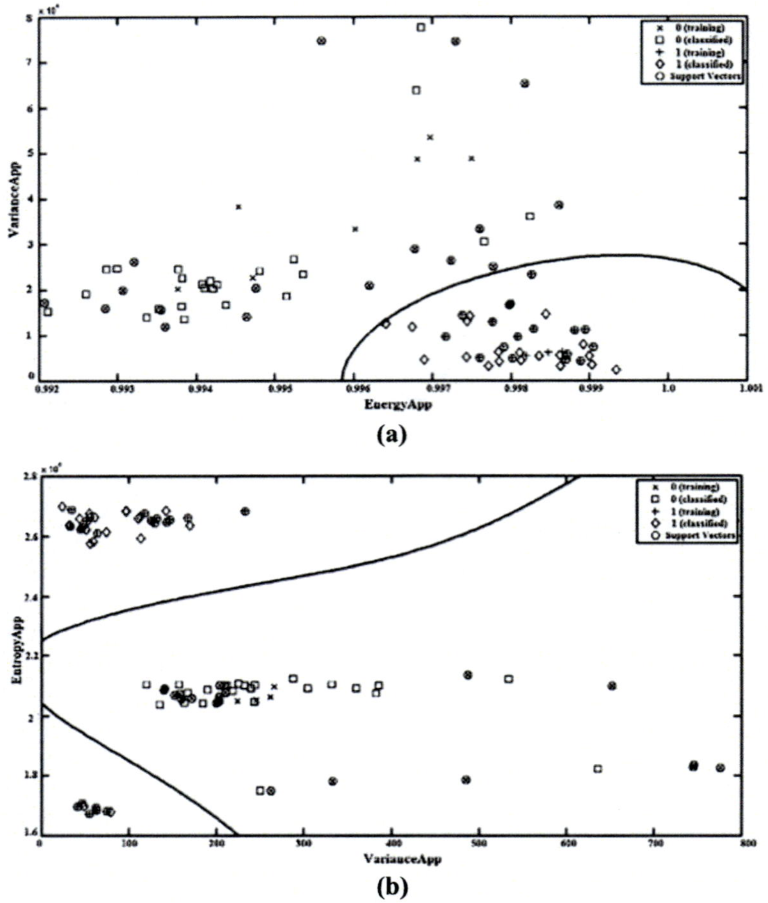

Fig. 6.17 SVM classification plots [43]. Reprinted from Journal of Manufacturing Processes, Volume 210, Nagraj N. Bhat, Kanchan Kumari, Samik Dutta, Surjya K. Pal, Srikanta Pal, Friction stir weld classification by applying wavelet analysis and support vector machine on weld surface images, 274–281, 2015, with permission from Elsevier

can be employed for both regression and classification tasks. Let us first understand the details of this model, followed by a discussion on the manufacturing example. As the name suggests, a decision tree model forms a tree structure for solving a problem. It is very much intuitive, the way human think. It is like asking a question, before making a decision. Refer to Fig. 6.18, a tree structure can be seen. The modelling begins from the 'root node' (shown in orange colour) and based on the dataset, it questions to form the tree. A 'node' (shown in blue colour) is the one that has inputs coming from the root node and also has an output. Finally, the 'leaf nodes' (shown in green colour) have inputs from a node. Let us consider a small example to understand

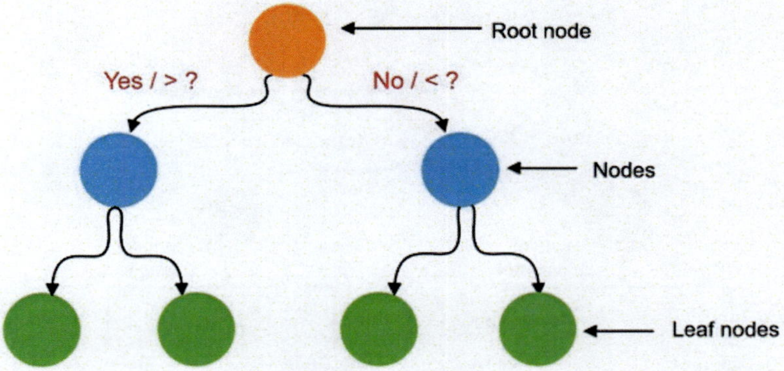

Fig. 6.18 A decision tree model

the concept. It is a dataset of a machining process that consists of two inputs i.e., temperature and surface roughness, and one output i.e., defect-free or a defective cutting tool. Thus, each row of the dataset has one value of temperature, one value of surface roughness, and the state of the cutting tool. A decision tree model, in this case, is for classifying the state of the cutting tool based on the available temperature and surface roughness data. For modelling, the decision tree can take either temperature or surface roughness. The question is how to decide this?

Let us consider, the decision tree takes temperature as the root node. This particular dataset will have only numeric values (defect-free can be considered as 0, and defective as 1). Keeping the temperature in the root node, the model investigates each row and determines the state of the cutting tool. Here the state of the cutting tool is the final output, so it will be termed leaf node. The model keeps on splitting the dataset to learn the state of the cutting tool based on the values of the temperature. This is referred to as pure subsets, wherein, the term 'purity' refers to the quality of splitting. If beyond a particular value of temperature, the state of the cutting tool is defective, then splitting will take a pure form. However, most of the real-world scenario is non-linear and thus result in impure splitting. To account for this, a decision tree model computes 'Gini impurity', and it is determined as shown in Eq. 6.3.

$$G_i = 1 - \sum_{k=1}^{n} {p_{i,k}}^2 \tag{6.3}$$

where $p_{i,k}$ is the ratio of k instances amongst the training instances in the ith node, and G_i is the Gini impurity. A node with $G_i = 0$ can be said to be pure. This means that for a particular split, all the training instances belong to the same class. Thus, to measure the impurity, the ratio is subtracted from 1, as shown in Eq. 6.3. Therefore, an effective splitting would result in a lesser division of data in a child node. This explanation for the case of temperature is shown in Fig. 6.19. For the sake of example, the splitting is considered to start with a value of 320, which forms the root node.

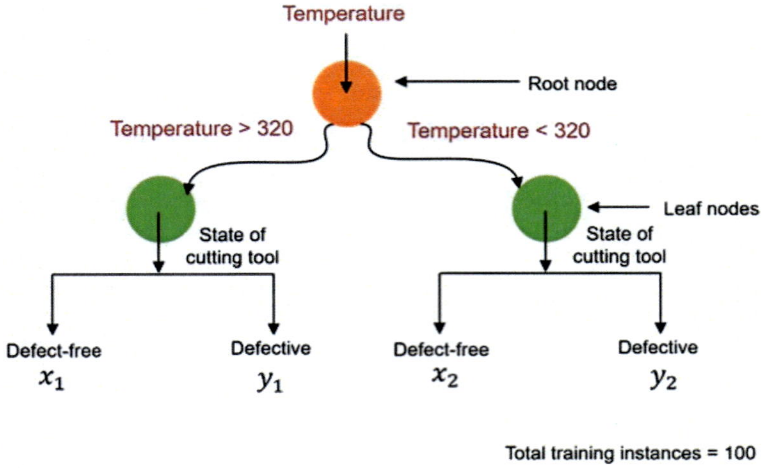

Fig. 6.19 Example tree structure

The first leaf node (left) checks the state of the cutting tool when the temperature is more than 320 and the second one (right) does the same for temperature less than 320. Considering the total number of training instances to be 100, the model lists the state of the cutting tool for this split. To measure the effectiveness of the split, G_i of the root node is determined. The same is shown in Eqs. 6.4, 6.5 and 6.6.

$$G_{i(\text{left})} = 1 - \left(\frac{x_1}{x_1 + y_1}\right)^2 - \left(\frac{y_1}{x_1 + y_1}\right)^2 \tag{6.4}$$

$$G_{i(\text{right})} = 1 - \left(\frac{x_2}{x_2 + y_2}\right)^2 - \left(\frac{y_2}{x_2 + y_2}\right)^2 \tag{6.5}$$

$$G_{i(\text{rootnode})} = \frac{x_1 + y_1}{100} G_{i(\text{left})} + \frac{x_2 + y_2}{100} G_{i(\text{right})} \tag{6.6}$$

where x_1, y_1, x_2, y_2 are the number of instances, $G_{i(left)}$ is the Gini impurity index of the left leaf node, $G_{i(\text{right})}$ is the Gini impurity index of the right leaf node, and $G_{i(\text{rootnode})}$ is that of the Gini impurity index of the root node. The Gini impurity index of the root node is a weighted average of the indexes of the leaf nodes. In this way, the model evaluates for each input parameter and the one with the lowest impurity forms the root node. Based on the selection of the root node (also called parent node), child nodes are formed. In this manner, the tree structure grows. An important parameter in decision tree modelling is the depth of the tree. The depth of a tree is specified as per Eq. 6.7.

$$N = 2^D \tag{6.7}$$

where D is the depth of the tree and N refers to the number of terminal nodes. Therefore, with a depth of 1, the number of terminal nodes will be 2, as shown in Fig. 6.19. However, as the depth of the tree increases, the terminal nodes increases, which thereby increases the complexity of the tree structure. Besides Gini impurity, informational entropy is also employed to determine the purity index of a node. Entropy can be defined as the degree of randomness and it is computed as per Eq. 6.8.

$$H_i = -\sum_{\substack{k=1 \\ p_{i,k} \neq 0}}^{n} p_{i,k} \log_2(p_{i,k}) \tag{6.8}$$

where H_i is the informational entropy. One question, at this point, on the index which is to be employed? Both Gini impurity and informational entropy measures the impurity index, however, the former makes in purer subsets, while the latter in the balanced ones. Also, Gini impurity is computationally less expensive than that of the informational entropy. As highlighted, the depth of the tree is one of the parameters of this model. Usually, if the depth of the tree is not fixed, then during modelling, the tree goes on expanding, and it models the training data perfectly, resulting in overfitting. This is also referred to as a non-parametric model. Therefore, to avoid overfitting of the model, the depth of the tree should be fixed and the same is passed during modelling. Thus, the model will generalize i.e., it will learn. There are other parameters to control as well, which includes, maximum nodes, maximum leaf nodes, samples for splitting, etc. In the following paragraph, let us implement a decision tree algorithm to a real-life manufacturing example.

An example is a classification of bearing health condition. For this, the experimental set up consists of an induction motor that is connected to a hydraulic power pack. This induction motor has two bearings, one on the drive-end and the other on the fan cover end. A vibration sensor (accelerometer) is fixed on the induction motor that acquires the vibration signal while the motor is operating. A picture of this set up is already shown in Chap. 2 (Fig. 2.61) and Chap. 3 (Fig. 3.15). The bearings selected for the experiments were of three different health conditions namely, jammed, broken cage, and healthy. The first two conditions resemble faulty bearings and the last one is for a healthy bearing. Time-domain features such as mean, RMS, standard deviation, maximum, and minimum were collected from the vibration signal for fault analysis. The objective is to develop a machine learning model, a decision tree in this case, for classifying the health of the bearing by using the aforementioned features of the vibration signal.

For modelling, the dataset consisted of 5 features of the vibration signal i.e., mean, standard deviation, RMS, maximum, and minimum of the signals acquired for different health conditions of the bearings. Three labels are included, which correspond to the health conditions, 'class = 0 for broken', 'class = 1 for healthy', and 'class = 2 for jammed'. The modelling was performed as discussed in this section. For the sake of understanding, two approaches are considered, first, a non-parametric model. This means there was no control over the parameters of the model.

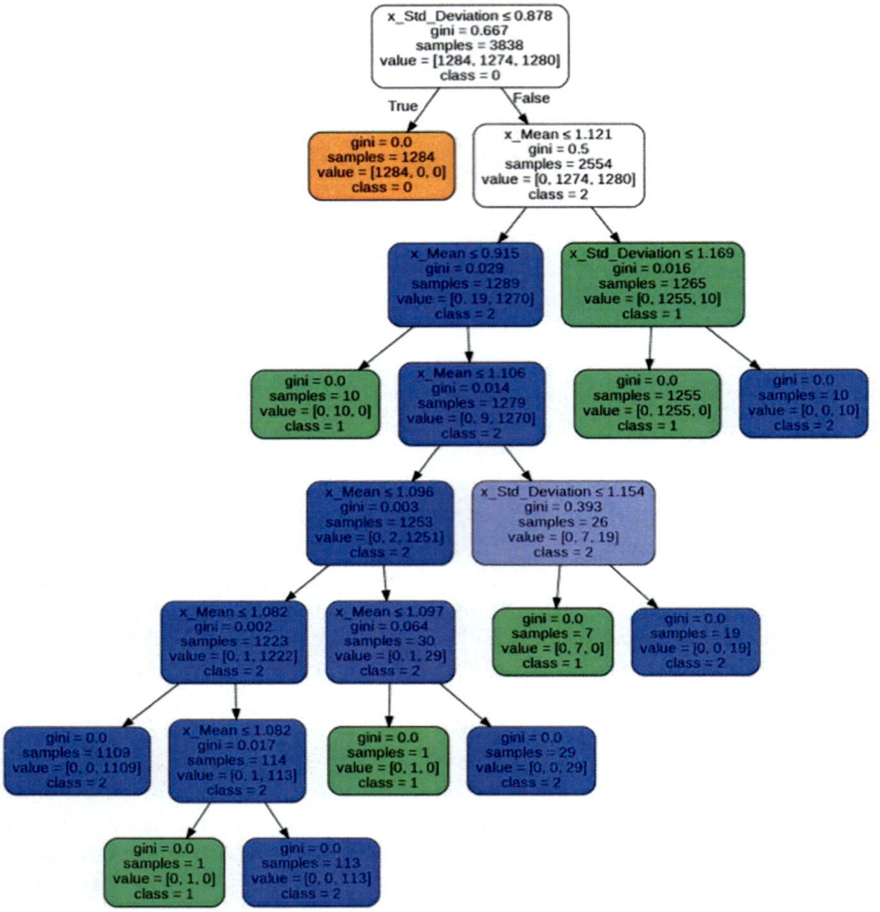

Fig. 6.20 Decision tree model for the example

Gini impurity is considered for determining the purity of the nodes. Figure 6.20 depicts the decision tree model for this case. It can be observed that out of the five features present in the dataset, the standard deviation is having the least impurity and thus, it is considered at the root node. The accuracy of this model is 99.69%. This might be a case of overfitting as there is no control over the depth of the tree. Moreover, a decision tree is considered a 'white box', which means that it is intuitive and interpretable. However, as the depth of the tree, and the number of nodes, and the leaf nodes goes on increasing, it becomes less explainable.

To address the above, the second approach considers controlling the depth of the tree. As an example, let us consider the maximum depth of the tree to be 3 and index to be entropy. In this manner, the model will be controlled and also the sets will be balanced because of the informational entropy measure. Figure 6.21 depicts the

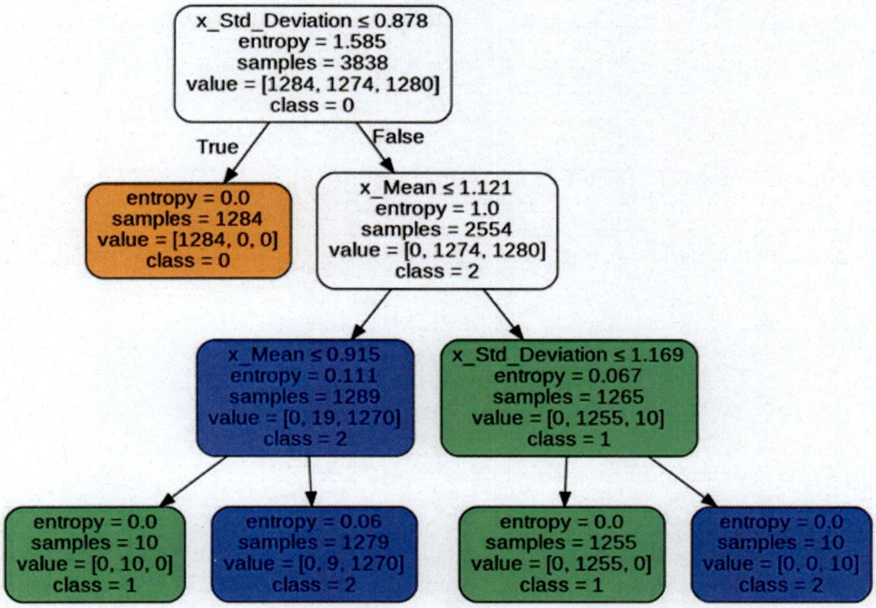

Fig. 6.21 Decision tree model for the example

decision tree obtained with this approach. As the depth is 3, the number of terminal nodes is 8, as shown in the picture. The accuracy of the model, with this approach, is 99.57. It can be observed that the classification is carried out with two features namely, standard deviation and mean of the vibration signal. This indicates that these two features will be on top of the list if ranked.

Thus, a decision tree is a white box model and it can capture non-linearity existing in data. Since the model has a non-parametric approach, it does not have any assumption on the distribution of the data. However, it can be prone to overfitting if its parameters are not controlled.

A Real-Life Example in Manufacturing—Random Forest

Random forest algorithm is a tree-based, supervised learning algorithm, which operates on the "ensemble learning" method that refers to a collection of models (decision trees) for learning a dataset, as presented in Fig. 6.22. The final prediction is either the majority or the weighted average of the sub-models. In a real-world scenario, it works similar to an individual referring to the advice of multiple experts before finalising a decision, and the final decision is based upon the majority. These sub-models are referred to as "base models". The ensemble learning consists of two families, one is bagging, and the other is referred to as boosting. The random forest model operates on the bagging method, which resembles the building of multiple independent

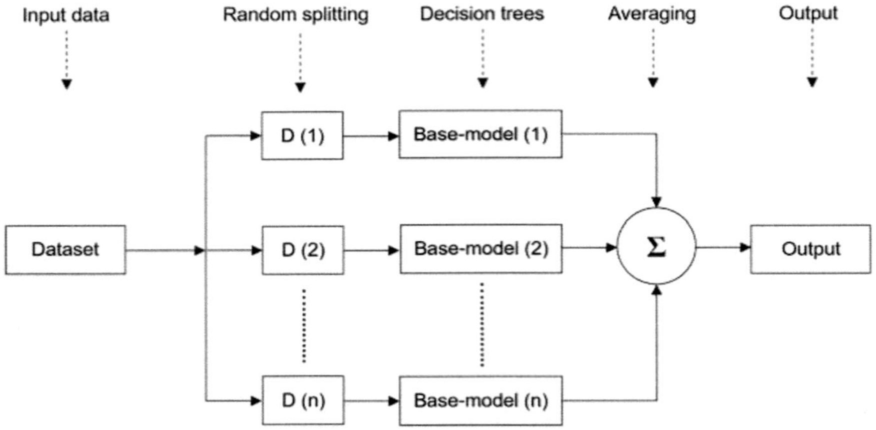

Fig. 6.22 Figure depicting the method of random forest model

estimators for prediction, and the final prediction is the average of all the predictions made by these estimators. One of the advantages of the random forest is its ability to reduce the variance without increasing the bias. Thus, a better result can be expected by using random forest without having a trade-off between the variance and the bias. Let us discuss the difference between random forest and decision tree. As discussed in the previous section, a decision tree generates a set of rules for learning on a given dataset. Random forest, on the other hand, build several decision trees with a given dataset and finally averages the individual results to produce the output. Thus, the number of trees becomes one of the important parameters during the modelling of a random forest.

A random forest model can be utilized both for regression and classification. It is advantageous in a way that it produces a good predictive result. But it is a black box model, not a descriptive one. It produces good results as it considers numerous decision trees, where each of them takes random samples from the dataset and produce the result. Thus, the chances of overfitting in a random forest are very rare. However, owing to these numerous decision trees, the algorithm slows down making it difficult for implementing in real-time applications. In the following paragraph, let us try implementing this model to a real-life example i.e., to predict the grain size of a welded sample.

This is a regression task. The prediction of grain size is important because welding is the last process in the sequence of manufacturing. Thus, it is highly essential to ensure that the weld-joint is error-free. This indicates the weldment must have the required mechanical and metallurgical properties and is free from welding defects. Owing to the heat generated in welding, the properties of the weldment differ from that of the parent material. Ideally, a weldment must possess similar properties across

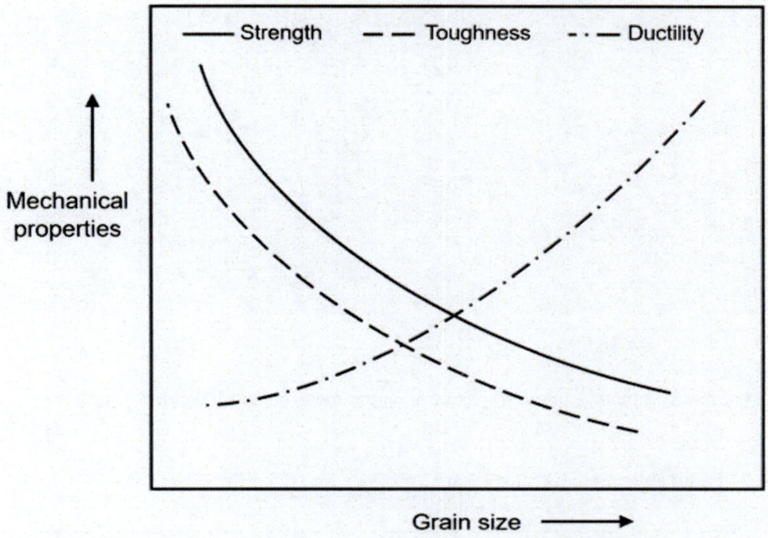

Fig. 6.23 Picture depicting the relationship between mechanical properties and grain size

the joint, i.e., across the zones of welding. However, this does not exist. A variation occurs from the weld zone to the heat-affected zone when compared with the parent metal. Grain size has been one of the important factors for ensuring the quality of a weldment.

Figure 6.23 shows the general relationship between the mechanical properties and the grain size. The mechanical properties namely, strength and toughness can be observed to be decreasing with an increase in the grain size. The variation in the grain size or growth of the grain is dependent on the heat input. New grains generate by absorbing each other as the temperature rises above the recrystallization temperature. The resultant grain structure depends upon the temperature and the time of exposure and usually can be fine or coarse. While a fine-grained microstructure is beneficial, the coarse grain size is undesirable for the mechanical properties and weldability. This marks the importance of exploring grain size in the weldments. Thus, in this example, a machine learning model i.e., a random forest model is built for predicting the grain size of a weldment by utilizing the welding data. This is crucial because of the close relationship of the grain size with the mechanical parameters.

For the acquisition of the data, a force sensor (piezoelectric dynamometer by *Kistler, 9124B1111*) was utilized for sensing the force signal. The sensor was mounted on the spindle of the FSW machine (a vertical milling machine by *BFW, V3.5*). A picture of the FSW machine fitted with the piezoelectric dynamometer has already been shown in Chap. 2, Fig. 2.42. AA6061 (an alloy of aluminium), was chosen as the base material. Aluminium is one of the popular metals because of its characteristics such as low density, high strength, high thermal and electrical conductivity, resistance to corrosion etc. The parametric combinations opted for welding have been tabulated

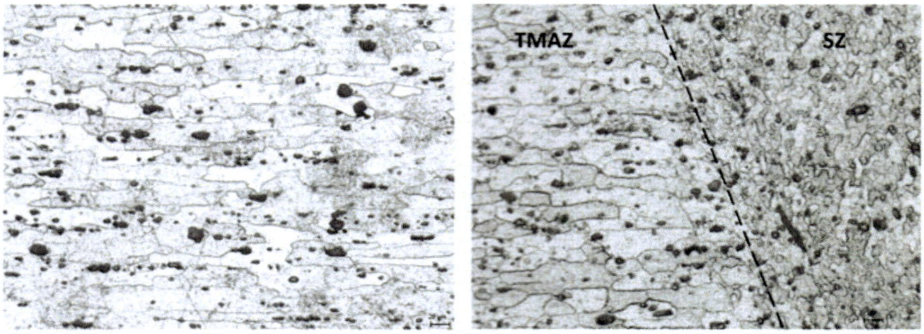

(a) (b)

Fig. 6.24 Picture depicting micrographs of, **a** parent material and **b** welded sample

Table 6.2 Parameters selected for welding

Expt. no	1	2	3	4	5	6	7	8	9	10	11	12
ω (rpm)	710			900			1120			1400		
v (mm/min)	50	160	200	50	160	250	50	80	200	63	80	315
α (°)	0, 1.5, and 2.5											

Expt.—Experiment, ω—Rotational speed, v—Welding speed, α—Tilt angle

in Table 6.2, which consisted of 36 numbers of experiments. To cover a wide range of parametric combinations, no proper experimental design was followed. While rotational speed (ω) was varied from 710 to 1400 rpm, welding speed (v) was varied from 50 to 315 mm/min, and tilt angle (α) from 0° to 2.5°. These 12 numbers of experiments shown in Table 6.2 were repeated for each of the values of α resulting in 36 experiments, Expt. no. 1–12 for 0°, Expt. no. 13–24 for 1.5°, and Expt. no. 25–36 for 2.5°. After fabrication, a knowledge base was created which consisted of the variation of input parameters and the data collected from the sensor. Post the weld fabrication, the average grain size was determined for all the welded samples.

The occurrence of recrystallization during deformation leads to grain refinement. The grains in the stir zone experience an extreme level of plastic deformation and heat generation because of the rotating tool and translating bed, which results in a remarkable change in the grain size. This is followed by thermo-mechanically affected zone (TMAZ) and heat-affected zone. Figure 6.24a, b depicts the micrographs of the parent material (AA6061) and welded sample, respectively. While larger sized grains can be observed for the parent material (80.03 μm), there has been a remarkable reduction in the grain size in the case of the welded sample (8.82–23.38 μm).

Figure 6.25 shows a 3D map of the distribution of the grain size obtained with the parametric combinations opted in this study. The grain size increased with increasing values of ω keeping v and α constant, and it decreased with increasing values of v keeping ω and α constant. As mentioned earlier, the grain size is affected by

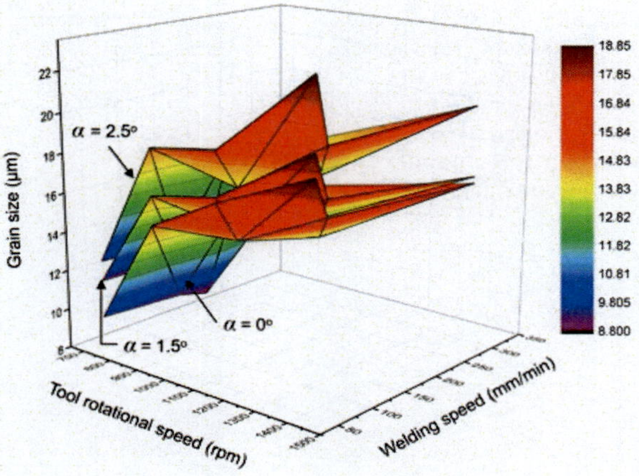

Fig. 6.25 Variation of grain size with opted parametric combinations

deformation and heat. With increasing values of ω, the degree of deformation and heat generation will be higher owing to the higher level of interaction between the tool and the workpiece. On the other hand, these factors reduce with increasing values of v keeping ω and α constant. Thus, high deformation and heat are probably the reasons behind grain growth. In such a case, the heat input to the weld-joint will decrease, thereby decreasing the grain growth (for ω being constant). Further, the variation in the grain size with change in the values of v was found to be smaller as compared to the change because of the values of ω, which suggests that v has a lesser effect on the grain size than ω.

One more observation which was noted is that at lower $\omega = 710$ rpm, with change in α from 0° to 1.5°, the change in grain size is 19.95%, whereas from 0° to 2.5° it is 27.73%. Although, at a high value of ω (1400 rpm), with change in α from 0° to 1.5°, the change in grain size is almost negligible, i.e., only 2.12% whereas from 0° to 2.5°, it is 24.51%. The possible reason for this variation could be the effect of ω which is sufficiently high for plastically deforming the workpiece. Thus, tilting the tool by 1.5° does not create much of a difference, however, with further tilting by 1° (i.e., 2.5°), there occurs a huge change.

Let us discuss the details of the modelling that has been performed on the welding data for predicting the grain size. The analysis began with understanding the available inputs for the target variable. Figure 6.26a–c show the variation of the obtained grain size values with ω and α; with v at selected values of ω and α; and with "feed per revolution" (FPR) at selected values of α, respectively. FPR is the ratio of the feed of the machine bed per revolution of the tool. This ratio will be higher with an increase in the feed of the machine bed per revolution of the tool. A non-linear relationship can be observed from these plots. Figure 6.26d presents the change in grain size values along with the processed force signal at various parametric combinations. The clusters of the obtained grain size values, as shown in Fig. 6.26d, can be observed to be having some correlation with the force data. However, the trend of the two, i.e., the force and grain size is not similar because of their behaviour difference in the case of

Fig. 6.26 Variation of grain size with, **a** rotational speed at various tilt angles, **b** welding speed at various tilt angles and rotational speeds, **c** FPR at various tilt angles, and **d** force values

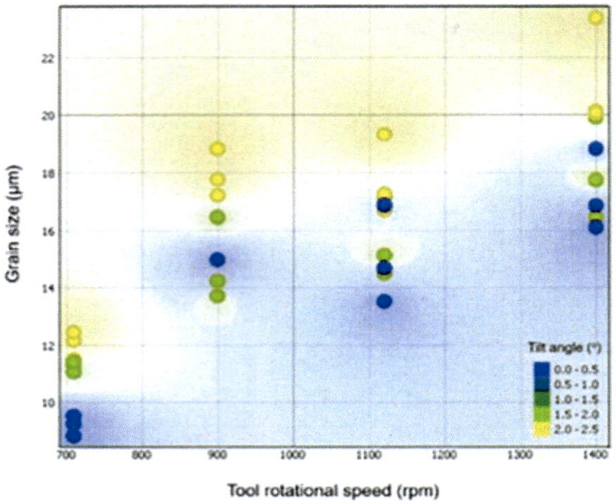

(a)

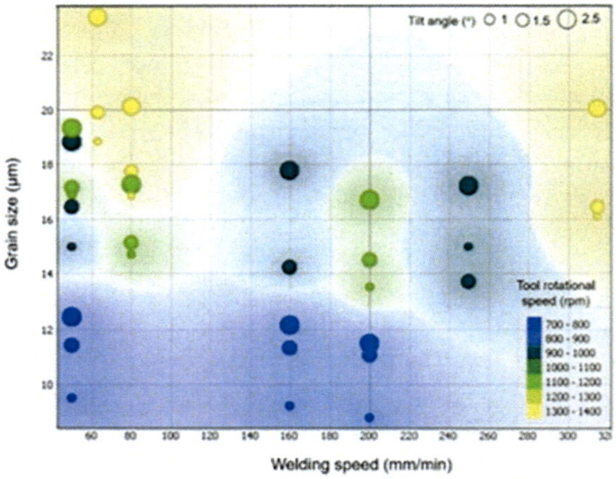

(b)

the input parameters of welding. The force is the reaction that reflects the amount of deformation the material is experiencing and will decrease with increasing ω or decreasing v, or the force will increase with low values of ω and high values of v.

To quantify the interaction among the variables and the target, the correlation coefficient was determined. This method, i.e., determining the correlation coefficient is a feature selection technique under the category of "filter method". "*Pearson correlation*" has been determined which is obtained by dividing the covariance of two variables with the standard deviation of the same. It is expressed in Eq. 6.9.

Fig. 6.26 (continued)

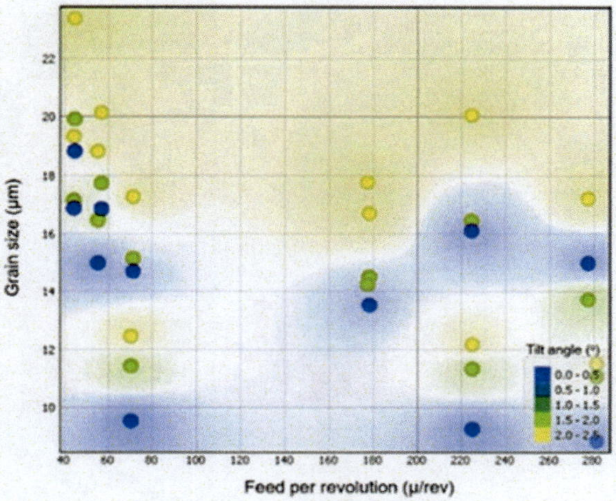

(c)

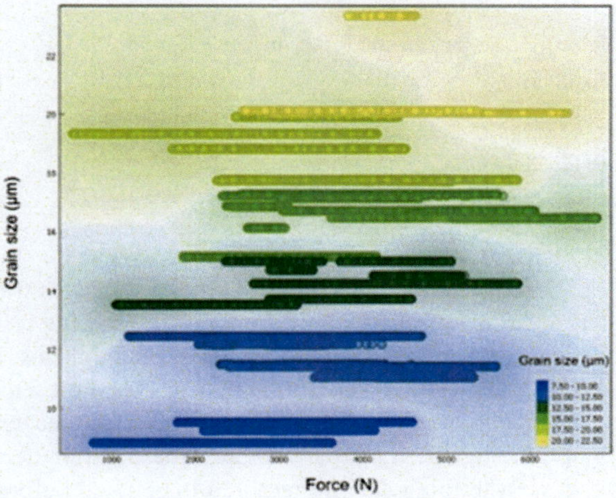

(d)

$$r = \frac{\sum(x_i - \overline{x})(y_i - \overline{y})}{\sqrt{\sum (x_i - \overline{x})^2 \sum (y_i - \overline{y})^2}} \tag{6.9}$$

where x_i denotes the values of the x variable, y_i denotes the values of the y variable, $\overline{x}$ is the mean of the values of the x sample, $\overline{y}$ is the mean of the values of the y sample, and r is the *Pearson correlation coefficient*. Figure 6.27 shows the colour map of the coefficients obtained. ω has the highest correlation in the positive direction with the grain size followed by force and α. v and FPR have a negative correlation.

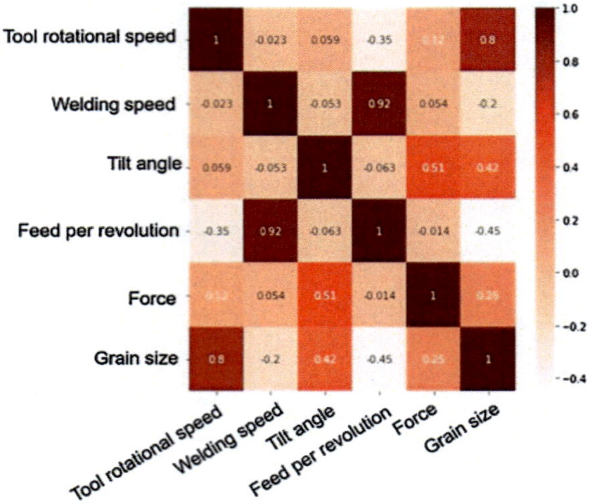

Fig. 6.27 Correlation map showing the relationship amongst the variables

Table 6.3 Testing dataset for the machine learning model

Expt. no	6	10	15	16
Parameters $(\omega\text{-}v\text{-}\alpha)$	900-250-0	1400-63-0	710-200-1.5	900-50-1.5
Expt. no	20	29	32	35
Parameters $(\omega\text{-}v\text{-}\alpha)$	1120-80-1.5	900-160-2.5	1120-80-2.5	1400-80-2.5

The dataset was divided into two halves, first for training and the other for testing. Out of 36 experiments, 8 of them constituted the testing part, and the rest 28 sets were used for training a random forest model. Thus, the split was 77% for training and 22% for testing. The testing part has been presented in Table 6.3.

In machine learning, one of the practices that help in achieving a good result is optimizing the hyperparameters. These parameters are those which govern the process of training and are required to be fixed before the beginning of the training of the machine learning model. One of the most important hyperparameters in a random forest is the number of trees considered in the forest. For this, a subset from the training dataset was randomly selected and the rows of the same were randomly shuffled. A threefold cross-validation technique was applied to this subset to determine the optimum number of trees, which were 200 in number. This result was then applied to the training dataset and the testing dataset. Figure 6.28 compares the actual and predicted values of grain size in the case of the testing dataset. The predictions of the model have been tabulated in Table 6.4, from which, the absolute average percentage error for predicting the grain size can be observed to be 4.97.

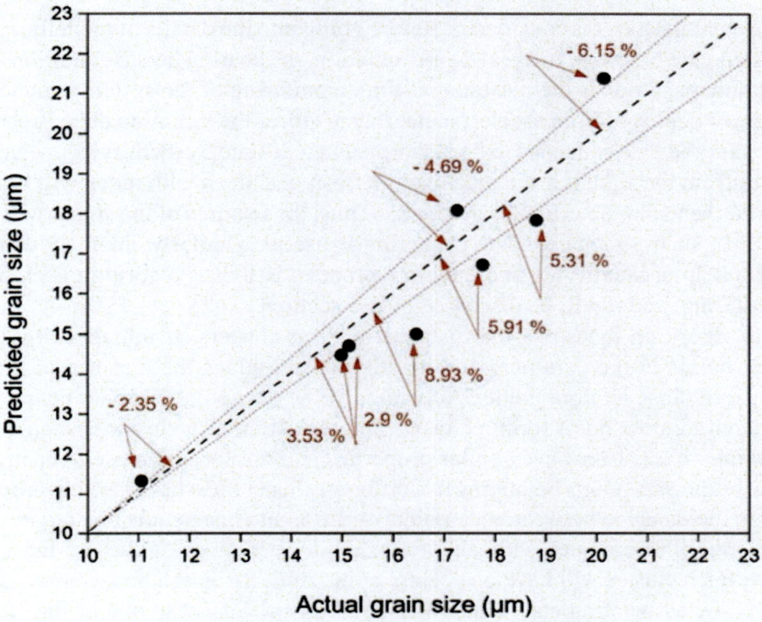

Fig. 6.28 Comparison between actual and predicted values in case of testing data

Table 6.4 Predicted values obtained from the model on testing data

Expt. no	Actual grain size (μm)	Predicted grain size (μm)	Error (%)
6	14.99	14.46	3.53
10	18.82	17.82	5.31
15	11.06	11.32	−2.35
16	16.46	14.99	8.93
20	15.14	14.7	2.9
29	17.76	16.71	5.91
32	17.26	18.07	−4.69
35	20.14	21.38	−6.15
Absolute average error (%)			4.97

A Real-Life Example in Manufacturing—Clustering

This section will bring the importance of unsupervised learning through a case study. As highlighted, in an unsupervised machine learning algorithm, the training data is unlabelled. This learning system is not for performing tasks such as regression or classification, rather, it is employed for clustering, anomaly and novelty detection, visualization and dimensionality reduction, etc. To illustrate this learning system, the bearing health dataset is considered. Before going into the details of modelling, let us understand the problem. In a real-world situation, the bearing health condition would be unknown and this is the reason, real-time monitoring of the system is inevitable. The sensor deployed is an accelerometer that acquires the vibration data. Processing of the acquired data is needed to be performed in real-time by identifying meaningful features from the signal. If the machine is a fresh one, the health index will be good, however, the health deteriorates gradually. Thus, the features of the signal will differ as well. In such a scenario, data clustering is useful. Similarly, anomaly detection can also help in identifying anomaly in a process. K-means algorithm is employed for clustering, and it will be discussed in this section.

This algorithm measures the distance to form clusters. It initializes by taking random points in the data map and iteratively determines the distance of the data points from those random points. Accordingly, the data points that are nearer to the considered random point form a cluster. The formation of a cluster is such that the data points in a cluster have similar properties and are located close to each other. Similarly, the data points belonging to a different cluster must have varying properties and also the distance between data points of different clusters must be large.

The algorithm initiates with, say, for example, k number of clusters for a given dataset. Each cluster will have one centre point, thus, there will be k centres. This is followed by associating each data point in the given dataset to one of the clusters, to which it is nearest. This is performed by calculating the squared distance of each data point from a centre point, and it is assigned to the one to which it is the nearest. This is followed by the determination of the centroid of each cluster. The objective function, which is being solved in this algorithm is expressed in Eq. 6.10.

$$J = \sum_{i=1}^{m} \sum_{k=1}^{k} w_{ik} \|x_i - \mu_k\|^2 \tag{6.10}$$

where x_i refers to the data points in the given dataset, μ_k refers to the centroid of the clusters. The approach followed by this algorithm is also referred to as the "*Expectation-Minimization*" problem, wherein, the expectation is for assigning the data points to the clusters, and minimization is for determining the centroids of the clusters.

6.7.1.3 Other Variants of Machine Learning—Reinforcement Learning

The reinforcement learning system works by rewarding. It is mostly employed to obtain the best possible behaviour or path in a particular situation. In this case, the data is unlabelled, so the model learns by itself. In this case, the model is referred to as an *agent*, as shown in Fig. 6.29. Another term in reinforcement learning is the *environment*, which is the world, in which, the agent performs the task. The learning begins with an initial *state* and it has multiple routes to approach the final destination.

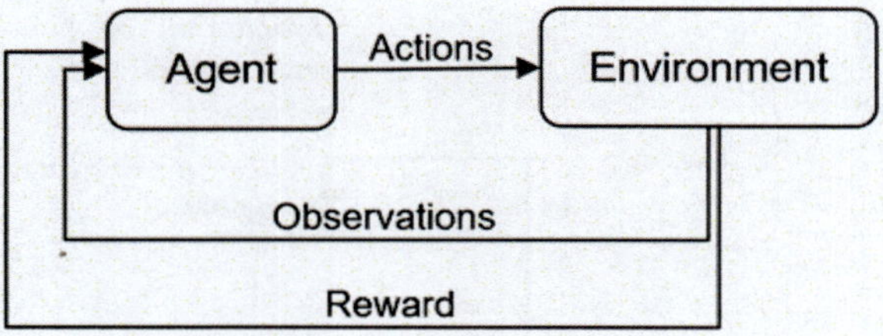

Fig. 6.29 Schematic picture of reinforcement learning

The training is based on the input, and the model returns a state, which is validated by the user by rewarding or punishing. In this way, the model continues to learn from its experiences what actions yield long-term cumulative rewards, and thus, the best solution is decided based on the maximum reward.

The reinforcement learning technique is best suited for robotics application where the best possible path from a start point to a final destination is to be learnt by the robot. In such a case, the robot may be fitted with a camera to acquire images of the path. The training instances will be the acquired images that are processed by the model to find the best possible path. The feedback to the model by the user enables it to learn the best path to be followed.

6.7.1.4 Batch Learning and Online Learning

This categorization is based upon the fact that whether a machine learning model can learn incrementally. For the first i.e., batch learning, the model does not learn incrementally. In this case, the model is trained offline using the entire available dataset. Upon successful modelling, the trained model is deployed in the working environment to predict/classify the new incoming data. Thus, the model applies what it has learnt from the training dataset. This sort of modelling is also referred to as offline learning. This type of learning requires generating a huge amount of data, and the modelling may take several hours. Most often, this sort of model is updated weekly/daily with a new dataset, which is performed offline. A schematic of this learning is represented in Fig. 6.30.

In contrary to the batch learning mode, an online learning system enables the machine learning model to learn incrementally. In this case, the data to the model is fed in small batches. The model keeps learning as the new data comes in. This sort of learning is more suitable for systems that generate data continuously. Particularly, for manufacturing operations, online learning will be more suitable for making predictions about the process conditions. Health is an index that never upgrades, it rather deteriorates as time passes by. This indicates the life of the mechanical components employed in a machine, which deteriorates with time. This will get reflected in the data, which would be acquired using various sensors depending upon the type of manufacturing process. Hence, the machine learning model that would be deployed in this case for performing monitoring/prediction, must learn and update incrementally. This will ensure the efficiency of the model in making useful predictions. One

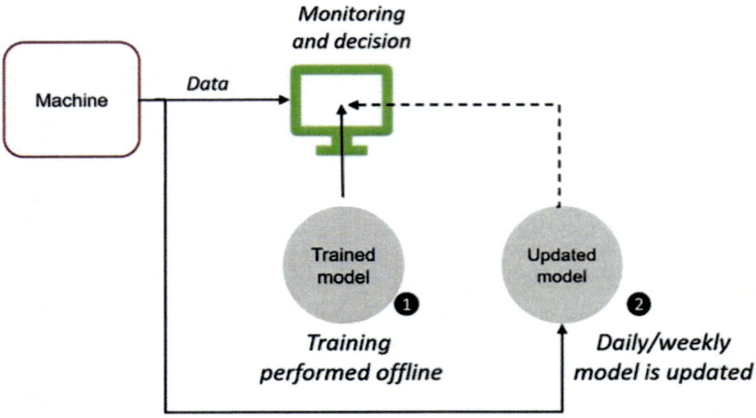

Fig. 6.30 Schematic picture of batch learning

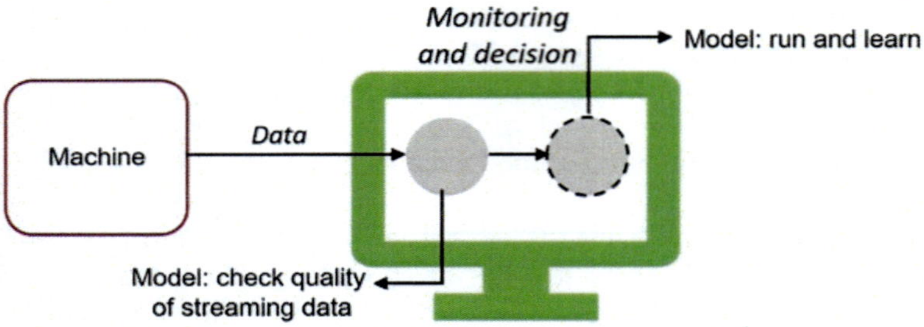

Fig. 6.31 Schematic view of online learning

challenge in online learning mode could be posed by the quality of the data entering into the model. Imagine a scenario where a sensor that is engaged in a machine for monitoring the process fails or starts behaving erroneously. In this case, the quality of the data that will go into the model will be poor and will result in wrong predictions. In such cases, the quality of the streaming data is required to be ensured before feeding it to the system. A schematic of this learning is represented in Fig. 6.31.

6.7.2 Artificial Neural Network

Artificial neural network (ANN) is one of the most popular machine learning algorithms. History explains the art of machine learning as a model of brain-cell interaction. This was told by *Donald O Hebb*, who is known as the *father of neural networks*. The algorithm consists of a set of interconnected neurons that transfer data and try to learn a system.

Let us first pay attention to the human nervous system before introducing the context of ANN. Do not get surprised, the section still aims to explain ANN only, however, this algorithm has something to do with the human nervous system. Thus, let us explore. The human nervous system consists of three things, first is the central nervous system, second is the peripheral nervous system, and third is the autonomic nervous system. The central one is the most complex and it contains the brain and marrow. The brain is composed of numerous neuron units connected via axons and dendrites and the connecting region between axon and dendrite is referred to as synapses. A representational image of the biological neuron is depicted in Fig. 6.32. The main function of the brain is to control the activity of a human, based on the processing of information. The processed information is passed on to the peripheral nervous system, which has contacts with the sensors and effectors. The autonomic system supervised the activity of the internal organs. Thus, the brain is the most crucial thing in the entire architecture of the nervous system. The basic element, which a brain consists of is a neuron. Thus, numerous neurons make up the brain.

So, what is happening inside the brain? Well, the so-called neuron receives a signal and produces a response. Whenever an electrical impulse arrives at the synapse, it activates the transmission of the signal via the dendrite to the cell. Thus, when numerous signals are accumulated in the cell via the dendrites as shown in the picture of the biological neuron, it causes the neuron to produce an output, which is the action of the axon. This interconnection forms the basis of neural information processing. *Donald O Hebb* also explained that when one of the cells repeatedly assists in firing another, the axon of the first cell develops synaptic knobs in contact with the soma of the second cell.

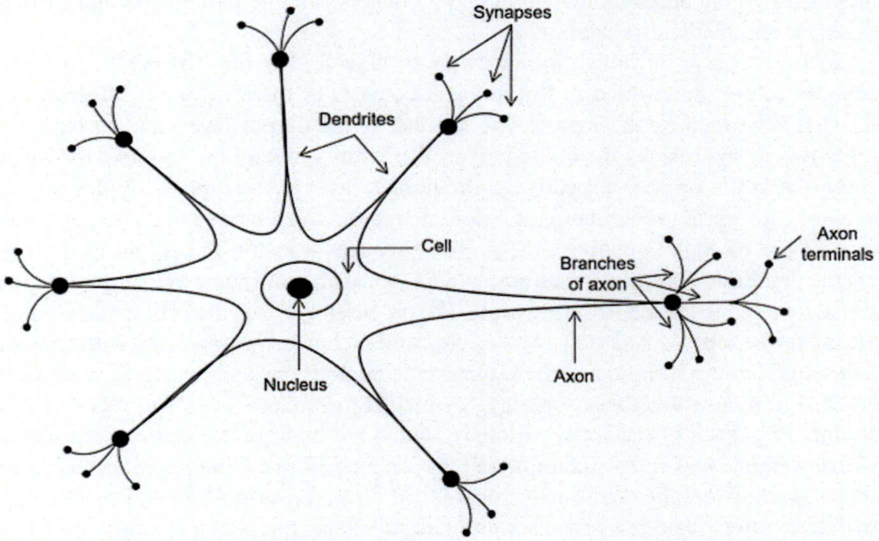

Fig. 6.32 Representation of a biological neuron

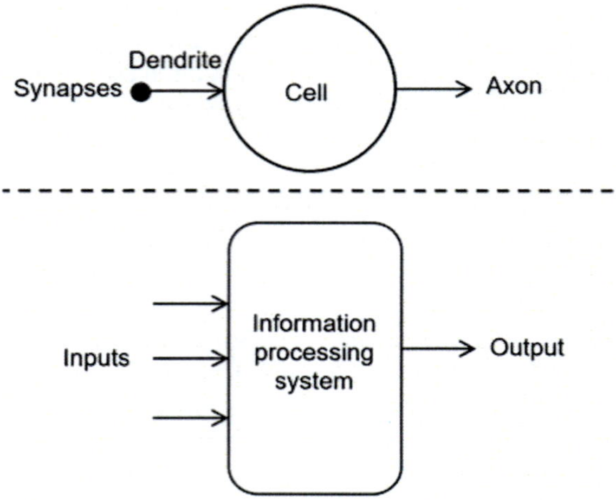

Fig. 6.33 Representation of an artificial neural network

This biological mechanism forms the core context of the ANN. Since it is attempted to be simulated artificially, it is titled so. Figure 6.33 depicts a representational picture of the ANN, which mimics the mechanism of the biological neuron. It consists of inputs, an information processing system, and output.

Very similar to the structure of the brain, ANN consists of multiple neurons, and that is one reason why these models come up with excellent results. Typically, their use for computations makes them referred to as neurocomputing. So, by now, it is clear why the discussion began on the human nervous system. Well, it is quite interesting to understand how engineering concepts can be understood and related to nature and medical-related fields.

With this basic introduction, let us try to dive further into the ANN, with the basic structure represented in Fig. 6.34. It consists of three layers of information, the first is termed as the input layer, the last is the output layer, and in between these two layers resides the hidden layer. While the first and the last take the input and provide the output, respectively, the hidden layer is the processing unit of the network. Recap the mentioning of external signal, which comes to the synapses in case of the biological neural system. The synapses transmit the signal to the cell via the dendrites. Similarly, the training data (consisting of inputs and outputs) is the signal here, which is fed to the network. The processing unit is the cell. In the picture, the inputs are represented as $x_1, x_2, \ldots, x_n$, and each input is associated with another unit termed as the weight, and the same is represented as $w_1, w_2, \ldots, w_n$. The input layer is going towards the second layer, which is the hidden layer, and it consists of neurons. So, each of the input variables, which are input to the neurons are scaled with a weight. Next is the summing unit, which multiplies the input variable with the corresponding weights and finally sums it up. Thus, the associated weight function will determine the output of the summing unit. Hence, the neural network, which is shown in Fig. 6.34, is trying to learn a function from the provided input variables by altering the associated weights.

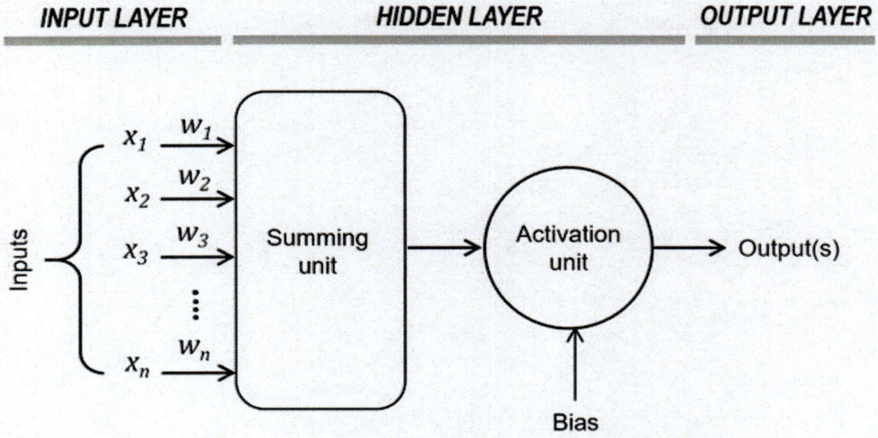

Fig. 6.34 Basic structure of an artificial neural network

Till this point, the discussion has given a shape of what a neural network is and the importance of data, which is fed to the network. The next important thing in modelling ANN is the activation function, as shown in the figure. Comparing with other machine learning models, neural networks also take data as input, which consists of instances of features and labels. Naturally, the model tries to learn from the corresponding features and labels fed to it. So, the question is why an activation function is included in the neural network? Well, the foremost reason is to address non-linear modelling. By this time, one must have got an idea that neural networks are complex, and this makes it popular for modelling highly non-linear data. This is fulfilled by employing an activation function. This function takes the weighted output of the neurons and decides what will be the input to the next neuron.

Let us visualize through an example, what a neural network will look like when it is employed for some computation. For this, let us consider the example of predicting the weld quality using features of arc signal with the help of an ANN [44]. Here are the details of the problem statement. The manufacturing activity involves the joining of two mild steel specimens using the gas metal arc welding technique. Conventionally, to determine the strength of the weld, one needs to extract a tensile sample from the weld, followed by testing the same in a universal tensile tester. Alternatively, in this work, a neural network is modelled to predict the joint strength of the welded sample by feeding features of the arc signal. During the experiments, the signal has been collected using a current sensor. The neural network designed for this purpose consisted of multiple neurons for feeding the inputs, multiple hidden layers, and one output neuron for predicting the weld tensile strength. Figure 6.35 depicts the architecture of the developed model. It consists of eight neurons in the input layer and the parameters as inputs are pulse voltage, background voltage, pulse duration, pulse frequency, wire feed rate, welding speed, RMS current, and RMS voltage. The hidden layer consists of multiple hidden layers (2 shown in the picture) with multiple

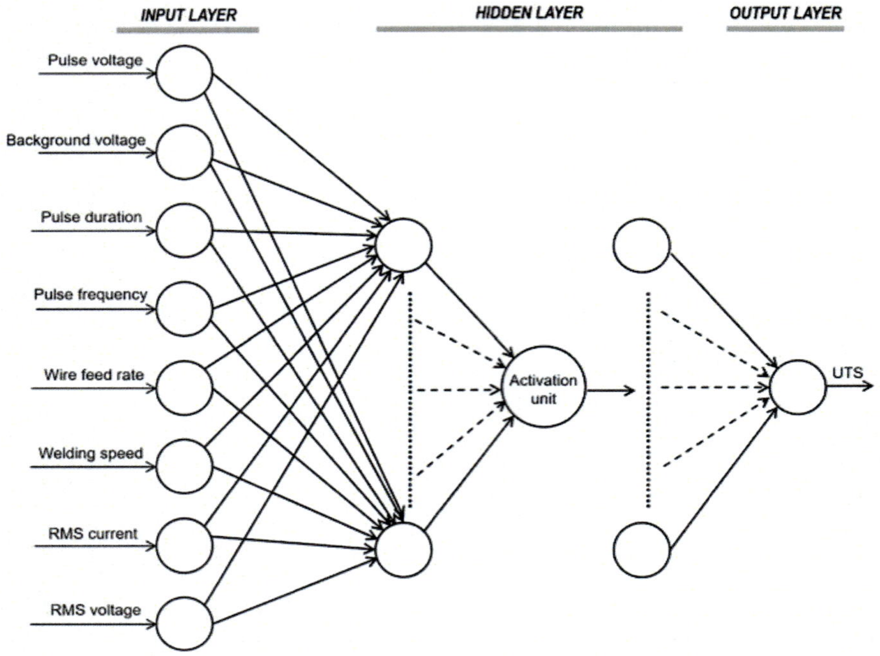

Fig. 6.35 Architecture of the developed artificial neural network

neurons in each layer, and finally, the output i.e., the tensile strength of the welded sample (shown as UTS).

Let us try to pose the research questions on the architecture shown above followed by a discussion on each of them. The research questions will help to understand neural networks and the associated terms better.

The first question could be the choice of input features. As discussed previously, inputs are those features, which have a certain correlation with the desired output. In the present problem, the objective is to predict the tensile strength of the welded sample, which is affected by the selected parameters. The second question is the selection of the number of hidden layers and the numbers of neurons in each hidden layer. This question in particular relates to the architecture of the neural network because the hidden layer forms the neural processing system. The distribution of this layer i.e., the number of layers and the number of neurons in each layer is the main deciding factor for the performance of the network. To begin the process of learning the network, the weights were initialized and were kept in the range of ± 0.9. Recap Fig. 6.34, where a bias is shown. Why this bias is required in the learning process? Well, the bias can be thought similar to that of the intercept in the case of a straight line. The output can be expressed as represented in Eq. 6.11, which is the sum of the input values multiplied with the corresponding weights and the bias.

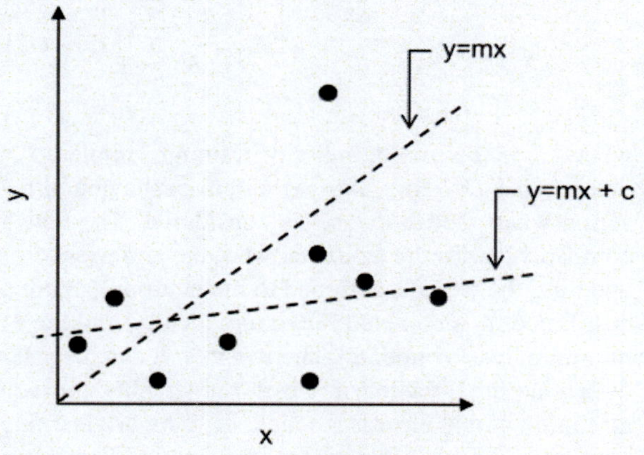

Fig. 6.36 Usefulness of bias

$$\text{Output} = \text{sum}(\text{input} \times \text{weights}) + \text{bias}$$

$$\Rightarrow \text{Output} = \sum_{i=1}^{n} x_i w_i + \text{bias} \tag{6.11}$$

Considering the input–output relationship, the so formed model will always then get trained over the points passing through the origin because of this direct dependency, as shown in Fig. 6.36. While this may hold good for an ideal situation, the real-world scenario is complicated and in such a case, the network will not be able to correlate the desired output with the selected inputs. However, the bias present in the network helps in generalising the input–output relationship.

For the example considered above, the bias selected for the input layer is 0, and 1 for the hidden and the output layers. Coming back to the second research question i.e., about the selection of hidden layers and hidden neurons in each layer, it depends upon the learning of the network. Because it is the information processing system of the network, one will try to build a network with the least possible error, and accordingly, the required layers and neurons will be selected. So, let us discuss the training algorithm.

One of the popular learning algorithms is backpropagation, and the neural networks trained with this algorithm are referred to as backpropagation neural network. The term backpropagation indicates the backward propagation of errors in the network. This learning algorithm works on the concept called gradient descent. First, let us begin with gradient descent followed by others. Gradient descent is a first-order iterative optimization algorithm for determining a local minimum of a differentiable function. The purpose is to find coefficients that minimize the error. A neural network begins the process of learning with random weights and while it learns, it attempts at minimizing the error of a cost function. In the considered example, where the objective is to predict the weld strength, the training objective is to minimize the mean square error (MSE). This MSE function is depicted in Eq. 6.12.

$$\text{MSE} = \frac{1}{N}\sum_{i=1}^{N}\left(T_i - O_i^k\right)^2 \tag{6.12}$$

where N is the total number of training samples, T_i is the target output of the ith dataset, and O_i^k represents the output of the neural network model on the kth interaction when the ith dataset is considered. The network after making a prediction compares it with the actual target value and tries to minimize the obtained error by updating the weights assigned to the neurons. Figure 6.37 depicts this idea, where a cost function is depicted diagrammatically, and the objective is to reach the global minimum loss by updating the weights. It can be observed that the true essence is to determine the direction in which the weights are required to be updated. Once the minimum global error is reached, the network learning is completed. Thus, for each prediction made by the model, the error is determined using the cost function and accordingly, the weights of the network are updated. This optimization technique is termed gradient descent, and it operates by calculating the gradients of the error concerning the weights. It can also be observed that by doing so, the network will reach a position after which further increasing or decreasing the weights will result in increasing the error.

Let us try to compute citing a small example to demonstrate the backpropagation, followed by further insights to the considered example of predicting the weld strength. Consider the neural network shown in Fig. 6.38, which consists of two neurons in the input layer, two neurons in the hidden layer, and one neuron in the output layer. Let us try to make the computations and explore the concepts. As discussed, the weights for this network are initialized and randomly set as shown in the picture. The computation of the network will begin in the forward direction, which has been shown in Eqs. 6.13, 6.14 and 6.15.

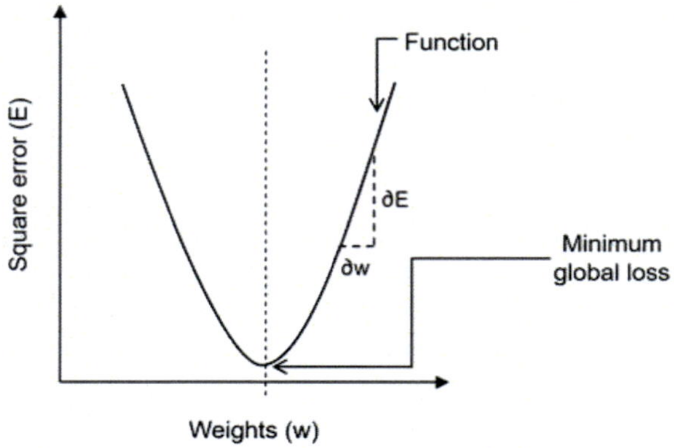

Fig. 6.37 Finding the global minimum loss of a function

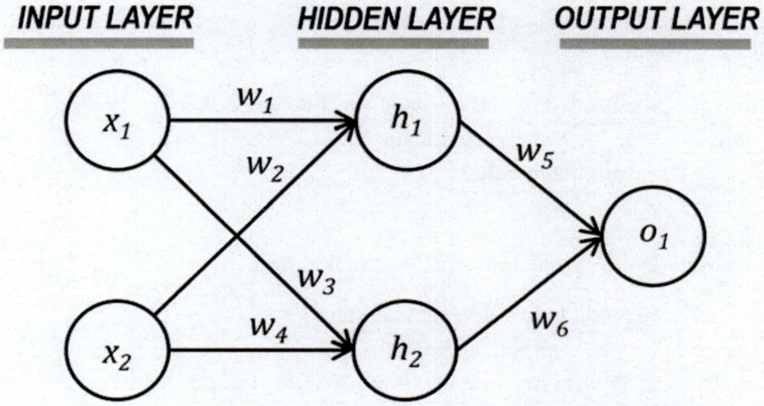

Fig. 6.38 An example neural network

$$\begin{bmatrix} h_1 & h_2 \end{bmatrix} = \begin{bmatrix} x_1 & x_2 \end{bmatrix} \begin{bmatrix} w_1 & w_3 \\ w_2 & w_4 \end{bmatrix} \tag{6.13}$$

$$o_1 = \begin{bmatrix} h_1 & h_2 \end{bmatrix} \begin{bmatrix} w_5 \\ w_6 \end{bmatrix} \tag{6.14}$$

$$\text{error} = \frac{1}{2}(\text{predictedvalue} - \text{actualvalue})^2 \tag{6.15}$$

where x_1 and x_2 are the inputs, h_1 and h_2 represent the hidden neurons, w_1, w_2, w_3, and w_4 are the weights, and o_1 is the output from the network.

Finally, the error is computed, which forms the basis for updating the weights assigned to the network. Equations 6.16 and 6.17 represents the method for updating the weights of the network.

$$w_j = w_j - \alpha\left(\frac{\partial \text{error}}{\partial w_j}\right)$$

$$\Rightarrow w_6 = w_6 - \alpha\left(\frac{\partial \text{error}}{\partial w_6}\right) \tag{6.16}$$

$$\Rightarrow \frac{\partial \text{error}}{\partial w_6} = \frac{\partial \text{error}}{\partial o_1} \times \frac{\partial o_1}{\partial w_6}$$

$$\Rightarrow \frac{\partial \text{error}}{\partial w_6} = \frac{\partial \frac{1}{2}(\text{predicted value} - \text{actual value})^2}{\partial \text{predicted value}} \times \frac{\text{predicted value}}{\partial w_6} \qquad \text{Using Eq. 6.15}$$

$$\Rightarrow \frac{\partial \text{error}}{\partial w_6} = 2 \times \frac{1}{2}(\text{predicted value} - \text{actual value}) \frac{\partial (\text{predicted value} - \text{actual value})}{\partial \text{predicted value}} \times \frac{\partial [w_5(x_1 w_1 + x_2 w_2) + w_6(x_1 w_3 + x_2 w_4)]}{\partial w_6} \qquad \text{Using Eqs. 6.13 and 6.14}$$

$$\Rightarrow \frac{\partial error}{\partial w_6} = (\text{predictedvalue} - \text{actualvalue}) \times (x_1 w_3 + x_2 w_4)$$

$$\Rightarrow \frac{\partial error}{\partial w_6} = (\text{predictedvalue} - \text{actualvalue}) \times h_2 \qquad \text{Using Eq. 6.14}$$

$$\Rightarrow \frac{\partial error}{\partial w_6} = \Delta h_2$$

(Considering (Δ = predictedvalue − actualvalue))

$$\text{Substituting in Eq.6.16}, \; w_6 = w_6 - \alpha \Delta h_2 \tag{6.17}$$

Based on the above, the weight, w_6 will be updated in the network, and similarly, the updated weights will be determined for the entire network. The term α in Eqs. 6.16 and 6.17 is termed the learning rate and it is a hyperparameter of the neural network. It is employed for updating the weights of the network and can be thought of as the value of the step-size for updating the weights. This is what is getting back-propagated i.e., based on the obtained error, the network updates the weights by propagating backwards. This is the idea behind the training algorithm.

Another important concept in the neural network is the use of momentum coefficient. Recap Fig. 6.37, where the picture of the cost function is shown and it was discussed that the objective is to reach the global minimum loss by updating the weights of the network. For a practical problem, the function might have several local minima points, as depicted in Fig. 6.39. In case you are interested to know what a local minima of a function is, then it refers to a point, where the value of the function is smaller than the value at the nearby points, however, it is possibly greater than a distant point. During the process of optimization, neural networks should not get stuck in the local minima points. To prevent this, a momentum coefficient is employed in addition to the learning rate. The term momentum physically refers to an impetus gained by a moving body. This coefficient value assists the neural network in reaching the global minima without getting overshoot.

Based on these theories, in the present case study, the learning rate was varied from 0.65 to 0.95 and the momentum coefficient from 0.3 to 0.75. Also, the neurons

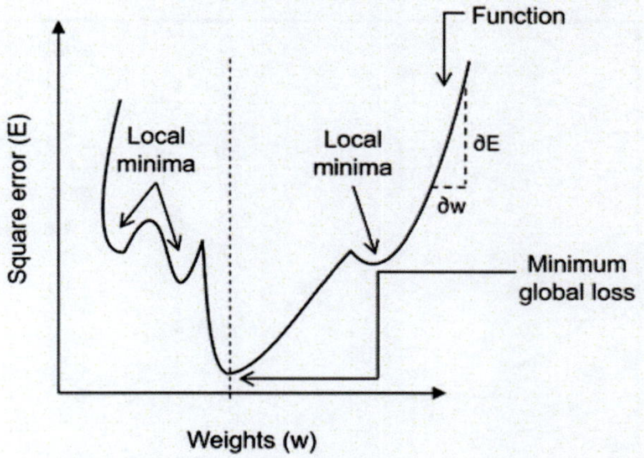

Fig. 6.39 Finding the global minimum loss of a function

in the 1st hidden layer were varied from 8 to 20 and that from 6 to 22 for the second hidden layer. To determine the optimum architecture of the neural network model, the MSE was found for each case. The dataset was divided into two, training and testing. So, MSE was determined for both these sets, and accordingly, the optimum architecture was selected, out of which, 8-13-8-1 was found to be the optimum. With this optimum architecture, the network was utilized for predicting the tensile strength for the test dataset. Figure 6.40 represents the outcome of the model on the test dataset, where the percentage error of prediction is shown for a few cases. The net average error of prediction is 3.542.

By now, discussions have been around various machine learning algorithms, their categories, ANN, and a few case studies. In the following section, other strategies of the analytics pipeline are discussed.

6.7.3 Adaptive Filtering

Let us look at adaptive filtering as automation of the matched filtering design and implementation. An adaptive filter is a system with a linear (usually) filter that has a variable parameter transfer function where these parameters can be adjusted according to an optimization algorithm. An adaptive filter can be designed using FIR or IIR filters and can have two basic kinds of adaptation algorithm—Least-mean-square (LMS) and Recursive-least-square (RLS) [45–47]. Adaptive filters can be thought of as a digital filter that self-adjusts its coefficients to minimize an error function, which is the distance between the reference or desired signal and the output of the adaptive filter. The optimization of the cost function is performed by employing the least-squares hence, the name LMS or RLS [48–50]. Interestingly, adaptive filters can be used to automate not only the matched filtering part but also the whole initial

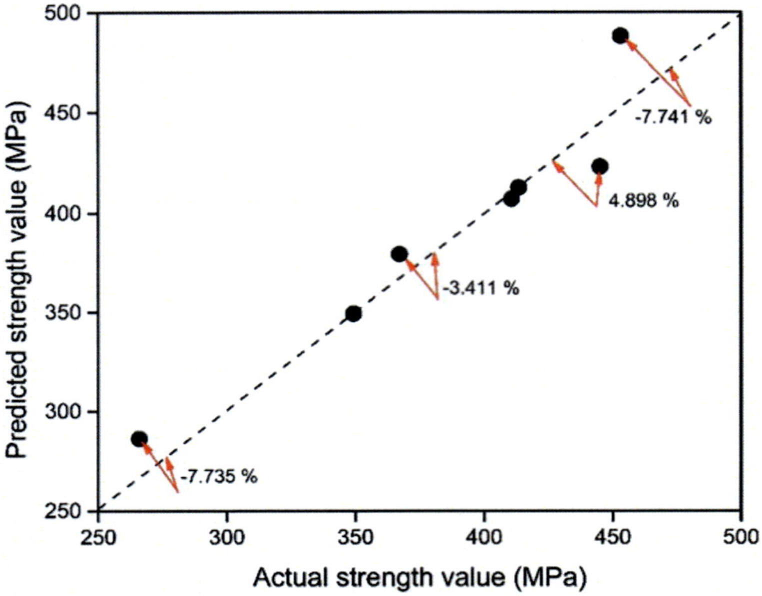

Fig. 6.40 Actual versus predicted values

pipeline of signal representation and feature extraction. This is achieved by using the adaptive filter as a *k* sample ahead predictor, allowing the filter to converge and then take the signal output (which represents the deterministic part of the signal) and the error output (which represents the stochastic part of the signal) as to orthogonal feature components that can be fed into the subsequent machine learning engine [51]. It should also be noted that though usually adaptive filters are used as linear filters, non-linear versions of the same are also available—they are called kernel adaptive filters (KAF) [52].

6.7.3.1 A Real-Life Example in Manufacturing—1

This section discusses a case study that aims at detecting the condition of an engine using the engine's vibration. This is a classification problem. The problem is to diagnose from engine vibration noise, whether a certain symptom exists or does not exist in an automotive subsystem. The standard dataset used globally for this kind of problem is FordA and FordB, both available from the University of California, Riverside (UCR) [53]. This data was originally used in a competition in the IEEE World Congress on Computational Intelligence, 2008. Each case consists of 500 measurements of engine noise and a classification. For FordA, the train and test datasets were collected in typical operating conditions, with minimal noise contamination. For FordB, the training data was collected in a typical operating condition, but the

Table 6.5 Accuracy of different approaches for engine vibration classification

UCR engine vibration dataset	Traditional signal processing pipeline (DTW) [54]	Automation using AutoML (Sect. 4.2) [55]	Automation using deep learning (Sect. 4.3) [56]	Automation using adaptive filters (Sect. 4.1) [51]
FordA	65.9	93.17	96.81	64.00
FordB	68.6	89.25	92.93	80.31

test data samples were collected under noisy conditions. The classification results of the state-of-the-art traditional signal processing approach using dynamic time warping (DTW) is presented in Table 6.5, along with the three different automation approaches presented in previous sections. It is seen that the deep learning-based automation performs the best, followed by AutoML approach and adaptive filter-based approaches. However, it should also be noted that the computation complexity of learning and inferencing is the highest in the deep learning-based approach, followed by AutoML and then adaptive filters. Hence, for different automation approaches, the accuracy and computation complexity can be treated as a trade-off.

6.7.3.2 A Real-Life Example in Manufacturing—2

This section discusses another case study that involves the detection of defects during the manufacturing of semiconductor wafers. The wafer data relates to semiconductor microelectronics fabrication. A collection of inline process control measurements recorded from various sensors during the processing of silicon wafers for semiconductor fabrication constitute the wafer database; each data set in the wafer database contains the measurements recorded by one sensor during the processing of one wafer by one tool. The two classes are normal and abnormal. This dataset is also available as part of UCR. Table 6.6 presents similar results for wafer quality classification—here it can be seen that all methods produce a very high level of accuracies. However, automation methods save significant development time as compared to traditional signal processing/manual pipeline building (a couple of days as compared to 2 months). The computation complexity of the different methods follows a similar trend (deep learning > AutoML > adaptive filter).

Table 6.6 Accuracy of different approaches for wafer quality classification

UCR wafer dataset	Traditional signal processing pipeline (DTW) [54]	Automation using AutoML (Sect. 4.2) [55]	Automation using deep learning (Sect. 4.3) [56]	Automation using adaptive filters (Sect. 4.1) [51]
	99.50	99.88	99.9	95.47

6.7.4 AutoML

Automated machine learning (AutoML) is the process of automating the signal processing pipeline till the machine learning block as depicted in Fig. 6.4. A similar approach is possible for the image/video processing pipeline too. AutoML was proposed as an artificial intelligence-based solution to the ever-growing challenge of applying machine learning algorithms [57]. AutoML can automate tasks such as feature engineering, model selection, and tuning of hyperparameters of the model. There are many AutoML tools available from major vendors, like AutoGluon from Amazon [58], Auto-sklearn available in Python [59], Neural Network Intelligence from Microsoft [60], CloudAutoML from Google [61], AutoAI from IBM Watson [62], and AutoML pipeline from Oracle [63]. However, while most of these works pretty good for model selection and hyperparameter tuning, they do not work well on diverse sensor signals (they are tuned more towards text, speech, and categorical data). Researchers have combined both signal processing based feature generation along with autoencoder based feature representation followed by either a simple grid-search based feature selection or a complete search space optimization based feature selection that tries to find out features that causally correlates with the labelled data using measures like Maximal Information Coefficient (MIC) [64], Minimum Redundancy Maximum Relevance (mRMR) [55, 65–67]. A variant of this kind of AutoML has been employed using a limited set of feature transformation and a limited set of machine learning algorithms [68].

6.7.4.1 A Real-Life Example in Manufacturing

To set an example, the CloudAutoML from Google is utilized for developing a machine learning model. The manufacturing set up consists of an induction motor and the objective is to investigate the health condition of the bearings fitted in the motor. This is the same set up whose picture is depicted in Chap. 2 (Fig. 2.61) and Chap. 3 (Fig. 3.15). A vibration sensor was engaged with the induction motor for acquiring the vibration signal. The bearings selected for the experiments were of five different health conditions namely, inner-race fault, outer-race fault, jammed, broken cage, and healthy. The first four resembles faulty conditions that were induced in the bearing and the last one was a healthy bearing. Various time-domain features were collected from the vibration signal for fault analysis. Those were mean, RMS, standard deviation, maximum, minimum, range, crest factor, form factor, kurtosis, and skewness. The objective is to develop a machine learning model for classifying the health of the bearing by using the aforementioned features of the vibration signal.

For the same, the Google AI platform, CloudAutoML is employed. The data was uploaded into the platform, which consisted of the time-domain features as the inputs and the corresponding health conditions of the bearing as the labels. The platform was instructed to build a classification model using the said data. Figure 6.41 depicts the outcomes of the developed model. It could be observed that the developed model

Fig. 6.41 Outcomes of the developed model: **a** accuracy and **b** feature ranking

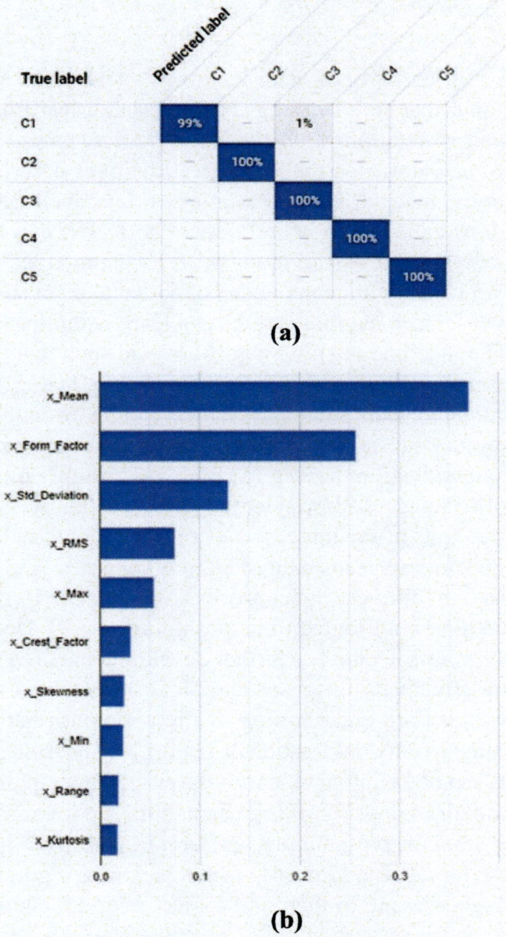

has high accuracy as it has accurately classified the health conditions of the bearing i.e., C1–C5, as shown in Fig. 6.41a. Figure 6.41b shows the ranking of the features considered for developing the machine learning model. The mean of vibration signal has the highest correlation and thus, it has been at the top, followed by crest factor, standard deviation, RMS, and so on. This sort of platform is beneficial in a way that it eliminates the hurdles in selecting a model. However, several things are not disclosed by the platform such as the name of the algorithm and related details of how it is selected, its optimization, and so on. It gives the details of the trained model and also allows the deployment of the trained model.

6.7.5 Deep Learning

Recall Fig. 6.8, where deep learning is shown as a subset of machine learning. Traditional machine learning techniques can learn only when the structure of the input–output relationship is known via feature engineering. The feature engineering part can be automated via AutoML techniques described above, but they have their limitations. Instead, if there is a lot of data available for training, a new class of algorithms called deep neural networks (DNN) can learn both the structure of the input–output relationship along with its parameters by loosely modelling the way our brain works with many layers of networked decision making [69]. Such learning is known as deep learning, which can learn complicated patterns in large amounts of data. The machine learning algorithms are such that they require domain expertise, feature engineering, and utilizing these, they learn. A deep learning algorithm, on the other hand, learn categories incrementally through its deep and wide architecture consisting of several hidden layers and neurons. For instance, considering a case of image classification, a deep learning algorithm learns about the lighter and darker areas of an image, followed by the lines, and finally, the shapes present in the image. These features of the image are learnt incrementally by the layers of the network.

Deep learning needs a lot of computing power and recent advances in computing systems including graphical processing units (GPUs) and deep learning accelerated chipsets have contributed to its proliferation [70]. Deep learning has shown a lot of promise in image data but further development is required in the field to apply deep learning effectively in sensor data. It is interesting to note that since deep learning automates end-to-end learning of input–output relationship, and its application is not limited to only automation of feature engineering and machine learning, but also automation of the initial blocks of the processing pipeline. This includes tasks such as signal conditioning, signal representation, and learning/inferencing, or image conditioning, image representation, and learning/inferencing [56, 71, 72]. Therefore, when compared with the traditional machine learning algorithms, deep learning algorithms are versatile owing to their end-to-end solution, contrary to the solution derived in multiple steps as in the case of machine learning algorithms. For instance, considering the case of classification of images, a deep learning algorithm, such as You Only Look Once (YOLO), takes the image as an input, splits it into several cells, wherein, each cell creates a bounding box for classification. In contrary to this approach, considering SVM, first, a bounding box detection algorithm is required that detects all the possible objects that have the histogram of gradients, and the same is then feed to the model.

Manufacturing specific applications of deep learning-based automation have been addressed using data augmentation via generative networks followed by Deep Belief Networks and Deep Feed-Forward Networks [73]. As highlighted and discussed, the deep learning models are very popular owing to their increasing performance with the size of data as compared to the traditional machine learning algorithms. This is diagrammatically represented in Fig. 6.42. The plot shows that with the increasing size of data, a deep learning model outperforms a traditional machine learning model.

Fig. 6.42 Performance of deep learning versus traditional machine learning algorithms

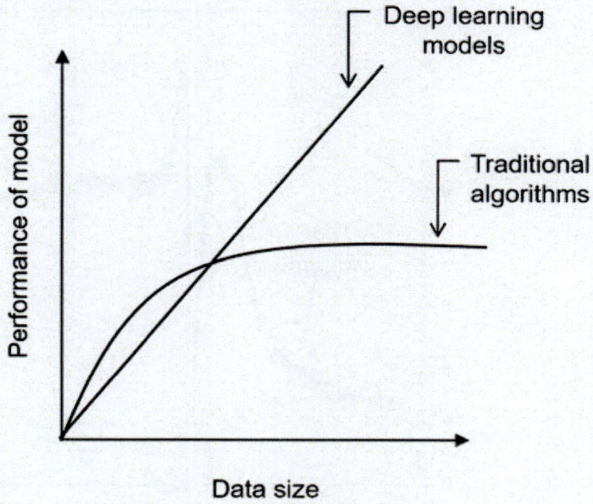

To visualize the same, let us perform a small exercise and deduce insights from the same.

6.7.5.1 Real-Life Example in Manufacturing—Deep Neural Network

As an example, the friction stir welding process is considered. The objective of using machine learning here is to predict the joint strength of the weld using data acquired from the sensors. This process has already been introduced in previous chapters, so, let us directly head towards the problem statement. The sensors engaged in the process were force, torque, and power. Several experiments were performed to create a knowledge base that consisted of the variation of parametric conditions, variation of force, torque, and power, and the corresponding tensile strength of the welds. The evolution of the force in the welding process has also been shown and discussed in the previous chapters. Thus, here, the evolution of torque and power signals is shown in Fig. 6.43a, b to help readers understand at the importance of these signals in monitoring the welding process.

From Fig. 6.43, the events of the welding process can be observed in the signals. The torque signal, in the beginning, shows a peak value, which is because of the initial torque induced in the spindle motor as the welding process begins. This indicates the start of the tool rotation [74]. As the tool plunges inside the materials to be welded, the torque again increases and during the welding phase, lesser fluctuations can be observed in the torque signal. Finally, as the tool retracts, the torque decreases, which marks the end of the welding process. These events could also be marked in the power signal, as shown in the picture. Both the signals correspond to a defect-free weld, and that is why the instance of the signals corresponding to the welding phase has

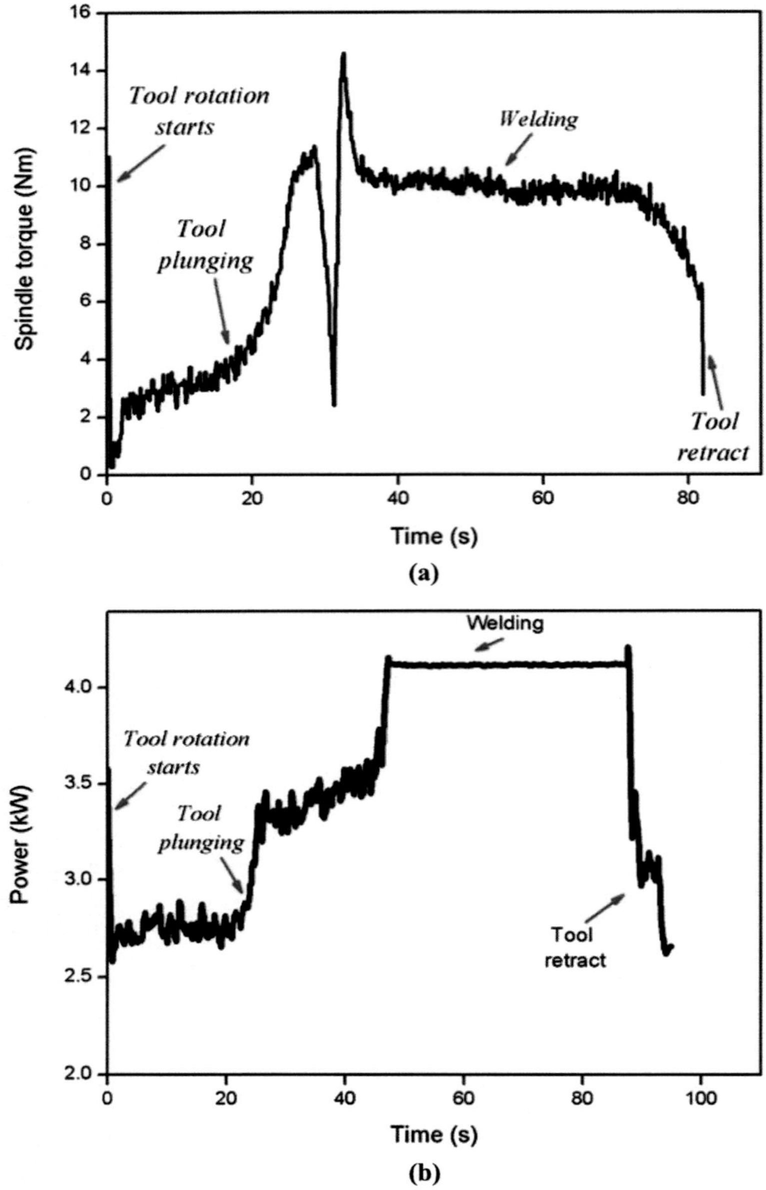

Fig. 6.43 Evolution of acquired signals: **a** torque and **b** power

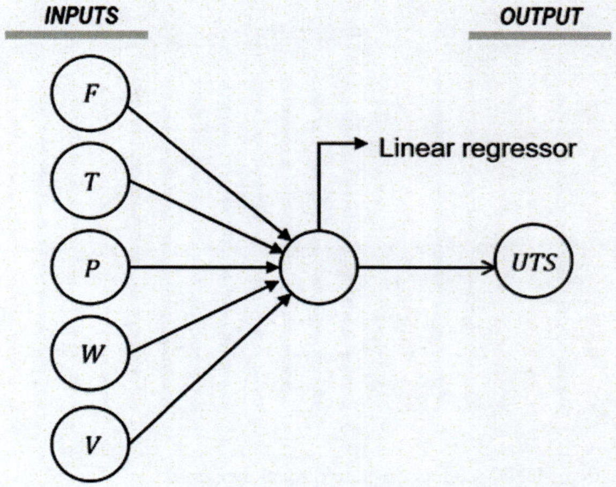

Fig. 6.44 Schematic of linear regression model considered for the example

lesser fluctuations. If the welding would be improper i.e., during a defective weld, the fluctuations will get reflected in these signals [75, 76].

To predict the tensile strength of the weld, two approaches are followed, first is using a traditional machine learning algorithm, and the second is using a deep learning model. In the first case, a linear regression model is considered. A schematic diagram of the model is depicted in Fig. 6.44. Linear regression is one of the oldest algorithms utilized for machine learning and, in this example, it is trained with the ordinary least square (OLS) method. OLS is utilized when multiple inputs are involved and it aims at minimizing the sum of squared residuals. The technique attempts at calculating the distance of each point from a regression line. The calculated distances are the errors that are squared and summed and are minimized during the training. In this example, multiple inputs are present, which are the sensors' data and welding parameters, and the output is a single value which is the tensile strength.

In the second case, a deep neural network is considered for predicting the tensile strength. As discussed, the ANN is popular owing to its complex computing structure that involves multiple hidden layers and neurons. Because of this, it can solve complex problems and is considered supreme at non-linear modelling. In this example, one such DNN is modelled that has 22 hidden layers, each layer with multiple neurons. A schematic diagram of this DNN architecture for this example is depicted in Fig. 6.45.

An optimization technique called genetic algorithm is considered for selecting the number of hidden layers and respective neurons. A genetic algorithm is a search technique that follows the principles of genetics and natural selection. It is utilized in finding optimal solutions to certain problems, which would otherwise need a whole lifetime to solve. The methodology starts with the selection of an initial population, each having their own set of chromosomes. The population refers here as the set of possible solutions to a specific problem. Similar to the process of natural genetics,

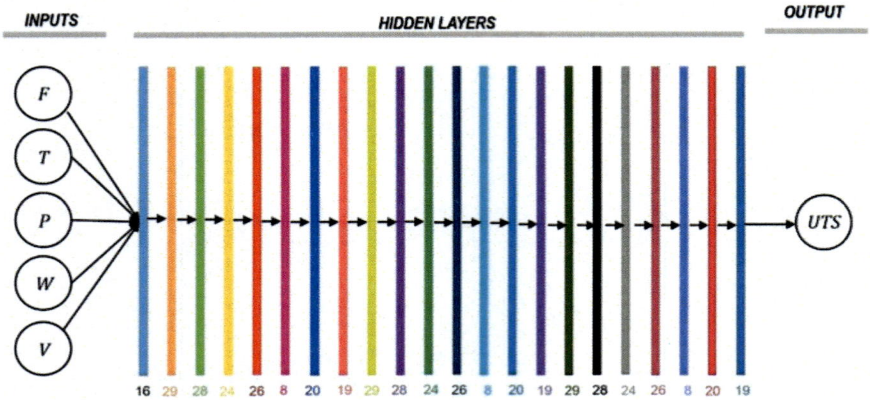

Fig. 6.45 Schematic of DNN model considered for the example

the solutions here also undergo recombination and mutation producing new children. Each individual is associated with a fitness value. A fitness function is defined to classify individuals of the population as fit or unfit. Based on the fitness of the individual, they have the chances of getting selected to produce offspring. The individuals are randomly selected through natural selection. The individuals with higher fitness value have higher chances of getting selected, and their chromosomes are interchanged with each other, the process is referred to as a crossover. As known from biology, the offspring do not have a complete copy of genes from their parents, thus, the mutation is performed where a random change in the chromosome is carried out. The offspring are again validated by using the fitness function, and if considered fit, then it replaces the lesser fit individuals from the population. For implementing the genetic algorithm technique to this example, a subset of the experimental data was considered. Accordingly, it resulted in multiple hidden layers and the corresponding neurons. With that architecture, the model obtained an accuracy of ~92%. Further, a few layers were added following a trial and error approach to increase the accuracy.

Figure 6.46 shows a bar chart depicting the accuracy obtained with the linear regression model and the DNN. It can be seen that the accuracy of the DNN is superior to that of the traditional machine learning algorithm. To visualize the importance of data size, as discussed, training data was divided into multiple sizes. These are 5000 (5 K), 10,000 (10 K), 15,000 (15 K), 20,000 (20 K), 25,000 (25 K), and 30,000 (30 K). The dataset consisted of 30,000 rows and it was divided into six parts. With fewer data i.e., 5000, the accuracy of the linear regression model can be seen to be more as compared to that of the DNN model. However, as the data size grows i.e., with the increase in the number of instances, the DNN model can generalize the problem better. This is why the accuracy of the DNN model increases. Also, it could be observed that the difference in the accuracy of the linear regression model with the increasing data size is not that significant. For instance, as the data size increased from 5000 to 10,000, the accuracy is increased from 78 to 81%. Similarly,

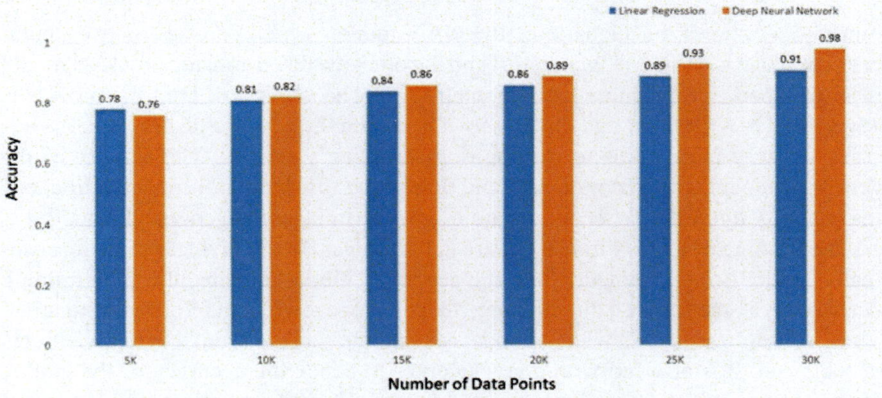

Fig. 6.46 Performance of linear regression model versus DNN

it increased from 81 to 84% with an increase in data size to 15,000 from 10,000. Thereafter, the accuracy is not improving significantly, it is like getting stabilized. However, with the DNN, a significant change in the accuracy can be observed with the increasing size of the data.

6.7.5.2 Real-Life Example—Generative Adversarial Network

This section will demonstrate the application of another deep learning algorithm, a generative adversarial neural network (GAN) in manufacturing [77]. Specifically, the algorithm is applied for augmenting the quality of images. For this example, first, the problem statement is defined and the details of the GAN follows next. Finally, the results obtained from the model are shown.

An image is one such information, which is highly driven by the conditions in which it is captured. It demands huge efforts for ensuring quality. This includes a high-quality camera system with associated accessories, proper illumination of the working area, presence of nearby objects, etc. All these things have been discussed in Chap. 4. The problem is that the industrial scenario, particularly the shop floor, is filled with several machines and systems that are working simultaneously and continuously. It also has the presence of humans who look after those machines, systems, and the work. Thus, it is very unlikely to get a high-quality image with all these challenges. Thus, a technique is required to augment the quality of the image. This forms one part of the problem statement. The other part is to employ a technique that can augment the image captured from a low-quality camera to bring it at par with that of high quality. This is aimed at cost savings. So, the augmentation of images forms an essential requirement for manufacturing applications because it desires both the parts of the problem statement posted here.

To address the aforementioned problem, a GAN model is developed. Let us explore the details of GAN followed by its implementation in this example.

Figure 6.47 shows a schematic of the GAN model, which consists of two neural networks, one is named as 'generator' and the other as 'discriminator'. A GAN model is an unsupervised machine learning technique. The activity of both the models is adversarial to each other and this is why it is named the adversarial network. Adversarial because the models try to oppose each other's activity. The activity of the generator model is to generate data and that of the discriminator model is to check the authenticity of the generated data, i.e., to discriminate the generated data.

The training of a GAN model occurs in two stages. In the first stage, the discriminator model is only trained, while the generator model remains idle. The training occurs only in the forward direction i.e., there is no involvement of backpropagation. The discriminator is feed with a real dataset and the same is trained to learn features of the same. It is also fed with a few samples to check the accuracy of the model. In the second stage, the generator model is trained, while the discriminator model remains idle. The generator model is fed with random input, which the model utilizes to generate new instances of data. In this case, predictions of the trained discriminator model are also employed to train the generator model. Thus, the generator model tries to fool the discriminator model by generating fake data sample. In this way, both the models learn.

In the present case study, a variant of the GAN model called progressive GAN is employed. This variant is popular for generating a high-quality image from a low-quality. In this modelling approach, the training begins by feeding a low-quality image to the GAN, and then progressively adding layers to the network for increasing the output size of the generator model is developed for augmenting the quality of an image. The benefit of employing a GAN model is that it does not need a huge dataset. In this example, the discriminator model was trained with a high-quality image, so that the model learns features such as pixel intensity and aspect ratio. The generator model is trained with the low-quality image, which is performed with the prediction of the discriminator model. Thus, new layers are added progressively to the generator model to generate an image that is equivalent to the one that has been employed to train the discriminator model. The low-quality image is acquired with the help of a webcam. This saves the cost of employing an expensive camera. As the low-quality image is fed to the generator model, it starts generating new instances, which are given as input to the discriminator model in real-time to classify them. This process continues until the output of the generator model is equivalent to the

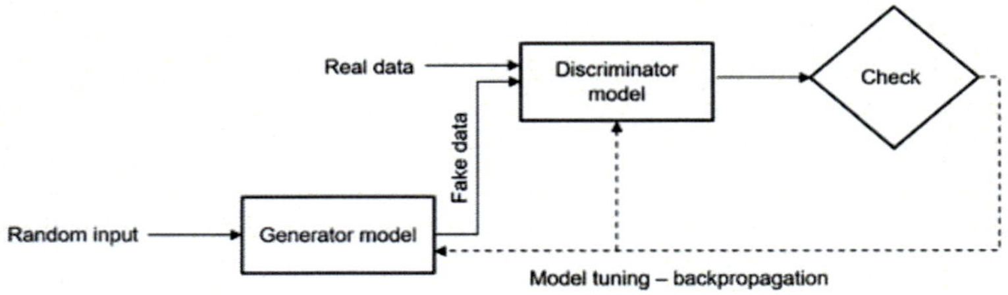

Fig. 6.47 Schematic of a GAN model

high-quality image. Figure 6.48 shows five images that include a low-quality image and the enhanced images generated in four iterations.

From Fig. 6.48, it can be observed that the original image is of low quality, which was captured by using a webcam. As highlighted, GAN eliminates the requirement of

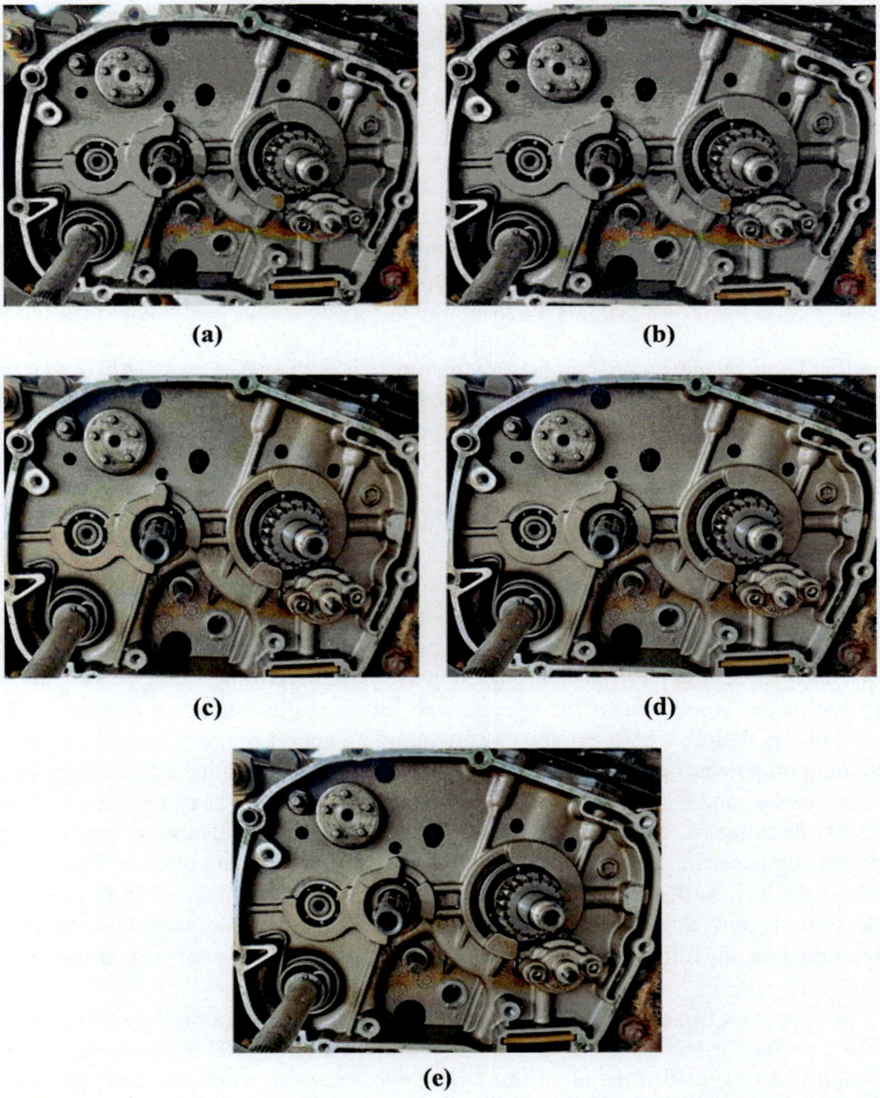

(a) (b) (c) (d) (e)

Fig. 6.48 Outputs of the progressive GAN model: **a** original image, **b** enhanced image after the first iteration, **c** enhanced image after the second iteration, **d** enhanced image after the third iteration, and **e** enhanced image after the fourth iteration a huge dataset as the generator model itself generates the data. Thus, after feeding the low-quality image to the developed GAN model, it iteratively enhances the image, which in this case is achieved in four iterations. The number of iterations depends on the quality of the input image and the desired output. One limitation of the developed model, at this point, is that it works offline. Also, the training time is very high. This requires focused research to automate the pipeline and make it work in real-time and include changes to reduce the training time.

One could ponder upon what could be the objective after augmenting an image in manufacturing? Owing to the industrial revolution, digitalization is at the edge of industries and as highlighted, it is of higher importance to the manufacturing industries. There is a vast open area in manufacturing for the application of machine vision. This possibly includes the automobile industries at the top, followed by others. This is so because the automobile sector involves many components, which are assembled. Thus, the quality of each component is ensured. Machine vision that involves the algorithms of image processing could be applied for quality assurance. In this case, the developed GAN model will be advantageous as it would result in a high-quality image. Applying machine vision to poor quality image does not make sense. Thus, in addition to automating the pipeline of the developed GAN model to operate in real-time, the second research requirement will be coupling the developed GAN with a machine vision-based system for quality assurance.

6.7.5.3 Real-Life Example—Siamese Neural Network

Siamese neural network is one of the recently introduced deep learning algorithms. The term 'siamese' means 'being a conjoined twin'. It is a one-shot learning technique, which means, that the algorithm requires only one instance of training data of one class for learning [78]. As discussed in the earlier sections, image classification is one of the popular tasks for which, deep learning algorithms are employed. In case of the YOLO, which employs a convolutional neural network, requires a large amount of labelled data for training. This is also referred to as standard classification, wherein, the model is feed with a large amount of data and the model outputs a probability distribution of each class. Contrary to this, the one-shot learning technique is promising because it requires only one instance of training data of each class. This eradicates the burden of creating a huge dataset for modelling, and it also makes the process of re-training easier. The siamese neural network employs this one-shot learning technique for image classification. A schematic of this network is shown in Fig. 6.49.

It consists of two convolutional neural networks, which have a similar architecture. The two-way arrow mark, in the picture, shows that the neural networks share the weights. As the title of the algorithm, these two neural networks form the Siamese twins or conjoined twins. Thus, they cannot be separated. Both neural networks have the same weights, parameters, and architecture. Thus, in the case of any parameter update in one of the neural networks, the same is also reflected in the other network. These networks are feed with the input image. As the image is fed to the networks,

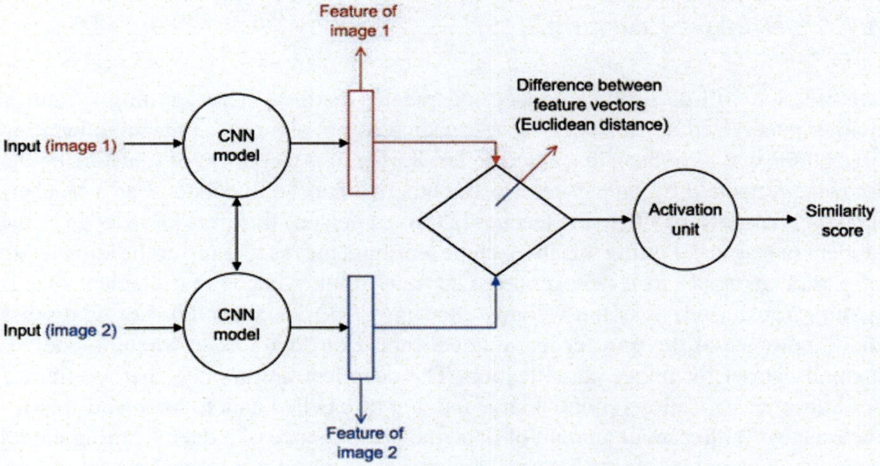

Fig. 6.49 Schematic of Siamese network

they start learning and generate feature vectors. In this case, two feature vectors are generated from the two convolutional neural networks, and the Euclidean distance between the vectors is determined. The loss function employed in a Siamese network is shown in Eq. 6.18.

$$L(s_1, s_2, y) = \alpha(1 - y)D_w^2 + \beta y \max(0, m - D_w)^2 \tag{6.18}$$

where s_1 and s_2 are the input images, y is a binary indicator that specifies whether the input images belong to the same class or not, α and β are constants, m is the margin, and D_w is the Euclidean distance between the feature vectors. It is computed as given in Eq. 6.19.

$$D_w = \| f(s_1; w_1) - f(s_2; w_2)\|_2 \tag{6.19}$$

where f is a function mapping the input image to the vector space via the neural network and w_1 and w_2 are the learned weights. The loss function is also referred to as contrastive loss. For training this model, image pairs are generated. The pair can consist of similar or dissimilar images; in practice, the dataset will include both for the training of the model. The image pair is termed as 'positive' in case of similar ones, and 'negative' in case of dissimilar images. Thus, when images (inputs) would be of the same class, then the similarity score will be higher as compared to the score of dissimilar input images.

6.7.6 Transfer Learning

This section will discuss the concept of transfer learning. This learning system is neither supervised nor unsupervised, it also does not signify the machine learning algorithms that have been discussed so far. Rather, it is a concept of transferring the knowledge gained by one system to the other. It can be understood as a teacher-student relationship, where the teacher (a trained person) transfers knowledge to the student (a learner). Putting this in machine learning, means to transfer the knowledge of a trained model to a new model. One may think what is that element that is getting transferred? It is the weights and bias of a trained model. Let us discuss the importance of the transfer learning concept. Consider a case, wherein, one has a small dataset for image classification. The deep learning models that are trained for image classification require a large amount of labelled data to learn and classify the images. With a small amount of data, the performance of a deep learning model cannot be guaranteed. In such a case, one can download a pre-trained model. A pre-trained model is one, which is already trained with some dataset, in this case, it would be an image classification model. This pre-trained model can then be employed for a different recognition task. The weights and bias of the model can be retained as learned features, and the training of the model can be resumed with the learned weights and bias. Thus, instead of training it from the scratch, which would have required a large amount of data as well as computational resources, the training would be faster owing to the transfer of knowledge.

As an example of the implementation of transfer learning, a case study is discussed in this paragraph. The objective of the study was to classify objects of cylindrical and rectangular shapes. This problem is essential to the manufacturing industries, wherein, jobs of different shapes are manufactured. In this case, after the fabrication of jobs, they are transferred to the warehouse, wherein, they are subjected to quality tests and finally to dispatch. In many places, a robot manipulator is employed for picking the fabricated jobs from the conveyor and put them in a storage box. But the challenge here is that the fabricated jobs may be of different shapes, and therefore, for the robot to pick a job, it should have a suitable gripper according to the shape of the object. This is achieved via integrating the robot with a camera, which acquires the image of the jobs, and the same is processed in real-time to detect its shape. For detection of the shape of the object, machine learning algorithms are deployed. Another challenge, in this case, appears with the deployment of a machine learning algorithm for image classification. For this purpose, various deep learning algorithms are available. However, the problem is that they require a large amount of labelled data. Such a large amount of labelled data was not available in the present example. To deal with this problem, the concept of transfer learning was utilized. Figure 6.50 shows a schematic of the approach followed for implementation of the transfer learning.

The pre-trained model was a YOLO model, which was trained with 200,000 number of labelled instances for image classification. The model consisted of 83 classes of images. In the present example, the number of images of objects belonging

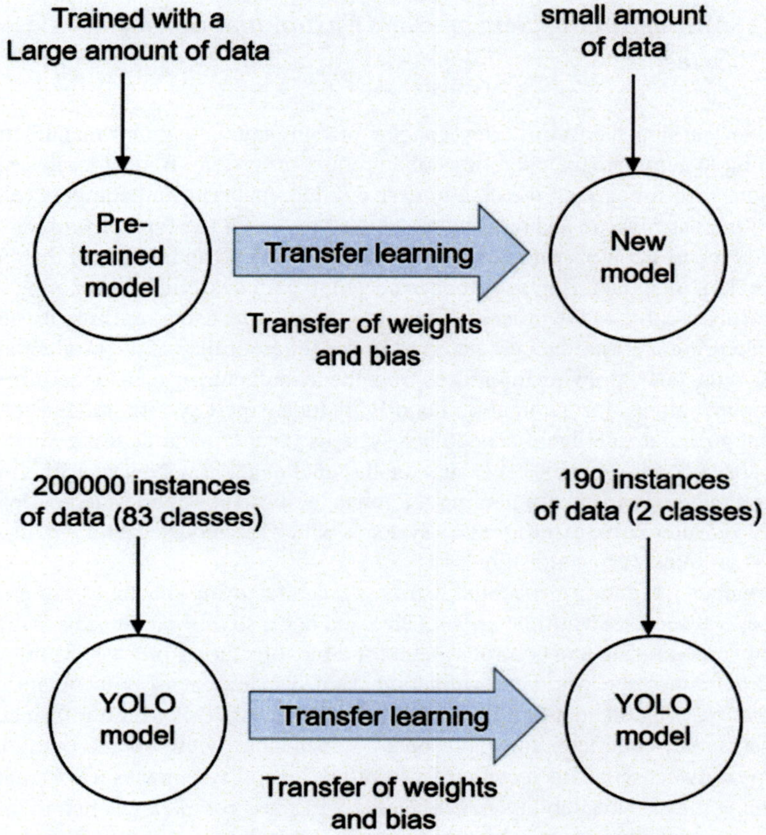

Fig. 6.50 Schematic for transfer learning and implementation in the present example

to cylindrical and rectangular shapes was only 190. The pre-trained model along with its weights and bias was utilized, wherein, the last layer was modified. As mentioned, the pre-trained model consisted of 83 classes, which was modified to have only two classes. These classes were cylindrical and rectangular. Thus, utilizing the weights and bias of the pre-trained model, a new model was built with only 190 instances of data, and a training time of 12 h. Considering the pre-trained model to have a uniform number of training instances for each class, it will contain ~2400 images of each class. In addition, it would have taken a huge amount of time for training, as it was built from the scratch. The same is reduced to only ~95 images of each class.

6.7.7 Automation of Sensor Data Fusion and Sensor Selection

Machine learning has been a key enabler of computer integrated manufacturing, intending to automate the operations of a production system. It includes a variety of techniques for processing the data to derive hidden patterns and complex relationship in the data. These techniques are selected based on the type of problem to be solved, type of the input and output data, and the volume and nature of the dataset. The modern manufacturing system is expected to gain flexibility by enabling online monitoring/control of the process. The major benefits of using machine learning in manufacturing are that the cost incurred in carrying out the offline product assessment is reduced by drawing inferences from the manufacturing data in real-time. The existing techniques for automation majorly include expert system utilized for decision support and intelligent scheduling systems on the manufacturing shop floor. They are becoming obsolete because of the challenges involved in gathering the required expertise for implementing a knowledge-based system. Machine learning has been an alternative to these expert systems, which can reduce the time-consuming process of knowledge acquisition.

To gather the data for machine learning, manufacturing machines are engaged with several sensors. Multiple sensors are used because of their varying abilities to develop a monitoring and control system for a manufacturing process. Multi-sensor fusion combines sensory data originating from disparate sensors to obtain useful information, which is not possible using a single sensor. This useful information is more accurate, complete, and represents a dependable 'view' of the entity or the system being sensed. This technique also offers several advantages associated with enhancing the data availability and authenticity in a less complex manner. In the case of smart manufacturing, it is primarily employed to improve the decision-making capability of predictive and prescriptive analytics built on such systems.

Over the recent past, several advancements in multi-sensor fusion have taken place and they can be classified into three categories namely, data-level, feature-level, and decision-level. The data-level fusion is also referred to as sensor-level fusion and in this case, the raw-sensors data is directly combined without performing any feature extraction. The combined data is treated as a single unit. Feature-level fusion includes the extraction of meaningful features from the data, which are combined to obtain a new feature. Finally, decision-level fusion employs the decision of multiple algorithms to come up with a combined decision.

In contrast to the hand-crafted feature design approach, a popular task for sensor fusion includes learning representations directly from the data streams. For this purpose, one of the techniques employed is referred to as transform learning. The techniques of AutoML and deep learning-based automation for feature engineering can be used to select relevant features or model input–output relationship when there are multiple streams of different sensors. This leads to automated sensor fusion models. The use of deep transform learning uses autoencoder based techniques to fuse multi-sensor data [79, 80]. As an additional benefit from the same feature selection approach, such systems also help to learn from the data, which sensors are useful in a given context, thereby leading towards sensor selection. This section will discuss the basic transform learning framework followed by the introduction of a novel transform learning framework and its implementation to a real-life example in manufacturing.

A transform learning framework can be schematically represented as depicted in Fig. 6.51. The original data is represented by X, and it contains m number of features of length N. The transform is represented as, $T \in \mathbb{R}^{K \times m}$, where, K is the number of atoms of the transform T. The corresponding coefficient is represented by matrix $Z \in \mathbb{R}^{K \times N}$. Thus, the relationship for the basic transform learning framework is expressed in Eq. 6.20 [80].

$$TX = Z \tag{6.20}$$

Let us compare this notion of transform learning with the neural network. A neural network architecture consists of three basic elements, an input layer, a hidden layer, and an output layer. As the data is fed to a neural network, the hidden layer starts learning the representations, or the relationship between the given inputs and the outputs. This is shown schematically in Fig. 6.52a. In contrast to this, Fig. 6.52b depicts a segregated representation in which representations are learnt between an input layer and a representation layer, and between the representation layer and the output layer. Representation learning refers to the learning that occurs between the input layer and the representation layer, as both the weights and the outputs are unknown.

This section will introduce a novel transform learning framework, named "*TransFuse*", which learns individual transform for each sensor's data and fuses them to learn a common transform. Thus, this transform learning approach performs a feature-level multi-sensor fusion. A schematic of the transform learning approach is shown in Fig. 6.53. In the case of the basic transform learning framework, the optimization problem is expressed as shown in Eq. 6.21 [80].

$$\min_{T,Z} \|TX - Z\|_F^2 + \lambda\left(\|T\|_F^2 - \operatorname{logdet} T\right) + \mu\|Z\|_0 \tag{6.21}$$

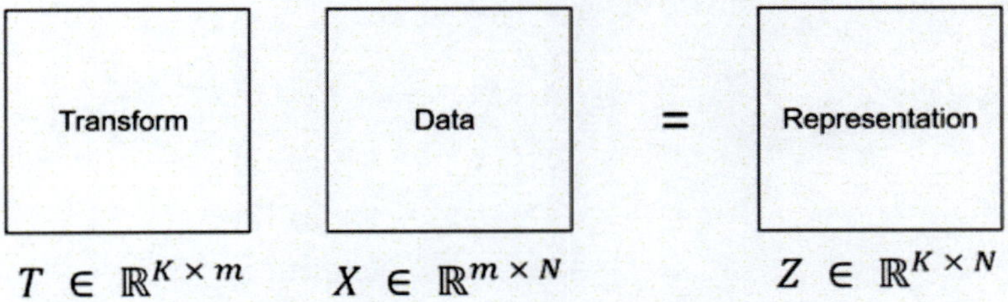

Fig. 6.51 Schematic of transforming learning framework

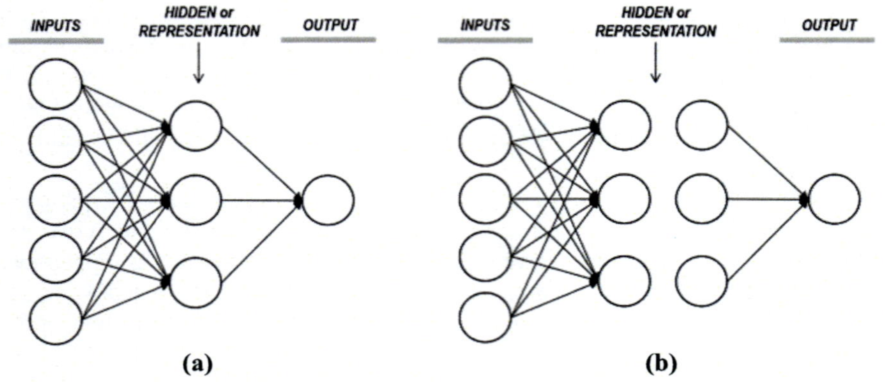

Fig. 6.52 **a** Learning in a neural network and **b** segregated representation

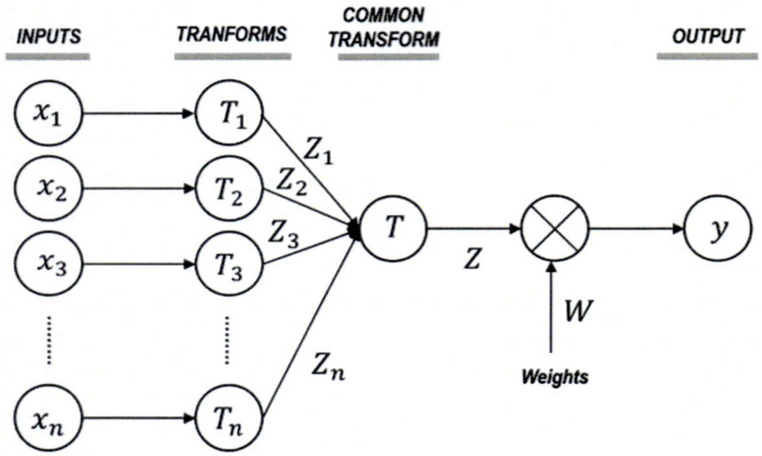

Fig. 6.53 Schematic of *TransFuse*—A transform learning approach

Considering a dataset consisting of data of three sensors, the optimization problem for the *TransFuse* is expressed in Eq. 6.22 [80].

$$\min_{T_1,T_2,T_3,T,Z_1Z_2,Z_3,Z,w} \|T_1X_1 - Z_1\|_F^2 + \|T_2X_2 - Z_2\|_F^2 + \|T_3X_3 - Z_3\|_F^2$$
$$+\lambda_1\left(\|T_1\|_F^2 - \log\det T_1\right)$$
$$+\lambda_2\left(\|T_2\|_F^2 - \log\det T_2\right)$$
$$+\lambda_3\left(\|T_3\|_F^2 - \log\det T_3\right)$$

$$+\gamma \left\| T \begin{bmatrix} Z_1 \\ Z_2 \\ Z_3 \end{bmatrix} \right\|_F^2$$
$$+\lambda\left(\|T\|_F^2 - \log\det T\right) + \alpha\|y - wZ\|_2^2 \quad (6.22)$$

where $X_1 \in \mathbb{R}^{m_1 \times N}, \ldots., X_n \in \mathbb{R}^{m_n \times N}$, $m_1, \ldots., m_n$ are the feature lengths of the individual sensors, the output is $y \in \mathbb{R}^{1 \times N}$, the individual transforms are $T_1 \in \mathbb{R}^{K_1 \times m_1}$, $T_2 \in \mathbb{R}^{K_2 \times m_2}$, $T_3 \in \mathbb{R}^{K_3 \times m_3}$, $T \in \mathbb{R}^{K \times (K_1+K_2+K_3)}$ are the transforms, $Z_1 \in \mathbb{R}^{K_1 \times N}$, $Z_2 \in \mathbb{R}^{K_2 \times N}$, $Z_3 \in \mathbb{R}^{K_3 \times N}$, $Z \in \mathbb{R}^{K_1 \times N}$ are the coefficients, regression weight is $w \in \mathbb{R}^{1 \times K}$, and λ, γ, α are the hyperparameters.

As an example, the *TransFuse* was employed for a regression task, where the objective was to predict the weld-joint strength by using information captured from the sensors. The process is friction stir welding and the sensors employed were force, torque, and power. The utility of these sensors has already been discussed in the preceding sections. Prediction of the weld quality can be achieved offline by using a sophisticated instrument; a universal tensile testing machine. However, to do so, a sample needs to be cut from the weld, which is destructive. Besides, this is not going to fulfil the objective of a digital twin application, where the idea is to monitor and control a process in real-time driving insights into the process and making profitable decisions. Alternatively, machine learning and artificial intelligence techniques can be leveraged for predicting the quality of the weld.

The raw sensors' data captured from force, torque, and power was fed to the *TransFuse* model. The model performs a feature-level multi-sensor fusion. The dataset consisted of sensors' data corresponding to 40 numbers of experiments, in which, welds were fabricated at different processing conditions. The dataset had 1000 samples, each of force, torque, and power, and they were split into windows of 50 samples. The results obtained from the model for both single as well as multi-sensor are shown in Table 6.7. The accuracy can be seen to be higher with multiple sensors as compared to the result obtained with the single sensor. Further, amongst the three signals, the force can be observed to be having higher accuracy followed by torque and power, respectively. In contrast to the deep learning-based techniques, the transform learning approach is computationally less intensive and complex as well. Thus, it provides higher flexibility.

Table 6.7 Weld quality prediction via automated multi-sensor fusion

Algorithm/performance measure	Using only force sensor	Using only torque sensor	Using only power sensor	Fusion of all three sensors
Mean-square error	0.0212 ± 0.004	0.0216 ± 0.005	0.0400 ± 0.005	0.0088 ± 0.002
Squared correlation coefficient (actual and predicted)	0.56	0.43	0.22	0.82

6.7.8 Automation of Compression and Partitioning for Edge Devices

As outlined, the huge computing power and memory required for deep learning-based systems create challenges in running them on memory and CPU constrained devices. The only way to handle such issues is model compression and partitioning. However, as with any other processing pipeline, manual compression and partitioning is time and effort consuming and require specialized skills. The compression can be automated by deploying novel student–teacher learning techniques and knowledge distillation, where the input–output of a pre-trained large sized deep neural network teacher model can be approximated by a smaller sized student model with some trade-off in model accuracy. Researchers have shown how the above techniques can be used in compressing DNN models used in image processing, whereas similar techniques can also be applied to compressing DNN models used in signal processing [81].

Model compression can take care of the memory constraints in edge devices, but it cannot cater for the CPU constraints for multi-processor systems. For that, model partitioning techniques will be required, that treats the DNN model as directed graphs and partitions it into multiple segments each of which can be run in a separate CPU core, different processor or accelerators like GPUs or neural compute sticks as is available in a given edge computing system [82, 83]. Researchers have shown how such a system can be used in object detection from camera images on an edge platform and also how such a system can be used on speech signals for automated speech recognition (ASR) on the edge [84, 85].

6.8 Summary

In summary, this chapter introduced the concept of artificial intelligence-driven automation of the signal/image processing pipeline for IIoT systems. The typical signal processing and image/video processing pipelines were introduced in the beginning. The discussions on different artificial intelligence-driven methodologies for automating such pipelines follows, which includes adaptive filtering, AutoML, and deep learning-based systems. Also, the automation of model compression and model partition have been discussed. Several IIoT/manufacturing use cases are also discussed.

6.9 Questions for Practice

1. *Define machine learning and its types. List the differences in each type of machine learning technique.*

2. *What is the difference between machine learning, deep learning, and artificial intelligence?*
3. *What is neurocomputing?*
4. *What is the importance of bias in machine learning?*
5. *You are the owner of a supermarket and want to use machine learning for better insights. Which machine learning technique will you use?*
6. *Differentiate between generative and discriminative model.*
7. *What is a kernel trick?*

References

1. Context. http://dictionary.cambridge.org/dictionary/english/context
2. Schintler LA, McNeely CL (2020) Encyclopedia of Big Data. Springer International Publishing, Cham
3. Nyquist's sampling theorem. https://www.cs.cf.ac.uk/Dave/Multimedia/node149.html
4. Olshausen BA Aliasing. http://www.rctn.org/bruno/npb261/aliasing.pdf
5. Resample. http://in.mathworks.com/help/signal/ref/resample.html?s_tid=gn_loc_drop
6. Matched filters. http://ee.eng.usm.my/eeacad/mandeep/EEE436/chp3.pdf
7. Hodge V, Austin J (2004) A survey of outlier detection methodologies. Artif Intell Rev 22:85–126. https://doi.org/10.1023/B:AIRE.0000045502.10941.a9
8. Grubbs FE (1969) Procedures for detecting outlying observations in samples. Technometrics 11:1–21. https://doi.org/10.1080/00401706.1969.10490657
9. Maddala GS (1992) Introduction to econometrics II. Macmillan Publishing Company
10. Frigo M, Johnson SG, FFTW: an adaptive software architecture for the FFT. In: Proceedings of the 1998 IEEE international conference on acoustics, speech and signal processing, ICASSP '98 (Cat. No.98CH36181). IEEE, pp 1381–1384
11. Daubechies I (1990) The wavelet transform, time-frequency localization and signal analysis. IEEE Trans Inf Theory 36:961–1005. https://doi.org/10.1109/18.57199
12. Kay SM, Fundamentals of statistical signal processing: practical algorithm development
13. Lin J, Keogh E, Lonardi S, Chiu B (2003) A symbolic representation of time series, with implications for streaming algorithms. In: Proceedings of the 8th ACM SIGMOD workshop on research issues in data mining and knowledge discovery—DMKD '03. ACM Press, New York, New York, USA, p 2
14. Hall DL, Llinas J (1997) An introduction to multisensor data fusion. Proc IEEE 85:6–23. https://doi.org/10.1109/5.554205
15. Liggins M II, Hall D, Llinas J (2017) Handbook of multisensor data fusion. CRC Press
16. Patra K, Pal SK, Bhattacharyya K (2007) Artificial neural network based prediction of drill flank wear from motor current signals. Appl Soft Comput J 7:929–935. https://doi.org/10.1016/j.asoc.2006.06.001
17. Bhat NN, Dutta S, Pal SK, Pal S (2016) Tool condition classification in turning process using hidden Markov model based on texture analysis of machined surface images. Meas J Int Meas Confed 90:500–509. https://doi.org/10.1016/j.measurement.2016.05.022
18. Poor HV (1994) An introduction to signal detection and estimation. Springer New York, NY
19. Kay SM (1991) Fundamentals of statistical signal processing, vol 1. Prantice Hall
20. Least squares estimation. http://stat.ethz.ch/~geer/bsa199_o.pdf
21. Maximum likelihood estimation. http://times.cs.uiuc.edu/course/410/note/mle.pdf
22. Bayesian minimum mean squared error. http://www.lx.it.pt/~mtf/learning/Bayes_lecture_notes.pdf
23. kalman filtering. http://stanford.edu/class/ee363/lectures/kf.pdf

24. Introduction to machine learning. http://alex.smola.org/drafts/thebook.pdf
25. Awad M, Khanna R (2015) Machine learning. In: Efficient learning machines. Apress, Berkeley, CA, pp 1–18
26. Ernest F-H (2008) Jess, the rule engine for the java platform
27. Proctor M (2012) Drools: a rule engine for complex event processing, pp 2–2
28. Zalta EN (2020) The problem of induction, the stanford encyclopedia of philosophy, Spring 202. Metaphysics Research Lab, Stanford University
29. Magnani L (2001) Abduction, reason and science. Springer US, Boston, MA
30. Derler P, Lee EA, Vincentelli AS (2012) Modeling cyber–physical systems. Proc IEEE 100:13–28. https://doi.org/10.1109/JPROC.2011.2160929
31. Operations research. http://mathworld.wolfram.com/OperationsResearch.html
32. Graves SC, Kletter DB, Hetzel WB (1998) A dynamic model for requirements planning with application to supply chain optimization. Oper Res 46:S35–S49. https://doi.org/10.1287/opre.46.3.S35
33. Martinez GS, Sierla S, Karhela T, Vyatkin V (2018) Automatic generation of a simulation-based digital twin of an industrial process plant. In: IECON 2018 44th annual conference of the IEEE industrial electronics society. IEEE, pp 3084–3089
34. Ramanath R, Snyder WE, Yoo Y, Drew MS (2005) Color image processing pipeline. IEEE Signal Process Mag 22:34–43. https://doi.org/10.1109/MSP.2005.1407713
35. Pal A, Mukherjee A, P. B (2015) Model-driven development for internet of things: towards easing the concerns of application developers, pp 339–346
36. Dey S, Jaiswal D, Paul HS, Mukherjee A (2016) A Semantic algorithm repository and workflow designer tool: signal processing use case, pp 53–61
37. Murphy KP (2012) Machine learning: a probabilistic perspective. The MIT Press Cambridge, Massachusetts London, England
38. Copeland M (2016) What's the difference between artificial intelligence, machine learning and deep learning? https://blogs.nvidia.com/blog/2016/07/29/whats-difference-artificial-intelligence-machine-learning-deep-learning-ai/
39. Pyle D, José, San C (2015) An executive's guide to machine learning. https://www.mckinsey.com/industries/technology-media-and-telecommunications/our-insights/an-executives-guide-to-machine-learning
40. Machine learning. http://mlclass.stanford.edu/#:~:text=What Is Machine Learning%3F,understandingofthehumangenome
41. Samad A (2019) What is machine learning. https://becominghuman.ai/what-is-machine-learning-d292114cc6ce#:~:text=ThefieldofMachineLearning,thatgovernalllearningprocesses%3F"&text=Machinelearningisbasedon,relyingonrules-basedprogramming
42. Le J (2021) 12 useful things to know about machine learning. https://www.kdnuggets.com/2018/04/12-useful-things-know-about-machine-learning.html#:~:text=ByJameLe%2MachinLearningambitiouproblemcanbetackle
43. Bhat NN, Kumari K, Dutta S et al (2015) Friction stir weld classification by applying wavelet analysis and support vector machine on weld surface images. J Manuf Process 20:274–281. https://doi.org/10.1016/j.jmapro.2015.07.002
44. Pal S, Pal SK, Samantaray AK (2008) Artificial neural network modeling of weld joint strength prediction of a pulsed metal inert gas welding process using arc signals. J Mater Process Technol 202:464–474. https://doi.org/10.1016/j.jmatprotec.2007.09.039
45. Madisetti VK, Williams DB (2018) The digital signal processing handbook, vol 3. CRC Press
46. Goodwin GC, Sin KSS adaptive filtering prediction and control
47. Simon H (1996) Adaptive filter theory. Prentice Hall, Upper Saddle River, New Jersey
48. Ingle VK, Kogon SM, Manolakis DG (2005) Statistical and adaptive signal processing
49. Honig ML, Messerschmitt DG, Adaptive filters: structures, algorithms and application. Springer, US
50. Jenkins WK, Hull AW, Strait JC et al (1996) Advanced concepts in adaptive signal processing. Springer, US, Boston, MA

51. Pal A, Ukil A, Deb T, et al (2020) Instant adaptive learning: an adaptive filter based fast learning model construction for sensor signal time series classification on edge devices. In: ICASSP 2020—2020 IEEE international conference on acoustics, speech and signal processing (ICASSP). IEEE, pp 8339–8343
52. Liu W, Prncipe JC, Haykin S (2010) Kernel adaptive filtering. Wiley, Hoboken, NJ, USA
53. Time series classification repository. http://www.timeseriesclassification.com/index.php
54. Lines J, Bagnall A (2015) Time series classification with ensembles of elastic distance measures. Data Min Knowl Discov 29:565–592. https://doi.org/10.1007/s10618-014-0361-2
55. Ukil A, Malhotra P, Bandyopadhyay S, et al (2019) Fusing features based on signal properties and TimeNet for time series classification. In: ESANN 2019 proceedings, European symposium on artificial neural networks, computational intelligence and machine learning
56. Ukil A, Bandyopadhyay S, Pal A (2020) Sig-R^2ResNet: residual network with signal processing-refined residual mapping, auto-tuned L_1-regularization with modified Adam optimizer for time series classification. In: 2020 international joint conference on neural networks (IJCNN). IEEE, pp 1–8
57. Thornton C, Hutter F, Hoos HH, Leyton-Brown K (2012) Auto-WEKA: combined selection and hyperparameter optimization of classification algorithms
58. AutoGluon: AutoML for text, image, and tabular data. https://auto.gluon.ai/stable/index.html
59. Auto-sklearn. https://github.com/automl/auto-sklearn
60. Neural network intelligence. https://github.com/microsoft/nni
61. CloudAutoML. https://cloud.google.com/automl
62. AutoAI. https://www.ibm.com/cloud/watson-studio/autoai
63. AutoML. https://docs.oracle.com/en-us/iaas/tools/ads-sdk/latest/user_guide/automl/overview.html
64. Maximal information coefficient. https://medium.com/@rhondenewint93/on-maximal-information-coefficient-a-modern-approach-for-finding-associations-in-large-data-sets-ba8c36ebb96b
65. Peng H, Long F, Ding C (2005) Feature selection based on mutual information criteria of max-dependency, max-relevance, and min-redundancy. IEEE Trans Pattern Anal Mach Intell 27:1226–1238. https://doi.org/10.1109/TPAMI.2005.159
66. Banerjee S, Chattopadhyay T, Pal A, Garain U (2018) Automation of feature engineering for IoT analytics. ACM SIGBED Rev 15:24–30. https://doi.org/10.1145/3231535.3231538
67. Ukil A, Sahu I, Puri C, et al (2018) AutoModeling: integrated approach for automated model generation by ensemble selection of feature subset and classifier. In: 2018 international joint conference on neural networks (IJCNN). IEEE, pp 1–8
68. Thokala NK, Kumar K, Girish Chandra M, Ravikumar K (2019) Long-term forecasting of heterogenous variables with automatic algorithm selection. pp 186–197
69. DNN. http://neuralnetworksanddeeplearning.com/
70. Developer blog. https://developer.nvidia.com/blog/deploying-a-scalable-object-detection-inference-pipeline/
71. Schwartz E, Giryes R, Bronstein AM (2019) DeepISP: toward learning an end-to-end image processing pipeline. IEEE Trans Image Process 28:912–923. https://doi.org/10.1109/TIP.2018.2872858
72. Rajbhoj A, Deshpande S, Gubbi J, et al (2019) A system for semi-automatic construction of image processing pipeline for complex problems. pp 295–310
73. Bapna A, Thokala N, Chandra MG, Kumar K (2018) Deep learning based tool wear monitoring in CNC machines. In: European conference on machine learning (ECML-PKDD workshop), Dublin, Ireland
74. Mishra D, Roy RB, Dutta S et al (2018) A review on sensor based monitoring and control of friction stir welding process and a roadmap to Industry 4.0. J Manuf Process 36:373–397. https://doi.org/10.1016/j.jmapro.2018.10.016
75. Mishra D, Gupta A, Raj P et al (2020) Real time monitoring and control of friction stir welding process using multiple sensors. CIRP J Manuf Sci Technol. https://doi.org/10.1016/j.cirpj.2020.03.004

76. Mishra D, Shree S, Gupta A et al (2020) Weld defect localization in friction stir welding process. Weld World. https://doi.org/10.1007/s40194-020-01028-5
77. Banerjee A, Mishra D, Nayak P, Pal SK (2020) Low-cost real-time machine vision based quality inspection system (Filed)
78. Lamba H (2019) One shot learning with siamese networks using Keras. https://towardsdatascience.com/one-shot-learning-with-siamese-networks-using-keras-17f34e75bb3d
79. Sahu S, Kumar K, Majumdar A, Chandra MG (2021) Deep transform learning for multi-sensor fusion. In: 2020 28th European signal processing conference (EUSIPCO). IEEE, pp 1996–2000
80. Kumar K, Majumdar A, Chandra MG, et al (2020) TransFuse: a transform learning based multi-sensor fusion framework. IEEE Sens Lett 1–1. https://doi.org/10.1109/LSENS.2020.3039300
81. Hegde S, Prasad R, Hebbalaguppe R, Kumar V (2019) Variational student: learning compact and sparser networks in knowledge distillation framework
82. Dey S, Mukherjee A, Pal A, Balamuralidhar P (2018) Partitioning of CNN models for execution on fog devices. In: Proceedings of the 1st ACM international workshop on smart cities and fog computing. ACM, New York, NY, USA, pp 19–24
83. Dey S, Mukherjee A, Pal A, P B (2019) Embedded deep inference in practice. In: Proceedings of the 1st workshop on machine learning on edge in sensor systems—SenSys-ML 2019. ACM Press, New York, USA, pp 25–30
84. Mondal J, Dey S, Mukherjee A, et al (2019) Edge acceleration of deep neural networks (demo). In: Proceedings of the 17th annual international conference on mobile systems, applications, and services. ACM, New York, NY, USA, pp 691–692
85. Dey S, Dutta J (2020) A low footprint automatic speech recognition system for resource constrained edge devices. In: Proceedings of the 2nd international workshop on challenges in artificial intelligence and machine learning for internet of things. ACM, New York, NY, USA, pp 48–54

Chapter 7
Digital Twin Application

Check for updates

7.1 Overview of This Chapter

This chapter introduces the concept of digital twin in details. The explanations given in Chap. 1 about building a digital twin model are revisited and the ideas are elaborated. Digital twins proposed in different routes of manufacturing are also discussed. These demonstrate the benefits of digital twins in manufacturing. The discussions also include a few schematic representations for a better understanding of the flow of the information. A few case studies are also included where digital twins for different processes are developed. Towards the end, several new concepts of Industry 4.0 are also elaborated and their importance in the development of a digital twin is highlighted.

7.2 The History Behind the Conceptualisation of Digital Twin

It dates back to the "*Apollo Program*" of the *National Aeronautics and Space Administration (NASA)* who attempted to mirror the life of a flying aircraft. In this program, *NASA* made two aircrafts identical to each other, in which one was on the earth and the same was utilized to mirror the life of the flying counterpart. The one left on the earth was termed as the twin of the other in space. This was idealized to help astronauts in space in taking correct decisions. The twinning was performed by simulating the space model on the ground. In this case, the twin is a prototype, which also resembles a physical asset, but was used for mirroring the life of another physical asset. Though this does not represent the true meaning and essence of a digital twin, the idea was in line with the reason one would like to build a digital twin.

S. K. Pal et al., *Digital Twin – Fundamental Concepts to Applications in Advanced Manufacturing*, Springer Series in Advanced Manufacturing,
https://doi.org/10.1007/978-3-030-81815-9_7

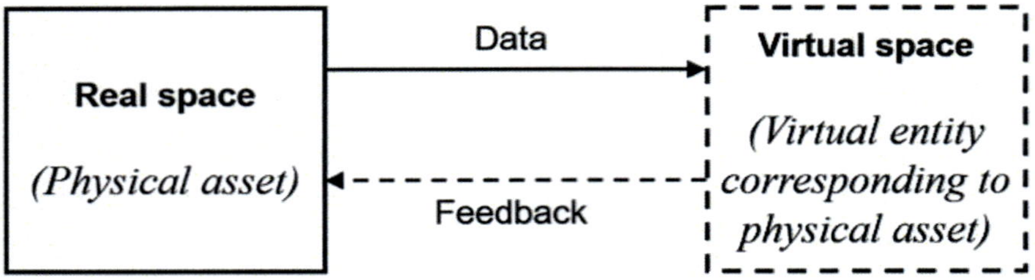

Fig. 7.1 Concept of digital twin

Later, the concept of virtual digital representations was proposed by *Michael Grieves* in the field of "*Product Lifecycle Management*" *(PLM)*. In this, the explanation was to develop a digital copy of a physical asset that can be used under real and simulated conditions to represent a physical asset. The conceptual model of digital twin by *Michael Grieves* consists of three components namely, real space, virtual space, and a connection between these two spaces. This is shown schematically in Fig. 7.1, where the real space depicts the physical asset and the virtual space refers to the virtual products, and the connection between them through the exchange of data.

The conceptual diagram shown above digitizes the virtual space or the twin by using digital expressions to build a virtual product corresponding to a physical one. It also establishes the fact of exchanging data between the two spaces. From these, it can be observed that the conceptualization of digital twin happened long back, sometime from 2005 to 2011. However, it is believed that research on digital twin was not so popular because of the limited and expensive resources for collecting data from a physical asset. Also, many systems were manually operated and thus, it was essential to prepare them for sensing data. Also, there were limited means and resources such as hardware, software, etc. for working on the real-time data. The network was another constraint as well.

Well, history has a vast list of things to tell about the inception of concepts. Not only for the case of the digital twin, but each of the concepts of Industry 4.0 might have been a funny dream to someone and the same would have evolved. At present, it is no more hype. Things are doable. One of the popular terms in this world of twins is the "*Airframe Digital Twin*". It is being developed by the US Air Force Research Laboratory [1]. They aim to solve the maintenance issues and predict the life of an aircraft in complex service environments. The *Airframe Digital Twin* is proposed as a hyper-realistic model of an airframe, consisting of a computational model and associated data of materials, manufacturing, etc. The *Airframe Digital Twin* can be utilized to simulate and judge whether the airframe meets the task condition. It is an integrated model of several sub-models. Let us revisit the concept of digital twin presented in Chap. 1 and elaborate the things for a clearer picture.

7.3 Concept of Digital Twin

Digital twin refers to the creation of a virtual model of a physical system [2, 3]. In particular, the term 'twin' refers to a replica. In manufacturing, the replica could be for a process or a machine. This could further be broken down into many sub-parts such as, within a process, one could be interested to identify and track a particular phenomenon. For instance, during the process of machining, one of the objective could be to track the health of the cutting tool. So, with the help of certain data, a digital twin of the cutting tool could be built. It also could be for the entire process. For instance, the objective may be to track the health of one component of a machine. Bearing in the induction motor is one such critical component, whose failure can lead to loss of production. Within a process, there occur various phenomena and tracking each one of them could be crucial. In that case, one could also build a digital twin for those phenomena. Otherwise, the objective may be to track the health of an entire machine or an entire process. The important thing to note here is the complexity. Referring to Fig. 7.2 shows a plot of digital twin and the complexity that is going to be involved. For the case of tracking just a particular phenomenon or the health of just one component, the complexity would be less. It would be high when the entire process or the machine is involved.

But why one should be concerned about the complexity? The idea of digital twin remains the same for different possible cases; this includes sensors, data, connectivity, information processing, making predictions, etc. The question is how to tackle this complexity? This is exactly where the low-level technicalities will come into the picture. This includes the right sensor to be chosen for acquiring information about a phenomenon, the right choice of technique for information processing, good connectivity between the nodes, etc.

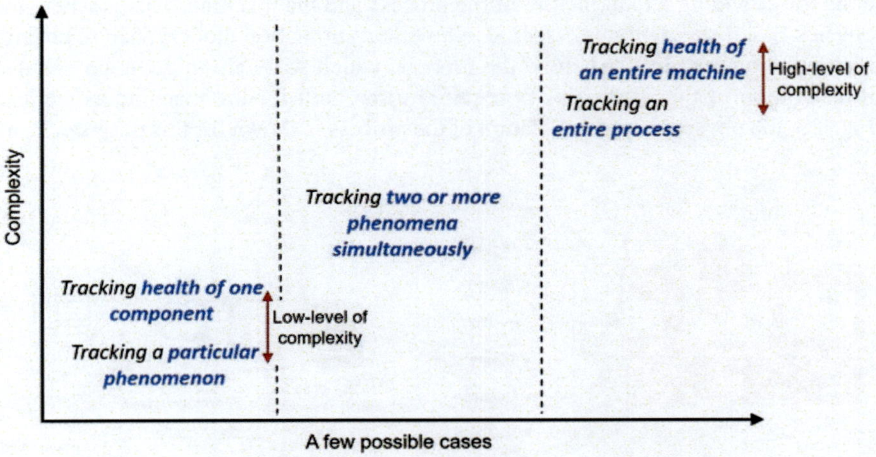

Fig. 7.2 DT and level of complexity

7.4 Realizing a Digital Twin Model

In Chap. 1, the digital twin has been pictorially represented (Figs. 1.8 and 1.9). A digital twin consists of two spaces, the physical space and the virtual space. Physical space consists of the physical machine, sensors, controllers, etc., and it represents the source of data generation and things, which are to be controlled. This space will consist of numerous sensors for acquiring the data from a process or about components of a machine. The collected information is sent to the virtual space in real-time via the internet. This space contains the applications, which will work on the received data to derive useful information out of it. Useful information as in the descriptive analyses of the physical machine or process. Real-time decisions will be taken based on the analyses, which will be an enabler of controlling the physical machine.

Let us dive into a more detailed understanding of how a realistic model for a practical environment can be built. The physical space is the reference point, as this will consist of the machine, process to be executed in that machine, etc. So, it means that it is the process or that machine that is the point of interest. For instance, let us consider the example of a machining process. A computer numerical controlled (CNC) machine must be there on the shop floor for this purpose. To capture the data, various sensors may be engaged with the CNC machine. Figure 7.3 depicts a CNC machine with sensors such as force, speed, position, temperature, vibration, and current. There may be other sensors also, depending upon the requirement. The data from these sensors will be useful for making predictions about the quality of the job being machined. It may also be useful for determining the state of the cutting tool employed for machining. Using this data, one could develop machine learning models to predict the job quality, state of cutting tool, etc. This will help to a certain extent, but, this is not where the capability of a digital twin ends. It aims for providing a total picture of the process and the machine. This is where the physics-based simulation is essential. Numerical simulation models need to be built incorporating the physics behind the process, which will help in getting a detailed investigation of the asset. This is required to be built for the machine as well. In Fig. 7.3, the physics-based simulation of the process is shown for this purpose. From

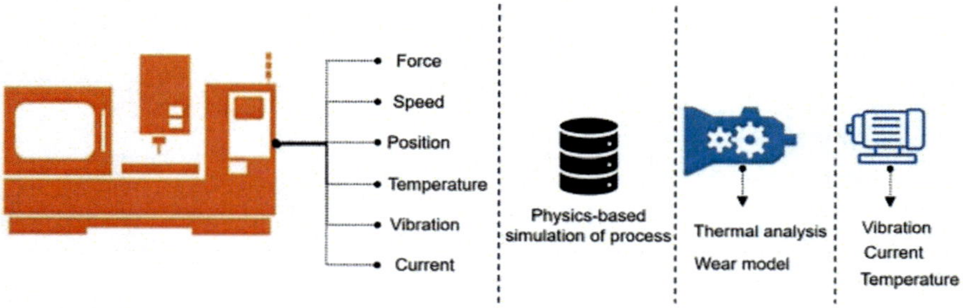

Fig. 7.3 Things involved in a digital twin model

the perspective of determining the health of the asset or the machine, simulations will be required for several components. For instance, the machine will be having a gearbox mechanism. To reduce the downtime of this machine, the state of the gearbox is required to be known a priori, which is one of the critical components. Physical sensors, which is also referred to as hard sensors, such as vibration can be fitted on the gearbox to determine the level of vibration. However, this will not be enough and will not suffice the need. For the details, a thermal or wear analyses will be required, which could be fulfilled by simulating the conditions. Similarly, the machine may also include motors, which are also critical components. Sensors such as vibration, current, and temperature can be engaged to identify the health of the bearings in the motor, load conditions, and so on. However, for more critical analyses, such as the evolution of electrostatic force or torque within the motor, simulation or inferences from a soft sensor will be required.

Needlessly, it should be noted that a digital twin will have several models to make the analysis and provide a holistic view of the component or the process, it is intended for. As there may be a need for 'multiple physics driven models' for one component, a digital twin is referred to as a multi-physics model. Also, the analytics involved in such a scenario will be of multi-scale manner, which gives it another attribute termed as 'multi-scale'. With these computations and information, a digital twin is expected to make useful predictions about the real-time performance and also it needs to state the remaining useful life (RUL) of the machine. This requires 'probabilistic simulations'. That is why, *NASA* has defined "*Digital Twin as an integrated, multi-physics, multi-scale, probabilistic simulation of an as-built vehicle or system that uses the best available physical models, sensor updates, and fleet history to mirror the life of its flying twin*" [4].

There a few things to note here. The objective mentioned is to make useful predictions in real-time, and also predict the remaining useful life (RUL) of the asset in real-time. How is this possible? Simulation runs for hours, consumes a lot of memory, needs high-tech specifications for computing, and other things. How to realize the real-time needs? For this purpose, a simulation model built for a digital twin application is a scaled-down model. It is so built that it can work in a real-time environment at lower computing requirements. Thus, the optimization of the model is a big challenge here. Once optimized and scale-down models are available, the next crucial thing is how to play with them? Anyway, since an asset may consist of multiple such models, one intelligent thing to follow is a *Doctor's Rule-out Test Approach (DROTA)*. Because of the involvement of multiple models, ensuring the timeliness of a digital twin model is challenging. It would be hard to maintain the system's outcomes in real-time, and the entire purpose may get lost. The *DROTA* idea may be useful. Let us understand this approach.

Consider the example of the machining process, which has been discussed above. Here, the importance of different simulations is brought for the complete diagnostics of the process and the machine. To implement the *DROTA* here, once the system (machine + process) is online, data should be fed to the model whose output is much more crucial and insightful to understand the behaviour of the machine. When a patient visits a doctor, and if the doctor is not sure about what could be the possible reason for such a health condition which the patient would be having, the doctor prescribes a few tests and mentions the priority as well. The intention of the doctor here is a structured and systematic approach to reach to actual disease or cause of that health condition. Similarly, for the machining process, maybe the temperature around the cutting tool and the workpiece could be the most important findings, then that should be simulated first. Then for ensuring the results of the first series of tests, a doctor may prescribe for the second test, before making a recommendation of his diagnosis. Similarly, the thermal analysis of the gearbox is of higher importance and should be performed first. If abnormalities are observed, the wear model could then be triggered. In this manner, all the models can be sequentially triggered and ultimately, decisions on the process and health of the machine can be made based on the outcomes of the models by the digital twin application. Further, the simulation models can be operated and run in various applications depending on the solvers required for the problem. All these are expected to deal with the timeliness of the digital twin application. These discussions are schematically represented in Fig. 7.4.

Figure 7.4 shows the two spaces i.e., physical and virtual. The physical space includes the CNC machine and the equipped sensors. The digital twin model, a collection of multiple simulations and machine learning models is the virtual space. The acquired sensory data is transmitted to the virtual space, where the models do work on them. Based on the outputs of the virtual space, feedback is sent to the physical space for controlling the physical asset. Thus, the process can be optimized to ensure the quality of the product. Since the digital twin can reveal the true condition of the physical asset, corrective measures can be taken at the right instant to prevent failure. Besides, the user can also optimize the next phase of production based on those predictions.

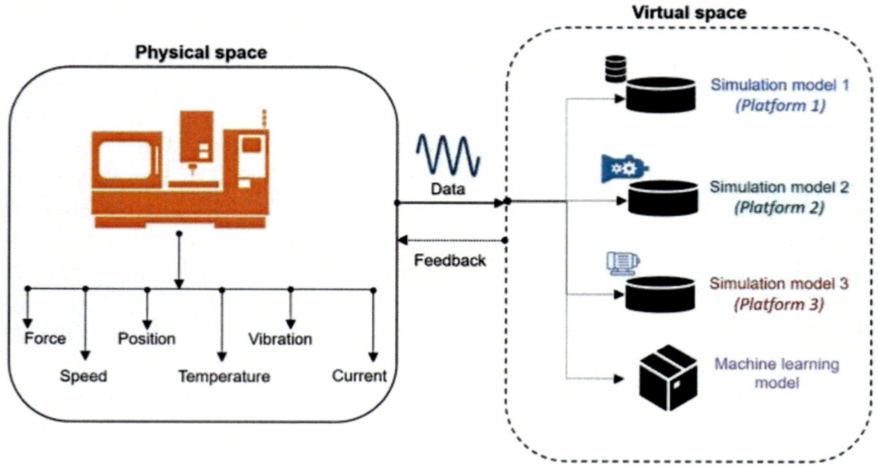

Fig. 7.4 *DROTA* for a digital twin application

Another need is the calibration of the simulation model, which attempts to deal with the accuracy of information. How a simulation model will be able to diagnose a process or a machine? The health will be deteriorating gradually and thus, the simulation model needs to be calibrated periodically or continuously by feeding the sensors' data to the model. In this manner, the model can match its computations to the actual machinery conditions. For instance, consider a simulation model of an induction motor, which is intended to perform a thermal analysis. In the ideal case, the simulation model is unaware of the vibration that is being generated in the motor during the actual running condition. Thus, the data from a hard sensor, vibration sensor, in this case, needs to be fed to the model, which will help it behave close to the condition of the physical asset. This is the actual essence of a digital twin; behaving as close as the condition of the physical asset.

7.5 Digital Twin as a Tool Throughout the Life Cycle

In Chap. 1, a digital twin has been mentioned to be a digital representation of a physical asset. It was not discussed when a digital twin comes into the picture or the life of a physical product. But the digital twin has been narrated to be serving throughout the life cycle of a product. In this section, discussions will be highlighted on this aspect i.e., digital twin as a tool throughout the life cycle of a product. Now, this is very crucial once the entire life cycle is being envisaged to be digitally driven, monitored, controlled, and optimized, and so on.

A product life cycle conventionally has four phases namely, *Introduction, Growth, Maturity,* and *Decline* [5]. The first phase i.e., *Introduction* includes awareness of product and identification of after-market opportunities. The second phase i.e., *Growth* highlights product branding and promotion strategies to enhance the marketability of the product. In the third phase i.e., *Maturity*, the product has established itself in the market and the only challenge is to deal with the market competition from similar products. In the final phase of *Decline*, the product becomes obsolete and requires significant remodelling and reformation to sustain itself in the market, if possible. This concept has affected the sustainability of the product and also its design and manufacturing. Therefore, digital twins are being introduced as an innovative way of designing and manufacturing a product. The following paragraph discusses the idea of using digital twin for this purpose.

It is a three-stage process namely, conceptual, design, and virtual verification [6]. In the first stage i.e., conceptual, a digital twin is more of like a simple design of a product. It necessarily includes a feedback form sort of thing to take feedback from other design engineers, customers, etc. The feedback is incorporated in the digital twin model for refining its design. Now, why this is important? Usually, the practice includes developing a prototype, where money, effort, and other resources are involved. With a digital twin, these resources can be saved. An important thing to note here is that the physical product is not available. It is the digital representation, which is available. So, the physical one is still inside the brain of the designer, who has built the model [7]. For example, if a designer is willing to develop a new car, the designer builds a model of the design, which is a computer-aided design (CAD) model. This model will describe how the car will appear. It will be a 3D model listing the components, dimensions of the components, grade of material, manufacturing processes, etc. So, in this stage, a digital twin is similar to a CAD model. A CAD model is responsible for product design, and it is heavily based on the designer's experience.

In the second stage i.e., design, the digital twin undergoes multiple simulations for ensuring the product performance. The simulations will depend upon the type of product that is being developed. Based on the results of the simulation, the product design can be optimized. Consider the manufacturing of a car, simulations will be required to ensure the structural rigidity of the vehicle, ability to withstand various road conditions, the behaviour of the vehicle in different loading conditions, etc. Accordingly, the designer can make changes to the design of the product. This is helping to eliminate building physical prototypes and also it will eliminate the use of infrastructure that would have been employed by the designer in making the calculations and inspections.

In the third stage i.e., virtual verification, things such as the manufacturing aspects, design for manufacturability, etc. are investigated. It is very important to analyse the sequence of manufacturing and also the processes involved in fabrication. Some processes may perform poorly than others and thus, the same also needs to be known beforehand. Also, the third stage includes dealing with the prediction of the product quality. Based on the outcomes, the feedback is sent to the previous stage, to make further changes to the model.

For manufacturing industries, be it process, chemical, refinery or others, they consist of numerous machines. The above-made discussion is for developing digital twin in the product design and manufacturing/installation stages, as shown in Fig. 7.5. What about the machines installed on the shop floor? In general, it can be said that a product has three stages such as designing, manufacturing and installation, and service period. Out of these three, the service period is the longest, where the physical asset runs uninterruptedly and provides its service. This is the major concern area of this textbook. Once a machine is installed and is ready to use, how one can make use of the digital twin strategy to monitor, control, optimize, and make other profitable decisions. The machining example discussed in the previous section is the one applicable.

These discussions justify the importance of digital twin as a tool throughout the life cycle of a product.

7.6 Digital Twins for Design and Manufacturing

This section discusses the digital twin models that have been proposed in the existing literature for the designing and manufacturing stage. The purpose of this section is to help readers in understanding digital twinning via practical examples.

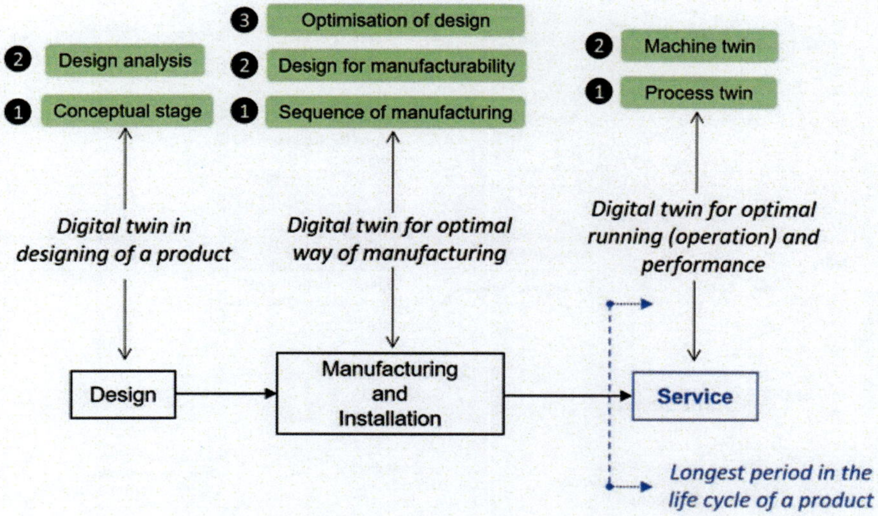

Fig. 7.5 Digital twin as a tool throughout the life cycle

7.6.1 Digital Twin for Iron and Steel Product Life Cycle

One example available in the existing literature is a digital twin model proposed for the iron and steel product life cycle [8]. The proposed digital twin model constitutes raw material acquisition to recycling of the scrap. A picture of this twin is schematically shown in Fig. 7.6. In this process, first comes the order stage, where the customer needs are identified, examined, and well-understood before the requirement by employing inventory and order models. Accordingly, a comparison between the current order and future requirements are forecasted. This helps in preparing for the upcoming orders and it also eliminates the burden of over-storage and wastage of raw materials.

After acquiring the materials, the design stage follows, which consists of models for generating the drawings, and other models are also involved for predicting the strength, heat transfer, residual stress, distortion, and fatigue to determine the strength, service-life, and optimum product design parameters. During the production stage, models for capacity requirement and material requirement planning are employed to determine the phasing-in and phasing-out of capacity, as and when required. It establishes the required coordination amongst the machines, materials, and men on the shop floor. Accordingly, the production activities are scheduled. Planning and scheduling in this stage are for the overall record and it gives feedback to the material purchasing stage. Based on the information, the purchase and logistics models plan for the procurement of materials, their logistics, and their expenses. The manufacturing stage includes various models for production processes viz. blast furnace, converter, casting, hot-rolling, etc., which aims at determining the total

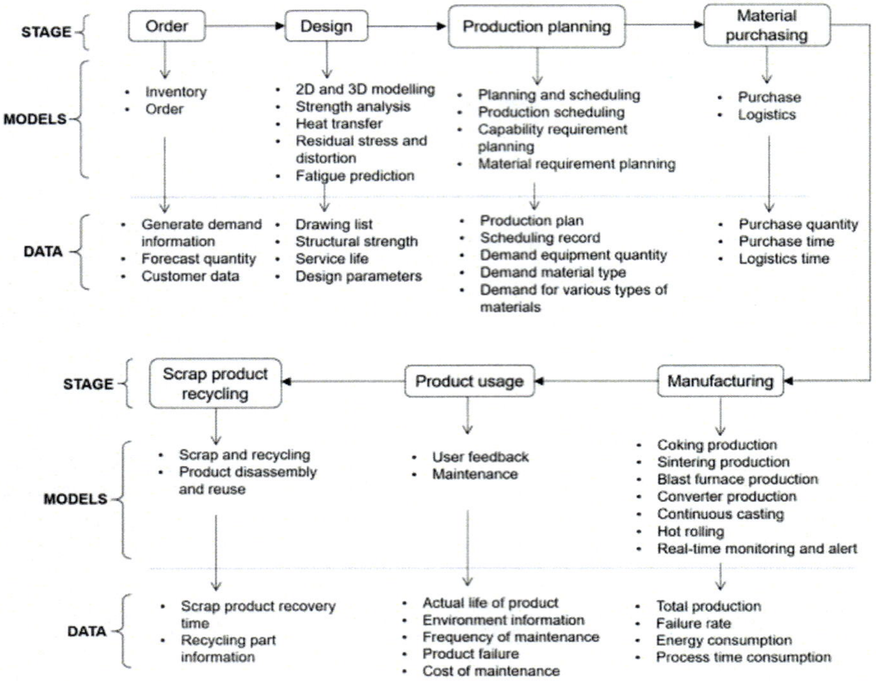

Fig. 7.6 Schematic picture of the twin proposed for iron and steel product life cycle

production rate, failure rate, energy consumption, and process time consumption. Additionally, the twin will have a product usage model for obtaining feedback from the customer to evaluate the product on its intended features during its service life, frequency of failure, maintenance, and cost incurred in those failure and maintenance etc. It will also have information of the geometric models, material properties, environmental data, historical data, etc. The life of the product can be predicted via this information. The same can be utilized to compare with the actual feedback received about the life and maintenance of the product from the customer. Accordingly, changes can be suggested to the design and manufacturing stages. A scrap recycling model is also included in the twin for finding out the scrap products. These models are required to have uninterrupted information flow between the physical and virtual entities, which will ensure product traceability.

7.6.2 *Digital Twin for Weld-Joint Life Cycle*

Researchers have stated the importance of digital twin modelling in developing proper planning of a welding process [9]. These days, a lot of research is being focused on the joining of dissimilar materials. The combinations of dissimilar materials mostly include steel to aluminium, steel to magnesium, copper to aluminium, etc. These combinations are popular owing to the unique characteristics of the two materials in combination. For example, steel is unique for its strength, hardness, and the high-strength low-alloy steels are even famous for their alloying properties. Aluminium is popular for its high strength to weight ratio, resistance to corrosion, weldability, etc. Thus, combining these materials by creating permanent joint results in a hybrid weld such that its unique properties can be leveraged for different applications. Majorly, the automobile sector is one of the areas, where applications of hybrid joints can be found and applied. One example is the *Honda Accord* passenger car, where the front sub-frame is manufactured from hybrid materials, it is a steel to aluminium welded joint. For the car, steel is providing strength and ductility to the vehicle components, and at the same time, aluminium on the other side is providing light-weight to the car. So, a digital twin can be useful in simulating the welding conditions for joining of these dissimilar materials to determine the optimum process conditions before actual welding. This aims to improve the quality of welding and productivity.

A digital twin model has been proposed for spot welding, which includes virtual simulations for the entire life cycle of welds [9]. This includes weld designing, optimization, planning, production, and evaluation. A schematic of this proposed twin along with its utility for each stage is depicted in Fig. 7.7.

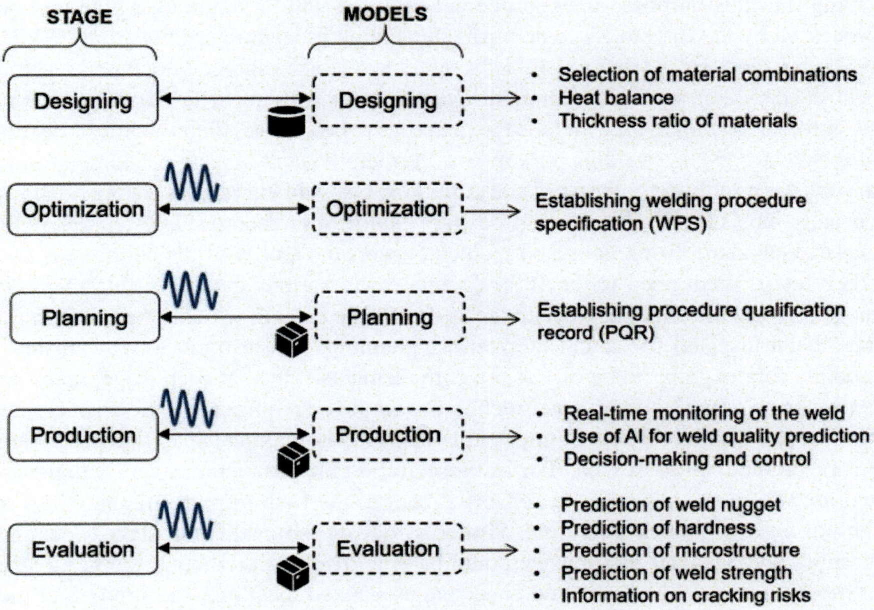

Fig. 7.7 Schematic picture of the twin proposed for spot welding

Let us understand this twin model in details. The first is weld designing, where the importance is to make an optimum design of the welds including material combinations, heat distribution, positioning of the welds, the thickness ratio of the materials to be welded and so on. Now, why this step i.e., weld designing is important? Industries attempt at employing newer materials to improve the efficiency of the product, and at the same time, reduce the cost. Before introducing new materials to fabricate a product, a large number of tests are required to be carried out. The testing not only involve different grades of material, but varying dimensions, and other design factors. So, if one could obtain information by simulating the conditions in a virtual scenario, then a lot of efforts could be saved. In this case of spot welding, it is crucial to understand the distribution of heat as the electrodes press the materials to be welded allowing the current to pass through them resulting in the fusion of those materials. So, outputs such as spot size, tensile strength, etc. become the deciding factors of the weld quality. Thus, simulations will be beneficial to visualize these things before actual welding. In the schematic diagram, the twin represents simulation models for achieving these things. This is followed by optimization of the welding design, where the optimized parameters' window is obtained. A thing to note here is that a parameter window cannot be generalized and the same is case-specific. For instance, the parameters' window obtained for joining aluminium to steel via spot welding technique need not be applicable for joining copper to aluminium via the same technique. Thus, manually, one would need to perform various experiments to define the window. Instead, the virtual simulation for optimizing window will prove to be cost-saving. For this purpose, an experimental database will be required, which will be processed by the twin to come up with the welding procedure specification (WPS). If you are wondering what a WPS is; it refers to a standardization document, which is utilized by industries to maintain the standards for welding [10]. For example, the WPS for spot welding may include the materials to be welded, their thickness, details of the electrode gun, welding parameters. Further, the next objective is to determine the optimum settings for the weld and evaluating the weld via various testing methods virtually. This is referred to as the procedure qualification record (PQR). Again, PQR is also a standardization document, which ensures the ability of the parameters and other details mentioned in the WPS document via various quality testing methodologies [10]. One could appreciate the saving in the capital, which otherwise would have been invested for actual fabrication, planning, and testing. So, for all these models, data is going to be one of the core elements. Thus, at each stage, machine learning and artificial intelligence techniques would be required to perform analytics on the data, which will help in improving the decision-making capability, predicting errors, predicting failures, etc. For instance, during the production stage, actual data acquired from the machine via sensors is sent to the twin to perform analytics for monitoring the weld in real-time. Similarly, during the evaluation stage, machine learning and artificial intelligence could be employed for predicting the grain size of the weld nugget, heat-affected zone, and associated locations, the strength of the weld, etc.

7.7 Digital Twins for Service

Digital twin modelling carries a huge potential for the service period of physical assets. This is because the physical asset spends an enormous amount of time during this period. Digital twins for this period can be broadly classified into two categories namely, a process twin and a machine twin. The first refers to a digital twin of a manufacturing process, which is performed by an asset and the second refers to a digital twin for the asset itself. Recall, Fig. 7.2, which discussed the complexity in the building of the digital twin application. Here also, as one goes on adding components for monitoring, the complexity will be increasing. Let us discuss a few case studies proposed by researchers around the globe for the service period.

7.7.1 Digital Twin for Critical Component—A Gearbox

Researchers proposed a digital twin model for monitoring the gearbox, which includes predicting the breakage of the tooth [11]. Here, the gearbox forms the physical asset. Failure is being monitored by using a vibration sensor. Vibration suffices the behaviour of an asset and that is why vibration monitoring is usually performed for indirectly assessing the health of machines. In this research problem, the vibration frequency spectrum was studied, and the same was observed to be depicting the varying level of energies and frequencies corresponding to different faults of the gearbox. Thus, a few features were extracted from the vibration signal.

Figure 7.8a represents a schematic of the twin developed for the health monitoring of the gearbox. The twin consisted of both a simulation model and a data-driven model. The simulation model was envisaged for reflecting the faults of the gearbox via the stress distribution across the tooth. Simulations were performed for understanding the contact stress on the tooth surface and bending stress on the root of the tooth. The simulated outputs considered were maximum contact stress, maximum bending stress, and the number of gear meshing. The linkage between the asset and the model is the data. The developed simulation model has been validated by correlating with the input and output torque values of the gearbox. For this purpose, a torque sensor is also installed on the gearbox from which, the experimental data was collected. Accordingly, a correction factor was calculated to compensate for the deviation between the measured and simulated values obtained from the gearbox and the simulation model, respectively.

So, the twin, in this case, is a machine learning model, which is trained with features of a hard sensor (vibration engaged on the gearbox) and a soft sensor (simulation model) for predicting the health of the gearbox. The outputs of the sensors have been shown in Fig. 7.8b, c, respectively. It is a neural network model built for predicting faults. Thus, it depicts the example of a hybrid digital twin model that is clubbing the data obtained from the hard and soft sensors.

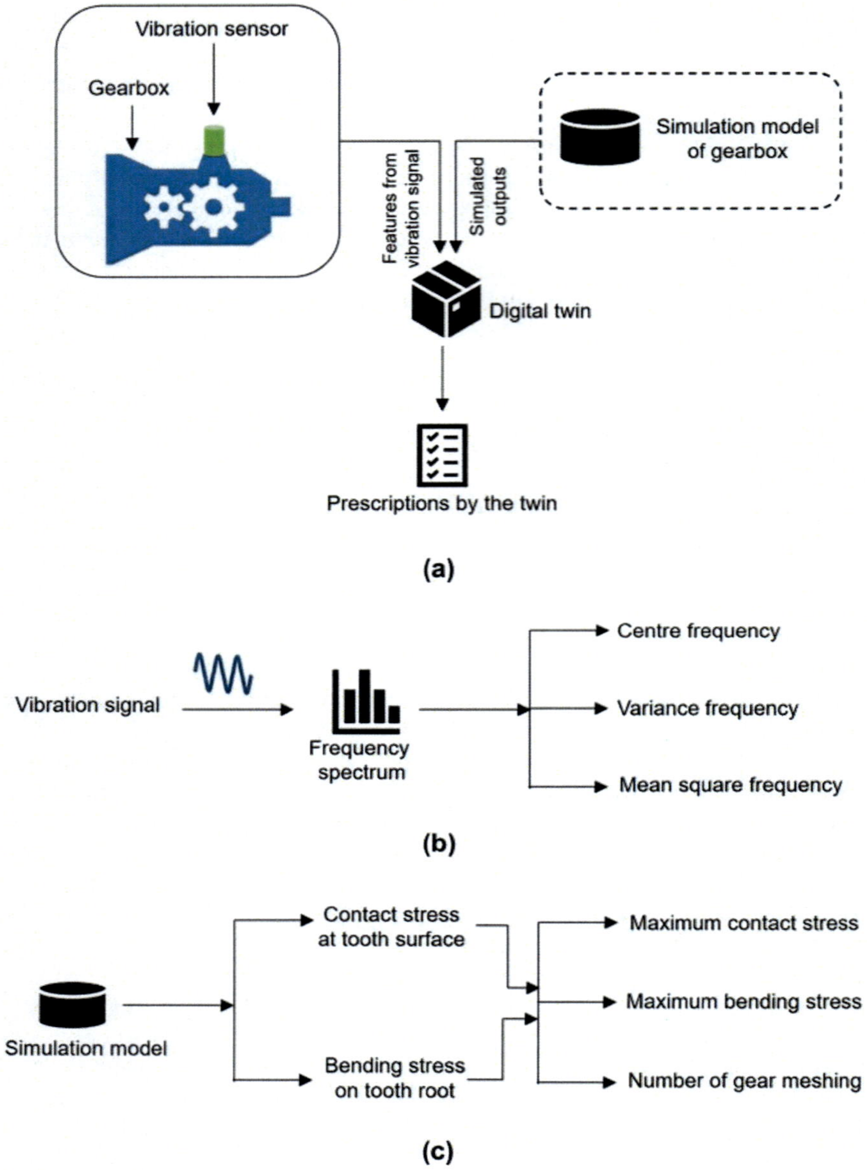

Fig. 7.8 Schematic picture of the twin developed for health monitoring of gearbox

The developed machine learning model i.e., the neural network consists of a single hidden layer. The proposed digital twin-based solution was compared with that of a data-driven solution in predicting the faults, wherein the accuracy of the digital twin model was found to be higher. It is believed that the vibration signal could be prone to interferences arising out of other components present in the gearbox, which is lowering the accuracy of fault prediction. This model may need to be made deeper when it will be employed for predicting more faulty cases. Also, the model is only prescribing the possible repair solution required for addressing a particular fault. More complexities will be involved, when the twin will send feedback to control the asset in real-time.

7.7.2 Digital Twin for Critical Component—A Voltage Converter Station

A digital twin platform named "*Pavatar*" has been developed for the inspection of the ultrahigh-voltage converter station [12]. This station converts direct current (DC) to alternating current (AC) from renewable energy sources like wind, solar, nuclear, etc. The station consists of a synchronous compensator dedicated to stabilizing the outgoing current by absorbing or generating reactive power, in response to the voltage fluctuations. A picture of the twin proposed for this voltage converter station is depicted in Fig. 7.9.

It consists of several sensors such as temperature, pressure, vibration, and rotation for sensing the conditions of the rotating equipment in the station, and other sensors such as temperature, humidity, noise, air quality, and liquid leakage for sensing the environmental conditions and other information. The twin forecasts the system errors, identify the faults, and diagnoses them. The collected data is envisaged to be utilized for detecting the eccentricity in rotating components, detection of specific events, and data visualization.

7.8 Challenges and Intelligence

As pointed out previously, the complexity involved in building a digital twin application primarily depends upon the problem statement. Just to recap, if one aims to track a phenomenon occurring during a process, the complexity is going to be less as compared to tracking the entire process. In this section, challenges associated with a digital twin application will be discussed, which are represented schematically in Fig. 7.10. The most basic need is preparing the physical asset for twining. This means the physical asset needs to be instrumented with different varieties of sensors. Next is coming up with the simulation models that will be a part of the virtual space. So, the model needs to be securely placed in a virtual environment i.e., in a cloud server,

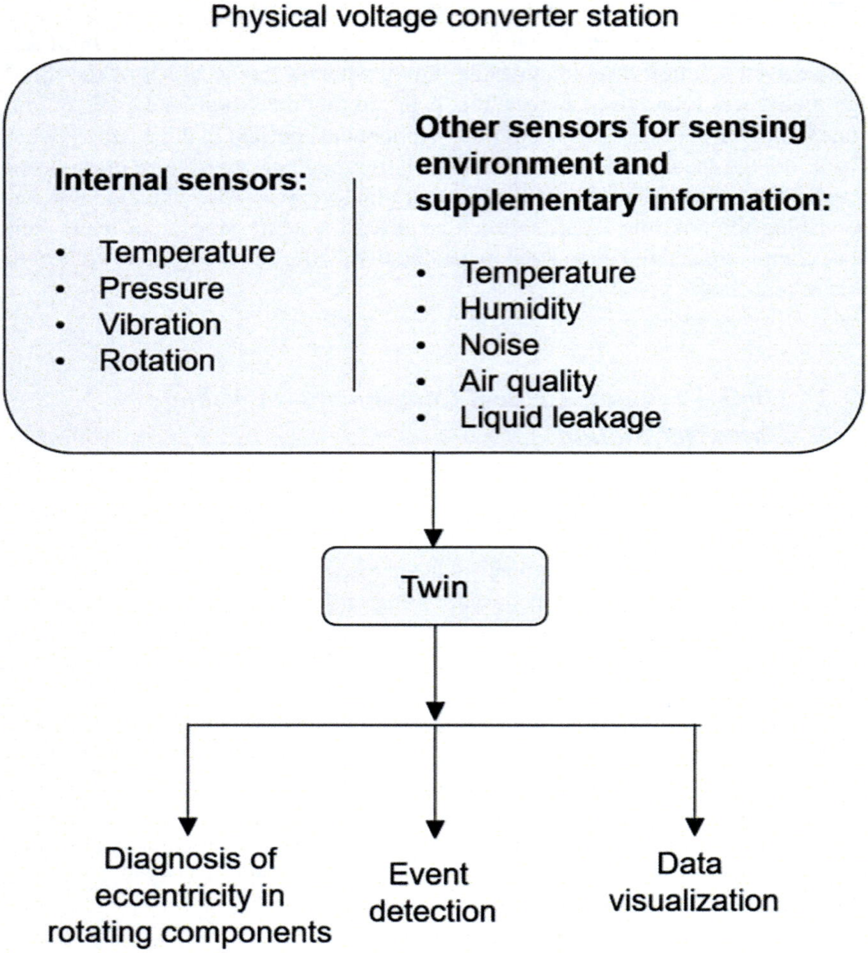

Fig. 7.9 Schematic picture of the twin proposed for monitoring of voltage converter station

or it also may be a service system on the shop floor or a service system within the group enterprise [13]. Integration of the model with the physical asset for enabling the real-time transfer of information is to be performed. The incorporation of big data analytics into the digital twin model is also essential to find hidden patterns in the data. Since, this process is going to generate a huge amount of data in various stages of a product, proper and logical management of the data is essential. The challenges discussed above are at a higher level. Within each of these, there may be several more issues, which needs to be delivered. For instance, in the instrumented physical asset, selection of the type of sensors, data acquisition and its management, multi-sensor synchronization, integration of edge devices for pre-processing, control

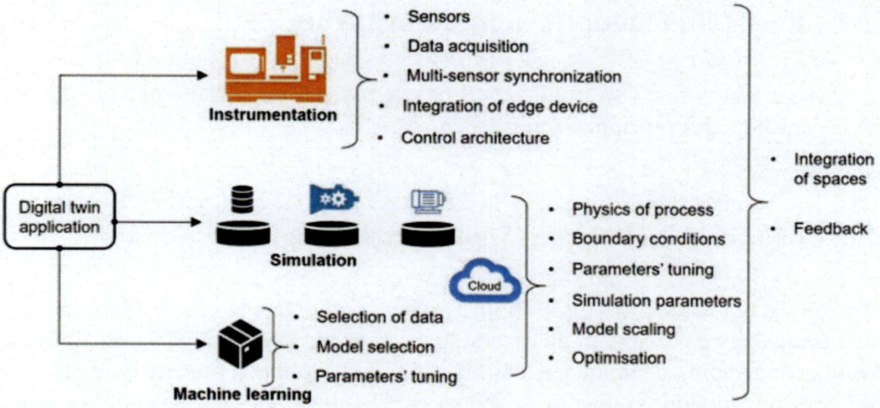

Fig. 7.10 Challenges in digital twin

architecture of the machine, etc. Similarly, for the simulation model, one needs to have a complete understanding of the physics of the process, associated boundary conditions, parameters' tuning, parameters to be simulated, creating a scaled-down model, optimization of the model, etc. Besides, one also needs to develop the machine learning models from the sensors' data as well as by using the simulated outputs.

A digital twin application must possess "*intelligence*". In the previous sections, it is discussed that the stages of a product are designing, manufacturing and installation, and service. During the design stage, the digital twin will serve as a tool for coming up with an excellent product. This is possible because of the realistic simulations that could be carried out, and also because of the feedback that would be made available from the end-users, feedback from other engineering departments, etc. So, it would not be in silos anymore. Instead, it is attempting at connecting the product from one end to the other across the value chain. In the manufacturing and installation stage, the digital twin will help in optimizing the design, generating the code for manufacturing, checking for design for manufacturability, sequence of manufacturing, etc. If any changes are required in the manufacturing sequence, then feedback could be triggered to the design department. Finally, in the service stage, to be more specific, if a manufacturing process is considered, a digital twin will aim at process monitoring and control, which will lead to the prediction of errors if any, and to track the health of the machine. However, the challenge lies in augmenting the intelligence of these models with the accumulated data with the aim of continuous improvement of the physical object. For example, a digital twin developed for a machine will gradually acclimatize with the various situations arising within the process. It also learns the response of the machine in such situations. Accordingly, it shall prescribe its feedback. This process of learning needs to strengthen with the passage of time.

7.9 Real-Life Examples in Manufacturing

In this section, a few case studies will be discussed, which show practical digital twins developed for manufacturing processes.

7.9.1 Process Twin Using Signal Processing

As pointed out in the previous section, a process twin refers to one build for tracking and controlling a process. In this example, friction stir welding (FSW), an advanced solid-state welding technique, is considered for developing a process twin [14, 15]. For this, a detailed introduction to the process is given first, followed by the nitty–gritty of the process twin.

7.9.1.1 Details of FSW

Friction stir welding is a solid-state joining process, which means, it does not involve melting of the base materials to be welded [16, 17]. For joining, a specialized tool is utilized, which is non-consumable. This tool is mounted on the spindle of the welding machine. Once the process begins, the tool rotates and is allowed to plunge inside the base materials. This results in the generation of frictional heat in the tool-workpiece interface regions. Because of the sustained deformation by the action of the tool and prevailing heat, the base materials get plastically deformed. This means they attain a stage, where they can flow. This is the very first stage of the process and thus the term friction justifies the title of the welding technique. Following this, the machine bed upon which, the base materials are placed, is allowed to traverse. Thus, there occurs a combined action, tool rotation and bed traversal. The combined action is responsible for stirring the deformed material. This could be visualized as a process of blending. This is why the second term is in the title of the welding technique. Finally, the weld joint is created because of the frictional heating and mechanical stirring. So, the major parameters involved in the process are tool rotational speed and welding speed. Besides these two, plunge depth and tilt angle are two more parameters. However, in most applications, they are kept constant.

7.9.1.2 Why Develop a Process Twin of FSW?

Welding comes at the last in the sequence of manufacturing processes. So, a product may include numerous components, which are fabricated via different routes of manufacturing. In the end, they are welded together for final assembly. In the context of welding, maintaining the product quality is crucial, or in simple words, the product should be free from defects. Thus, automation along with monitoring and control of welding is inevitable. Friction stir welding, because of its solid-state nature, renders numerous benefits. To list, let us start with the tool utilized for welding. It is non-consumable, thus, it does not add extra weight to the weldment. Besides, it also makes the process cost-effective unlike the numerous filler rods employed for joining in the case of conventional welding technique. Next comes the process variables, which are very few in FSW, thus, it makes the process easy. Coming to the ability of the process, since it works in solid-state, dissimilar materials could also be welded. Also, the welding defects, which occur because of solidification are eliminated. Other benefits include low distortion and low residual stress, absence of fumes, arc which make the process environmentally friendly. With all these advantages, it has a high potential for the manufacturing industries. However, if welding defects occur, it degrades the weld strength of the material. A bad quality weld is highly undesirable in industries as it indirectly increases the production cost. There are numerous destructive and non-destructive post-processing methods for analyzing the weld quality. But these offline methods are not economically optimal for industries as defective welds must be eventually discarded. This also accounts for a huge wastage of the welding material. Thus, online monitoring and control of the process are required for subsequently improving the weld quality after a defect has occurred during welding. Thus, with continuous monitoring and control, the process can be made more reliable.

7.9.1.3 Experiments for Developing the Twin

For the twin, the physical machine is a CNC-FSW machine *(WS004, ETA Technology)*, the picture of which is shown in already shown in Chap. 2 (Fig. 2.31) during the discussion on sensor electronics. The machine is capable of fabricating linear welds only. For welding, a user needs to feed the input parameters to the machine. Recapping the challenges that were discussed above, the first is the instrumentation of the machine. This FSW machine is instrumented with a load cell, which acquires force and torque signals, and a power sensor is also connected externally, which acquires the power signal. An environment was created using LABVIEW and PYTHON to acquire data from the sensors mentioned above. Since, multiple sensors were engaged in the process, for synchronization, a multiplexer is employed.

7.9.1.4 Development of the Process Twin

This twin developed for the FSW process is purely driven by data. This data corresponds only to the sensors' data i.e., force, torque, and power. The physics-based simulation is not involved in this case. The objective of the twin is to track the process in real-time i.e., to monitor the weld quality and also control the process if the quality is found to be poor. Figure 7.11 depicts a schematic of how in this case the physical and the virtual spaces are connected. The instrumented FSW machine comprises the physical space. The data from the sensors is acquired in the computer connected to the machine. The acquired data is sent to the virtual space in real-time i.e., a cloud

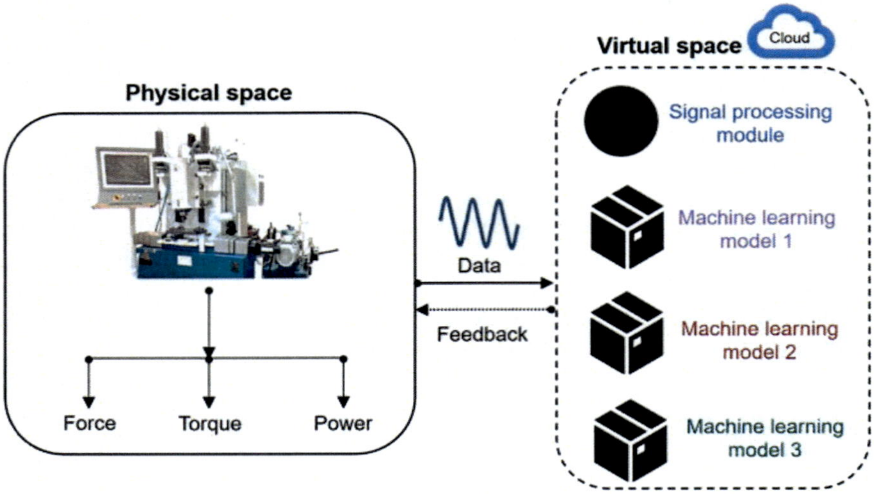

Fig. 7.11 Physical and virtual space for FSW process twin

server *(Meghamala; the private cloud of IIT Kharagpur)*, which is the twin. The twin consists of four distinct, but connected modules namely, a signal processing module and three machine learning models. The analysis performed by the twin is sent back to the physical space i.e., the welding machine in real-time as feedback for process control. The three machine learning models are for predicting the weld quality, predicting the modified parameters, and predicting the health of the tool.

The data acquired from the sensors is transferred to the virtual space in real-time. For acquiring the data from the sensors, a data acquisition card is used. The physical machine (FSW machine) is having a Windows operating system. The virtual space *(Meghamala)* is having Linux operating system. Thus, to establish communication between the friction stir welding machine and *Megahamala*, a third device was utilized. This device remotely accessed the FSW machine and connects the same to the cloud server. Figure 7.12 schematically depicts the communication established between the physical and virtual spaces. The internet connectivity of the FSW machine and the third computer was achieved through cellular 4G communication. For accessing the FSW machine via the third computer and subsequent transfer of data, TCP/IP application protocol is employed. Here, the *WebSockets* was used for data transfer, which takes the IP and port number of the devices to communicate. In the third computer, the IP and port number of the FSW machine were specified and the two started communicating. The cloud server was also accessed through this computer using the same protocol i.e., by providing the IP and port number of the server. Once the data transfer occurs in real-time, analysis was carried out in the virtual space, and feedback was triggered to the physical space. Let us see and understand the computations that were performed in the virtual space.

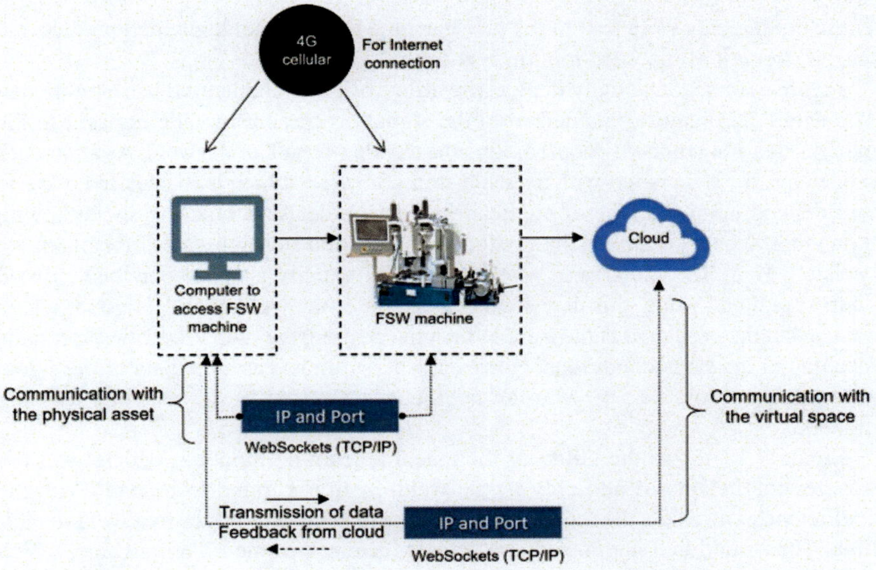

Fig. 7.12 Communication between physical and virtual spaces

As mentioned previously, the purpose of the twin was to monitor and control the occurrence of a defect in the welding process and another aim was to prescribe the health of the welding tool. Let us discuss the modules involved in the virtual space and their operations. So, the first is the signal processing module and its operation is described in Fig. 7.13. Once the data arrives at the cloud server, batches of data are formed. In this case, a batch of 64 samples is formed for analysis. This batch of data i.e., 64 samples of force, torque, and power signals are analysed via discrete wavelet transform and the D1 i.e., the wavelet coefficients at the first level were determined. In the chapter on signal processing (Chap. 3), it was observed that a majority of the information of the raw signal is contained in the initial levels. Thus, to ensure the timeliness of the entire communication, only the D1 coefficients were selected.

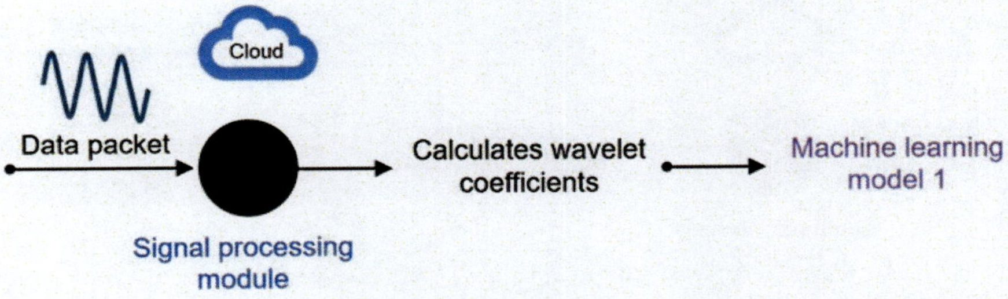

Fig. 7.13 Signal processing module of the twin

These coefficients were sent to the first machine learning model, which predicts the tensile strength of the weld-joint in real-time.

Figure 7.14 schematically depicts the utility of the first machine learning model. The detail coefficients obtained from the signal processing module are sent to the machine learning model 1 for predicting the tensile strength of the weld. As a standard procedure, the twin is fed with a tensile strength value, which is the desired one and will depend upon the type of application. Thus, a decision box is associated with the model, which compares the predicted and desired strength values for making a decision. Here, the decision is whether the weld quality is 'good' or 'bad'. Based on the predicted value, this decision is taken, and if the weld quality is good, which means that the predicted value is more than the desired, the twin prescribes to run the machine in the same conditions. However, if it is not the case, the data of the signal processing module i.e., the wavelet coefficients are sent to the machine learning model 2.

Figure 7.15 shows the utility of the machine learning model 2, which takes the wavelet coefficients, if and only if, the weld quality is found to be bad. With the coefficients, it predicts the modified parameters to achieve a defect-free weld in real-time. These modified parameters are sent as feedback to the FSW machine, which resembles the physical space.

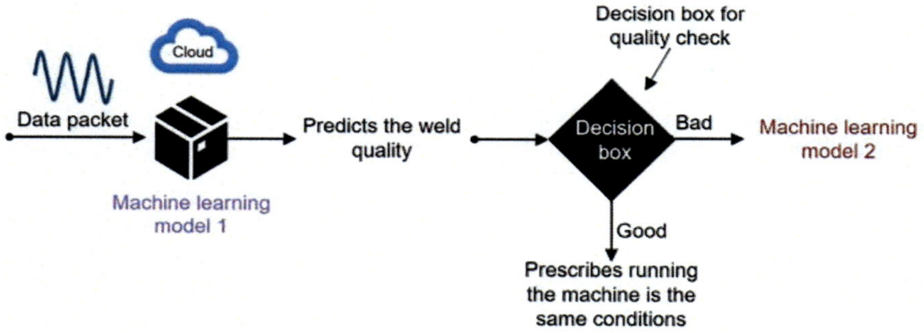

Fig. 7.14 Machine learning model 1 of the twin

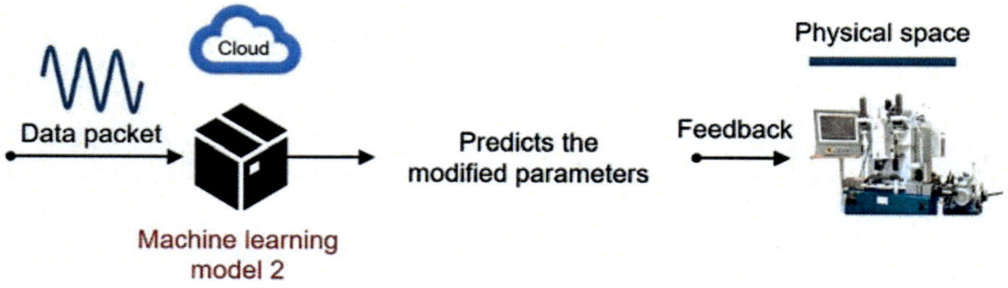

Fig. 7.15 Machine learning model 2 of the twin

The first two machine learning models (1 and 2) are artificial neural networks (ANNs) with multiple hidden layers. The input to these models is the wavelet coefficients obtained at the first level of decomposition of three signals i.e., force, torque, and power. For modelling, the backpropagation technique was utilized which employs gradient descent rule for optimization. The backpropagation technique searches for the minimum error and accordingly the weights of the network are considered. To optimize the hidden layers of these models, a genetic algorithm was employed. A genetic algorithm is a search-based optimization technique, which is based upon the principles of genetics and natural selection.

Other than these two machine learning models, the third one is for predicting the health of the tool employed for welding. As pointed out previously, the tool in FSW does not get consumed in the welding. However, it plunges inside the base materials to be welded. To understand the tool better and the problem which is being posed, let us stay here in this section to visualize it. Figure 7.16 depicts a picture of this tool, where the features of the tool namely pin and shoulder can be observed.

The pin is the part, which plunges into the base materials, and thus, the height of the pin is selected based on the thickness of the materials to be joined. The shoulder rotates over the surface. Because of these two features, the heat is generated in the thickness direction and surface of the material, respectively. It is for sure that the material out of which the tool is fabricated has superior properties in terms of at least strength and melting point as compared to the materials to be welded. There might

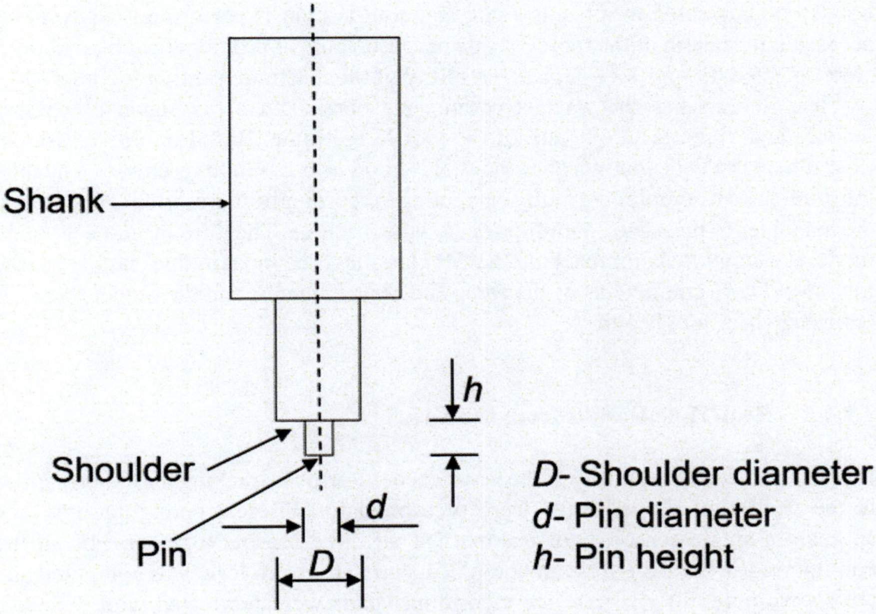

Fig. 7.16 Picture of FSW tool

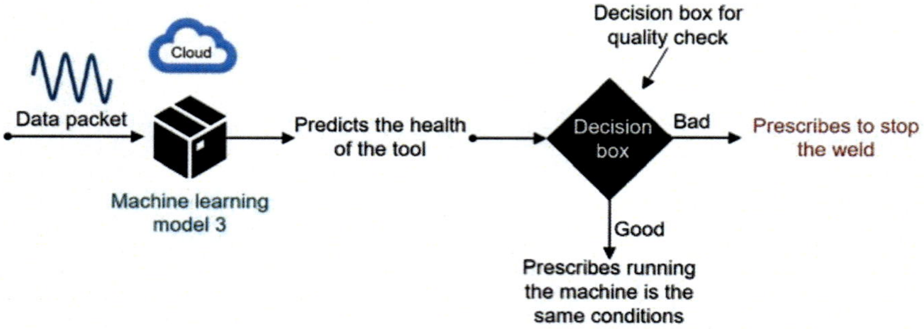

Fig. 7.17 Machine learning model 3 of the twin

be other properties as well. Strength is important because the tool has to deform the materials to be welded. Similarly, because of the involvement of heat, the tool should not get plasticized as well. This is the concern. Despite all these, a part of the tool may break or may get damaged during welding. As mentioned previously, the plunge depth is specified which is passed to the machine as a parameter to understand the depth to which the spindle has to travel in the vertical direction. So, if the spindle fails, or if there occurs some problem with the spindle, maintaining the vertical direction will be difficult. Thus, determining the health of the tool is crucial, which will further fulfil the benefit of performing real-time monitoring and control. In a way, this could be also be attributed to a health twin, as here the twin is performing analysis for accessing the health of the tool. For this purpose, tools of different health conditions were selected. Figure 7.17 depicts the utility of the machine learning model 3.

This model takes the wavelet coefficients obtained via the signal processing module and classifies the health of the tool in real-time. Based on this classification, it makes a decision, whether the tool is good or bad. Good resembles a healthy tool while bad resembles the faulty categories. Accordingly, it prescribes either to run the machines in the same conditions or stop the machine. The third machine learning model is a support vector machine (SVM) classifier. The input to this classifier is the first level DWT coefficients of the force and power signals, and the model forecasts the category of tool health.

7.9.1.5 Real-Time Results from the Twin

Figure 7.18 depicts the utility of the wavelet coefficients in tracking the abnormalities in the weld-joint. Experiments were performed for different configurations such as using a single sensor, two sensors, and all three sensors together. For all the configurations, the sum of detail coefficients up to the third level was computed and are shown here. To visualize the algorithm, welds were fabricated with defective welding conditions so that it was bound to obtain a defective weld.

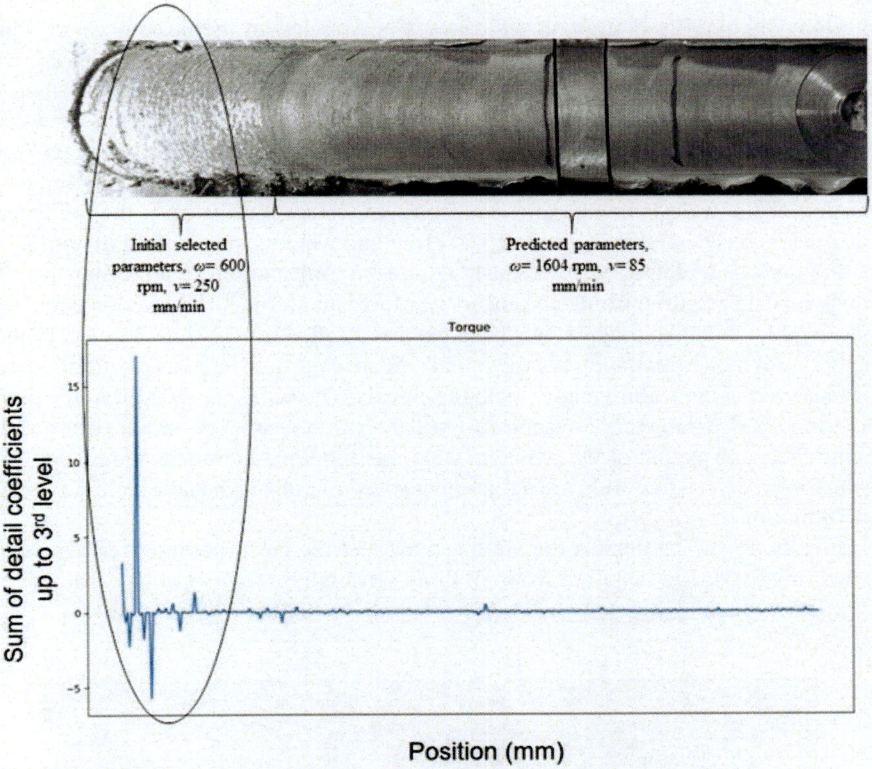

Fig. 7.18 Weld fabricated with the process twin

Figure 7.18 depicts a weld sample and the feature map, where it can be observed that for the defective portion of the weld, which is the initial portion, the signal shows high fluctuations, which is absent in the subsequent portion that corresponds to a defect-free weld. For this case, the force signal is considered. The initial parameters selected for welding were tool rotational speed of 600 rpm and welding speed of 250 mm/min. From the plot, the utility of the signal processing module can be visualized. It identifies the occurrence of welding defect in real-time. If welding defects are found, the strength will be reduced. This is why, the machine learning model is fed with the wavelet coefficients, which are having a high correlation with the defects, to predict the strength of the weld. From the plot, it can be seen that after the initial defective portion, the defect-free portion is there. This is possible because the weld was fabricated with the twin working aside. Let us understand this.

As the welding initiated, the sensors' data was fed to the twin, where the signal processing module decomposes the incoming signal. It calculates the wavelet coefficients of the incoming signals in real-time and sends them to the machine learning model 1 for predicting the strength. As the weld is a defective one (i.e., the starting portion), the predicted strength will always be lesser than the desired value. The desired value is always greater than 90% of the strength of the base material. Of course, the desired value may vary from application to application. As the predicted value of weld strength was lesser than the desired one, the wavelet coefficients were sent to the machine learning model 2, which predicted the modified parameters for achieving a defect-free weld. This can be seen from the picture as the subsequent portion of the weld is completely free from any defect. In this case, the predicted parameters were rotational speed of 1604 rpm and welding speed of 85 mm/min.

Figure 7.19 depicts another weld sample and the corresponding feature map, which involved data acquired from two sensors i.e., force and torque. Here also, it could be observed how the welding defects have been controlled in real-time because of the activities of the process twin. For the initial defective portion, large fluctuations could be observed in the feature map of both the signals. The same got reduced as the twin sent modified parameters for achieving a defect-free weld. The initial parameters being the same as that of the previous weld, the modified parameters predicted for achieving a defect-free weld are a rotational speed of 1684 rpm and a welding speed of 78 mm/min.

Finally, Fig. 7.20 depicts the weld sample and the corresponding feature map, which involved data acquired from all three sensors i.e., force, torque, and power.

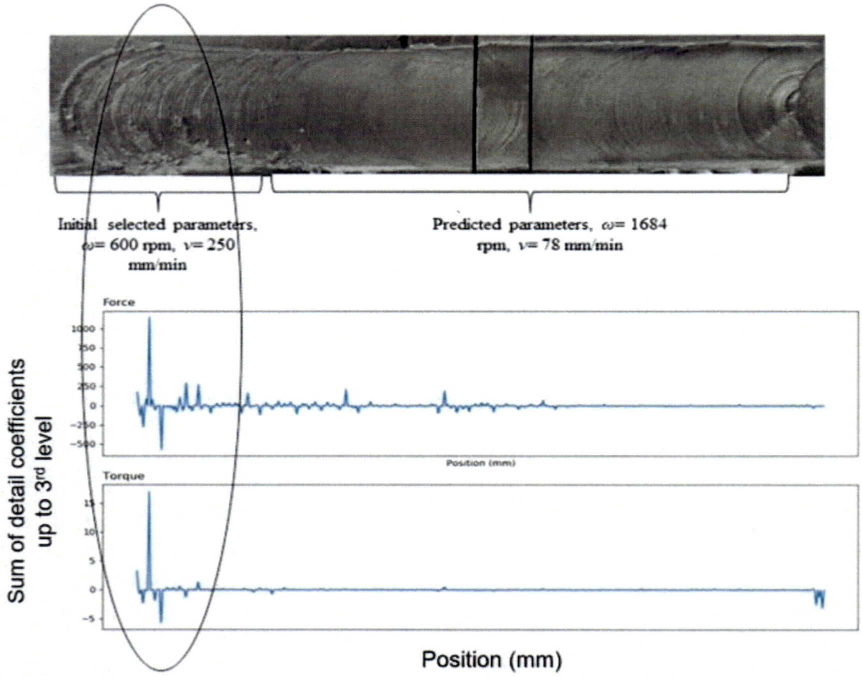

Fig. 7.19 Weld fabricated with the process twin

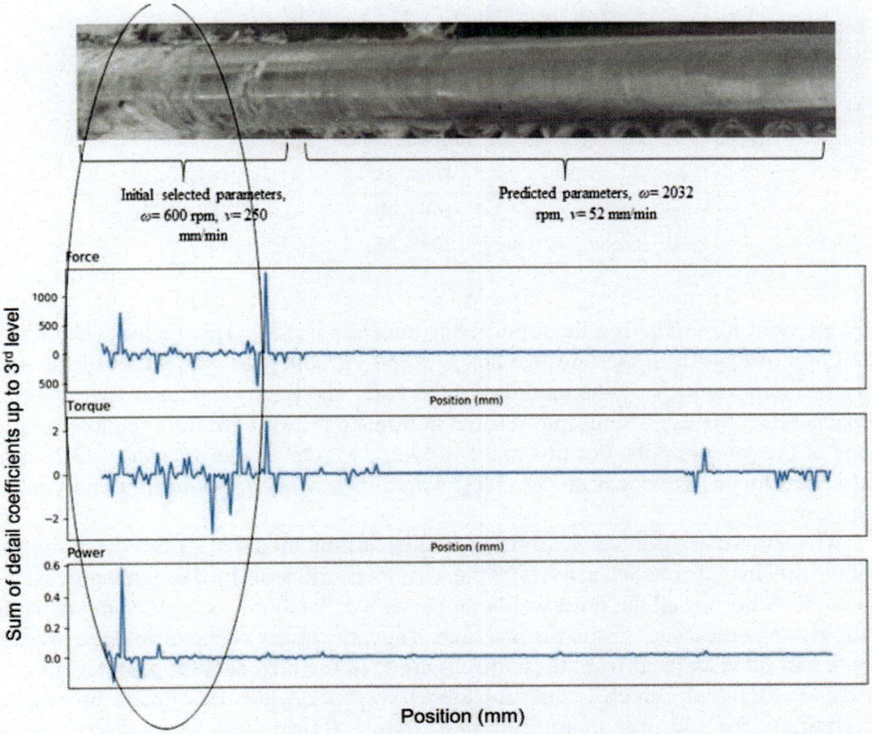

Fig. 7.20 Welds fabricated with the process twin

The control of the weld from defective to defect-free in real-time could be observed from the figure. The initial parameters for this weld were also the same as those selected for the previous two welds. However, the modified parameters predicted for achieving a defect-free weld, in this case, are a rotational speed of 2032 rpm and a welding speed of 52 mm/min.

One could wonder, why different combinations of sensors are attempted. In the first case, it is only a force sensor-based real-time monitoring and control, the second uses force and torque, while the third uses all three sensors. It is to present that the information or the accuracy of the predicted value increases with more amount of information. For the three welds discussed above, the fabrication was initiated with similar conditions, however, when the twin was fed with data corresponding to those combinations, the accuracy of the one predicted with all three sensors is the highest as compared to the others. This was known by determining the strength of the weld of the defect-free portion, as shown in Table 7.1.

Of course, the time for prediction will be more when multiple sensors are considered, but accuracy is also important. For the current problem, the total time taken

Table 7.1 Multiple versus single sensors

Experiment #	Combination selected	Predicted parameters (tool rotational speed (rpm), welding speed (mm/min))	Ultimate tensile strength (MPa)
1	Force	1604, 85	187.1
2	Force and torque	1684, 78	188.44
3	Force, torque, and power	2032, 52	194.19

by the twin for predicting the modified parameters is ~8 s. This includes the transmission of data from the physical space to the virtual space, formation of the data packet, data analysis by the modules of the twin, and finally feedback to the virtual space. Also, the use of multiple sensors will make the system more reliable in case one of the sensors fails. For this process twin, a graphical user interface (GUI) was also developed, as shown in Fig. 7.21, which displays the monitoring and control information.

The SVM classifier i.e., the third machine learning model classifies the condition of the tool in real-time when fed with the wavelet coefficients by the signal processing module. Since for all the three welds shown above, the tool was healthy, the twin did not give any message to stop the machine. Thus, the utility of the developed process twin can be visualized from these discussions. In the next section, another process twin is discussed, which is also for the FSW process but uses image processing techniques for real-time monitoring and control. Before that, let us discuss a few queries related to this process twin.

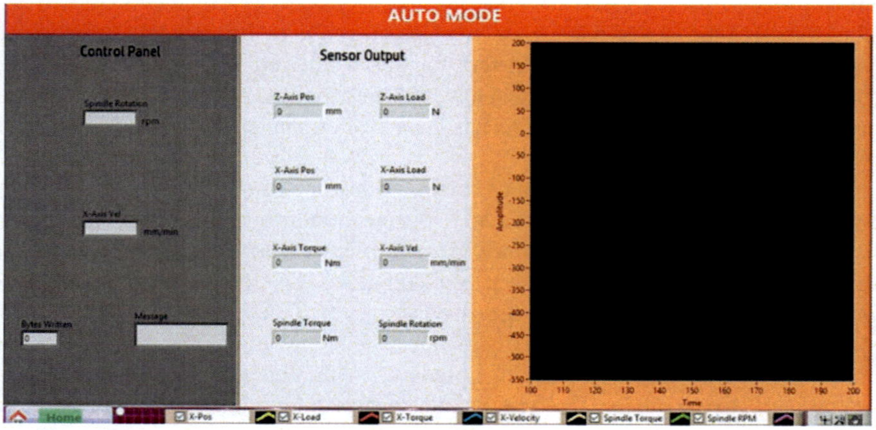

Fig. 7.21 GUI developed for process monitoring

7.9.1.6 Queries Related to the Twin

There could be a few research questions related to this process twin. The following sentences discuss them.

The first is why even a cloud server is considered for this twin? As pointed out previously, the twin may reside in the cloud or even in a local system in the enterprise itself. However, when multiple machines will be there on the shop floor, the local systems will not be sufficient for the processing of the data. Thus, a cloud server has been intentionally considered for addressing the scalability problem. One could introduce an edge-based service system in this architecture. For this purpose, a part of the twin will reside in the edge system, and the remaining part will reside in the cloud server. The signal processing module, machine learning module 1 and 3 could be kept in the edge system for continuous monitoring of the weld quality and health of the tool. In case, the quality is found to be poor, then only the cloud server is going to receive the data.

The second question could be the necessity of using three machine learning models. This has been performed so that the deployment of the twin could be successfully achieved. Instead of making a very heavy model, one could distribute the data traffic by building multiple models. The one that will reside in the edge system will only predict the tensile strength. Thus, the data need not be sent to the cloud server. This will help in ensuring timeliness.

The third question is why neural networks are considered for modelling? Neural networks are one of the best machine learning algorithms for non-linear modelling. As observed in the aforementioned discussion, the wavelet coefficients were able to detect the occurrence of welding defects. However, they did not have a linear correlation with the tensile strength values. However, it was important to consider the wavelet coefficients as they rightly detect the welding defects, which ultimately corresponds to the strength of the weld. Thus, to develop a reliable and robust model, neural networks were built for predicting tensile strength and modified parameters.

One could even point out, why even consider the machine learning models? This forms the fourth research question. One may advocate that via the experiments that are performed, the optimum parameters are already known. So, whenever the twin encounters an anomaly through the signal processing module, the optimum parameters can directly be fed. Why route the data through machine learning models? If the optimum parameters are only fed, the twin as such will not have any intelligence. This twin can serve various applications. One thing to note is that experiments cannot be performed for every combination of parameters, as they are costly. Different industries will have varying strength requirements owing to the product. In such a case, a machine learning model will help in predicting the desired parameters required for the desired strength value. Besides, the health of the machine is not going to remain the same. It will deteriorate with time and thus, the sensors' data will depend upon the health condition. The reliability of the twin will also depend upon the health.

7.9.1.7 Betterment of the Twin

Finally, how can one improve this twin? From the discussion held on the digital twin in this chapter, this twin does not incorporate any physics-based model. It only relies on the data-driven models, which are operating solely on the hard sensors' data. The control of the process is also being performed depending upon the tensile strength of the weld only. To scale the twin and improve its performance, one can make use of the physics of the process to simulate the conditions. The FSW process involves heat and plastic deformation for creating the welds. This means dynamic metallurgical changes are occurring during the heating, deforming, and stirring of the material during welding. Thus, in addition to the strength of the weld, grain size, hardness, strain, etc. are also playing a major role. A data-driven approach is robust and is providing a lot to automate the process and also make the twin happen, but it is expensive as it requires a large amount of training data. Simulation can help ease out this problem. Other factors such as grain size, hardness, strain, etc. could be simulated by involving the physics of the process. Once a successful simulation model is built, the same can be utilized for simulating various processing conditions, and by doing so, a large amount of data can be generated. This can be utilized to develop machine learning models. Also, the scaled-down version of the simulation model can be plugged into the twin for generating real-time results of the welding process.

7.9.2 Process Twin Using Image Processing

In this section, another process twin will be discussed which uses image processing techniques for real-time monitoring and control of the FSW process [18]. In the previous section, the said welding technique has been discussed in details, so, the same has not been repeated in this section. Directly, the development process of the twin has been discussed below.

7.9.2.1 Overview of the Twin

The developed twin monitors the surface defects in the FSW process by using a 3D laser scanning of weld bead for feedback and taking the requisite control action in real-time by changing the process parameters to overcome the upcoming defects. The 3D laser scanning technology employed is faster and more accurate/precise 3D reconstruction algorithm. This technology uses a thin laser line and a USB camera, both kept still concerning the machine, for visualizing the relatively moving laser plane, to the moving weld surface over the surface during the welding. A schematic diagram of the system is shown in Fig. 7.22.

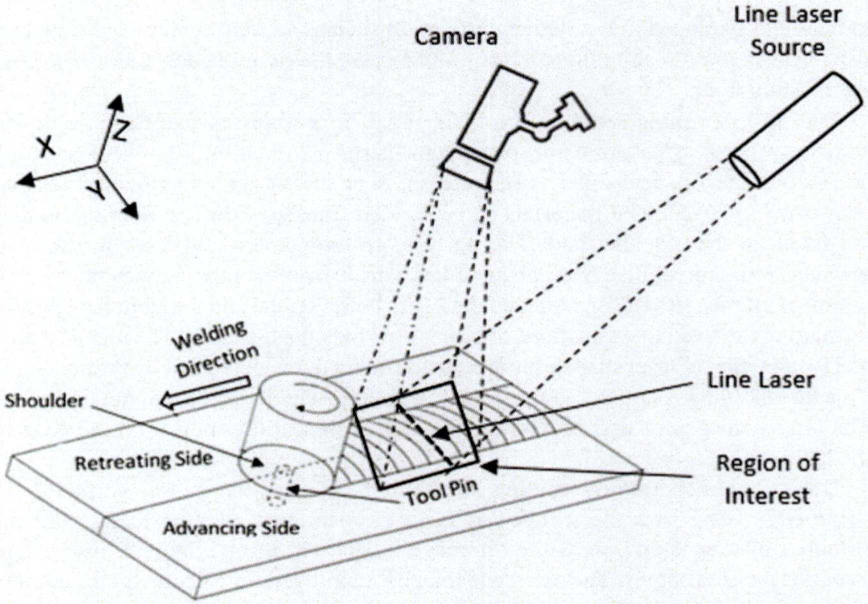

Fig. 7.22 Schematic representation of the process monitoring system

7.9.2.2 Details of the Process Twin

The process twin involving image processing for monitoring and controlling FSW process is schematically represented in Fig. 7.23. It consists of the FSW machine constituting the 'slave computer' and the developed twin with the image processing module as the 'master computer'. Here, the image processing module refers to the laser scanning setup, which has been shown in Fig. 7.22. For this problem, there is no deployment of the things in a cloud environment and the developed twin resides in a computer locally available in the space, which is being referred to as the master computer. As pointed out during the previous discussion, the digital twin model needs

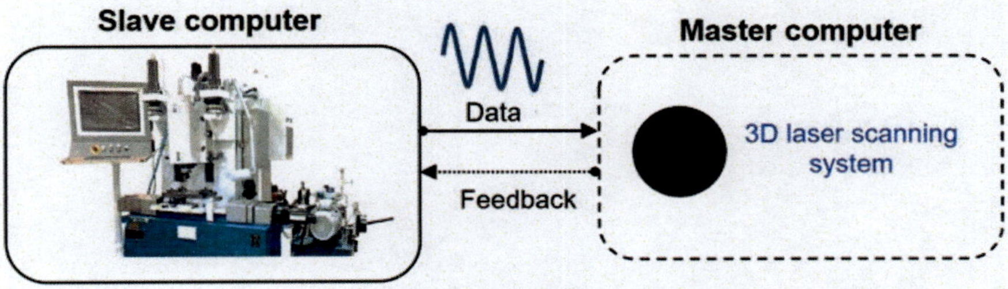

Fig. 7.23 Schematic of the process twin utilizing image processing

to be securely placed in a virtual environment such as a cloud server, or it may be a service system on a shop floor. Thus, in this case, the twin resides within a system on the shop floor.

The laser scanning setup as shown in Fig. 7.22 comprises five items: a digital camera, a laser stripe projector, two planar surfaces, a calibration checker-board, and a computer for real-time reconstruction. A picture of the experimental setup is shown in Fig. 7.24. The materials to be welded translates during welding as they are fixed on the machine bed. The position of the camera and laser projector is always fixed, projecting a fixed plane of laser stripe over the moving weld surface. A region of interest (RoI) is specified manually in both vertical and horizontal planes for estimating the fixed line equations of laser stripe over these planes. The identification and localization of laser stripe edge are carried out in the selected RoI by thresholding. This thresholding is applied to the grayscale image, which results in a binary image. The binary image consists of white pixels that correspond to points on the edge of the laser stripe.

During the inception of welding, the camera captured the image with the laser stripe edge lying over the weld bead surface, positioned past the location of the coloured marker is chosen as the reference image. After this, the real-time reconstruction process begins. The reference image is employed for calculating the amount of distance travelled in X and Y directions by computing the difference between the corresponding centroid of X and Y coordinates of the identified coloured markers in current and reference images. The positions of coloured markers are extracted based on the thresholding of coloured channels of the image.

For each white pixel, 3D reconstruction process is carried out to create a 3D surface profile of the weld cross-section. As the laser plane and the camera position are fixed, the reconstructed surface coordinates are spatially translated in the direction of the

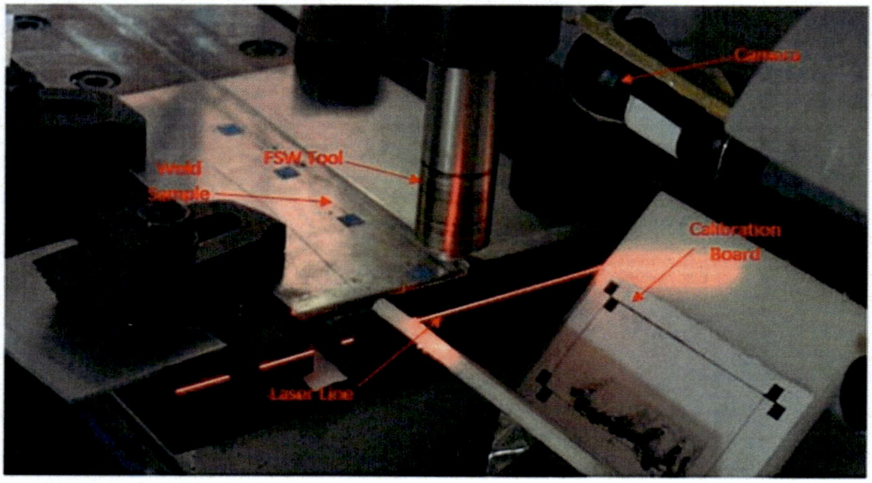

Fig. 7.24 Picture of the experimental setup

Fig. 7.25 Point cloud data of a weld surface

weld motion by the amount of triangulated distance. These shifted 3D coordinates are iteratively stored for each instance of a weld, thus, a 3D point cloud of the weld surface is generated in real-time. The point cloud data of a weld surface obtained using the above-stated algorithm is shown in Fig. 7.25.

Using this, the exact position and depth of defects were measured concerning a smooth weld surface. The 3D triangulation and reconstruction algorithm was modified as per the dynamics and relative motion between the camera, weld surface, and the laser plane. Finally, based on the occurrence of a surface defect during welding, the welding process parameters were altered by communicating the new parameters to the machine computer through Ethernet. For communication between the slave and master computer, TCP/IP communication protocol was utilized using an Ethernet wire. As mentioned, the master computer is utilised for real-time 3D reconstruction of weld surface and providing feedback for changing the welding parameters to the slave computer, which is the welding machine.

The entire algorithm was implemented in MATLAB. A GUI for controlling the FSW machine was developed in the LABVIEW environment. This GUI was used to receive the altered parameters from MATLAB and reflect them in the actual welding process. The TCP/IP server was hosted on MATLAB for transmitting modified weld parameters upon inquiring about a defect on the weld surface. A TCP/IP client block was set into the listening mode in LABVIEW for acquiring the data sent from the server. The TCP/IP server hosted on MATLAB sent two different data sequences on detecting a weld defect. The first data received on the machine computer was used as an identifier to figure out whether the next data would correspond to which of the welding parameters. The next data was the actual value of the welding parameters to be sent to the welding machine using shared variables to overcome upcoming defects. This algorithm was implemented using flat and stacked sequence structures in LABVIEW. Thus, surface defects were removed in real-time by the above feedback mechanism.

7.9.2.3 Real-Time Results from the Twin

Figure 7.26 depicts the picture of a sample welded with the developed process twin. The initial parameters fed to the machine were a rotational speed of 600 rpm, a welding speed of 80 mm/min, and a plunge depth of 0.05 mm. It could be observed from the picture of the welded sample that the selected parametric conditions resulted in a defective weld, which has been annotated as 'Defective region' in the picture. The twin analyses the data in real-time, and sends feedback to the welding machine, which resulted in a defect-free weld. The same has been annotated as the 'Defect-free region' in the picture. This shows the utility of the twin in monitoring and controlling the process. A picture of the developed GUI is shown in Fig. 7.27.

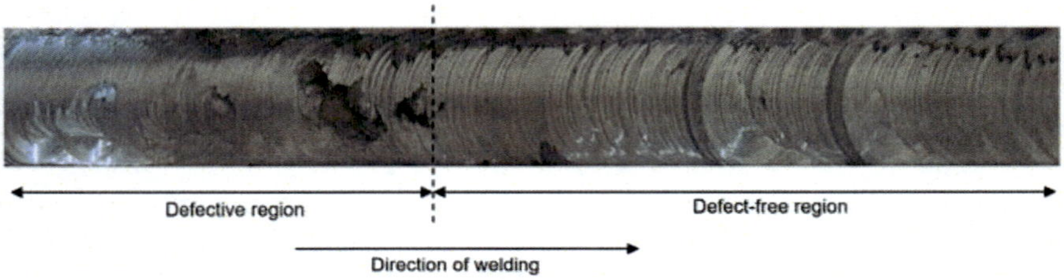

Fig. 7.26 Point cloud data of a weld surface

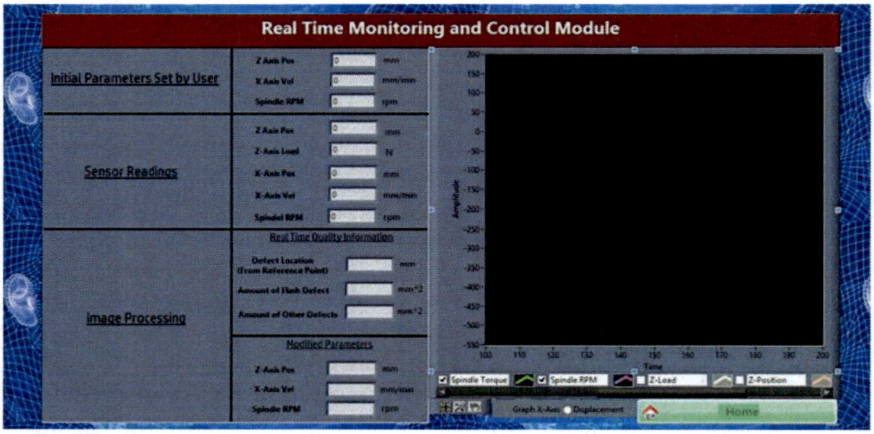

Fig. 7.27 Picture of the developed GUI for the twin

7.9.3 *Machine Twin*

This section will discuss the development of a machine twin from a health monitoring perspective. Though the twin, which will be discussed in this section, is not fully loaded with all features of a digital twin application, it aims at highlighting the importance of clubbing data and physics-driven approaches. For this, let us discuss the physical space of the digital twin application, followed by the inclusion of sensors and simulation information.

7.9.3.1 Overview of the Twin

The physical space consists of a hydraulic power pack, which was built as a prototype. A hydraulic power pack is a general-purpose utility system, which is employed by CNC-enabled machines for energizing themselves. The major components of a hydraulic power pack are a tank that holds the hydraulic fluid, a pump and motor arrangement for energizing the fluid, actuators for actuating end components, and piping for transporting the fluid to and fro between the tank and actuator. A picture of this hydraulic power pack is shown in Fig. 7.28.

Since the hydraulic power pack consists of numerous devices, it is referred to as a system. Thus, a system-level digital twin will be required to be developed for the hydraulic power pack. Recall Fig. 7.2 discussed the complexity of a digital twin model concerning the tasks to be performed. The complexity could be visualized from this hydraulic power pack system. However, for demonstration, a unit-level

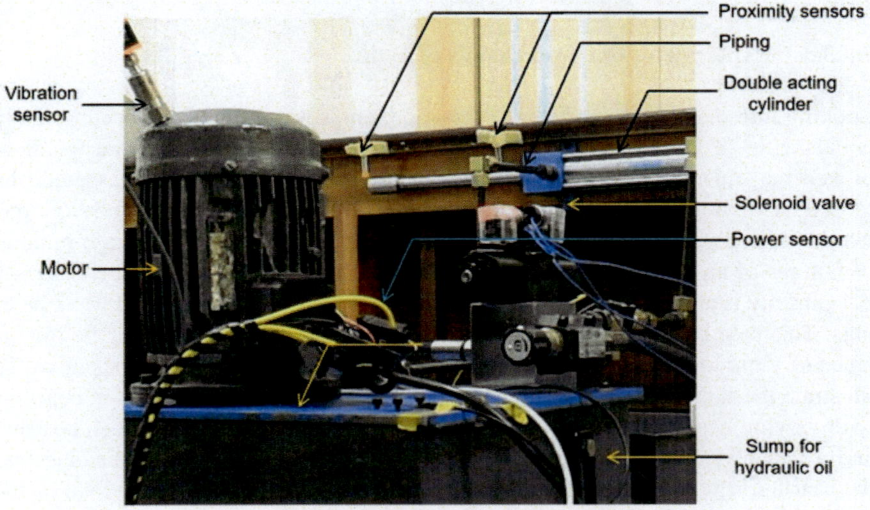

Fig. 7.28 Picture of the hydraulic power pack

digital twin application has been attempted to build in this case study. Within the system and unit-level digital twin, there also lies a machine-level digital twin. The machine-level twin is proposed to be built for a machine, and the unit-level indicates a particular component within a machine. Thus, the complexity of a digital twin increases from unit to system level. In the present case study, a unit-level digital twin is built for monitoring the induction motor fitted in the hydraulic power pack. Further, the specific objective within the induction motor is to monitor the health of the bearing fitted inside the motor. Induction motors are one of the important assets of industries as they are included in almost all types of machines. This rotary machine's health monitoring is important and crucial because its shutdown would cause large damage to the machine it is associated with.

7.9.3.2 Sensors Fitted to the Machine

Multiple sensors are fitted on to the hydraulic power pack, which includes vibration, power, flow rate, temperature, turbidity, and proximity. The proximity sensor is for controlling the actuation of the double-acting cylinder. A vibration sensor is mounted on the motor via a stud for capturing the vibration signature of the motor during its operation. Similarly, a power sensor is also connected with the motor for acquiring the motor power signature. These two sensors i.e., vibration and power are engaged for analysis of the health condition of the bearings in the motor. Other sensors such as flow rate, temperature, and turbidity were connected in the piping section for acquiring the variation in the fluid flow rate, fluid temperature, and inclusion of foreign particles in the fluid, respectively.

7.9.3.3 Experimentation and Details of Twin

Looking into the induction motor as a component, several faults can occur namely unbalancing of the rotor component, damage of bearings, faults in the winding, broken bar, misalignment, eccentricity, etc. Reasons for these damages could be looseness of bolts, variation in speed, variation in the load of the motor, over-temperature, insufficient lubrication, and so on. Out of all these, a large number of failures of an induction motor is because of the damage caused to the bearings. Eccentricity is another fault that arises because of the misalignment and weak bearings. This case study attempts to deal with these two faults. In Chap. 2, the importance of vibration signal was discussed from the health monitoring perspective. To illustrate the same, Fig. 7.29 depicts the vibration signature acquired from bearings with varying health conditions. This includes a jammed bearing, a broken bearing, and a healthy one. In the jammed bearing, foreign particles were embedded into the bearing, which makes it difficult to freely rotate. Because of this, the load on the motor increases. Practically, this happens because of the wear and tear of the machine components. In the case of the broken bearing, the cage of the bearing was broken and a few balls i.e., the rolling elements were removed. The healthy one is a fresh

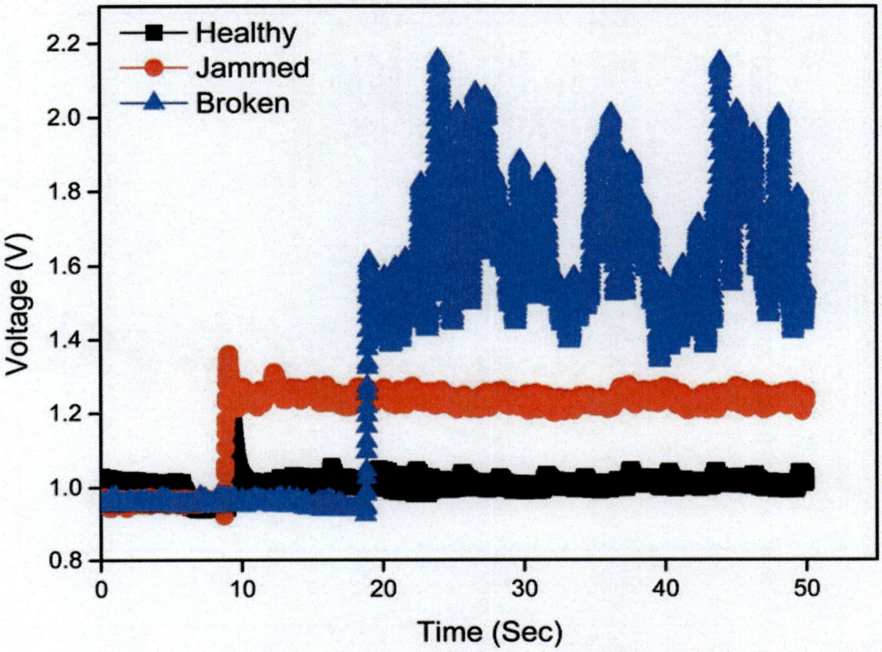

Fig. 7.29 Variation in vibration signature with varying health conditions of bearing

bearing. Each of these bearings was fitted in the induction motor and the vibration signature was collected.

For experimentation in this example, five bearings were selected, one is having an inner race fault, the second with a fault in the outer race, the third is a jammed bearing, the fourth is a broken bearing, and the last one is with a healthy condition. The first four bearings are faulty as the faults are induced in them. Experiments were conducted by operating the hydraulic power pack with these bearings, one at a time. These bearings were fitted at the fan-end of the motor. For all the cases, data was acquired from vibration and power sensors. Figures 7.30 and 7.31 show the raw signal collected from vibration and power sensors at a sampling rate of 2 kHz. Both of these raw signals correspond to the healthy bearing.

7.9.3.4 Data-Driven Analysis

This section will elaborate on the analyses performed using the sensors' data for investigating the condition of a bearing. To initiate the analysis task, the signals were studied in the time domain and various features were extracted from them. These features were mean, RMS, standard deviation, variance, crest factor, form factor, skewness, kurtosis, maximum, minimum, range, and sum. Here, the objective was

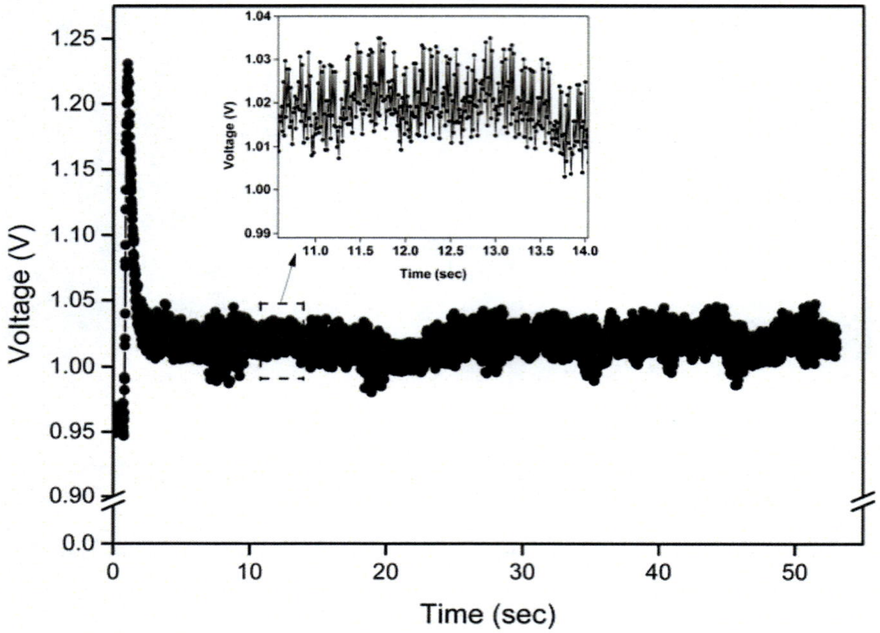

Fig. 7.30 Raw signal-vibration sensor

to develop a machine learning model that can classify the bearing faults. For the same, feature engineering was carried out. Figure 7.32 shows the correlation matrix obtained from the features of the power signal, which was computed to determine the inter-correlation existing between the features. If the features would be having a high correlation between them, this indicates that they will be carrying redundant information. Thus, feeding these features into a machine learning model will not be useful. Rather, extraction of meaningful and unique features is more important. The correlation matrix shown in Fig. 7.32 is obtained by using the Pearson correlation method, which is computed as expressed in Eq. 7.1.

$$r = \frac{\sum(x_i - \overline{x})(y_i - \overline{y})}{\sqrt{\sum(x_i - \overline{x})^2 \sum(y_i - \overline{y})^2}} \tag{7.1}$$

where x_i denotes the values of the x variable, y_i denotes the values of the y variable, $\overline{x}$ is the mean of the values of the x sample, $\overline{y}$ is the mean of the values of the y sample, and r is the Pearson correlation coefficient. From the matrix, it can be observed that some of the features are having a high correlation coefficient with other features. Say, for example, the standard deviation is having a high correlation with other features.

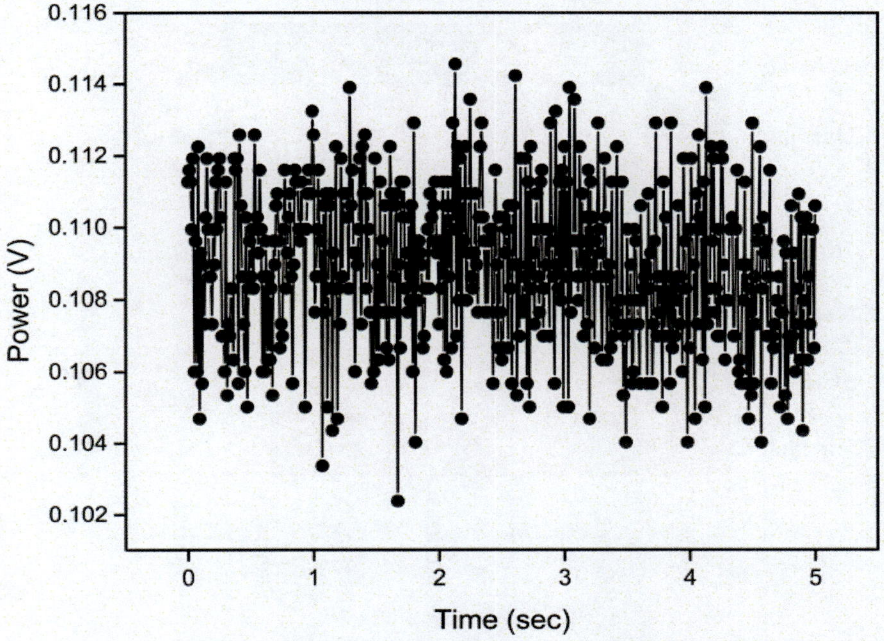

Fig. 7.31 Raw signal-power sensor

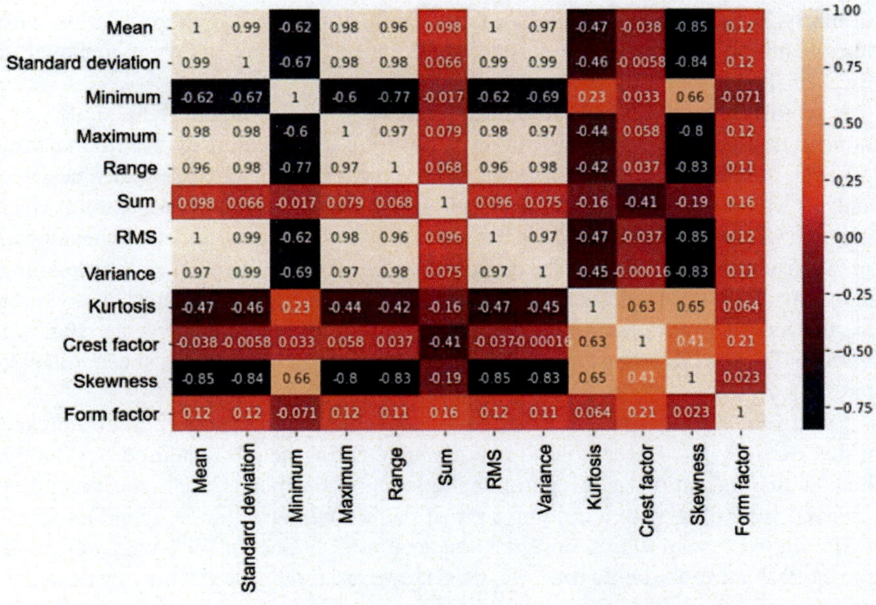

Fig. 7.32 Correlation matrix for features of the power signal

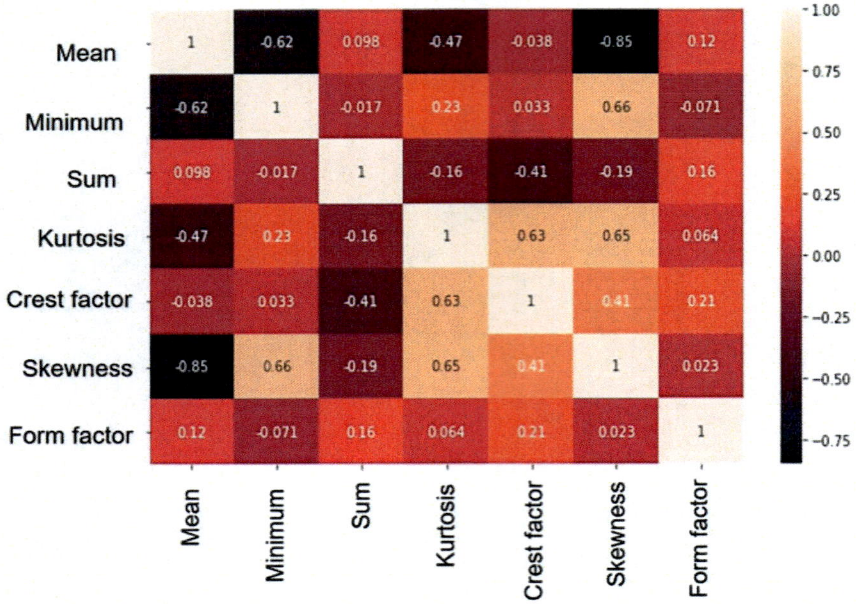

Fig. 7.33 Correlation matrix for selected features of the power signal

Similarly, maximum, range, and variance are also having a high correlation with other features. Thus, they are dropped, and the final feature matrix is depicted in Fig. 7.33.

It can be observed that out of the 12 features extracted from the signal, only 7 are found to be meaningful. Though this analysis reduces the volume of data for machine learning, it retains meaningful information, which is crucial and essential. The same analysis was also performed for the vibration signal and the meaningful features were selected. Using the above-selected features, a machine learning model was developed for classifying the condition of the bearing. For this purpose, several machine learning algorithms were selected, such as SVM, decision tree, and random forest. The results i.e., the accuracy and training time of the machine learning algorithms are tabulated in Table 7.2. The learning from a single sensor as well as multiple sensors is also reported.

From Table 7.2, it can be observed that the machine learning algorithms employed in this example have been able to classify the bearing health conditions. Amongst them, with information captured from a single sensor i.e., power, the highest accuracy achieved is 99.07% with a training time of 0.7 s. The algorithm is a random forest model. In the case of vibration signal, the accuracy obtained is 99% with a decision tree model having a training time of 0.06 s. However, for the case of the power signal, the decision tree model takes only 0.04 s for training with an accuracy equivalent to that obtained with the vibration signal. In all the cases, the accuracy achieved

Table 7.2 Comparison of machine learning algorithms for classification

Algorithm	Single sensor				Multiple sensors	
	Power		Vibration		Power and vibration	
	Accuracy (%)	Training time (s)	Accuracy (%)	Training time (s)	Accuracy (%)	Training time (s)
SVM	93	0.87	88	0.99	96	0.29
Decision tree	99	0.04	99	0.06	100	0.03
Random forest	99.07	0.70	98	0.62	100	0.5

with the power signal is more as compared to that obtained with the vibration signal. Also, the training time is less in the case of the power signal, except for one case, i.e., in the random forest model, the training time is 0.7 s for the power signal, which is 0.6 s for the vibration signal. However, the accuracy is higher with the former. Interestingly, when the features of both power and vibration signals (multiple sensors) are employed for machine learning, the accuracy is higher than the case of a single sensor. This is so because more information is available for learning about a phenomenon. Thus, the data from the multiple sensors get fused in the algorithm and is utilized for learning about the target feature. Indeed, to say, a digital twin application will require information from multiple sensors to accurately mirror the activities of the physical asset and predict its RUL.

7.9.3.5 Data-Driven and Physics-Based Simulation Analyses

This section will demonstrate the usefulness of the combined approach i.e., data-driven along with a physics-based approach for a digital twin application. For this, two bearings are considered, one is a jammed bearing and the other is a healthy bearing. Figure 7.34 depicts the workflow of the proposed twin. It begins with a preliminary analysis of the bearing fault by analyzing the vibration and power signatures. To explore the condition of the motor and analysis of the predicted faults by the machine learning model, a simulation model is run.

In the picture, "*platform 1*" is mentioned, which refers to a simulation platform, and in this case, it is a COMSOL-based simulation. The simulation takes the current values as input and the outputs are electromagnetic torque and electromagnetic forces in X and Y directions. "*Platform 2*" is also proposed for simulation of rotor speed, rotor current, and stator current. Recall the *DROTA*, which has been discussed previously; the vibration and power signature are tracking the health condition of the bearings. In the case, the machine learning model detects a fault, data can be sent to "*platform 1*" for fault analysis.

The simulation is designed in the COMSOL environment for the analysis of the motor [19]. It takes certain inputs about the construction of the motor such as supply frequency, the number of turns in the stator winding, phase span of winding, rotor

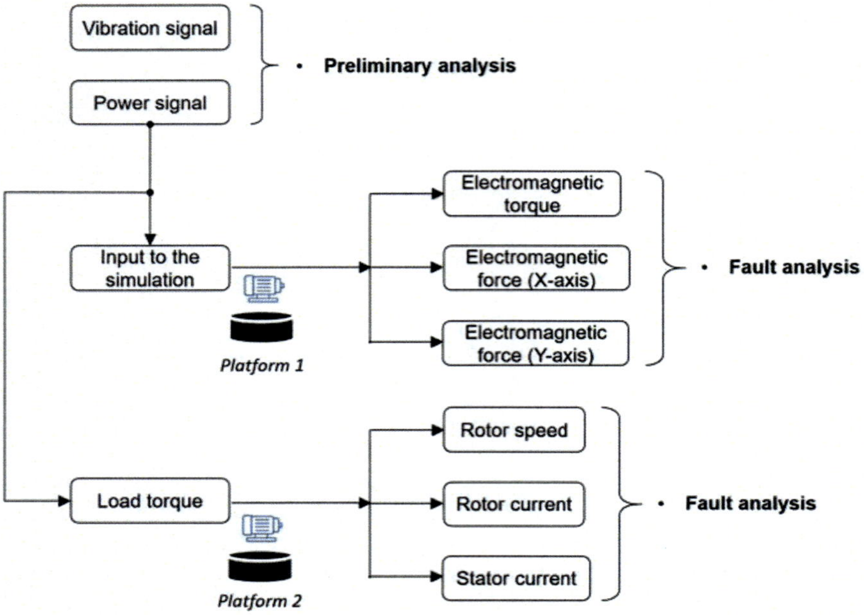

Fig. 7.34 Proposed overview of the twin

length, radial misalignment, and time-dependent current input in the coils of the motor. Out of these parameters, a few were taken after analyzing the construction of the motor, and data from the power sensor which was converted at a static assumed voltage supply of 230 V and a delta circuit for the induction motor. The power sensor's output was found equivalent to the phase supply current reading, which was made using a multimeter device during experimentation. Figure 7.35 shows the mesh prepared for the simulated entity.

The electromagnetic simulation is carried out on the 2D cross-sectional model of the induction motor, which is performed in three steps. The first is a magnetic field analysis in the rotor–stator structure, the equations for this are represented in Eqs. 7.2, 7.3, 7.4, and 7.5.

$$\nabla \times H = J \tag{7.2}$$

$$B = \nabla \times A \tag{7.3}$$

$$E = -\frac{\partial \mathrm{A}}{\partial \mathrm{t}} \tag{7.4}$$

$$J = \sigma \mathrm{E} \tag{7.5}$$

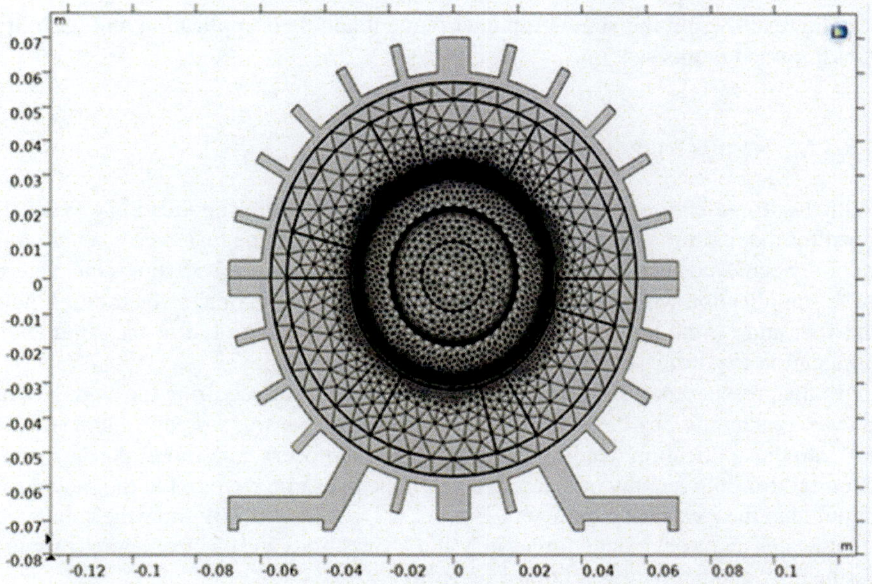

Fig. 7.35 Mesh prepared for the simulated entity

where ∇ indicates divergence, H indicates magnetic field intensity, J indicates the current density, B is for magnetic flux density, A is magnetic vector potential, E is the applied electric field, and σ indicates the conductivity of the material.

The second is the evaluation of the current density for the stator coils, which is represented by Eq. 7.6.

$$J_e = \frac{N I_{\text{coil}}}{A_{\text{coil}}} \text{e}_{\text{coil}} \tag{7.6}$$

where N is the number of turns, I_{coil} is the current flowing in the coil, e_{coil} is the induced voltage, and A_{coil} is the cross-sectional area.

Finally, the electromagnetic forces and torque are calculated based on the rotor–stator misalignment after performing the basic magnetic field analysis and current density evaluation. Equations for the same are represented in Eqs. 7.7 and 7.8, respectively.

$$F = \int_{\partial\Omega} \text{dnTds} \tag{7.7}$$

$$\tau = \int_{\partial\Omega} \text{d}(\text{r} - r_0) \times (\text{nT})\text{ds} \tag{7.8}$$

The results from the simulation environment and their application and usability are discussed in this section.

7.9.3.6 Results from the Simulation Model

With the aforementioned information, a COMSOL simulation application was developed for the real-time feeding of the information. Though the real-time evaluation is not fully achieved, a processing pipeline has been built. Figure 7.36 shows the picture of this application window, where on the top left side of the image, the motor inputs are fed, and on the bottom side, the real-time signal can be fed to the simulation application for further analysis.

In the present work, the power sensor was engaged for acquiring the signature of power consumption by the motor. The acquired data, that was stored in a local file, is fed into the application window for simulating the process conditions. As data were acquired for both healthy and jammed bearings, the same were fed to the model for simulating the electrical parameters. Figures 7.37, 7.38, and 7.39 show the simulated outputs i.e., electromagnetic forces in X and Y directions and electromagnetic torque for healthy and jammed conditions, respectively.

From the plots of electromagnetic forces and torque, a clear distinction between the bearing conditions can be observed. This data obtained from the simulation can be employed for modelling a machine learning model for the classification of the bearing conditions in real-time. The data gathered from the sensors, like in this case, the vibration and power can help in a preliminary analysis, while the physics-based simulation can benefit in getting a lot more information. Figure 7.40 shows the simulation output window, where the plots of electromagnetic torque and magnetic flux density can be seen.

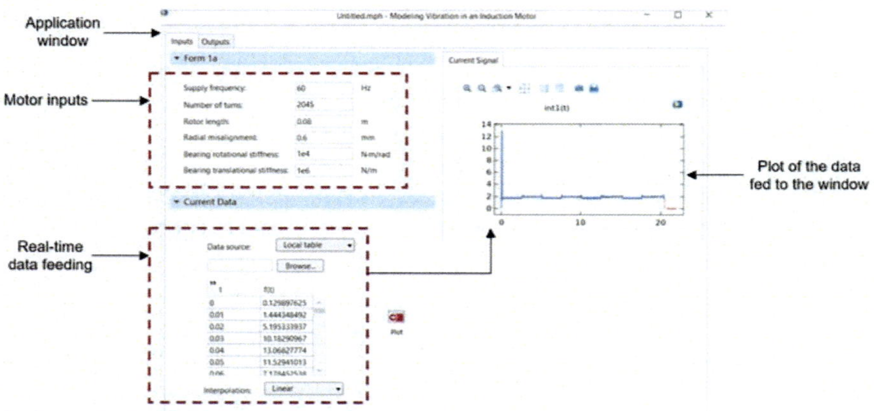

Fig. 7.36 COMSOL simulation application window

(a)

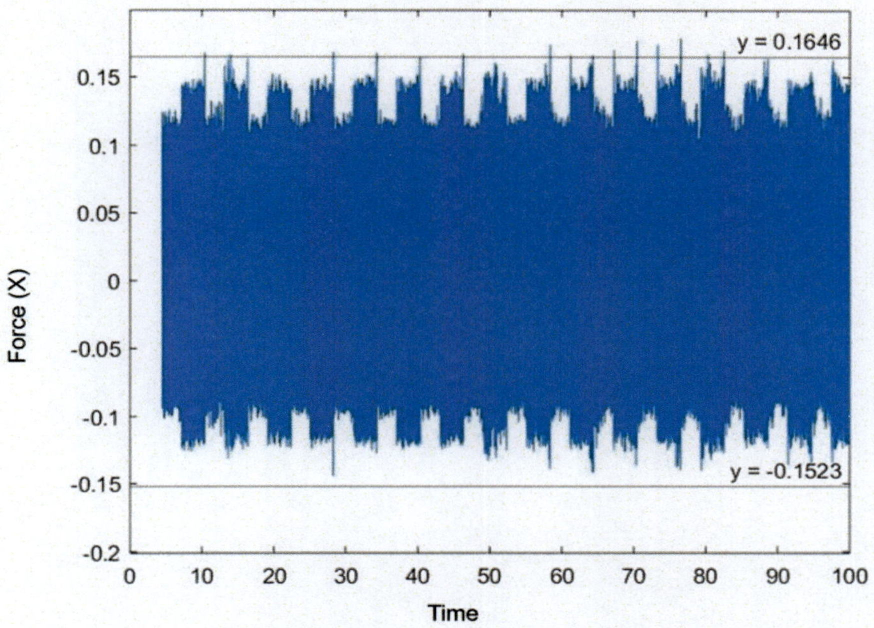

(b)

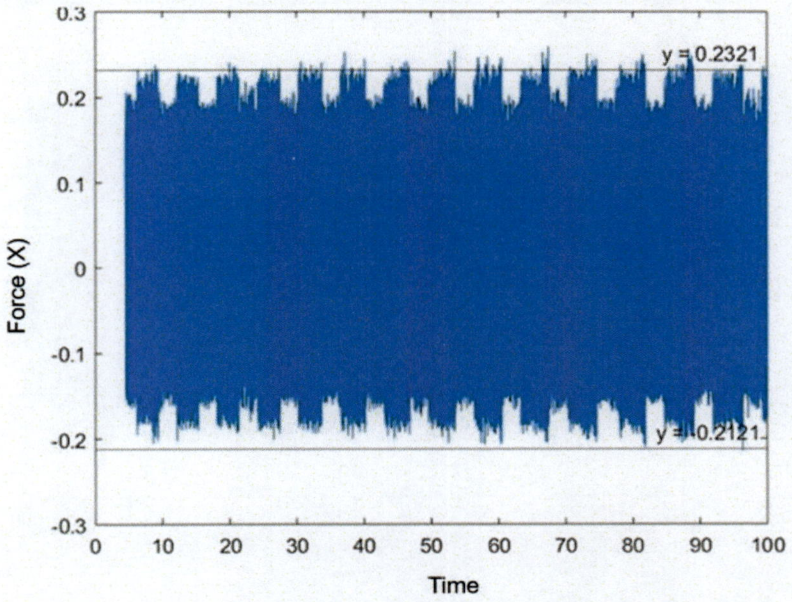

Fig. 7.37 Variation of electromagnetic force (X-direction) for **a** healthy bearing and **b** jammed bearing

(a)

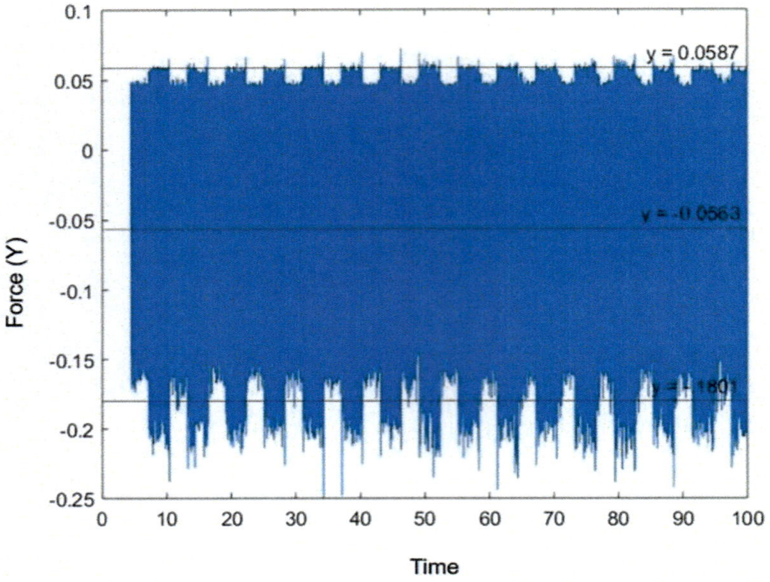

(b)

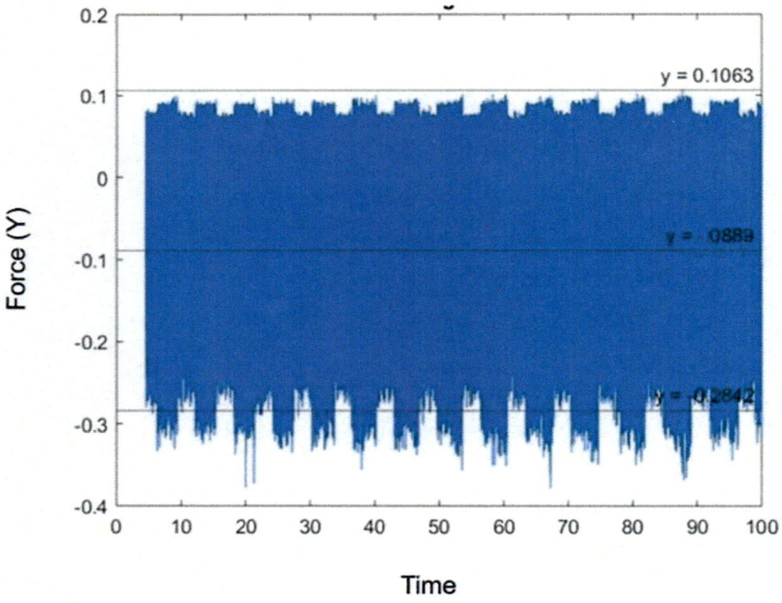

Fig. 7.38 Variation of electromagnetic force (Y-direction) for **a** healthy bearing and **b** jammed bearing

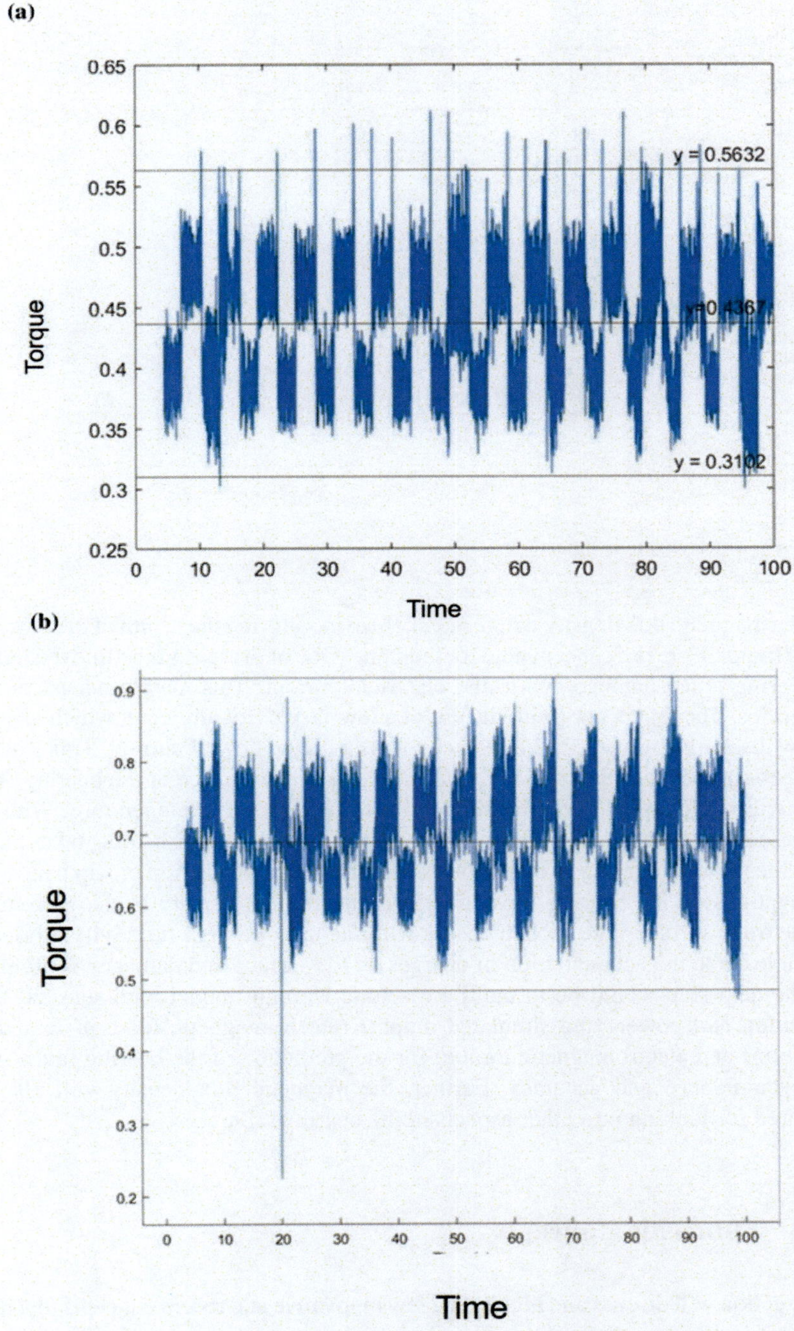

Fig. 7.39 Variation of electromagnetic torque for **a** healthy bearing and **b** jammed bearing

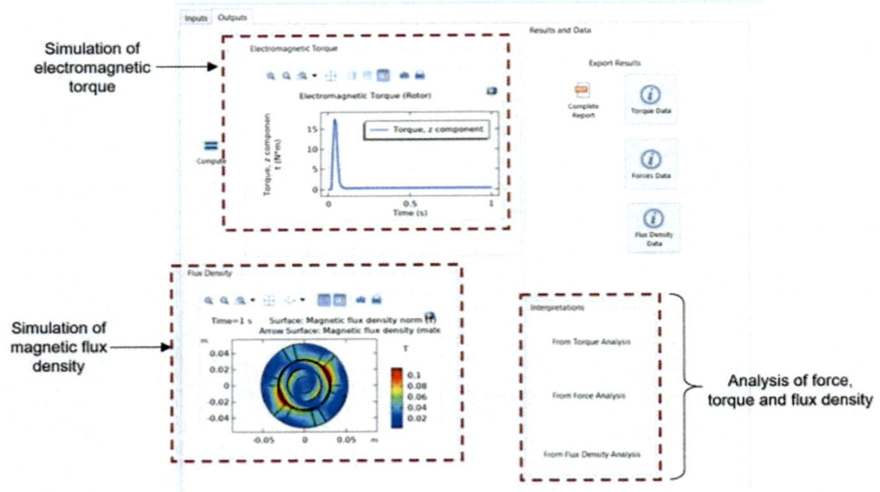

Fig. 7.40 Simulation output window

The magnetic flux density will further help in monitoring the health of the bearings of the motor. Here is a concept note for the same. One of the reasons behind the failure of bearing is the damage due to the electrical current. This mainly occurs in two categories. The first occurs with the use of a low-resistivity lubricant, which leads to a slow discharge through the bearing elements with the flow of current. This leads to the decomposition of the lubricant and corrosion of the surface of the bearing. This affects the fatigue life of the bearing and also increases the temperature. With the increase in temperature, thermal stresses are induced and surface heating takes place, which leads to tempering of the surfaces resulting in the formation of slip bands and corrugations on the bearing. Besides, the magnetic flux density is also developed on the track surface. The second occurs with the use of a high-resistivity lubricant, which leads to the accumulation of charges on the surface and damage via arcing.

The next step would be to build a machine learning model with sensors' data (vibration and power) and simulated outputs (electromagnetic force in X and Y-directions and electromagnetic torque) for the prediction of the bearing faults with more confidence and accuracy. Further, the magnetic flux density will also be explored for looking into other aspects of the bearing failure.

7.10 Futuristic Concepts

This section will discuss and highlight a few innovative and recent concepts that have been put forward by researchers. The objective is to give a philosophical note on the possible deployment of innovative concepts in digital twin application.

7.10.1 *Federated Learning in Digital Twin*

The machine learning approach resembles the training of a global model on a single server with data from various devices. Things discussed in this textbook are like sending the data to a server over the network and training a model, which could then be deployed for making useful predictions. This process has been referred to as centralized learning. It consists of a centralized server and multiple clients.

On the other hand, federated learning, introduced by Google, enables client devices to collaboratively learn a shared prediction model while keeping all the training data on their respective devices. It refers to a distributed learning approach. So, it is like, sending copies of a global model to multiple devices, which are the clients. Each client has its data, they do not send the data to the server. Once the copy of the global model is received, it is trained with the local data. The weights of the model are updated via local training in each of the client devices and the trained models are sent back to the server. The server aggregates the weights and improves the global model. Thus, the data, which is utilized for training the model is not revealed, hence, it is secured. Therefore, in this learning system, machine learning is distributed over mobile devices instead of computing on a large centralized server. It allows faster deployment with lower latency and lowers power consumption. It also helps in eliminating the need of storing data in a cloud or a centralized server, thus, ensuring the privacy of the data. So, why it could be an important aspect of a digital twin application? In IIoT applications, a large amount of data is expected to be generated each second from various manufacturing devices. This data (time series) will have a lot of spatial and temporal redundancy information. As digitalization will increase, billions of devices will be connected, thus, network congestion may occur with the increasing traffic. This may affect the timeliness of the models. Manufacturing applications are very much crucial and sensitive to time. Also, industries are concerned about the security of their data. For this purpose, federated learning could be leveraged.

Consider a case, as shown in Fig. 7.41. It consists of a centralized server with a global model, which has been trained on some previously collected data to the various mechanical devices. Each machinery collects data from various sensors in a real-time monitoring environment and trains their local models which is the global model. Then, each device uploads its local models to the centralized server at the end of each training process. A digital twin stores the real-time information about the health of each manufacturing device. A device having better health than another would quite naturally provide less error-free information regarding the process it has performed. So, in federated learning, to improve the learning accuracy and provide the best global model, the parameters uploaded by the nodes (individual devices) would have greater weight in aggregation.

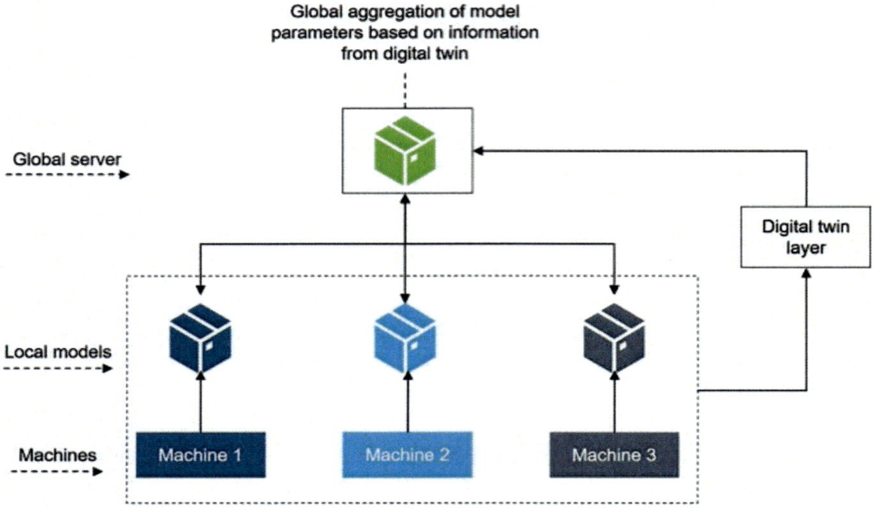

Fig. 7.41 Federated learning approach for digital twin application

7.10.2 Blockchain in Digital Twin

As pointed out throughout this textbook, digital twins are the replicas of physical objects with their functions, data, and computation capabilities. A digital twin interacts with the physical objects in real-time ensuring synchronization of data for mirroring its life and make useful predictions. The thing to note is that the spaces i.e., the physical and the virtual exchange a lot of data, functions, and analytics to replicate the physical system, which attracts various intruders and hackers to corrupt the data.

To protect the privacy of the user and enhance data security, a blockchain-empowered digital twin can be used. Blockchain provides various features such as data immutability, decentralized storage, peer to peer communication, and transparency to refine the concept of a digital twin. Blockchain can be used as an essential tool for the sharing of data and values between the physical system and digital twin. As discussed in previous chapters, digital twins need various intermediaries for data management, which also perform analytics, and are very prone to cyber-attacks. With blockchain technology, a digital twin can be strengthened i.e., the data can be made more secured certainly by using hash functions. Also, owing to the unalterable nature of the blockchain, it will ensure that a digital twin transaction history remains ineradicable. This attribute not only brings a remarkable level of trust but also provides data immutability and integrity. Since, with the blockchain, data can be stored globally and accessed securely without using any third party, data tracking for auditing and record-keeping becomes easy.

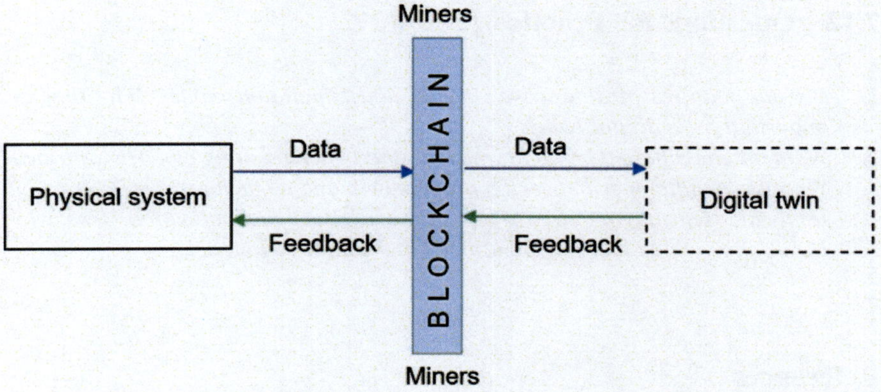

Fig. 7.42 Blockchain for digital twin application

To put forward the philosophical note, consider a scenario, where fake digital twins are present. These twins prove their authenticity by altering the digital certificates. These fake digital twins may corrupt the data and the protection of digital twin data is crucial. In such a scenario, the use of blockchain to issue and store digital certificates ensure the legitimacy and the identity of digital twins. The schematic shown in Fig. 7.42 depicts a blockchain-driven digital twin system. Instead of routing the data directly from the physical system to the digital twin, it is routed via a blockchain. Owing to the immutability feature of the blockchain, the data acquired from the physical system cannot tamper. Similarly, the other features add to security, transparency, decentralized storage, etc., which ensures a higher efficacy of the system.

7.11 Summary

This chapter discussed the nitty–gritty of digital twin application in details. Starting from a historic to present day's point of view, the digital twin has been explained. Discussions have also been around how digital twin can be thought of as a tool throughout the product life cycle. A few examples of digital twins in manufacturing have been explained and discussed as well. The concept of the hybrid digital twin is discussed in length along with the associated challenges and features to be possessed by a digital twin application. Three real-life case studies have been discussed, where process twins have been developed for real-time monitoring and control of a welding process and a machine twin for health monitoring of bearings of an induction motor. The results of those twins have also been shown. Finally, a few futuristic concepts are discussed that could be leveraged in developing robust and accurate digital twins.

7.12 Questions for Practice

1. *List the features of data-driven model and simulation model. Why they are important for a digital twin?*
2. *What are the different levels at which digital twin modelling can be performed?*
3. *What is the difference between a process twin and a machine twin?*
4. *Is machine learning required to be a part of digital twin application? If so, why?*
5. *How can machine learning be utilized in a simulation model?*

References

1. Tuegel EJ (2012) The airframe digital twin : some challenges to realization, pp 1–8
2. Roy RB, Mishra D, Pal SK et al (2020) Digital twin: current scenario and a case study on a manufacturing process. Int J Adv Manuf Technol 107:3691–3714. https://doi.org/10.1007/s00170-020-05306-w
3. Haag S, Anderl R (2018) Digital twin—Proof of concept. Manuf Lett 15:64–66. https://doi.org/10.1016/j.mfglet.2018.02.006
4. Glaessgen E, Stargel D (2012) The digital twin paradigm for future NASA and U.S. air force vehicles. In: 53rd AIAA/ASME/ASCE/AHS/ASC structures, structural dynamics and materials conference 20th AIAA/ASME/AHS adaptive structures conference 14th AIAA, pp 1–14. https://doi.org/10.2514/6.2012-1818
5. Cao H, Folan P (2012) Product life cycle: the evolution of a paradigm and literature review from 1950 to 2009. Prod Plan Control 23:641–662. https://doi.org/10.1080/09537287.2011.577460
6. Tao F, Sui F, Liu A et al (2019) Digital twin-driven product design framework. Int J Prod Res 57:3935–3953. https://doi.org/10.1080/00207543.2018.1443229
7. Qi Q, Tao F (2018) Digital twin and big data towards smart manufacturing and Industry 4.0: 36° comparison. IEEE Access 6:3585–3593. https://doi.org/10.1109/ACCESS.2018.2793265
8. Xiang F, Zhi Z, Jiang G (2018) Digital Twins technology and its data fusion in iron and steel product life cycle. In: 2018 IEEE 15th international conference on networking, sensing and control
9. Zhang W, Welding simulation & digital twins. https://www.swantec.com/welding-simulation-digital-twins/. Accessed 20 May 2020
10. Mishra D, Pal SK, Chakravarty D (2021) Industry 4.0 in welding, pp 253–298
11. Tao F, Zhang M, Liu Y, Nee AYC (2018) Digital twin driven prognostics and health management for complex equipment. CIRP Ann 67:169–172. https://doi.org/10.1016/j.cirp.2018.04.055
12. He Y, Guo J, Zheng X (2018) From surveillance to digital twin: challenges and recent advances of signal processing for industrial internet of things. IEEE Signal Process Mag 35:120–129. https://doi.org/10.1109/MSP.2018.2842228
13. Zhang H, Liu Q, Chen X, et al (2017) A digital twin-based approach for designing and decoupling of hollow glass production line. IEEE Access 1–1. https://doi.org/10.1109/ACCESS.2017.2766453
14. Mishra D, Pal SK, Gupta A, et al (2020) A system for real time monitoring, prediction and control of weld quality in friction stir welding (Filed)
15. Pal SK, Roy RB, Singh AK, et al (2018) Cloud based remote manufacturing with machine learning based real time control (Filed)
16. Mishra D, Roy RB, Dutta S et al (2018) A review on sensor based monitoring and control of friction stir welding process and a roadmap to Industry 4.0. J Manuf Process 36:373–397. https://doi.org/10.1016/j.jmapro.2018.10.016

17. Mishra D, Sahu SK, Mahto RP, et al (2019) Friction stir welding for joining of polymers. Springer Singapore, pp 123–162
18. Pal SK, Khan AR, Ranjan R, et al (2018) Real time surface defect analysis and correction in friction stir welding process by image processing
19. Induction motor in 2D, COMSOL. https://www.comsol.co.in/model/induction-motor-in-2d-14487